Electrophoresis '83
Advanced Methods
Biochemical and Clinical Applications

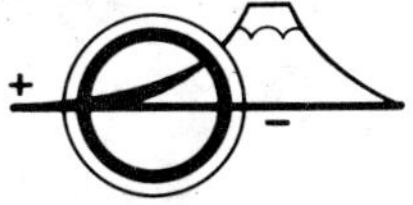

Electrophoresis '83

Advanced Methods
Biochemical and Clinical Applications

Proceedings of the International
Conference on Electrophoresis
Tokyo, Japan, May 9–12, 1983

Editor
H. Hirai

Walter de Gruyter · Berlin · New York 1984

Editor

Hidematsu Hirai, M. D., Ph. D.
Professor of Biochemistry
The First Department of Biochemistry
Hokkaido University School of Medicine
Sapporo Hokkaido
Japan

CIP-Kurztitelaufnahme der Deutschen Bibliothek

Electrophoresis . . .: advanced methods, biochem. and clin.
applications; proceedings of the Internat.
Conference on Electrophoresis. – Berlin; New York: de Gruyter
NE: International Conference on Electrophoresis
1983. Tokyo, Japan, May 9 – 12, 1983. – 1984.

ISBN 3-11-009788-5

Library of Congress Cataloging in Publication Data

International Conference on Electrophoresis (4th: 1983: Tokyo, Japan)
Electrophoresis '83.

Bibliography: p.
Includes index.
1. Eoectrophoresis--Congresses. I. Hirai, Hidematsu. II. Title. [DNLM:
1. Electrophoresis--Congresses.
W3 IN174M 5th 1983e / QU 25 159 1983e]
QP519.9.E434I57 1983   574.19'283   83-26342

ISBN 3-11-009788-5

PREFACE

The third meeting of the International Society of Electrophoresis was
held on May 9-12, 1983 in Keio Plaza Hotel, Shinjuku, Tokyo. Over 300
researchers from 19 countries participated. Just before this meeting,
the 33rd Annual Meeting of the Society of Electrophoresis (Japan),
which was established in 1951, had been held and many members of the
Japanese Society had the opportunity of getting to know researchers
in the field of electrophoresis from foreign countries. We greatly
appreciate that the International Society made this possible.

In the international meeting Dr. R.C. Allen gave an excellent lecture,
as the opening address, and Drs. B.J. Radola and M. Kitamura held
special lectures. In the oral sessions 62 papers were presented, in
the poster sessions 60 papers were demonstrated, and 6 Round-table
Discussions were held with very active participation. We also had several
commercial seminars, exhibitions of a large amount of equipment, and
of many reagents and books.

I was very pleased that we were able to come together to talk, both
scientifically and privately. I am sure many of the participants were
able to make new friends, which is an important purpose of the meeting.
I sincerely appreciate the cooperation of all those attending towards
making the meeting a success.

In these proceedings the papers are printed in the order of presentation:
general, two-dimensional electrophoresis, cell electrophoresis
(including free-flow electrophoresis), isoenzymes, isoelectric focusing,
isotachophoresis, affinity electrophoresis, and applications.

The publication of the proceedings could not have been accomplished
without the cooperation of all the authors. The editors express their
sincere thanks for their efforts.

The editors would also like to thank the staff of Walter de Gruyter,
the Japanese Ministry of Education, the Tokyo Metropolitan Government
and the Society of Electrophoresis (Japan) for their valuable suggestions
and help.

The 3rd International Electrophoresis Society Meeting was supported
financially by many individual members of the Society of Electrophoresis
(Japan) and by many pharmaceutical companies and producers of instruments.
Among these were the following foundations:

      Japan EXPO Memorial Foundation

      Japanese Society for the Promotion of Science

      Kashima Foundation for Promotion of Science

      Toyo Rayon Foundation for Promotion of Science

      Naito Memorial Foundation for Promotion of Science

      Shimazu Foundation for Promotion of Sciences and Technology

      Yoshida Foundation of Science and Technology

      Asahi Glass Foundation for Promotion of Industrial Technology

Finally, I would like to express my cordial thanks to Dr. N. Hashimoto,
Chairman, and to all the members of the organizing committee for their
tremendous efforts over the last two years.

Tokyo, February 1984   Hidematsu Hirai

Contents

Two-Dimensional Electrophoresis

CELL ELECTROPHORESIS

APPLICATION

LIST OF CHAIRMEN

Opening Speech

H. Hirai (Japan)

Plenary Lecture

Y. Tsuchiya (Japan), S. Akai (Japan)

General

O. Vesterberg (Sweden), K. Shimao (Japan)

Two-Dimensional Electrophoresis

S. Hanash (U.S.A.), M. Ohashi (Japan)
V. Neuhoff (F.R.G.), T. Inoue (Japan)
M. Bier (U.S.A.), N. Okuyama (Japan)

Cell Electrophoresis

W. Schütt, (G.D.R.), K. Ohkawa (Japan)
B. K. Shenton (U.K.), N. Hashimoto (Japan)
M. Wioland (France), T. Matsuhashi (Japan)

Isoenzyme

B. J. Radolla (F.R.G.), M. Yoshida (Japan)
M. Kitamura (Japan), T. Kanno (Japan)

Isoelectric Focusing and Isotachophoresis

P. G. Righetti (Italy), T. Horio (Japan)
A. Chrambach (U.S.A.), S. Kobayashi (Japan)

Affinity Electrophoresis

S. Hjerten (Sweden), K. Takeo (Japan)
T. C. Bøg-Hansen (Denmark), K. Taketa (Japan)

Application

R. C. Allen (U.S.A.), Y. Endo (Japan)
T. Kawai (Japan), Y. Sakagishi (Japan)

ELECTROPHORESIS - ITS PRESENT AND FUTURE ROLE IN BIOLOGICAL AND BIOMEDICAL
RESEARCH

Robert C. Allen, Departments of Pathology and Laboratory Animal Medicine,
Medical University of South Carolina, Charleston, SC 29425 USA

Introduction

I would like to take this opportunity to thank Professor Hirai, the
President of the Japanese Society of Electrophoresis, and the Organizing
Committee for their most kind and gracious invitation to me to present the
opening address at this most historic first meeting of the Electrophoresis
Society. It is indeed an honor and a pleasure to be here today to participate in this momentous event. This is a rare opportunity to expand the
exchange of information between two sides of the world and, I trust, the
beginning of new and meaningful cooperative relationships. I can only
hope that what information that I may impart at the beginning of this
meeting will be obsoleted by that which follows in the next three days.

The title of this talk, on reflection after I submitted it to Professor
Hirai, is a bit presumptuous and certainly presents an impossible task to
completely cover such a broad subject in the alloted time, or to condense
it into the requested number of pages for the proceedings book. I will
therefore, in the main, limit my remarks and illustrations of specific
points to the area of acrylamide gel electrophoresis of proteins.

The practical application of the practice of electrophoresis in the field
of biology and medicine may be credited Tiselius (1) and Konig (2) who
independently and almost simultaneously reported the separation of human
serum proteins by electrophoretic methods. Since that time, only some 46
years ago, we have seen and experienced an explosion of methodology for an
ever increasing resolving and informational capacity. Yet, when we carefully analyze the sequence of events leading to the accomplishments of

4

today, it becomes rather apparent the methods have evolved from the
theories presented by Kohlrausch (3) as early as 1897 as expanded by
Kendall and Krittenden (4) in the 1920s and Longsworth (5-7) in the 1940s
and 1950s on moving boundaries. The development of starch gel electrophoresis by Smithies (8) provided the first work wherein size was combined
with charge to improve resolution. This was quickly refined by Poulik (9)
with a moving boundary to further improve resolution. However, it remained for Ornstein and Davis (10) to utilize the controllable sieving
characteristics of acrylamide described by Raymond and Weintraub (11) with
a moving boundary system to provide the first really high resolution system
in which resolution was controllable by theory. The use of SDS by Weber
and Osborn (12) for molecular weight determination and its modification by
Laemmli (13), have provided us with the ingredients that are available to
us today in the use of charge-size separations.

There is a second aspect of charge that has played a major role in our
present ability to separate macromolecules, that is the isoelectric oint
of each macromolecular species, or its point of zero charge. Isoelectric
focusing was first actually demonstrated by Ikeda and Suzuki (14) in 1912
and was extended in 1929 by Williams and Waterman (15). However, it remained for Kolin (16,17), Svensson (now Rilbe) (18-20) to fully develop
this theory and for Vesterberg (21) in 1967 to bring it to a practical
application. Yet, as Hjelmeland and Chrambach (22) have recently described
this too is a special case of the moving boundary theory.

In this brief review it becomes quite obvious that major advances in the
separation of macromolecules by electrophoretic methods have centered on
only two of their characteristics; that of charge and size. Thus our
colleagues in this field have made the major contributions in electrophoresis which has progressed from a resolution capability of five to six
serum proteins to over 1000 with present two-dimensional methods based on
only these two attributes. Of course the utilization of various enzyme
immunological, and special staining techniques provide additional parameters of function and structure,

Present Applications

With all of the essential ingredients now in place let us examine the
practical and investigational uses to which we can apply electrophoresis
methodology. First let us look at the resolution capabilities and limitations of the individual and combined methods most commonly in use today,
using the more conventional discontinuous moving boundary methods as
applied in disc and SDS-PAGE types of electrophoretic separations. At
best only 200-400 theoretical plates are available in polyacrlamide gel
systems. Thus, there is at present a definite limitation in the number
of proteins that one may expect to resolve in, for example, SDS-PAGE.
Increasing the gel length does not increase the expected theoretical
plates due to the undisciplined behavior of proteins which diffuse with
the increasing separation times, particularly over long separation distances. While this can be controlled to a degree by producing ultrathin
starting zones by stacking with moving boundaries and by altering subsequent unstacking limits, it would appear that major improvements in this
technique may not be soon forthcoming. On the other hand, in a chromatographic column of 70 cm in length using 5u beads the theoretical plates
are in the order of 1000 to 10,000 and by doubling the length one may
increase these by approximately 1.5 times. The question then arises,
should or would we be better off to discard electrophoresis in favor of
such methods as HPLC for the purpose of molecular weight separations. A
similar argument for charge separations could also be made for the use of
chromatographic columns.

In the case of isoelectric focusing, the situation is somewhat more favorable for the electrophoretic approach. Here as may be seen in Fig. 1 the
resolution is contingent on the voltage gradient. Where theory predicts
(23,24) that for a four-fold increase in the voltage gradient will produce
a two-fold increase in resolution and indeed as is illustrated, at least
a 1.8-fold increase is achieved. However, to use a voltage gradient of
500V/cm as in this example; ultrathin-layer gels of 125-200u in thickness
run under special cooling conditions using a "Cold Focus" apparatus with a
special Beryllium oxide cooling plate (MRA, Corp.) are necessary. Such

gels are of an advantage for analytical work and may be run in less than
30 min. at up to 750V/cm (25).

A similar relation of gel length to resolution also exists as shown by
Giddings and Dahlgren (23) and Rilbe (24). Thus, a four-fold increase in
the length of a gel will also produce a two-fold increase in the resolution at the same voltage gradient. Here one is limited, not only by the
heat dissipation capability of the instrument, gel **vo**lume and ionic
strength of the buffer, but also by available power supplies. In the
example shown in Fig. 1 the distance between electrode wicks is 5.4 cm.

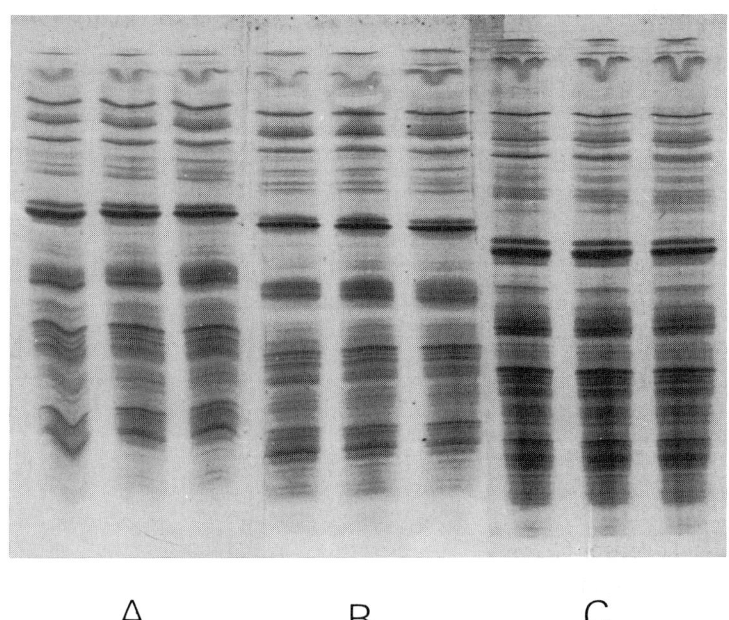

Figure 1. Separation of 1.0ul of a 0.1 percent solution
of Rohament P. Separations were carried out on   pH 3-7
Servalyte with a gel thickness of 125u for 450 V/h with
maximum voltage gradients of 125 V/cm (A), 250 V/cm (B),
500 V/cm (C). Separation times were 52 min. (A), 39 min.
(B), 29 min. (C). The gels were stained with diammine
silver (25).

To double the resolution by increasing the separation the separation
distance to 21.6 cm would require a power supply capable of delivering
2500 volts. To combine both methods to achieve a four-fold increase in
resolution would require at 500V/cm, not a 2500 volt power supply, but one
capable of delivering 10,000 volts. Presently such power supplies are not
commercially available, nor are safety disigns of present apparatus
adequate for such voltages. The present state of the art would seem to
limit this approach to furthering the resolution potential of isoelectric
focusing for the moment.

Combined Methods

Early reports by Smithies and Poulik (26) followed by inumerable other
studies culminating more recently with those of Felgenhauer and Hagedorn
(27) and Manabe et al. (28) have utilized non-denaturing two-dimensional
electrophoresis to solve the dilema of resolution and elucidation of maximal information in biological systems. The usefulness and limitations of
this technique are readily apparent. On the other hand, the almost
simultaneous reports of Klose (29), O'Farrell (30) and Scheele (31) utilizing dissociating isoelectric focusing in the first dimension and denaturing SDS-PAGE in the second dimension have expanded the information
potential obtainable in complex mixtures of macromolecules from some 200
bits of information to 1000-2000 bits.

Anderson and Anderson (32) utilized this technique in their ISO-DALT
system and have developed from this the concept of the human "Protein
Index", or the potential of a macromolecular map of any biological species.
Dunn and Burghes (33) have recently reviewed this field, in a most compre-
hensive manner, pointing out the advantages and disadvantages of many of
the approaches toward this aim. This concept has excited much interest in
these techniques as a clinical tool that is, perhaps, a bit premature.

8

Young and Tracy (34) have concluded, in the most recent journal of Electrophoresis, that there does not appear to be an immediate application of this
technique in the clinical laboratory for the direct examination of body
fluids. However, they feel that it has great potential as a tool in the
search to link specific proteins with a particular disease. By virtue of
its non-specificity, it permits a broad-based search for possible direct
genetic mutation changes, as are being studied by Dr. Neel's group, offers
a tremendous potential when studying the father-mother-child triad.   I
trust that this approach will be more fully illucidated in the following
presentations by this group during the meeting.

While procedures such as 2-D may be totally acceptable to the biologist in
the persuit of answers to research questions, our clinical colleagues require rapid, cost effective information that is of primary diagnostic
value; (predictive and not retrospective information as is presently the
case with most two-dimensional studies). A further, present constraint to
the use of such techniques, is that the information is still quite complex
and not in a form familiar to most of our clinical colleagues. Thus, a
clinician is quite comfortable and satisfied with a confirmatory Alpha 1-
Antitrypsin phenotype in a suspected case of infantile cirrhosis, which
may be obtained in 45 minutes by isoelectric focusing. Yet, he is abso-
lutely uninterested as well as perplexed by an exquisite multi-parameter
separation which takes several days to complete, although this may provide
a total molecular profile on the patient with invaluable attendent diagnostic and prognostic information.

Again, we must look at what the advantages and disadvantages of such a
technique offer to the field of diagnostics. First of all are we able to
get all of the potential information that is present in a serum sample
from a 20 ul sample, or are we looking only at the surface of an iceberg?
Present detectability limits with the best reported silver stains are
about 0.03ng per mm square of gel or 15-50 ug/dl. Thus many of the minor
proteins, which may be of biological or diagnostic significance, may be
undetectable with the present two-dimensional technique. An indication

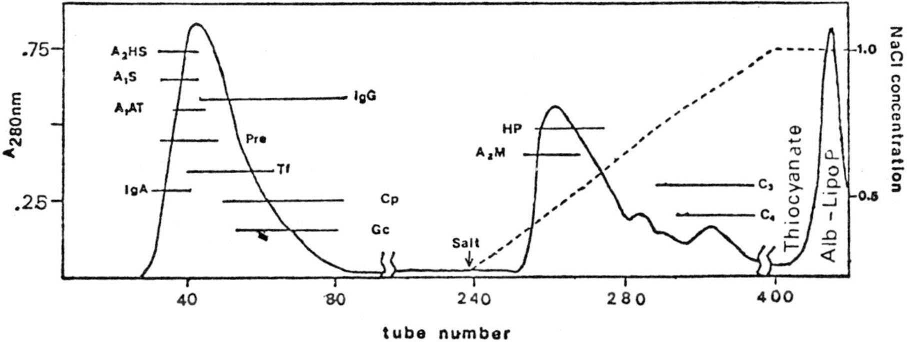

Figure 2. Elution profile of an Affi-Gel Blue column to which 10ml of dialysed plasma from a single individual was applied. Unbound proteins were eluted with 0.03M Na-phosphate buffer at pH 7.0 at 25^{0}C, followed by a salt gradient and Ammonium Thiocyanate. Proteins listed in each fraction were determined by fused rocket immunoelectrophoresis.

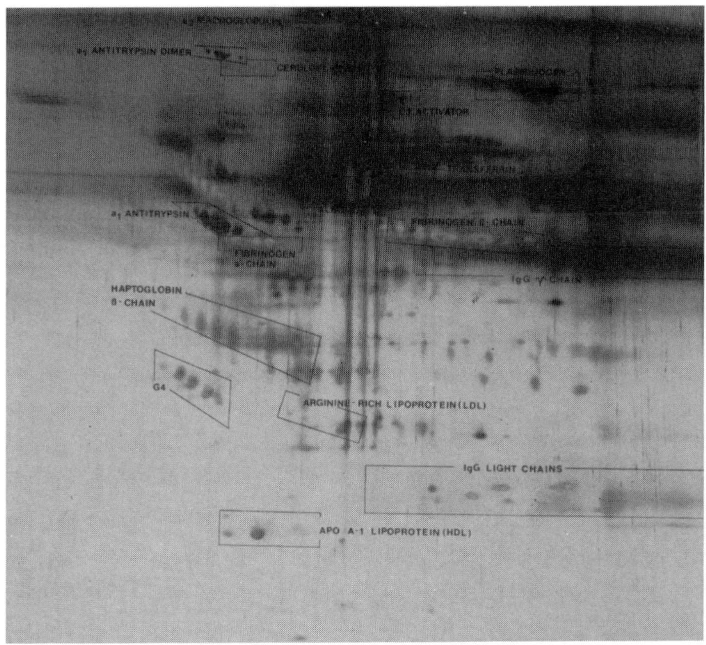

Figure 3. A 2-D ISO-DALT system separation of human plasma.

of this being the case is shown in the following example where a multi-parameter separation was carried out using pseudo-ligand affinity chromatography in combination with the ISO-DALT system.

Multi-Parameter Separation

The use of dimensions to describe the sequence of separation steps appears be be no longer adequate, especially when more than three dimensions or characteristics of macromolecules are employed. For the sake of simplicity in describing the following illustration, the term parameter is employed to signify that more than three characteristics of a macromolecule are being employed in the separation and demonstration thereof. To determine if the present information obtained in a two-dimensional system represents a complete picture of the plasma proteins present, a joint study between Dr. Arnaud's, Dr. Sammons' and my laboratory was undertaken. We wished to determine, if following pseudo-ligand affinity chromatography on Affi-Gel Blue columns (Bio-Rad) as described by Arnaud and Giannaza (35) and by this parameter and charge as reported by Allen and Arnaud (25) the expected additional plasma proteins could indeed be demonstrated. Briefly, 10 ml of dialyzed plasma was passed through an Affi-Gel Blue column with the unbound fractions being sequentially released with 0.03M phosphate buffer at pH 7.0, followed by elution of bound components with a NaCl salt gradient from 0.03 to 1.0M, with a final elution of the very strongly bound albumin and lipoprotein with ammonium thiocyanate. Fractions were pooled into 58 groups of ten tubes each and lyophilized. Each fraction was then diluted to contain 1.0mg protein/ml and was assayed by fused rocket immunoelectrophoresis against 27 antisera, by high voltage ultrathin-layer isoelectric focusing and by two-dimensional electrophoresis in an ISO-DALT system. Two dimensional separations were stained with the color silver stain of Adams and Sammons (36). Preliminary results from six of the 58 fractions analyzed are shown in the following figures.

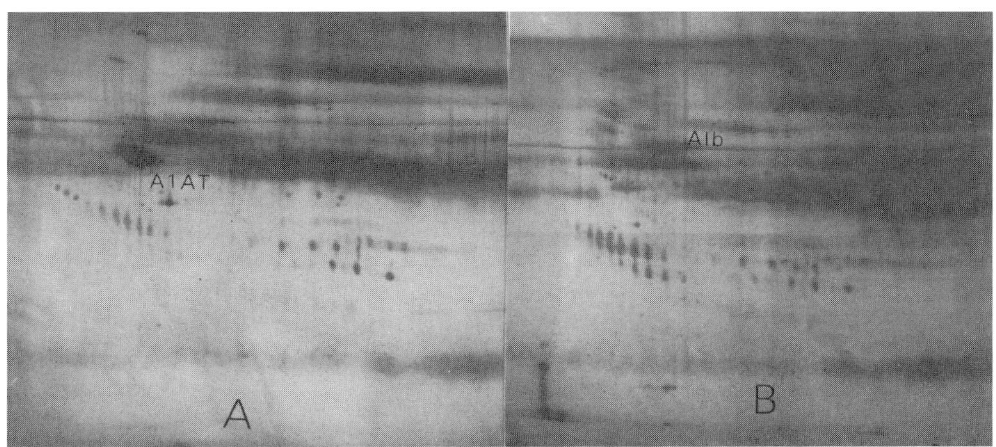
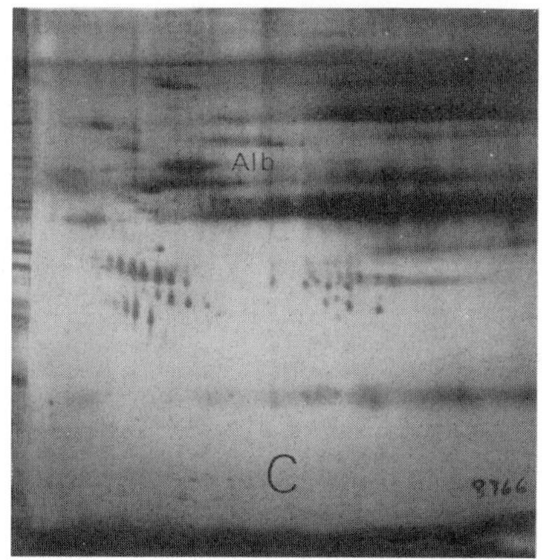

Figure 4. Lyophilized eluates pooled from each ten tubes at a concentration of 1.0mg/ml subjected to 2-D electrophoresis in an ISO-DALT system and stained with silver (36). Panel A resulted from tubes 30-39, Panel B from tubes 80-89 and panel C tubes 210-219 from the 0.03M phosphate elution step Fig. 1.

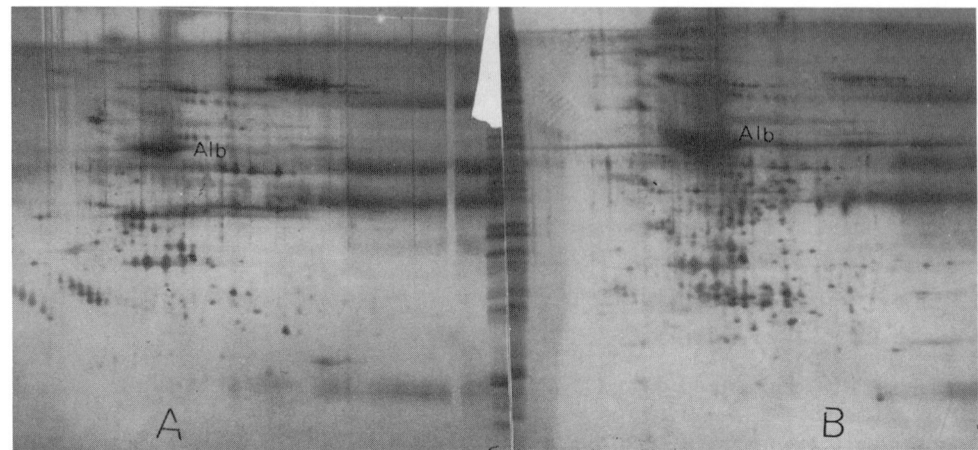
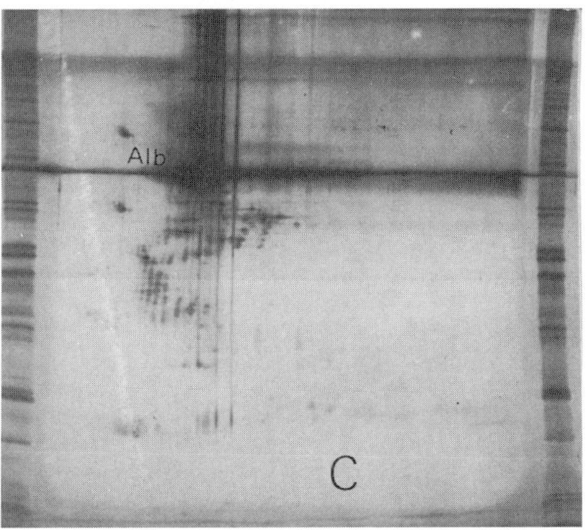

Figure 5. Salt gradient elution of tubes 300-309 panel A and 310-319 panel B. Concentration separation and staining are similar to Fig. 4. Panel C is the Ammonium Thiocyanate eluate treated similarly and consists mainly of Albumin and Lipoproteins.

Future Trends

These data above illustrate that preliminary fractionation of plasma with resultant enrichment of each fraction, followed by concentration, allow the demonstration of many more proteins and subunits than can be obtained with ordinary two-dimensional techniques based on charge and size alone. In conjunction with the colored silver stain used in this preliminary study one is able to utilize four parameters in the separation process to obtain information in greater depth. Obviously, the use of immobilized specific lectins, antibodies, etc. can tailor a given system to meet the specific needs of a particular investigation. The availability of high performance liquid chromatography (HPLC) now offers a rapid first fractionation procedure wherein now minutes rather than hours are required to fractionate plasma or other biological fluids. It would appear that in the future that such combined techniques will be that ones that will allow such concepts as the "Human Protein Index" to be successfully accomplished; with the demonstration of the 10,000 or more subunits predicated by a number of authors. The major point to be made here is that we have too long attempted to use only one technique to answer many of our problems at the molecular level and that it is now time for us to broaden our horizons beyond charge and size separations alone.

Also utilized in the above example was the highly sensitive colored silver stain which provides an important additional parameter of the proteins under study. Stain technology, largely borrowed from the histochemists, has played a major role in the development of electrophoretic methodology and advances in its use in the last thrity years. However, other than microdensitometry in the ultra violet spectrum, spectral characteristics of proteins have been little studied. The recent studies of Möller et al. (37) on the evaluation of stained and unstained electropherograms by photoacoustic spectrscopy, hopefully portend new developments in the elucidation of protein characteristics. Other such possibilities, with the availability of fiber optics, are nitrogen laser excitation at 330nm and Rhamen spectroscopy. The letter should be theoretically capable of demonstrating overlapping proteins as evidenced by double spectral images.

All of these possibilities are further feasible with the availability of ever more powerful micro computers, whose cost is no longer a prohibitive factor.

Why then, am I standing here today discussing the already in place technology of the analytical chemist, and why have we not taken greater advantage of such instrumentation in the field of electrophoresis? It would appear that the answer of this last question is a historical part of the field of electrophoresis itself. We have, in the main, developed and built our own equipment at a rather modest cost. Similarly, the cost of most commercially available equipment is also rather modest in comparison with other laboratory instrumentation. This has led to the dilema we face today in advancing the field. That is that the return on investment and market for this class of equipment is too small to attract the R&D investment by industry to carry the field to its full potential in biology and in biomedicine. Unfortunately large industrial firms with the manpower and expertise to do this type of development are not interested in a market of under $100,000,000. Yet, they will eagerly produce a new automated clinical analyzer using 30 year-old chemistries, which offer nothing new to the patient or to science, when they can project a 3-400,000 dollar market. This fact of life is no way aimed as a condemnation at the many companies producing electrophoresis equipment who have made so many major contributions to the field. We should not ask, to paraphrase Shakespear, "2-D or not 2-D" or, "out damned spot"; but continue as we are doing here to improve communication and the dissemination of information so that the full potential of electrophoresis will be recognized and utilized by our biological and biomedical colleagues, no matter the cost of the instrumentation required.

References

1. Tiselius, A.: Trans. Faraday Soc. $\underline{33}$, 93 (1937).

2. Konig, P.: in Acts and Words of the 3rd Congress of South American Chemists, $\underline{2}$, Rio de Janero, 334 (1937).

3. Kohlrausch, F.: Ann. Phys. Chem. $\underline{62}$, 209-239 (1897).

4. Kendall, J. and Krittenden, E.D.: Proc. Soc. Nat. Acad. Sci. USA, $\underline{9}$, 75-78 (1923).

5. Longsworth, L.G.: J. Am. Chem. Soc. $\underline{67}$, 1109-1119 (1945).

6. Longsworth, L.G.: in, Electrochemical Constants, National Bureau of Standards Circular $\underline{254}$, U.S. Government Printing Office, Washington, DC pp. 59-68 (1953).

7. Longsworth, L.G.: in, Bier, M. Ed. Electrophoresis, Academic Press, New York, pp. 91-178 (1959).

8. Smithies, O.: Biochem. J. $\underline{61}$, 629 (1955).

9. Poulik, M.D.: Nature, $\underline{180}$, 1477-1479 (1957).

10. Ornstein, L. and Davis, B.J.: Preprinted by Distillation Products Industries, Division of Eastman Kodak Co., Rochester N.Y. (1962).

11. Raymond, S. and Weintraub, L.: Science, $\underline{180}$, 711 (1959).

12. Weber, K. and Osborn, M.: J. Biol. Chem. $\underline{244}$, 4406-4412 (1969).

13. Laemmli, U.K.: Nature, $\underline{227}$, 680-685 (1970).

14. Ikeda, K. and Suzuki, S.: Patent 1,015-981 (1912).

15. Williams, R.R. and Watermann, R.E.: Proc. Soc. Exptl. Biol. and Med. $\underline{27}$, 56-59 (1929).

16. Kolin, A.: J. Chem. Phys. $\underline{22}$, 1628-1629 (1954).

17. Kolin, A.: Pro. National Acad. Sci. $\underline{41}$, 101-110 (1955).

18. Svensson, H.: Acta Chem. Scand. $\underline{15}$, 325-341 (1961).

19. Svensson, H.: Acta Chem. Scand. $\underline{16}$, 456-466 (1962).

20. Svensson, H.: Arch. Biochem. Biophys. Suppl. $\underline{1}$, 132-140 (1962).

21. Vesterberg, O.: Acta Chem. Scand $\underline{21}$, 206-216 (1967).

22. Hjelmeland, L.M. and Chrambach, A.: Electrophoresis $\underline{4}$, 20-26 (1983).

23. Giddings, J.C. and Dahlgren, H.: Sep. Sci. $\underline{6}$, 345-456 (1971).

24. Rilbe, H.: Ann. N.Y. Acad. Sci. $\underline{209}$, 11-22 (1973).

25. Allen, R.C. and Arnaud, P.: Electrophoresis $\underline{4}$, in press.

26. Smithies, O. and Poulik, M.D.: Nature $\underline{177}$, 1033 (1956).

27. Felgenhauer, K. and Hagedorn, D.: Clin. Chim. Acta, $\underline{100Y, 121-132}$ (1980).

28. Manabe, T., Hayama, E. and Okuyama, T.: Clin. Chem. $\underline{2}$ Part 2, 824-827 (1982).

29. Klose, J.: Humangenetic, $\underline{26}$, 231-243 (1975).

30. O'Farrell, P.H.: J. Biol. Chem. $\underline{250}$, 4007-4021 (1975).

31. Scheele, G.A.: J. Biol. Chem. $\underline{250}$, 5375-5385 (1975).

32. Anderson, N.G. and Anderson N.L.: J. Autom. Chem. $\underline{2}$, 177-179 (1980).

33. Dunn, M.J. and Burghes, A.H.M.: Electrophoresis $\underline{4}$, 97-116 (1983).

34. Young, D.S. and Tracy, R.P.: Electrophoresis $\underline{4}$, 117-121 (1983).

35. Arnaud, P. and Gianazza, E.: Febs Letters $\underline{137}$, 157-162 (1982).

36. Adams, L.D. and Sammons, D.W.: in, Allen, R.C. and Arnaud, P. (Eds.) Electrohporesis '81, Walter de Gruyter, Berlin, pp. 167-180 (1981).

37. Möller, U., Köst, H.P., Schneider, S. and Coufal, H.: Electrophoresis $\underline{4}$, 148-152 (1983).

HIGH RESOLUTION ISOELECTRIC FOCUSING: NEW APPROACHES TO ANALYTICAL AND PREPARATIVE SEPARATIONS OF PROTEINS AND ENZYMES

Bertold J. Radola

Institut für Lebensmitteltechnologie und Analytische Chemie,
TU München, D-8050 Freising-Weihenstephan, FRG

1 Strategy of high resolution isoelectric focusing

High resolution isoelectric focusing differs from traditional
techniques in resolution, speed and capactiy. Differences in
pI values of 0.001 to 0.01 pH are resolved by variations of
the method, and the number of zones found in one dimension can
be of the order of 100 to 120. High resolution is attained
under steady-state conditions in minutes rather than in hours
or days that were necessary previously. In practice, a compro-
mise often has to be made with regard to resolution and separation time. Diminished time requirements result also from
rapid visualization that may be accomplished in minutes for
protein location with dyes and the sensitive silver stain,
as well as for enzymes with new zymogram techniques. An
attractive feature of high resolution isoelectric focusing is
its capacity, which, in analytical applications, allows analysis of 25 to 300 samples per focusing unit depending on
cooling surface, sample volume and electrode arrangement (1).
High resolution preparative isoelectric focusing achieves the
resolution indicated above in terms of resolvable pI differences and number of zones also under conditions of high total
sample load. Resolution in isoelectric focusing depends on the
design of apparatus, anticonvective stabilization of the pH
gradient, parameters of the separation process (field
strength and shallowness of the pH gradient), and properties
of the separated substance (diffusion coefficient D, and
mobility slope du/dpH in vicinity of pI). In the first part
of this report the strategy of high resolution isoelectric
focusing will be discussed. In the second part specific tech-

niques for analytical and preparative separations of proteins and enzymes will be described.

1.1 Apparatus

Numerous apparatus for isoelectric focusing have been described differing with respect to (i) mode of operation: discontinous (2), continuous (3) and recycling (4), and (ii) characteristics of the pH gradient which may be either continuous, e.g. in gels and density gradients, or segmental, e.g. in multi-compartment electrolysers (5) or apparatus similar to that in zone convection focusing (6). Apparatus with a segmental design are inferior to those with continuous pH gradients.In the former,it has been estimated that hundreds of segments would be necessary to achieve a comparable resolution (6). In gels and density gradients,resolution may be lost through post-focusing segmentation of a continuous pH gradient, e.g. by gel slicing or fraction collection with resultant remixing of zones already separated in situ. Apparatus operated in a discontinuous mode are most widely used, both in analytical and preparative applications. Continuous and recycling apparatus offer advantages in industrial applications for preparative separation of hundreds of gram quantitities.

Horizontal gel-stabilized apparatus provide the optimal system for high resolution isoelectric focusing. (i) In horizontal gel layers cooling is most favorable due to a high ratio of cooling surface to the total gel volume.Horizontal flat-bed apparatus excel over systems with cylindrical geometry for either density gradients or in gels giving more efficient heat dissipation and versatility. Vertical flatbed systems, popular in other electrophoretic techniques, offer no advantage in isoelectric focusing. (ii) Movable electrodes afford increased flexibility in the choice of separation distance, and are easily adjustable to the number of analyzed samples. (iii) Controlled humidity is important for analytical and preparative separations at high field strength. This aspect has not been given sufficient consideration thus far. (iv) Power supplies capable of yielding 3000 to 6000 V are indispensible for high resolution focusing. For work in this voltage range the safety features of the commercially available focusing apparatus must be reevaluated. (v) Rapid fixation converting the diffusable focused species into insoluble precipitates in analytical applications, and rapid detection of proteins in preparative separations are essential for maintaining the resolution achieved during the focusing procedure. Gel slicing devices or fractionation grids are incompatible with high resolution focusing because of dividing the layer arbitrarily into a number of segments inevitably leads to remixing of the separated zones.

1.2 Anticonvective stabilization of the pH gradient

Density gradients and gels are most frequently used. Density gradient columns have been employed in early experiments (2) and continue to be applied widely, by inertness rather than any inherent merits. A recently described cross-sectional column overcomes some of the shortcomings of the original technique by allowing rapid prefocusing over its short dimension to be followed by focusing over the longer dimension after turning the column (8). With this design it is possible to substanstially reduce focusing time but it does not overcome the drawback of remixing, the separated zones during the elution process.
Gels offer the best anticonvective stabilization. Isoelectric focusing requires a non-restrictive gel because molecular sieving will retard migration of proteins, resulting in prolonged focusing times. The migration velocities decrease as proteins approach their pI and any molecular sieving effect will not only retard the migration but may appear to halt it

entirely, giving rise to fallacious pI values (8). However, a steady-state need not necessarily be attained because many proteins are sufficiently separated under non-equilibrium conditions. The following gels, arranged according to decreasing restrictiveness, are suitable for anticonvective stabilization: polyacrylamide gels for proteins with molecular weights up to 500 000, agarose for molecules up to several millions and particles 30 to 80 nm in radius (9), and granulated gels which may be used for all molecular sizes but are unsuitable for cells.

1.2.1 Polyacrylamide gels

Compact polyacrylamide gels dominate in analytical applications of isoelectric focusing but they have failed to attract much interest in preparative separations (10). Gels of low total monomer and crosslinker concentrations are usually employed in analytical isoelectric focusing, and gels composed of 5 % T and 3 % C_{Bis} (crosslinked with N,N'-methylenbisacrylamdide, cf. ref. (11)) have become particularly popular (12). A drawback of polyacrylamide gels in preparative applications is the difficult recovery of separated proteins (8). Proteins that are electrophoretically extracted from polyacrylamide gels may also be contaminated with non-proteinaceous impurities.

1.2.2 Agarose

Agarose with low electroendosmosis has been proposed as an alternative to polyacrylamide gels in analytical isoelectric focusing of high molecular weight proteins (13,14). In a few reports, agarose has been also used for preparative isoelectric focusing (15-17). High resolution, easy handling, non-toxicity and absence of molecular sieving for high molecular weight proteins are some of its advantages. Although the purified or charge - balanced agaroses are claimed to fulfil many requirements of a good anticonvective support, there is evidence of certain disadvantages. A severe degree of surface flooding in the cathodal part of the gel, water transport to both electrodes resulting in distorted pH gradients, protein loss into the water accululated on the gel surface and protein trailing at the edges of the gel are some of these shortcomings (15,18,19). To overcome these drawbacks a composite agarose-Sephadex matrix was developed (19). Higher field strength than in agarose could be used resulting in improved resolution. In analytical experiments photopolymerized composite agarose (0.5 %) - polyacrylamide gels (2.5 %) proved superior to each of the single gels for the analysis of crude tissue extracts containing a wide molecular range (20). Preparative isoelectric focusing of immunglobulins was improved by adding 0.5 % non-crosslinked polyacrylamide to 1 % agarose (21).

1.2.3 Granulated gels

Horizontal layers of granulated gels of the Sephadex or Bio-Gel type were introduced for anticonvective stabilization of the pH gradient with the intent of overcoming some of the limitations of both the density gradient technique and compact polyacrylamide gels (22). Granulated gels have also been used for preparative isoelectric focusing in columns (23) and in continuous flow configurations (3), but the horizontal systems offer distinct advantages over vertical, closed systems. In preparative applications, granulated gels excel over other gel matrices in a number of properties: (i) high load capacity; (ii) quantitative elution of the focused proteins from the gel; (iii) simple handling; (iv) absence of molecular sieving for high molecular weight proteins; (v) availability, partly in a prewashed form, with well defincd chemical and physical properties. The inertness of granulated gels towards biopolymers under a wide range of conditions is well established due to their widespread use in gel chromatography. Drawbacks to granulated gels have been reported. Preparation of a gel bed with optimum consistency has been considered difficult (16,17) or laborious (24). Mixtures of Sephadex and Pevikon (25) (a co-polymer of polyvinyl chloride and polyvinyl acetate) and Pevikon alone (24) were suggested as possible supports. Inferior resolution and poor printing properties are shortcomings of Pevikon-containing layers.

Sephadex G-200 is the gel of choice for most applications. It exhibits the best load capacity and handling properties, the highest water regain and is most economical (10). Sephadex G-200 was not practical previously due to unsatisfactory printing properties(22)but new printing techniques (10,26), overcome this limitation. Sephacryl S-200, prepared by covalently cross-linking allyl dextran with N,N'-methylenebisacrylamide, can be handled as conveniently as Sephadex G-200 and has also good printing properties,but resolution is much inferior. The higher G-numbered Sephadex gels may contain up to 10 % of free dextran which could interfere with some detection methods and contaminate the eluates. The enzymatically resistant polyacrylamide gel, Bio-Gel P-60, is potentially useful and superior to Sephadex in work with crude preparations of cellulases and hemicellulases (10). All granulated gels must be extensively washed with distilled water before use to remove charged solutes interfering with the formation of the pH gradient. Optimum results are obtained with gels with a dry bead diameter of 10 - 40 μm ("Superfine" or minus 400 mesh).

1.2.4 Rehydratable gels

Until recently granulated gels were prepared as wet layers on a glass plate or in a trough (22); they could not be stored and usually were prepared just prior to use. Rehydratable gel layers (26) would have the advantage of allowing storage. Preparation of rehydratable layers is simple. After spreading the gel suspension of the correct consistency over a support, the gel is dried in air. The dry gel firmly adheres to the support, is mechanically stable and can be preserved indefinitely. Instead of glass plates or troughs the rehydratable gel layers are preferably prepared on a plastic film. Best results are obtained with 100 μm polyester films (Mylar D, Du Pont) treated with alkali to impart hydrophilic properties to the film (27). Two commercially available supports (GelBond for agarose from Marine Colloids, and Gel-Fix from Serva) are also suitable. Rehydratable gels can be prepared with carrier ampholytes which, due to their hygroscopic properties, ensure the residual moisture necessary for storage. Even greater versatility is provided by preparing "empty" gels without added carrier ampholytes, but supplemented with 1 to 2 % glycerol. Before use the rehydratable gels, containing carrier ampholytes, are sprayed with an amount of water, calculated from the volume of the gel layer. Empty gels are sprayed with an appropriate solution of carrier ampholytes. Any formulation of carrier ampholytes supplemented if necessary with such additives as urea, can be used for rehydration.

1.3 Field strength

Analytical and preparative isoelectric focusing has been carried out thus far at rather moderate field strength mainly because of difficult heat dissipation. Typical values were: 20 to 50 V/cm in density gradients (2), 20 to 40 V/cm in gel rods (8), and 20 to 100 V/cm in gel layers (12,22). Although field strengths as high as 150 to 300 V/cm with power outputs of 0.3 to 0.5 W/cm^2 have been applied in 1 to 2 mm gels (28) they have not been widely used in practice. As a result of low field strengths traditional isoelectric focusing required long focusing periods. High field strengths offer two advantages: (i) sharpening of zones with resultant improved resolution, and (II) shorter focusing time. Long focusing times were repeatedly held to be a shortcoming of isoelectric focusing (7,17).

Extended residence times of labile substances at extreme pH
values or close to their pI values incur the risk of inactivation and artefacts. By reducing the thickness of the gel layer,
field strengths of 100 to 1000 V/cm can be applied in ultrathin
gels (28), and 100 to 500 V/cm in preparative separations (10,
26) improving resolution and drastically shortening focusing
times.

1.4 Manipulation of the pH gradient

There are several approaches towards improved resolution by
flattening the pH gradient. (i) Selection of narrow range carrier ampholytes. These are either commercially available or
can be prepared by fractionation of the commercial products by
preparative isoelectric focusing (22,28). (ii) Increased separation distance. For longer separation distances the gradient
is flattened linearly and resolution improved if focusing is
carried out at the same field strength. For a 40 cm separation
distance, pH gradients are flattened from 0.15 pH/cm for wide
range carrier ampholytes to 0.025 pH/cm for 0.5 to 1 pH ranges,
respectively. (iii) Cascade focusing combines in a two-step
or multistep procedure, prefractionation of the sample with a
fractionation of the carrier ampholytes. In the first step the
sample is focused at a high load and with lower resolution in
a steep pH gradient. In the subsequent step(s) parts of the
gel layer, enriched with the components of interest, are transferred to a prefocused narrow range pH gradient. The carrier
ampholytes, transferred with the sample, flatten the pH gradient and greatly improve resolution. (iv) Addition of separators. Single or multiple amphoteric substances added in large
amounts (5-50 mg/ml) to carrier ampholytes flatten the pH gradient in the vicinity of their stead-state positions. At present the manipulation of the pH gradient has to be conducted in
a systematic but empirical manner, because our understanding
of the mechanism of the pH gradient formation is still insuf-

ficient for predicting the pH gradient from the pK values of
the separators (8). (v) Buffer isoelectric focusing. With some
buffer mixtures in cylindrical polyacrylamide gels, using a
14 cm separation distance, the pH gradient was flattened to
0.02 t0 0.04 pH/cm (29). By addition or deletion of buffer constituents the course of the pH gradient may be manipulated.
(vi) Local increase of gel volume (30) or concentration (31)
of carrier ampholytes. Both approaches have been described for
analytical isoelectric focusing but with 0.2 - 0.3 mm layers
they could also be useful in preparative separations. (vii)
Continuous displacement and pH of the anolyte. The pH gradient
can be flattened by suitable choice of anolyte and catholyte,
which are chosen so that they are within the pH range of the
gradient (8). (viii) Immobilized pH gradients. With the aid of
Immobilines (LKB) the most shallow pH gradients can be created
improving resolution and increasing the distance between separated zones (32).

2 Ultrathin-layer isoelectric focusing

2.1 Preparation of ultrathin gels

For analytical high resolution separations ultrathin-layer isoelectric focusing is 50 to 250 µm gels (27,33) is the method
of choice. The method evolved from thin-layer isoelectric
focusing in 1 to 2 mm gels but it is more than just an alternative to the traditional technique. It combines high resolution, speed, versatility and reagent economy with simplicity
of operation (27). It was recognized early that ultrathin-layer
isoelectric focusing is most attractive with 50 to 100 µm gels
polymerized on a suitable support. The function of the support
is to protect the ultrathin gel during all steps of preparation,
separation and visualization (34). Polyester films and glass
plates pretreated with methacryloxypropyl-trimethoxysilane are

increasingly used. Two commercial supports, GelBond PAG (Marine Colloids) and Gel-Fix (Serva) are also available for the preparation of polyacrylamide gels. These supports provide adequate adherence of ultrathin polyacrylamide gels of a standard composition (5 % T, 3 % C_{Bis}) but binding may be less reliable in presence of 8M urea, detergents, and/or increased gel thickness. For some applications silanization of polyester films pretreated with "Prime Coat 1200" provides a better coating (35).

Ultrathin gels are prepared by various techniques: (i) flap technique (27,34), (ii) thin-layer capillary technique, (iii) sliding technique (36), and (iv) clamp technique (33). The flap technique is the simplest and most versatile. Polyacrylamide gels, agarose and composed gel matrices can be prepared without special equipment, down to a gel thickness of 20 µm (28). The thin-layer capillary technique which has been applied to the preparation of 1 mm layer between horizontal plates is adaptable to the preparation of 100 to 200 µm gels when operated in slanted position. For the sliding technique a special mold is commercially available (Macromold , LKB).

2.2 Miniature ultrathin-layer isoelectric focusing

Ultrathin-layer isoelectric focusing has been described for 1 to 3 cm (28,34), 5 cm (37) and 12 to 25 cm (27,34) separation distances. For the most frequently used 10 to 12 cm separation distance, isoelectric focusing in 5 % T, 3 % C_{Bis} gels requires 2000 to 3000 Vh to attain the steady-state judged by coalescence of samples migrating from both electrodes (27). These Vh products can scarcely be achieved in separation times shorter than 1 to 2 h, depending on pH range and power input. The main difficulty in shortening the separation time is that distorted patterns are observed when either prefocusing or focusing are carried out at high initial field strength. In an attempt to shorten focusing time miniature ultrathin-layer isoelectric focusing was developed (28). On 3 cm gels in pH 4 to 9 Servalyt carrier ampholytes and at final field strengths of 400 to 800 V/cm the total focusing time, including prefocusing, is only 10 min, and even ferritin ($M_r \geq$ 465 000) reaches a steady-state. On 1 to 2 cm gels the separation time can be further reduced but at the expense of resolution due to pH gradient suppression. Even on 1 cm gels with a total focusing time of only 2 min the resolution of a mixture of marker proteins was better than in a density gradient column operated under optimal conditions (38).

A major advantage of isoelectric focusing over short separation distances is the considerably smaller gel volume per sample. On 10 to 12 cm gels, the focused samples usually occupy 1 cm tracks resulting in 50 to 100 µl gel volumes per sample in 50 to 100 µm layers. On minature gels 3 to 4 samples (0.2 to 0.4 µl) per cm are applied as droplets with resultant gel volumes of a few µl per sample. In miniature systems the gel volume per sample is reduced by a factor of up to 1000 to 2000 when compared with the traditional 1 mm gel layer. Gel volumes in the range of a few µl have been known so far only from microelectrophoresis in capillaries (39).

2.2 Resolution

In ultrathin gels 2 to 8 zones per mm are found depending on the properties of the analyzed sample, separation distance, and steepness of the pH gradient and final field strength (27, 28,37). In wide pH range carrier ampholytes 0.02 to 0.03 pI differences are resolved. Resolution is better in commercially available narrow range carrier ampholytes and can further be improved by isolation of ultranarrow ranges of carrier ampholytes by preparative isoelectric focusing (28). An alternative approach is analytical cascade focusing in which a prefractionation of carrier ampholytes and sample is followed by focusing in narrow or ultranarrow pH range carrier ampholytes. In cascade focusing additional advantage can be taken from pH gradient flattening (30) by carrying out the first focusing step in a 0.5 mm gel layer and the second step in 50 μm layers. These approaches allow the detection of pI differences of 0.003 to 0.005 pH.

2.3 Rapid protein staining

Staining of proteins after isoelectric focusing in gel media has a troubled history of initial failures and incremental improvements (40). For many years the undesirable interaction of carrier ampholytes with most protein dyes was a worry and resulted in excessively long destaining times. Destaining for several days was not unusual in the beginnings of isoelectric focusing in gels. By natural selection some dyes were discarded, either because of difficult background destaining (e.g. Amido Black 10B) or unsatisfactory sensitivity (e.g. Bromophenol Blue, Fast Green or Ligth Green). Two dyes of the triphenylmethane family came into general usage, namely Coomassie Brilliant Blue G-250 and R-250. In an attempt to overcome long destaining, several procedures were described excelling by relative rapidity, however, at the price of decreased sensitivity. Staining of proteins was dramatically improved through the introduction of 50 to 100 μm gels (27). Staining of proteins in ultrathin gels offers considerable advantages when compared with traditional 1 mm gels. Using suitable carrier ampholytes and dyes (e.g. Serva Violet 17 or 49) the total time required for fixation, staining and complete destaining of the background is reduced to 10 to 15 min. This is a major achievement that places protein staining after isoelectric focusing in the domain of rapid staining known so far only from electrophoresis in cellulose acetate membranes. Usually hydrated gels are stained. A further shortening of staining and destaining time to only a few minutes is possible when dry gels are stained (40).

Over the past few years there has been an increasing number of
reports describing sensitive staining of proteins with silver
or silver diamine complexes (41). Most of the silver staining
methods have been applied to the traditional 0.8 to 3 mm gels
and only a few reports decribe staining of 100 to 300 µm gels
(37,41). Silver staining methods use either an aldehyde or
mild oxidation step before the silver reaction. Complete
washing out of the reagents, particularly of glutaraldehyde,
is a critical step in obtaining a clear background. Since
washing is more efficient in ultrathin gels, the total time
for silver staining can be reduced to only 3 to 5 min. In a
simplified silver staining method, the 50 µm gels are fixed
for 1 min in 20 % trichloroacetic acid and treated with 0.25M
glutaraldehyde in 0.5M Na_2PO_4 (warmed up to 80 ^{O}C) for 5 to
10 s. After a single washing with 10 % ethanol for 20 to 30 s,
the gels are stained with silver diamine for 2 min, and finally reduced in citric acid-formalin. Gels polymerized on silanized polyester films and glass plates are both suitable for
silver staining.

2.4 Enzyme visualization

Using conventional immersion or overlay techniques for enzyme
visualization the resolution is partly lost as a result of
diffusion during the extended incubation necessary for many
enzyme reactions. Considerable amounts of enzymes may be expected to leach from 50 - 100 µm gels placed in aqueous solutions
(40). Therefore the strategy for enzyme visualization in
ultrathin gels has had to be redesigned (42). In an attempt to
optimize visualization for different enzymes four approaches
were used. (i) Immersion. For a few enzymes substrate solutions
from conventional techniques may be used. Thus, peroxidase is
detected within 1 - 3 min with o-toluidine and urea peroxide
by immersion or overlayering with substrate solution. For most
enzymes, solutions with substrate concentrations employed in

conventional enzyme visualization (43) are inadequate because, at the low substrate levels, leaching of enzyme is more rapid than color development. Increasing the substrate concentration by a factor of 10 to 20, accelerates the reaction and color development with greatly reduced enzyme loss. Alcohol dehydrogenase and lactate dehydrogenase are detected within 2 min as intensively stained, sharp zones. The reaction is stopped by washing with 50 % ethanol. (ii) Membrane printing. The paper print technique previously used for enzyme visualization in granulated gels (22) is unsuitable for ultrathin gels due to the coarse structure of the paper which is incompatible with the subtle pattern in the gel. An alternative was found in dimensionally stable polyamide membranes which are pretreated with buffer and impregnated with the substrate solution. The impregnated membranes are rolled over the focused gel and the gel-membrane sandwich is incubated at elevated temperature (60 – 80 $^{\circ}$C) until dry. A permanent record is thus obtained. (iii) Salting out-membrane printing. Glycosidases are detected by salting out the enzymes, with 90 % $(NH_4)_2SO_4$ in the focused gel, prior to printing with a substrate-impregnated polyamide membrane. Fluorogenic 4-methylumbelliferyl-derivatives may be used for a variety of substrates for enzyme location with high sensitivity. (iv) Printing with ultrathin agarose. "Empty" 100 to 200 μm agarose layers are prepared with the flap technique and can be conveniently stored until use. Substrate gels are obtained by equilibrating the ultrathin layer for 1 to 5 min with solutions containing either low or high molecular substrates, buffer, and additives. After equilibration the ultrathin agarose is rolled over the focused polyacrylamide gel and incubated for 1 to 2 min at elevated temperature. The excess of substrate and water is removed by pressing the agarose with filter paper. Phosphoglucomutase and proteases are conveniently located with this technique. Ultrathin agarose printing excels over conventional agarose overlay techniques due to higher resolution, speed, versatility, simplicity, convenient documentation and lower consumption of reagents.

3 High resolution preparative isoelectric focusing

The potential of isoelectric focusing for preparative separations was recognized early (2) but applications have remained modest. Depending on the scale of fractionation all preparative isoelectric focusing techniques can be classified into two categories. (i) Techniques for laboratory scale fractionation of milligram quantities with one gram as an upper limit; only in a few applications are 0.5 to 1 g amounts of proteins actually separated (10). (ii) Techniques for large scale fractionation of gram amounts (and more) with the potential of industrial applications (3-5). Both groups of techniques comprise a plethora of diverse systems but most do not seem to have been applied due to unsolved practical problems. The two most widely used methods of preparative isoelectric focusing comprise density gradient columns (2) and layers of granulated gels (22). The former technique has remained unchanged over the past years and so have its shortcomings (7). Although granulated gels of the Sephadex and Bio-Gel series have shortcomings they afford the most challenging potential for preparative isoelectric focusing. High resolution isoelectric focusing in layers of granulated gels (10,26) has evolved from the previously described technique (22,44) through a number of modifications: (i) Resolution is improved by using high field strengths (100 - 500 V/cm) in thin (0.2 - 1 mm) instead of the thick (2 - 12 mm) gel layers employed previously. (ii) Loading is limited to 1 g because of inherent limitations in heat dissipation in thicker gel layers required for larger amounts. (iii) Increased flexibility is achieved by using dry, rehydratable gels (26) on polyester films instead of wet gel layers on glass plates or in troughs. (iv) Rapid focusing with resultant short residence times of the separated samples in the gel layer is possible using shorter separation distances, extended prefocusing and cascades, or a combination of these approaches. (v) Detection of proteins and enzymes is improved by new print techniques utilizing cellulose acetate membranes (10), trichloroacetic

acid impregnated paper strips (26) or high resolution enzyme
visualization techniques (42). The time for detection by these
techniques is shortened to a few minutes.

3.1 Load capacity

In order to compare isoelectric focusing in systems employing
a different geometry, pH gradient and/or other forms of anticonvective stabilization, load capacity is defined as the
amount of protein (mg) per milliliter of focusing volume (44).
Load capacity is calculated by dividing the total load by total
volume of the gel layer. The appearance of straight zones is
used as the criterion in determining the highest permissible
protein load capacity. Overloading results first in irregular
zones ,which with additonal protein, cause the major zones to
decay into droplets. The irregularities of the major zones in
some parts of the gel layer usually have no detrimental effect
on zone definition of minor components in other parts of the
gel. Thus, load capacity for total protein is much higher when
minor components are to be separated from an excess of major
components rather than when all protein zones have to be well
defined. Load capactiy has been determined for natural and
artificial mixtures of proteins as well as for a single protein
and carrier ampholytes of different pH ranges using 0.2 to
1 mm layers of Sephadex G-200 and Bio-Gel P-60. The highest
loads are attained for protein mixtures with a uniform distribution of protein zones over a wide pH range, e.g. crude
tissue extracts. The decisive parameter for preparative systems is the amount of material to be fractionated which, depending on the load capacity of the system, requires a specific focusing volume. In the early experiments with granulated
gels, thick (2 mm) layers were used, and subsequent separations
also employed thick layers of up to 12 mm (22,44). The notion
that preparative isoelectric focusing requires thick layers
became established. Recent work appears to make the thick layer concept

obsolete (10,26). In high resolution preparative isoelectric focusing thin layers afford the following advantages:(i) higher field strengths due to more efficient heat dissipation; (ii) better resolution as the result of higher field strength and absence of skew zones; (iii) shorter focusing time; (iv) easier and more rapid gel preparation; (v) better moisture control; (vi) reduced cost due to lower consumption of carrier ampholytes, buffer and gels; (vii) improved recovery and higher concentrations of the recovered proteins; (viii) reduced risk of inactivation of labile substances, e.g. as a result of chelating activity of carrier ampholytes. High resolution preparative isoelectric focusing should be carried out at as high a load level as possible. The optimum capacity is 3 to 5 mg per ml of gel bed volume with 10 to 20 mg/ml being well tolerated;although this is dependent on the specific properties of the separated material. An increase of total load over one gram can be achieved by selecting an apparatus with a larger surface area of the cooling plate (1), or, preferably, by applying a two-step cascade with prefractionation of up to several grams in the first step, followed by high resolution isoelectric focusing of selected parts of the gel in the second step.

3.2 Resolution

The excellent resolution of analytical isoelectric focusing is a challenge for any preparative focusing method. Whereas high field strength are being increasingly applied in analytical experiments (28,37) preparative isoelectric focusing has been carried out thus far at rather moderate field strengths, mainly because of difficulty in heat dissipation. By reducing the thickness of the gel layer, field strengths of 100 - 500 V/cm can be applied in preparative isoelectric focusing with resultant improved resolution and drastically shortened focusing times (10,26). In prefocused gels over a 5 - 10 cm separation distance the residence time of the sample is shortened to only

20 - 40 min under steady-state conditions for granulated gels. The most conspicuous effect of high field strength is improved resolution. Proteins differing by only 0.01 to 0.15 pH are resolved on 40 cm gels using a pH 4 to 6 gradient (10). This resolution has been achieved previously only with the analytical system. In a two-step preparative cascade of a crude fungal enzyme 0.003 - 0.005 pI differences were resolved in gels with 20 cm separation distance (26). In preparative experiments two components should not only be visibly resolved but they should also be amenable to elution from the gel layer by a simple slicing technique. The main argument for using longer separation distances is that zones can be handled more easily on elution.

3.3 Detection of proteins

Prior to recovery, the focused proteins and enzymes must be located in the gel layer by detection techniques that should be rapid, simple and preferably non-destructive. Speed is important because keeping the gels without or at reduced voltage will broaden the zones as a result of diffusion, an effect less conspicuous with long separation distances. The gel layer may be rapidly frozen if this is compatible with the separated material. There are several approaches to locating proteins and enzymes in horizontal gel layers. (i) Transparent zones. At high protein loading the major components are visible in the gel layer after focusing as transparent zones due to changes in refraction relative to the surrounding gel (10). This permits rapid visual identification, pI determination and direct isolation by gel slicing. (ii) Membrane and paper printing is the most versatile technique for protein location. Originally, paper prints were obtained with chromatographic papers (10). The drawbacks of chromatographic paper for printing are: diffuse zones resulting from the coarse structure of paper relative to the gel matrix, limited applicability to some gel matrices, e.g. Sephadex G-200, and long visualization time. These shortcomings are overcome by cellulose acetate membranes (10) or trichloroacetic acid-impregnated paper (26). Ponceau S instead of the more sensitive triphenylmethane dyes (27,40) is preferred for staining at high protein load. With membrane printing the total time required for fixation, staining and destaining requires only 2 to 3 min. Destaining depends on the chemical properties of the carrier ampholytes (27) and is most rapid for Servalyt. Narrow (1 - 2 cm) strips are sufficient for printing with only small amounts of proteins being removed. Membrane printing is nondestructive but the trichloroacetic acid from the paper exerts an fixative effect on the proteins in the gel layer. The stained prints are a convenient document which can be preserved easily and evaluated densitometrically. Location of radioactivity in the print by a strip scanner has been reported (10). (iii) UV densitometry (44) is more sensitive than zone transparency but less sensitive than staining of a print and requires an expensive instrument. (iv) A topographic method is based on the fluorescence of Servalyt carrier ampholytes in a paper print (45,46). (v) Fluorescence. Without printing, proteins can be visualized in the gel layer with 8-anilino-1naphtalene sulfonic acid by spraying a water solution of the reagent on the gel surface (47). (vi) Enzyme visualization. Substrate-impregnated papers (10), dimensionally stable polyamide membranes or 100 to 200 μm ultrathin agarose layers (42) containing a high concentration of the substrate and coupling dyes, can be used for enzyme location. (vii) Activity determination in eluates. In those cases in which visualization reactions are not available, the enzyme activity has to be determined in eluates of gel segments, with some unavoidable zone remixing within a single segment.

3.4 Recovery

Recovery in preparative isoelectric focusing will depend on a number of factors, which are related either to the proper separation including elution from the gel or to additional steps necessary for removal of the carrier ampholytes or concentration of the isolated fractions. Elution from granulated gels is simple, rapid, and quantitative. A loss of recovery at this step is negligible in comparison with elution from compact polyacrylamide gels (8). The recovery of isoelectrically homogeneous proteins was studied in preparative refocusing experiments for which protein recovery of 85 to 92 % was found (38). For crude protein mixtures recoveries of 80 - 90 % were determined by eluting all proteins simultaneously from gel strips removed lengthways from the layer (44); this approach gives a more reliable estimate than procedures in which protein recovery is calculated by summation of the protein content of individual isolated fractions (48). Determination of total activity in a portion of the gel layer is therefore a means of checking inactivation inherent to the separation process. Recovery may depend strongly on load capacity (49); for pronase E at loads from 0.5 to 10 mg per milliliter of gel suspension, recovery fo activity increases with increasing load from 14 to 80 %. The chelating properties of the carrier ampholytes have been implicated as resulting in the dependence of enzyme recovery on the ratio of enzyme to the carrier ampholytes. Since extreme pH values during focusing may cause denaturation contact with extreme pH values is avoided by application of the sample at a sufficient distance from the electrodes. Components that are focused at extreme pH values should be protected by efficient temperature control and a short focusing period. The risk of denaturation can be at least partially avoided by establishing a pH gradient by prefocusing in the absence of the sample. The residence time of the sample can thus be substantially reduced, in some instances at the expense of not reaching steady-state conditions.

Dialysis, electrodialysis, ultrafiltration, salting out, gel chromatography, ion exchange chromatography, hydrophobic interaction chromatography, and two-phase extraction with n-pentanol have been suggested for the removal of carrier ampholytes. A recently described technique is based on electrophoresis of carrier ampholytes through a dialysis membrane into a filter paper sheet soaked with buffer (10). The proteins are retained by the membrane and can be recovered with nearly 100 % yield. The technique is simple, flexible with respect to gel volume and processing of multiple samples. The technique can be easily used with most commercially available equipment for flat-bed isoelectric focusing with buffer vessels of sufficient capacity.

References

1. Radola, B.J., in: Modern Methods in Analytical Protein Chemistry, (Tschesche, H., ed.), in press, Walter de Gruyter, Berlin 1983.

2. Vesterberg, O.: Methods Enzymol. 22, 389-412 (1971).

3. Fawcett, J.S., in: Isoelectric Focusing (Catsimpoolas, M., ed.), pp. 173-..., Academic Press, New York 1976.

4. Bier, M., Egen, N.B., Allgyer, T.T., Twitty, G.E., Mosher, R.A., in: Peptides: Structure and Biological Function, (Gross, E., Meinenhofer, J., Eds.), pp. 79-89, Pierce Chemical, Rockford 1979.

5. Jonsson, J., Rilbe,H.: Electrophoresis 1, 3-14 (1980).

6. Valmet, E.: Sci. Tools 16, 8-13 (1969).

7. Jonsson, M.,Ståhlberg, J., Fredriksson, S.: Electrophoresis 1, 113-118 (1980).

8. An der Lan, B., Chrambach, A., in: Gel Elctrophoresis of Proteins: A Practical Approach, (Hames, B.D., Rickwood, D., eds.), pp. 157-187, IRL Press, London 1981.

9. Server, P., Hayes, S.J.: Electrophoresis 3, 80-85 (1982).

10. Frey, M.D., Radola, B.J.: Electrophoresis, 3, 216-226 (1982).

11. Hjertén, S.: Arch. Biochem. Biophys.Suppl. 1, 147-151 (1962).

12. Vesterberg, O.: Biochim. Biophys. Acta 257, 11-19 (1972).

13. Rosén, A., Ek, K., Aman, P.: J. Immunol. Methods 28, 1-11 (1979).

14. Saravis, C.A., Zamcheck, N.: J. Immunol. Methods 37, 315-323 (1980).

15. Ebers, G.C., Rice, G.P., Armstrong, H.: J. Immunol. Methods 37, 315-323 (1980).

16. Chapuis-Cellier, C., Arnaud, P.: Anal. Biochem. 113, 325-331 (1981).

17. Cantarow, W., Saravis, C.A., Ives, D.V., Zamcheck, M.: Electrophoresis 3, 85-89 (1982).

18. Thompson, B.J., Dunn, M.J., Burghes, A.H.M.: Electrophoresis 3, 307-314 (1982).

19. Manrique, A., Lasky, M.: Electrophoresis 2, 315-320 (1981).

20. Quindlen, E.A., Mc Keever, P.E., Kornblith, P.L., in: Electrophoresis '81 (Allen, R.C., Arnaud, P. eds.), pp. 539-548, Walter de Gruyter, Berlin 1981.

21. Mc Lachlan, R., Cornell, F.N., in: Electrophoresis '82, (Stathakos, D., ed.), in press, Walter de Gruyter, Berlin 1983.

22. Radola, B.J., in: Isoelectric Focusing (Catsimpoolas, N., ed.) pp. 119-171, Academic Press, New York 1976.

23. O'Brien, T.J., Liebke, H.H., Cheung, H.S., Johnson, L.K.: Anal. Biochem. 72, 38-44 (1976).

24. Harpel, B.M., Kueppers, F.: Anal. Biochem. 104, 173-174 (1980).

25. Otavsky, W.I., Bell, T., Saravis, C., Drysdale, J.W.: Anal. Biochem. 78, 302-307 (1977).

26. Flieger, M., Frey, M.D., Radola, B.J.: Electrophoresis, in press.

27. Radola, B.J.: Electrophoresis 1, 43-56 (1980).

28. Kinzkofer, A., Radola, B.J.: Electrophoresis 2, 174-183 (1981).

29. Nguyen, N.Y., Chrambach, A.: J. Biochem. Biophys. Methods 1, 171-187 (1979).

30. Altland, K., Kaempfer, M.: Electrophoresis 1, 57-62 (1980).

31. Låås, T., Olsson, I.: Anal. Biochem. 114, 167-172 (1981).

32. Bjellqvist, B., Ek, K., Righetti, P.G., Gianazza, E., Görg, A., Westermeier, R., Postel, W.: J. Biochem. Biophys. Methods 6, 317-339 (1982).

33. Görg, A., Postel, W., Wesermeier, R.: Anal. Biochem. 89, 60-70 (1978).

34. Radola, B.J., in: Electrophoresis '79, (Radola, B.J., ed.) pp. 79-94, Walter de Gruyter, Berlin 1980.

35. Burghes, A.H.M., Dunn, M.J., Dubowitz, V.: Electrophoresis 3, 354-363 (1982).

36. Ansorge, W., De Maeyer, L.: J. Chromatogr. 202, 45-53 (1980).

37. Allen, R.C.: Electrophoresis 1, 32-37 (1980).

38. Radola, B.J.: Ann. N.Y. Acad. Sci. 209, 127-143 (1973).

39. Neuhoff, V., in: Electrophoresis '79, (Radola, B.J., ed.)

40. Frey, M.D., Radola,B.J.: Electrophoresis 3, 27-32 (1982).

41. Allen, R.C., in: Elektrophorese Forum '82, (Radola, B.J., ed.), pp. 40-52, München 1982.

42. Kinzkofer, A., Radola, B.J.: Electrophoresis 4, in press, (1983).

43. Harris, H., Hopkinson, D.A.: Handbook of Enzyme Electrophoresis in Human Genetics, North-Holland, Amsterdam 1976.

44. Radola, B.J.: Biochim. Biophys. Acta 386, 181-195 (1975).

45. Bonitati, J.: Biochem. Biophys. Methods 2, 344-356 (1980).

46. Bonitati, J.: Electrophoresis 3, 326-331 (1982).

47. Merz, W.E., Hilgenfeldt, U., Dörner, M., Brossmer, R.: Hoppe-Seyler's Z. Physiol. Chem. 355, 1035-1045 (1975).

48. Delincée, H., Radola, B.J.: Eur. J. Biochem. 52, 321-330 (1975).

49. Radola, B.J.: in Isoelectric Focusing (Arbuthnott, J.P., Beeley, J.A., eds.) pp. 182-197, Butterworth, London 1975.

A SERUM ENZYME ANOMALY: BINDING OF ENZYMES WITH IMMUNOGLOBULINS

Motoshi Kitamura

Department of Clinical Chemistry,
Toranomon Hospital and Okinaka Memorial Institute
for Medical research,
Toranomon 2-2-2, Minato-ku,
Tokyo, 105, JAPAN

1. Introduction

The general abnormalities in enzyme titer or pattern which
occur in disorders should not be called anomalies. Rather, an
anomaly is the name given to those abnormalities which can not
be explained clinically. In many cases, such special abnormalities are seen as unreasonable data, for example laboratory
mistakes. However, if they are not regarded as "mistakes" but
examined further, such scrutiny often may lead to new and
unexpected findings.

For example, in our clinical laboratory, a serum sample
with a markedly low lactate dehydrogenase (LDH) activity of 77
IU/L (international units per liter) was found. As the clinical
meaning of low LDH activity is unknown, by simply reporting the
data, nothing would have been understood. However, the low
activity was clearly abnormal. The results of examining the LDH
isozyme by electrophoresis showed that the isozyme pattern was
entirely different from that of normal serum. Only one LDH-5
isozyme was detected. When the LDH in the erythrocytes of the
same patient was examined, it was also found that LDH was
composed of just one LDH-5 isozyme.[1]

This case, reported in 1971, was the first instance of
H-subunit LDH deficiency. To date, no other cases of this
extremely rare hereditary enzyme abnormality have been
reported. This H-subunit LDH deficiency, encountered by

chance, gives us the two following important concepts. One is that extremely rare disease-states may be hidden in electrophoretic abnormalities found by routine clinical testing; the other is that in the isozyme pattern, individual genetic information is expressed in the enzyme as a protein which exists in minute quantities.

In this communication, among the various enzyme anomalies the anomaly linked with immunoglobulin (Ig), the enzyme-immuno-globulin complex is reviewed. This research is thought to have begun around 20 years ago with the report in 1964 by England's Wilding. In the early stages of research, this was thought to be extremely rare, but when isozyme analysis was implemented as a routine test and specimens of numerous patients were analyzed, report of anomalies linked with immunoglobulin rapidly followed one after another.

It is interesting to note that the first report of each anomalous enzyme is limited to Europe or Japan. [2-10] There are some Japanese findings not cited in foreign papers, since not all Japanese research is published in a western language, but the discoveries of alkaline phosphatase (ALP) in 1975 through to that of acid phosphatase (ACP) in 1982 [10] were, in fact, made in Japan. In the past 20 year period from 1963 to 1982, a total of over 300 papers were published, including 149 papers on amylase. More than a half of these papers were reported in Japan. One characteristic of the work is that it is not conducted as a academic work in the research laboratory, but rather comes from the clues of anomalies encountered during routine work done in hospital clinical laboratories.

II. Detection of Enzyme-Ig Complex

The anomalies found during serum enzyme analysis can be roughly divided into two categories, i.e., unusually high levels of serum enzyme activity, and abnormally fluctuating patterns in electrophoretic analysis. In the former, the most widely known phenomenon is the unreasonably high serum amylase value coupled with low urinary excretion of the enzyme. The activity of serum and urine amylases is the most common clinical measurment in the diagnosis of pancreatic and parotid disorders. When these disorders have been excluded, the main reason for the contradictory high level of serum amylase may come from the link with immunoglobulin. Wilding's discovery of amylase-Ig complex was no doubt discovered in the same way.[2]

In one celiac disease patient with no abnormality in the pancreas, salivary gland, or kidney, hyperamylasemia continued for 8 years. When the extremely low value of amylase in the urine, in other words low renal clearance of amylase, was found, Wilding imagined that the amylase became a macro molecule and could not be excreted in the urine. In order to verify this, gel filtration was carried out and it was shown that this amylase was much larger than normal. From the observation that the variation of serum r-globulin correlated well with that of serum amylase, Wilding speculated that the amylase was bound to r-globulin.

Table 1 shows the laboratory findings for two cases with aspartate aminotransferase (AST) anomaly [6,7] and a case with ACP anomaly[10] discovered in our laboratory. The clue to the discovery of the AST and the ACP immunoglobulin complexes was the abnormally high level of AST or ACP, despite otherwise normal laboratory findings. LDH anomaly is also often found when LDH is slightly or even fairly high while all other liver function tests are normal.

The abnormality in the electrophoretic pattern is, in other words, the abnormality of the isozyme pattern. Isozyme analysis is carried out as routine work in many Japanese hospitals. Accordingly when a strange zymogram is obtained, it is easy to see through more detailed tests that this is the result of the link with Immunoglobulin. Isozyme analysis is routinely done on ALP, LDH, and more recently amylase and creatine kinase (CK), thereby increasing the number of reports of anomaly in the clinical laboratory.

Table 1. Serum Enzyme Levels of Two Cases with AST Anomaly and A Case with ACP Anomaly Found in Our Laboratory

Enzyme		M.S.,52,F	S.T.,46,F	M.H.,39,M	(Normal Range)
		IgGκ	IgGκ	IgAκ	
AST	K.U	**202**	**163**	15	(7- 24)
ALT	K.U	**13**	**9**	10	(3- 25)
LDH	Hill U	236	161	161	(121-223)
ALP	K.A.U	8.7	3.8	4.8	(2.8-8.4)
γ-GT	IU/l	7	18	17	(4- 50)
AMY	S.U/dl	58	65	97	(42-127)
CK	IU/l	53	67	86	(-100)
CHE	ΔpH	1.3	1.9	1.0	(0.7-1.6)
ACP	K.A.U	–	–	**12.0**	(0.3-0.9)

Kanemitsu of Kurashiki Central Hospital reported the first CK-Ig complex. [5,11] In response to the needs of clinicians, Kanemitsu introduced to the laboratory CK isozyme analysis by agar gel electrophoresis for diagnosing myocardial infarction. Because the diagnosis of myocardial infarction was the direct purpose, 600 cases of serum with high serum LDH activity were collected, and the tests were conducted to see what kind of CK isozyme patterns they showed. One in every 100 specimens exhibited an unusual electropherogram with an anomalous CK-MM band. This was in 1978, when the existence of immunolobulin-linked LDH, amylase, and ALP was already known; thus by gel filtration it was shown that CK with abnormal CK zymograms were macro molecules. Moreover, by enzyme immunoelectrophoresis, it was demonstrated that all 5 cases were bound to IgA-lambda type immunoglobulin. All patients of the high molecular mass CK were patients with carcinoma. The well-known discovery of ALP-Ig complex by Nagamine [4] also came from the abnormality of an ALP electropherogram. This work is valuable research study for the clinical laboratory which daily examines numerous blood samples from patients.

It was in the late 1970's that the number of reports

concerning enzyme-Ig complex increased. In 1976 the Japan Society of Electrophoresis held a symposium entitled "LDH Anomaly Discovered through Electrophoresis".[12] Nagamine, one of the participants, reported the high occurrence of LDH-Ig in chronic hepatitis and other hepatic disorders, suggesting the link with disease.

In 1981, in response to a proposal of Professor Hirai of Hokkaido University, Kanno of Hamamatsu Medical College organized a research group to investigate enzyme-linked immunoglobulins in Japan. This research group received a scientific research grant from the Ministry of Education. In its first year, the group conducted a survey to ascertain the vast number of unreported enzyme anomalies in Japan. Questionnaires were sent to 240 selected hospitals, of which 105 responded. Some of the results of the survey[13] are given below.

III. Occurrence of Enzyme-Ig Complex in Japan

Apart from enzyme-linked immunoglobulins, the subtype of common immunoglobulins as the components of serum proteins in healthy adult subjects exhibits the following features. For the heavy chain, γ-chain (IgG) accounts for the overwhelming majority, (IgA) for around 1/6, whereas μ(IgM) and δ(IgD) are minimal. For the light chain, the kappa type slightly dominates the lambda type, but the difference is not great.

On the other hand, the distribution of immunoglobulins bound to enzymes is clearly different from that of the serum immunoglobulins as given in Table 2 which shows the types of the immunoglobulins linked to 7 different enzymes. As there are very few cases of alanine aminotransferase (ALT), AST, and ACP, any conclusion regarding the frequency of these types of linked immunoglobulins is premature. However, in the case of other enzymes, a clear difference in distribution pattern was observed. For the heavy chain, the frequency of IgA linked to

48

Table 2. Distribution of Types of Enzyme-Linked Immunoglobulins Found in Japan

	IgA				IgG				IgM				IgA+IgG				IgA+IgG-IgM				Total
	K	L	K+L	ND*	K	L	K+L	ND	K	L	K+L	ND	K	L	K+L	ND	K	L	K+L	ND	
AMY	35	32		15	5	11							1	1	4						104
LDH	158	7		46	45	21	12	13				1	6	6		1			2		318
ALP					7	31	6	3	1												48
CK		7	2	11	4	3		1							2	1					31
AST					2	4									1						7
ALT					4	1	2	9													16
ACP	1																				1

* ND:Not determined

amylase is much greater than that of IgG, a trend also seen for
LDH and CK. In the 48 reported cases of immunoglobulin linked
with ALP, almost all were IgG and the lambda type in light
chain was dominant. For the light chain, the Kappa type was
involved in all cases of IgA linked with LDH.

Table 3 shows the classification of immunoglobulins from
241 cases whose disorders were ascertained. The most numerous
were liver followed by malignancy, circulatory, and lung
disorders, etc. The findings that a substantial number,
35cases, was detected in a healthy population, as listed in the
bottom column in the table, is especially notable.

In autoimmune disease IgG accounts for over 80% of the
immunoglobulins. After taking into account the average value
of LDH-linked IgG presented in Table 2, this indicates that IgG
is very frequently involved in autoimmune disease. The
distribution of IgG is also quite large in cases with malignant
tumours and pulmonary disorders, of which 1/2 are pulmonary
tuberculosis. The difference in the distribution patterns
between the cases with these disorders and healthy population
is that IgG is uncommon, and IgA accounts for the majority of
Ig in healthy persons. This will be discussed later in more
detail. As another characteristic, the co-existance of IgA and
IgG in malignant tumours tends to be greater than in other
disorders.

A number of cases of enzyme-immunoglobulin complex has
been thus merely collected, but the frequency is not known.

Table 3. Disorder of Cases with Enzyme-Immunoglobulin Complex

Disorder	Total	IgA	IgG	IgA+IgG	IgM	IgA+IgG+IgM
Liver	40	19	19	1	1	
Malignancy	36	17	12	7		
Circulatory	34	23	9	2		
Lung	18	6	9	2		1
Autoimmune	17	3	13			1
Endocrine	13	9	4			
Biriary tract	7	6	1			
Stomach	4	3	1			
Others	37	27	9	1		
Total	206	113	77	13	1	2
Healthy	35	30	4	1		

According to the Ministry of Health and Welfare's classification of disorders in the nation's population, cerebral apolexy and malignancies head the list, while pulmonary disorders ranked fifth and autoimmune diseases are scarce. Consequently the conspicuous appearance of IgG in liver and autoimmune disorders suggests a significant relation between the IgG immunoglobulin and these disorders.

IV. Enzyme-Ig Complex and Disease

Table 4 shows the results from liver and autoimmune disorders. There were many cases of the complex detected in liver cirrhosis and chronic hepatitis. This was also true in disorders such as ulcerative colitis and rheumatoid arthritis where it is thought that an abnormality of immunity is closely related to the disease itself. In ulcerative colitis, a very rare disorder, 6 examples of the complex were found, 5 of which were LDH-IgG complex.

It is quite some time since the relation between

ulcerative colitis (UC) and ALP immunoglobulin was first reported. Suzuki noticed the appearance of an abnormally slow moving ALP isozyme in UC patients when ALP was fractionated into isozymes using agar-gel electrophoresis. This was named ALP-VI.[14] The fact that this extra-band is frequently detected in UC was studied from various angles by Israel's Streifer[15], England's Qirbe[16], and Japan's Miki[17]. From all these authors' work, it became evident that the decrease or disappearance of the extra-band is related to the progress of UC, and also that ALP-VI is very closely related to ALP-II of hepatitic origin, and in UC, ALP-II is transformed into ALP-VI. In 1977, Kano demonstrated that ALP-VI is an ALP-Ig complex.[18]

Table 4.   LDH-Linked Immunoglobulins
in Liver and Autoimmune Disorders   1981, Japan

Disorders	Total	IgA	IgG	IgA +IgG	IgM	IgA +IgG +IgM
Liver disorders						
Liver cirrhosis	13	4	8	1		
Hepatitis	11	5	6			
Chronic hepatitis	10	8	2			
Acute hepatitis	2	1	1			
Subacute hepatitis	2		2			
Posttransfusion hepatitis	1				1	
Hepatic fibrosis	1	1				
Autoimmune disorders						
Ulcerative colitis	6	1	5			
Rheumatoid arthritis	6	2	4			
Sjögren's syndrome	2		2			
Hashimoto's disease	1		1			
Lupoid hepatitis	1		1			
SLE	1					1

Table. 5. Occurrence of Enzyme-Immunoglobulin Complexes in Patients with Ulcerative Colitis

	LDH						ALP						Not detected
	IgA			IgG			IgA			IgG			
	κ	λ	ND*	κ	λ	ND	κ	λ	ND	κ	λ	ND	
Leroux-Roels 1981 N:20		1			1						3		15
Tozawa 1983 N:76					3	5					11		57

* ND:Not determined

As results of the work focusing on UC accumulate, a disorder whose cause was formerly unknown, may be added to the growing list of those associated with the ALP-Ig complex as shown for 1 out of 20 cases of Leroux-Roels[9] and 8 of 76 cases of Tozawa[20], shown in Table 5. The upper column is for 20 cases from Leroux-Roels, and the lower from Tozawa's 76 cases. These are of great interest in that both found the immunoglobulin of the Ig Complex to be an IgG-lambda. Another interesting fact is that these two research groups also found LDH-Ig complex in UC patient's serum in addition to the ALP-Ig complex. Except for the one case of IgA in Belgium, the results in both countries are remarkably similar. By adding both complexes, ALP and LDH, one out of every 4 UC patients had an enzyme-Ig complex. UC is a very rare disease, and ALP-Ig complex is not found as often as described above.

In Tozawa's other research[21], screening by immunoelectro-syneresis was done on the serum of 13,000 patients who visited the hospital, and 33 cases of ALP-Ig complex were found. This is a frequency of 0.25% compared to 15% among the cases of UC, or 60 times the rate of occurrence in a population unaffected by UC. Moreover there is a high degree of frequency of LDH-Ig complex as well. In other words, although the reason for the relation is unclear, there is no doubt that there is a close relation between UC and the enzyme-Ig complex. However, the possibility of interference of liver disorder, rather than simply UC, should be considered, since fatty liver and pericholongitis often occur together with UC; ALP-II is the principal ALP linked with Ig in UC-serum, i.e.: it is of hepatitic origin, and there are many instances of liver disorders among disorders where LDH anomaly is found.

One more point about CK must be added regarding the relation between Ig-linked enzymes and disease. Table 6 shows the disease in 31 patients with CK-Ig complex examined in Japan, along with the type of heavy and light chains of Ig.

Table 6. Disease of Patients with CK-Immunoglobulin Complex

	Total cases	IgG		IgA			G+A	Not identified
		κ	λ	κ	λ	κ+λ	κ+λ	
Malignant disease	11	1			5	2	1	2
Muscular dystrophy	8							8
Myositis	2							2
Diabetes mellitus	2		1		1			0
Ulcerative colitis	1	1						0
Others	2	2						0
Not reported	5		2		1		1	1
	31	4	3	0	7	2	2	13

From this two distinct features can be seen. One is the prevalence among certain diseases, with many cases found in malignant tumours and muscular diseases. The other is that IgA-lambda is dominant especially in malignant tumours. The investigators in this country[5,11,22] demonstrated that the subunit of CK linked with Ig is M, in other words the most common form of the link is CK-MM-IgA complex.

Regarding the prevalence in certain diseases it is necessary to take into consideration that many of the patients

who visit the hospital are cancer patients and that there are
many muscular diseases such as muscular dystrophy among the
patients whose serum CK, in particular CK isozyme, are
measured, in other words the possibility of correlation with
the number of patients investigated. Even if the possibility
is not discounted, the results of the survey in Japan raise an
interesting problem, since the characteristics shown in this
table are entirely different from those reported in Europe and
the U.S. In reports from those countries, the Ig linked with
CK in the serum is IgG, and the connection between the
appearance of the complex and the disease is not recognized.

Germany's Stein[23] examined in detail many CK anomalies
detected by electrophoresis, or the so called macro CK, and
identified two anomalies. One was the CK-BB isozyme linked to
Ig, and the other was the macro-CK thought to be of
mitochondrial origin. The former IgG-linked CK-BB was found
often in older subjects with no correlation with disease, while
the latter macro-CK not linked to Ig was found in cases bearing
malignant tumour. Results of the close correlation between
the mitochondorial macro-CK and malignancy have accumulated.
However, the problem now is to discover the reason for the
different results in Japan and the other countries.

V. LDH-Ig complex in Healthy Population

Some important findings of our investigation related to
disease are presented below. Table 7 shows the results of the
work of Tsutsumi and Nagamine[24] regarding the occurrence of
enzyme-linked immunoglobulin in a healthy population. They
examined the sera of 14,543 volunteer blood-donors at the Red
Cross Hospital in Fukuoka, and found 21 cases of LDH-Ig
complex, or 0.14%. The ratio of men to women was 4 to 3.
Without exception, all 21 cases were the IgA-Kappa type. The
age dependency of the frequency of occurrence shows that there
was a higher or frequency in younger age groups, i.e., 0.4% in

Table 7.   Occurrence of LDH-IgA Kappa Complex
                         in Healthy Population

Age	Number of healthy subjects examined	Number of cases with complex	Male	Female	Frequency %
~ 20	3,494	2	1	1	0.06
21 ~ 30	3,884	7	5	2	0.18
31 ~ 40	2,523	10	5	5	0.40
41 ~ 50	2,156	1		1	0.05
51 ~ 60	1,685	1	1		
61 ~	801	0			0
Total	14,543	21	12	9	0.14%

Tsutsumi & Nagamine , 1982

the 30's (1 in 250) and 0.18% in the 20's (1 in 550).

In the distributions of the nationwide enzyme-linked immunoglobulins presented in Table 2, the LDH-Ig complex, IgA accounts for 66% with kappa the majority of the light chains. Thus, there is a strong possibility that these LDH-Ig complexes are unrelated to a specific disease.

Figure 1 shows the distribution of LDH-Ig complex by age group, obtained from a nationwide questionnaire. The dotted line represents the distribution curve of patients in Japan, according to an inquiry by the Ministry of Health and Welfare. In this figure IgG is shadowed for easy comparison to IgA. When comparing the patient distribution-curve with the distribution of cases with LDH-Ig complex, IgG complex cases are more prevalent in the older age group, while the IgA complex cases are clearly more prevalent in the younger age groups. The heavily outlined bar graphs represent patients, and the others are reported cases of healthy subjects. The distribution of IgA complex in healthy subjects predominants in the 20's and 30's age brackets.

These results suggest that there are at least two types of mechanisms for the formation of the enzyme-Ig complex.   In

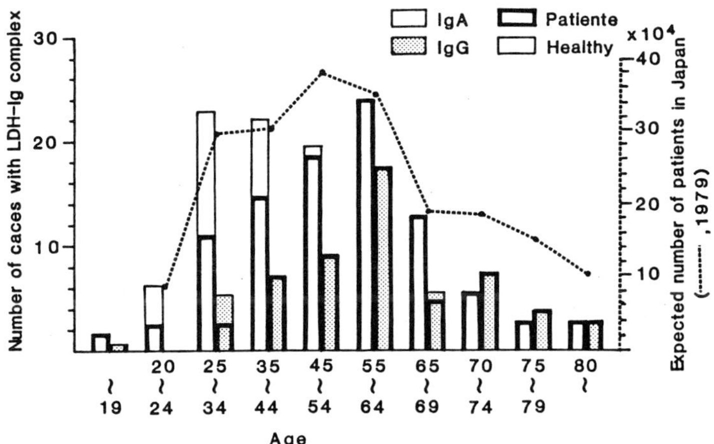

Fig. 1. Distribution of LDH-Ig Complex in Japanese Population

other words, one is the relation with a disorder for some
reason or other, and the other is spontaneous occurrence
probably unrelated to disease. As for LDH, the IgA-kappa type
complex occurs, especially in younger persons at a rate of 1 in
200 to 1 in 500. IgA is overwhelmingly involved in the amylase
Ig complex; moreover there is no clear link with disease but it
is possible that this Ig-complex occurs spontaneously as is the
case for LDH-IgA. No long-term follow-up studies to determine
whether or not disease occurs later on these cases with the
enzyme-Ig complexes probably unrelated to disease have been
appeared. The further investigation on this line is necessary.

VI. Complex Production and Mode of Binding

     Whether or not the Ig linked to enzyme is an atuo-
antibody, and whether or not that complex is an immune complex

is of great interest. When Ganrot[3] first reported a patient
with LDH-Ig complex in 1967, the fact that this case was
affected with lupoid cirrhosis suggested the possibility that
an immune complex was present. Experiments to demonstrate this
possibility were carried out fairly early in Japan and much
data are being accumulated.

The link of Fab to antigen is the most useful leading
evidence for immune complex. Imoto of the Kobe Shinko
Hospital[25] was the first to demonstrate the link with Fab in
LDH-Ig complex, followed by the Keio University's Kano.[26]
After that, a few cases where the enzyme-Ig complex was
demonstrated to be a Fab complex were reported. Investigators
in Japan have been interested for quite some time in whether or
not Ig is an auto-antibody. In these studies, Imoto, who found
the patient with lupoid hepatitis also demonstrated that the
patient's İgG linked to LDH is monoclonal in nature. There is
little room for doubt that this Ig is an auto-antibody.

Demonstration of the link with Fab is done, as is well
known, by hydrolizing Ig with papain or pepsin, and then
staining the enzyme activity of the product on the
precipitation-band produced by the anti-Fab antibody, or by
separating Fc from the product using protein A Sepharose and
determining the enzyme activity of the supernatant. It is
hoped that the Enzyme-Ig will be purified and that these
experiments will be pursued.

An even more interesting phenomenon in Imoto's case is
shown in Fig. 2. The photogram is the results of mixing this
patient's serum (B) with that of other hepatitis patients (A)
and conducting electrophoretic analysis. The proportions of
the mixtures are also shown. From the first, this patient's
serum showed only single band in the position of LDH-4. When
it was added to other serum it linked first with the LDH-2
fraction and the band disappeared, then it linked with LDH-1
and -5, and finally linked with all LDH isozyme fractions to
become only a single mobile band in the LDH-4 position. It is
well known that LDH is a tetramer composed of H and M subunits,

and the sequence of affinity for this LDH isozyme, from LDH-2, -5 equal -1, to -3, cannot be explained from the combination of the subunits.

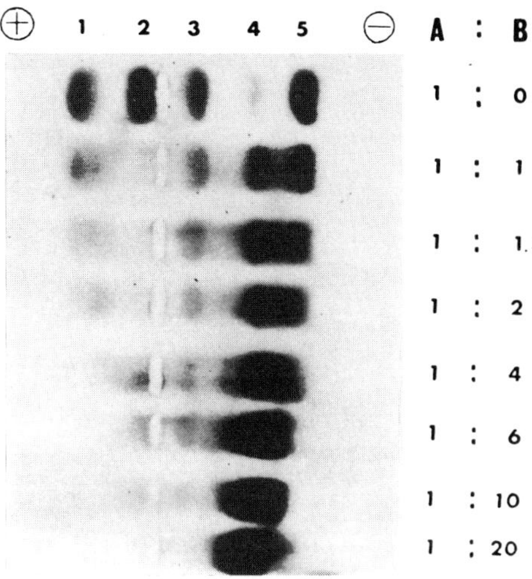

Fig. 2.   LDH isozyme patterns of mixtures of normal (A)
and the patient's (B, Imoto's case) sera

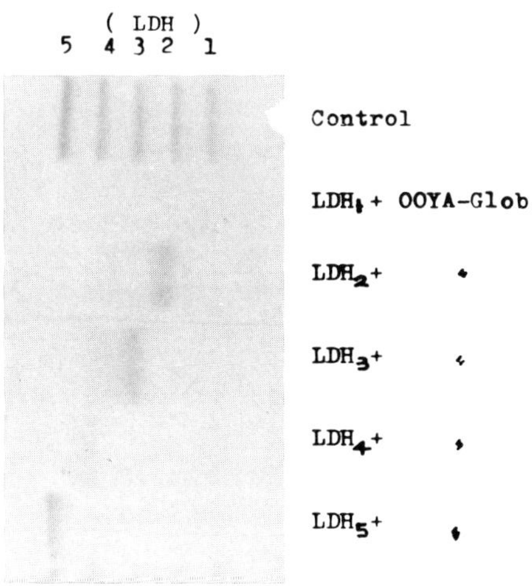

Fig. 3   Formation of complex between
LDH and immunoglobulin

Sugita and Yakata[27] of Niigata University found a mixed LDH-Ig complex, IgG and IgA, added five purified LDH isozymes to the complex, and investigated whether or not re-combination occurs (Fig. 3). The patient's Ig linked with all 5 isozymes. This complex formation caused retardation of the zymogram, and changed the mobility to the LDH-4 position. By Sephadex-G 200 gel filtration it was proven that a macro-molecule brought about the change in mobility. Such results indicate that the LDH antigenic determinant might be not related to the H, M subunit but is common to both.

Kuwa[28], of our laboratory, investigated the formation of LDH-Ig complex from the isozyme-standpoint. The Ig investigated was IgA linked to LDH. After spliting the complex by treatment at pH 3.4, serum proteins including the LDH-binding IgA-K were fractionated into 13 fractions by gel chromatography, at the same pH. Fraction numbers 5 to 8 contained the Ig. Each of the 5 LDH isozyme preparations was added to the fractions through gel filtration, which were then analyzed electrophoretically. In the fractions containing IgA a change in the mobility of LDH-2 and LDH-3 was observed and re-combination was demonstrated. No reaction with LDH-1, LDH-4, or LDH-5 was observed. Thus our results also exhibit a phenomenon which cannot be explained from the subunit structure of LDH.

Table 8.   Molecular Weight of Enzyme-Immunoglobulin Complex

Enzyme	M.W. of Enzyme	M.W. of Ig Complex	Method Used
AST	ca. 94,000	>800,000	Sephadex G-200
ALT	116,000	250,000	Sephadex G-200
AMY	56,000	150,000 ~ 800,000	Sephadex G-200 or Ultracentri.
ACP	100,000	500,000	Sephacryl S-300
ALP	140,000	480,000	Gradient Gel PAGE.
CK	80,000	250,000 ~ >800,000	Sephadex G-200
LDH	140,000	280,000 ~ >800,000	Sephadex G-200

Another important datum as to how enzymes and Ig link is the molecular mass of the complex. We examined by gel chromatography the molecular masses of the LDH-Ig complexes from the 10 cases. The masses of 5 samples of LDH-IgAκ complex were all in the range of 290,000 to 310,000 daltons, indicating that the enzyme and Ig linked in a molecular ratio of 1:1.

However, referring to various reports, there is a large variety in the molecular mass from different laboratories (Table 8). Both IgA and IgG have a molecular mass of about 150,000 daltons, so in samples of ALT, CK, and LDH, complexes with a 1:1 ratio may exist. Also a 2:1 ratio of ALP to Ig can be assumed, but the significance of molecular mass of over 800,000 daltons in ACP complex is unclear at this stage. It appears there are various forms of enzyme-Ig complex, whose structure will have to be characterized.

VII. Detection and Identification of the Complex

Material is needed in order to meet the above challenge.

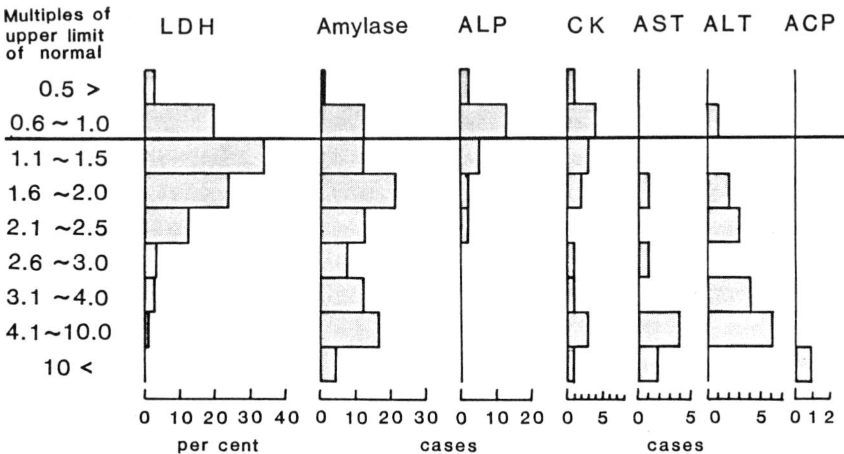

Fig. 4   Distribution of Serum Enzyme Activity in Japanese Patients with Enzyme-Ig Complexes

Consideration of the methodology for finding enzyme-Ig
complexes in routine checks in the daily clinical laboratory is
given below. The clue to Wilding's finding of amylase-Ig
complex was hyperamylasemia, as mentioned in the beginning. As
for other enzymes as well, a higher than normal value of total
activity in serum containing Ig-complex is often experienced.

Fig. 4 shows the distribution of levels of 7 different
serum enzymes in patients with enzyme-Ig complex found until
now in Japan. With 1.0 as the upper limit of normal value, the
multiples are shown in the left hand column. ALP is
distributed somewhat differently, while the other enzymes are
for the most part higher than normal, with some being extremely
high. This figure is the total of various facilities
nationwide, and as it is known that the normal reference value
for enzyme activity differs among laboratories due to technical
reasons, a more accurate comparison is needed.

Fig. 5 shows a more detailed comparison of the enzyme
activity of patients with 3 different enzyme-Ig complexes, and
of the normal reference values. LDH and amylase are the
results of our investigation, while ALP is from the results of
Tozawa[20] who investigated 1,300 patients by method of immunoelectrosyneresis. In the figure, the total enzyme activity of
each complex is classified according to its Ig-class. The
shaded squares on the left show the range of normal reference
values as the mean plus minus 2SD. Approximately 1/4 of the
ALPs are in the normal range, but there is very little LDH and
amylase in the normal range and then only close to the upper
limit. All tend toward high levels. The serum levels of
amylase and ALP in particular differ greatly individually.[29]
Thus it might be possible to generalize that there are no cases
of low levels of ALP, and in Ig complex the serum enzyme is
elevated.

Of course, the sera showing such enzyme activity are those
of patients with a variety of diseases, and not all enzyme
activity originates in the Ig complex. Thus it is difficult to
directly connect the abnormally high serum enzyme to the

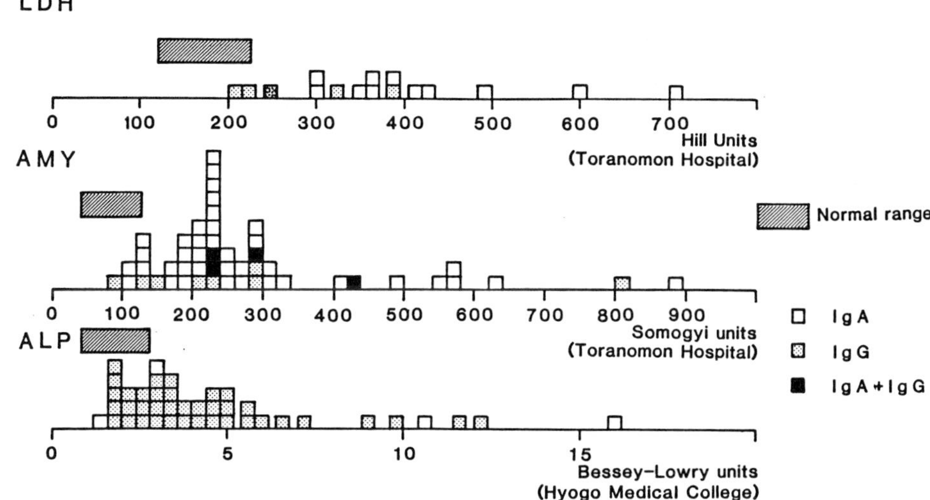

Fig. 5   Levels of Serum Enzyme Activity in Patients
with Enzyme-Ig Complex

formation of Ig complex. However, as mentioned before, in the
21 cases of LDH complex among 15,000 healthy subjects at the
Fukuoka Red Cross Hospital, the fact that all 21 showed high
serum LDH values [24] may support this conclusion.

Among the data for serum enzyme with a higher than normal
level, the existence of a higher frequency of enzyme-Ig complex
led us to develop a more effective Ig-complex screening-method.
Long[30] of Pennsylvania detected Ig-amylase complexes in 3,
or 5.8%, of 51 samples of patient's sera with values of over
200 Somogyi Units/dl. Now that determination of serum enzymes
and their isozyme have become routine, this kind of screening
procedure should be undertaken. Many patients' samples are
collected in the clinical laboratory, and it is natural that
these include many with abnormally high levels related to
diseases. Masuda[31] in our laboratory excluded cases with
serum creatinine of 2.0 mg/dl and urea nitrogen of over 30mg/dl
from among cases of hyperamylasemia, and in 8.3%, amylase-Ig

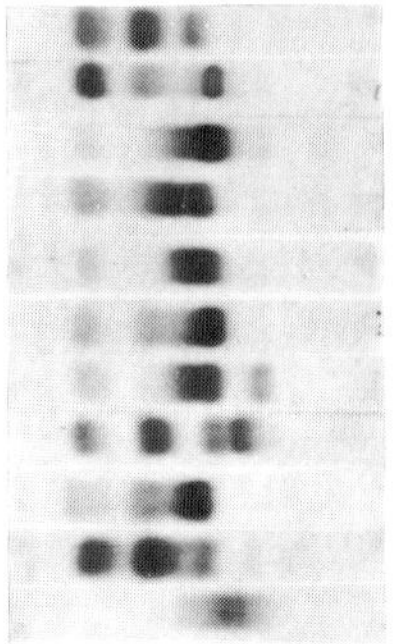

Case			LDH Enzymogram (+) 1 2 3 4 5 (−)	LDH* Hill U	AST Karmen U	ALT Karmen U	Immunoglobulin
Normal				−	−	−	−
M.O	M	71		244	23	6	IgAκ, IgGκ
G.O	M	66		499	−	−	IgAκ
S.A	F	47		328	14	11	IgAκ
T.A	F	43		499	11	4	IgAκ
M.Y	F	68		353	7	3	IgAκ
M.T	F	50		299	12	3	IgAκ
T.N	M	63		302	175	12	?
H.T	M	56		204	112	150	?
M.A	M	63		176	24	29	?
Y.A	M	54		237	8	4	IgGκ

* Normal range 126 – 202 Hill Unit

Fig. 6   10 Cases of LDH-Binding Immunoglobulin
found in 10 Months

complex was detected.

With respect to LDH, 9 years ago Kuwa[28] selected patients' samples where LDH was high while AST and ALT were normal. In the disease states showing this kind of serum enzyme pattern, there were some cases of patients with leukemia undergoing therapy. Excluding those, some 300 examples of serum were selected, and by agar gel electrophoresis 6 anomalies were found, or a frequency of 2%. Fig. 6 shows the zymograms and other laboratory findings of a total of 10 cases, the above 6 plus 4 others found by isozyme analysis ordered by clinicians. In 10 months of routine work, Kuwa found that it is not difficult to find various LDH anomalies. As mentioned before, the work in our laboratory concerning the binding form of LDH-Ig complex used the LDH anomalies thus is found as experimental material.

After screening for the enzyme anomalies in this way, the size of the molecules is ascertained, and then finally they are identified immunologically to determine whether they are Ig. Table 9 shows a practical comparison of the 4 methods

Table. 9   Method for Identification of Enzyme
Linked Immunoglobulins

Method (Assay material)	Sensitivity	Operation	Stability
Immunoprecipitation (S)*	◎ *	X	◎
Immunoelectrophoresis (P)	△	X	X
Immunofixation electrophoresis (P)	○	△	○
Immunoelectrosyneresis (P)	◎	◎	△
Immunoprecipitation (P)	◎	○	◎

* S: Supernatant , P: Precipitate

*Nonstandard abbreviations and symbols used; S; Supernatant;
P; Precipitate: ◎ , very good; ○ , good; △ , acceptable;
X , not acceptable.

generally used for measuring molecular size. It can be seen
that reliability and simplicity do not always co-exist. The
most reliable method is sucrose density-gradient centrifugation,
but as facilities, running costs, and operation present
problems, thin-layer gel-filtration method is most widely used.
This method's weak point is that enzymes such as amylase have
affinity with the supporting media resulting in retardation,
and thus molecular size is not accurately measured. Column gel
filtration would be a practical method if accuracy were
improved, especially in the range of high molecular mass.

A method using anti-serum has been used in the past for
identifying immunoglobulins, and is still in use. Antigen-
antibody complexes themselves have enzyme activity, so those
enzymes can be stained or measured. This technique has been

extensively investigated in Japan, and several methods have been developed. A comparison of some features of these various methods is shown in Table 10. Other methods in use include detection by autoradiography of the anti-enzyme antibody to

Table 10   Methods for Measurement of Molecular Mass

Method	Accuracy	Repro-ducibility	Recovery of Sample	Operation	Cost
Sucrose density-gradient centrifugation	◎	○	◎	X	X
Column gel filtration	△	◎	◎	○	○
Gradient PAGE	○	◎	X	○	○
Thin layer gel filtration	X	△	△	◎	◎

* see legend to Table 9

which a small amount of isotopelabelled enzyme has been added,

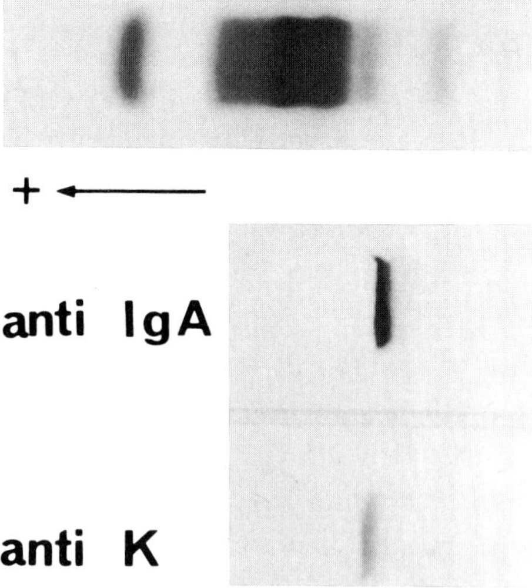

Fig. 7   Demonstration of Identification of enzyme-linked
         immunoglobulin by the method of enzyme-immunosyneresis

serum LDH zymogram without antisera is shown at the top
of the figure.

as well as removing immunoglobulins by affinity chromatography
using Protein A Sepharose. Immunoelectrosyneresis and
immunoprecipitation, however, are looked upon as the most
useful methods with a wide range of applications.
Immunoelectrosyneresis is also called counter-current
immnoelectrophoresis, and performed under standardized
conditions such conditions as the serum and antiserum move in
the opposite directions and the immuno-precipitation occurs.
Fig. 7 shows a photogram of the LDH immunoelectrosyneresis-
staining pattern developed by Nagamine[32], showing a clear
stain of the precipitates formed with anti-IgA and kappa
antisera. In this method, many materials are applied to one
sheet of cellulose acetate membrane and screening can be done
simultaneously. It can thus be used as a mass screening
method. Tozawa's new method[33] where the enzyme activity in
the precipitate is measured, is also very useful.
Standardization of methodology for accurate identification of
enzyme-linked immunoglobulins is being actively investigated as
one of the main objectives of our research group.
What has been presented so far regarding the enzyme-Ig
complex, principally of research done in Japan, can be
summarized as follows:

1. Regarding the occurrence of the complex, the enzyme-
    immunoglobulin complex is detected in most of the routine
    serum enzyme constituents. Frequencies are found to be
    from 0.1% to 0.5%, or more.
2. The enzyme-Ig complex is one of the most frequently encoun-
    tered enzyme anomalies, when total activity is elevated or
    the enzymogram exhibits an abnormal pattern.
3. It seems likely the enzyme complex may be one of the
    immuno-complexes. This has been confirmed in certain cases.
4. The frequency of heavy and light chain immunoglobulins

constituting the complex was found to be different from that of serum immunoglobulin. For example, in the LDH-Ig complex from a healthy population, all LDH-binding immunoglobulins, without exception, were the IgA-kappa type.

5. The clearest association between complex formation and a disorder was seen in the relation between alkaline phosphatase and ulcerative colitis. In CK, LDH and other enzymes, a distinct relation with disease was also suggested.

6. An explanation of the mode of binding between enzyme and immunoglobulin remains to be found. The isozyme specificity of ALP-, CK- and AST-binding IgA has been demonstrated, but for LDH, a clear explanation is not possible in view of the intriguing finding of a molecular binding ratio of 1:1.

The main purpose of this paper is not to present results of research but rather to raise and discuss some problems. There are no clear answers to the relation between the enzyme-Ig complex and disease: whether it is in fact an immune complex, why it possesses enzyme activity, why the enzyme activity in the serum rises, or what kind of molecule-level is involved in the complex formation.

The complex formation of serum enzyme and immunoglobulin, which 10 years ago was thought to be an extremely rare phenomenon related to some particular disease states, is now realized to be not so uncommon. If the enzyme-Ig complex is in fact an immunecomplex, it would be the simplest moiety in the immunecomplex and material for this investigation could be easily obtained. An important and interesting subject, that of resolving the structure and the formation mechanism of the complex, remains as a problem for the future.

References

1. Kitamura, M., Iijima, N., Hashimoto, F. and Hiratsuka, A.:
    Clin. Chim. Acta, 34, 419-423 (1971)
2. Wilding, O., Cooke, W.T. and Nicholson, G.I.: Ann. Inter.
    Med., 60, 1053-1059 (1964)
3. Ganrot, P.O.: Experientia, 23, 593 (1967)
4. Nagamine, M. and Ohkuma, S.: Clin. Chim. Acta, 65, 39-46
    (1975)
5. Kanemitsu, F., Katayama, N., Sasaki, R., Kawahishi, K.
    and Mizushima, J.: Jap. J. Clin. Pathol, 26(sup), 78 (1978)
6. Itoh, K., Nakajima, M., Kuwa, K., Nakayma, T. and Kitamura,
    M: Igaku-no Ayumi, 105, 233-235 (1978)
7. Nakajima, M., Itoh, K., Kuwa, K., Nakayama, T. and
    Kitamura, M.: Gastroent. Jap., 15, 330-336 (1980)
8. Konttinen, A., Murros, J., Ojala, K., Salaspuro, M., Somer,
    H. and Rasanen, J.: Clin. Chim. Acta, 84, 145-147 (1978)
9. Kajita, Y., Majima, T., Yoshimura, M., Hachiya, T.,
    Miyazaki, T., Ijichi, H. and Ochi, T.: Clin. Chim. Acta,
    89, 485-492 (1978)
10. Sakugi, F., Tsukada, T., Nakayama, T., Kitamura, M. and
    Shitan, U.: Physico-chem. Biol., 26, 290 (1982)
11. Kanemitsu, F., Kawanishi, I. and Mizushima, J.: Physico-
    chem. iol., 24, 301-308 (1981)
12. Kitamura, M., Omoto, K., Nakayama, T., Saga, M., Tanaka,
    F., Amino, N., Hayashi, C., Miyai, K., Nagamine, M., Kano,
    S. and Kuwa, K.: Physico-chem. Biol., 21, 169-214 (1977)
13. Kanno, T., Nagamine, M., Kano, S., Gomi, K., Yakata, M.,
    Sugita, O., Kajita, Y., Ishida, M., Kitamura, M. and
    Nakayama, T.: Physico-Chem. Biol., 26, 411-449 (1982)
14. Suzuki, H. Yamanaka, M. and Oda, T.: Ann. N.Y. Acad. Sci.,
    166, 811-819 (1969)
15. Streifeler, C., Schnitzer, N. and Harell, A.: Clin. Chim.
    Acta, 38, 244-246 (1972)
16. Qirbi, A.A. and Mass, D.W.: Clin. Chim. Acta, 60, 1-6(1975)
17. Miki, K., Suzuki, H., Ino, S., Niwa, H., Oda, T., Sugiura,

M. and Hirano, K.: Jap. J. Gastroent., 73, 162-168 (1976)

18. Kano, S., Takeshita, E., Kanno, T., Asakura, H. and
    Taniyama, M.: Physico-chem. Biol., 20, 325 (1977)

19. Leroux-Roels, G.G., Wieme, R.J. and DeBroe, M.E.: J. Lab.
    Clin. Med., 97, 316-321 (1981)

20. Tozawa, T., Satomi, M. and Shimoyama, T.: Saishin-Igaku, in
    press (1983)

21. Shibata, H., Tozawa, T., Taishi, K., Hayashi, K., Morita,
    S., Satho, H. and Okasaka, R.: Electrophoresis '83, Tokyo
    (1983)

22. Yuu, H., Ishizawa, S., Takagi, Y., Gomi, K., Senju, O. and
    Ishii, T.: Clin. Chem. 26, 1816-1820 (1980)

23. Stein, W., Bohner, J., Krais, J., Muller, M., Steinhart, R.
    and Eggstein, M.: J. Clin. Chem. Clin. Biochem. 19, 925-930
    (1981)

24. Tsutusmi, Y. and Nagamine, M.: Physico-Chem. Biol., 26, 49
    (1982)

25. Imoto. S., Uchita, K., Ota, M., Yoshida, M., Inoue, K.,
    Takatuki, K. and Yamasawa, I.: Jap. J. Gastroent., 71,
    1249-1255 (1974)

26. Kano, S., Kanno, T. and Saga, E.: Physico-Chem. Biol., 19,
    47-48 (1974)

27. Sugita, O. and Yakata, M.: Physico-Chem. Biol., 22, 151-156
    (1978)

28. Kuwa, K., Nakayama, T. and Kitamura, M.: Physico-Chem.
    Biol. 21, 209-214 (1977)

29. Kitamura, M. and Nishina, T.: Practical Clinical Chemistry,
    Suppl. ed., P.132 Ishiyaku Shuppan, Tokyo (1982)

30. Long, W.B. and Kowlessar, O.D.: Gastroenterology, 63,
    564-571 (1972)

31. Masuda, H., Tsukada, T., Nakayama, T. and Kitamura, M.:
    Physico-Chem. Biol, 27, 62 (1983)

32. Nagamine, M. and Okochi, K.: Physico-Chem. Biol, 27, 15
    (1982)

33. Tozawa, T., Taishi, K., Kuwahara, J.: Physico-Chem. Biol.,
    26, 243 (1982)

General

HIGH PERFORMANCE ELECTROPHORESIS (HPE)

Stellan Hjertén

Institute of Biochemistry, University of Uppsala, Biomedical Center,
P.O.Box 576, S-751 23 Uppsala   Sweden

Abstract

A method which is the electrophoretic counterpart of HPLC is described.
The electrophoresis tubes are made of ordinary glass (not quartz), which
much facilitates the handling of the equipment. These tubes transmit UV-
light (used for detection of low-molecular weight substances, proteins,
nucleic acids and viruses) because they have extremely thin walls (around
0.1 mm), which also allows very efficient cooling, particularly since the
diameter of the glass tubes is very small (0.05-0.4 mm). High field
strengths can thus be used without significant thermal deformation of the
zones, thereby permitting fast separations (5-50 min) which also
minimizes diffusional zone broadening. The detection system is that used
in the free zone electrophoresis equipment (1) or that in a home-built
gel scanner (2) (without moving the electrophoresis tube). A more
convenient apparatus with a liquid-cooled electrophoresis tube is being
constructed.

Introduction: The Analogy between HPLC and HPE

When one considers the parameters and the equations that govern the
transport of solutes in electrophoresis, chromatography and
centrifugation one can recognize considerable analogies among the
separation mechanisms in these methods. One can therefore expect that any
of these three techniques has a counterpart in the two others. In this
paper I shall describe the electrophoretic counterpart of high
performance liquid chromatography (HPLC), termed high performance

Electrophoresis '83
© 1984 Walter de Gruyter & Co., Berlin · New York

electrophoresis (HPE). Such a method should exhibit the same characteristic features as HPLC, namely

1) high resolution
2) short run times
3) capability to detect small amounts of material, particularly with the aid of UV-light
4) direct monitoring of the solutes (no time-consuming staining).

This can be realized by:

1) performing the runs in buffer alone or homogeneous gels such as agarose or polyacrylamide to avoid zone broadening due to macroscopic irregularities in the capillary structure of the electrophoresis medium

2) using glass tubes of a small inner diameter (0.05-0.4 mm) and a wall thickness of only 0.1-0.2 mm as electrophoresis chambers for efficient removal of the Joule heat generated to suppress the thermal deformation of a zone (high field strengths can accordingly be employed)

3) employing a highly UV-sensitive detector

4) scanning the electrophoresis tube or recording the solutes as they electrophoretically pass a stationary detector.

Equipment

The electrophoresis tubes were drawn to the dimensions desired after heating ordinary glass tubes in a Bunsen burner. The tubes obtained should have extremely thin walls (0.1-0.2 mm) to permit the passage of ultra-violet light thus eliminating the need to use quartz tubes. The tubes are filled with buffer alone or a gel supporting medium is cast in them. As the sample substances migrate electrophoretically in the

tubes they pass a stationary UV-beam striking a photomultiplier connected
to a recorder (Fig. 1). A peak on the recorder chart is thus obtained for

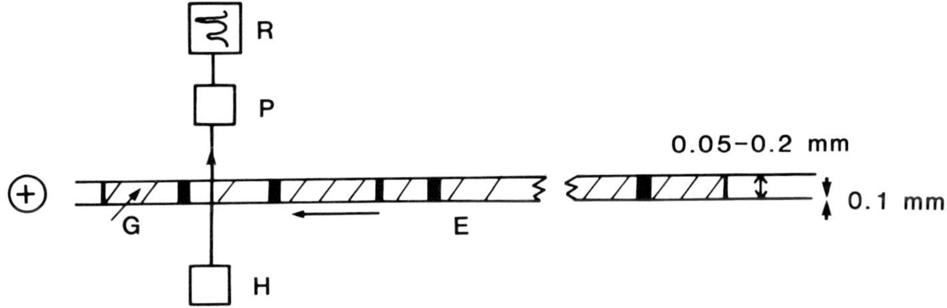

Fig. 1. An outline of the equipment for high performance electrophoresis
(HPE)

        E = thin-walled, narrow electrophoresis tube
        G = gel or buffer alone
        H = hydrogen lamp
        P = photomultiplier
        R = recorder

each UV-absorbing electrophoresis zone. The UV detection system used is
that of the free zone electrophoresis equipment (1) or a home-made
apparatus (2) for gel scanning (without moving the electrophoresis
tube). All experiments have been performed without actively cooling the
electrophoresis tube. A more sensitive and more convenient equipment with
a liquid-cooled electrophoresis tube is, however, under construction.
With this apparatus the duration of a run can be considerably reduced.

Results

Unless otherwise stated all experiments described below were done in 0.1 M
Tris-HAc, pH 8.6, as electrophoresis buffer and in a polyacrylamide gel
of the composition T = 6%; C = 3% as supporting medium (these parameters
are defined in ref. 3). The migration distances have varied from 6.5 to
8 cm, the voltage between 1100 and 3000 volts (corresponding to field
strengths of 75-200 volts/cm), and the current between 0.4 and 3.1 mA.
The volume of the sample has been 0.01-0.1 µl and the amount 0.01-5 µg.

The sample was applied by layering and was transferred into the gel by electrophoresis for some minutes at a <u>low</u> voltage (200 volts) to avoid broadening of the starting zone due to thermal convection. The monitoring of the solutes was done at 280 nm, except for the experiment shown in Fig. 3, where 265 nm was used.

Figs. 2 and 3 show HPE of artificial mixtures of pH indicators and aromatic carboxylic acids, respectively. The analyses were completed within 5-10 min. The duration of a run is thus comparable to that in an HPLC experiment.

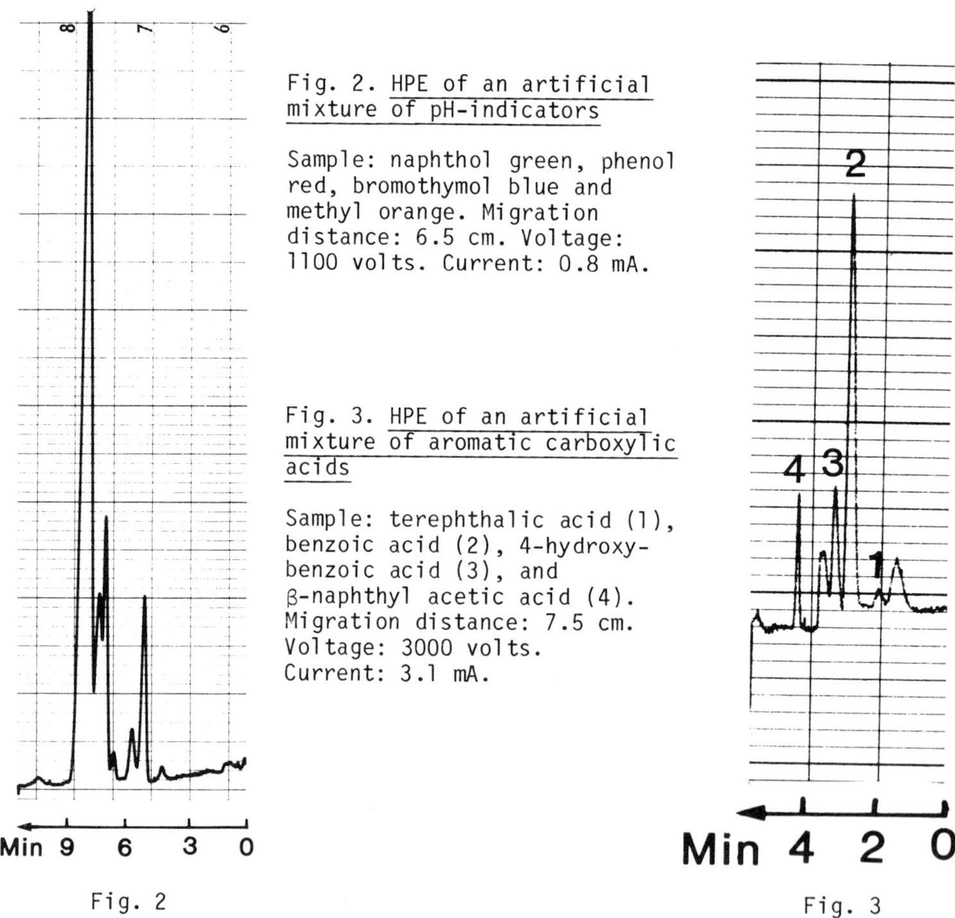

Fig. 2. <u>HPE of an artificial mixture of pH-indicators</u>

Sample: naphthol green, phenol red, bromothymol blue and methyl orange. Migration distance: 6.5 cm. Voltage: 1100 volts. Current: 0.8 mA.

Fig. 3. <u>HPE of an artificial mixture of aromatic carboxylic acids</u>

Sample: terephthalic acid (1), benzoic acid (2), 4-hydroxy-benzoic acid (3), and β-naphthyl acetic acid (4). Migration distance: 7.5 cm. Voltage: 3000 volts. Current: 3.1 mA.

Fig. 2

Fig. 3

An example of HPE of membrane proteins is shown in Fig. 4. The run was
performed in SDS in a discontinuous buffer system according to Neville
(4).

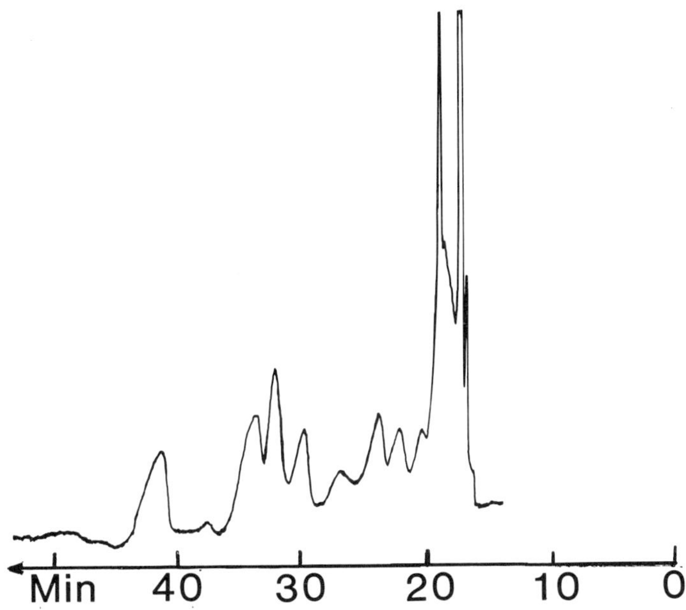

Fig. 4. HPE of membrane proteins of Acholeplasma laidlawii

   Migration distance: 8 cm. Voltage: 1200 volts. Current at the
   start: 0.6 mA.

HPE has been used in our laboratories to study the homogeneity of a
sample of tRNA from E. coli. A 0.4% non-sieving agarose gel was used
as supporting medium. The analysis, which required only 7 min (Fig. 5),
shows that the sample was electrophoretically homogeneous. The
symmetrical narrow peak indicates that all of the different species of
tRNA have the same ζ-potential.

For analysis of particles as large as TMV (tobacco mosaic virus) gels of
high porosity are required, for instance 0.2% agarose. An example is
given in Fig. 6. The peak corresponding to TMV is relatively broad

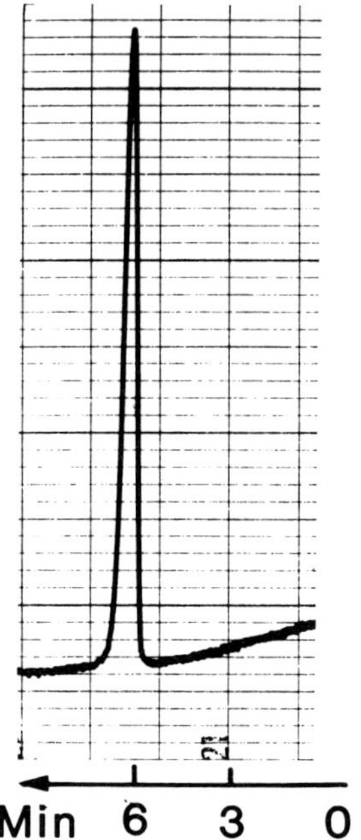

Fig. 5. HPE of tRNA from E. coli

Migration distance: 8 cm. Voltage: 1200 volts. Current: 0.8 mA.

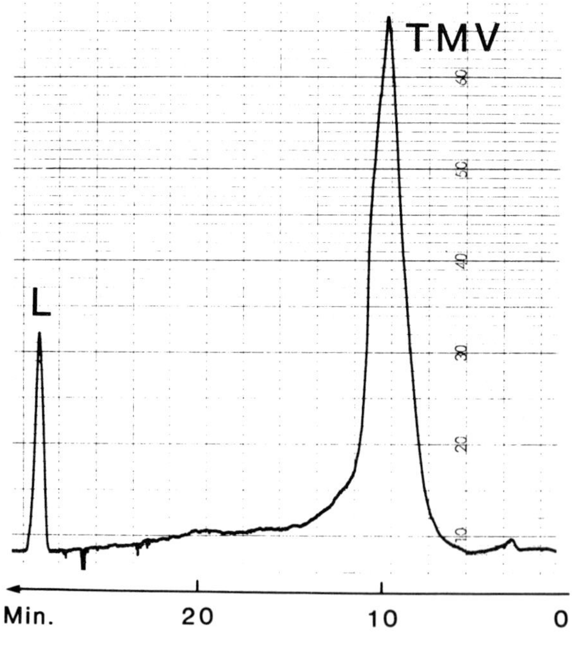

Fig. 6. HPE of tobacco mosaic virus (TMV)

Migration distance: 8 cm. Voltage: 1200 volts. Current: 0.8 mA.

compared to the last peak (L) in the electropherogram, probably due to some size-sieving in the agarose gel of TMV rods of different lengths.

Discussion

In the previously described free zone electrophoresis method (1) the runs were also performed in narrow bore tubes, although the diameters of the tubes were considerably larger (1-3 mm) than those used here. That equipment permits detection of the zones either by the same technique as described in this paper or by scanning the tube in UV-light. The latter technique is of ocurse preferable to the former but did not function properly with the small dimensions of the electrophoresis tube used in this study. The detection technique described herein, whereby the zones pass a stationary UV-beam, has also been used by Everaerts and others (5) for displacement electrophoresis and is a general method for monitoring substances in HPLC.

Polyacrylamide gel electrophoresis in narrow bore tubes has been described for instance by Grossbach (6), Hydén (7) and Neuhoff (8), and isoelectric focusing and electrophoresis in thin slabs of polyacrylamide by Delincée and Radola (9), Görg **et al.** (10), and Ansorge and de Maeyer (11). These techniques give very high resolution in short run times since high field strengths can be employed with minimal thermal deformation of the zones. The very thin-walled electrophoresis tubes employed in this investigation should be even better as to these parameters. Another advantage of the thin-walled glass tubes is that they permit passage of UV light, which in turn means that proteins, nucleic acids, viruses and other molecules and particles can be detected directly without staining, which is a prerequisite for an electrophoresis method which purports to be the electrophoretic equivalent of HPLC. Recently Jorgenson and DeArman Lukacs have described the separation of low molecular weight compounds in free solution in 1 m long glass capillaries by electroosmosis or by a combination of electroosmosis and electrophoresis (12). The detection system was based on prelabelling of

the solutes with fluorescamine and measurement of their fluorescence with
an on-column detector.

In the experiments presented here we have most often used 0.1 M Tris-HAc,
pH 8.6, as electrophoresis buffer. Those interested in extremely fast
separations should, of course, choose a more dilute buffer. However, at
very low ionic strengths one can obtain asymmetric peaks even with
homogeneous substances, particularly low molecular weight compounds. The
risk can be minimized by choosing the buffer such that the electrolyte
ion having the same sign as the solute also has a similar mobility (13).
A further decrease in the duration of a run can be achieved by
liquid-cooling of the electrophoresis tube (such equipment is under
construction).

Severe electroendosmosis, which occurs when the run is performed in
buffer alone, can be eliminated with the aid of methylcellulose in the
buffer (14) or as a coating on the glass of the electrophoresis tube
(1).

The popularity of HPLC as an analysis method for both low and high
molecular weight substances is rapidly growing. One reason for this is
certainly the fact that the analysis result is obtained by direct
UV-measurements (and not by staining) at the same time as the solutes
leave the column (point 4 in the Introduction). If an analogous detection
method is used, as herein, for electrophoresis it is very likely that
electrophoresis will increase in popularity even more rapidly than HPLC,
at least for the analysis of biopolymers, since in many cases these
macromolecules can be considerably better resolved by electrophoresis
than by HPLC. Such a detection method requires access to electrophoresis
tubes which are a) cheap (since they can be used only once for disc
electrophoresis and in a limited number of gel electrophoresis runs in
single-buffer systems); b) UV-transmitting; c) preferably not brittle
like quartz tubes (which is of importance since the tubes must be very
thin-walled for efficient dissipation of the Joule heat); d) more
hydrophilic than plastic tubes to ensure good adhesion of gels. Ordinary

narrow, thin-walled glass tubes fulfil these requirements. Therefore - and also for reasons mentioned previously - the HPE technique described herein will probably favourably compete with HPLC as an analytical method at least for biopolymer separations in sieving gel media.

Acknowledgements

The author is much indebted to Mrs. Karin Elenbring for skilful execution of the electrophoresis experiments. The work has been supported by grants from the Swedish Natural Science Research Council.

References

1. Hjertén, S.: Chromatog. Rev. 9, 122-219 (1967).
2. Fries, E., Hjertén, S.: Anal. Biochem. 64, 466-476 (1975).
3. Hjertén, S.: Arch. Biochem. Biophys., Suppl. 1. 147-151 (1962).
4. Neville, D.M.: J. Biol. Chem. 246, 6328-6334 (1974).
5. Everaerts, F.M., Beckers, J.L., Verhegen, Th.P.E.M.: Isotachophoresis, Theory, Instrumentation and Application, pp. 153-170, Elsevier, Amsterdam 1976.
6. Grossbach, U.: Biochim. Biophys. Acta 107, 180-182 (1965).
7. Hydén, H., Bjurstam, K., McEwen, B.: Anal. Biochem. 17, 1-15 (1966).
8. Neuhoff, V.: Arzneimittel-Forschg. (Drug. Res.) 18, 35-39 (1968).
9. Delincée, H., Radola, B.J.: Anal. Biochem. 90, 609-623 (1978).
10. Görg, A., Postel, W., Westermeier, R.: Anal. Biochem. 89, 60-70 (1979).
11. Ansorge, W., de Maeyer, L.: J. Chromatogr. 202, 45-53 (1980).
12. Jorgenson, J.W., DeArman Lukacs, K.: J. Chromatogr. 218, 209-216 (1981).
13. Hjertén, S.: in Topics in Bioelectrochemistry and Bioenergetics (ed. G. Milazzo), Vol. 2, pp. 89-128, John Wiley and Sons, Chichester 1978.
14. Hjertén, S.: Arkiv für Kemi 13 (16), 151-152 (1958).

SOME RECENT CONCEPTUAL ADVANCES IN MOVING BOUNDARY ELECTROPHORESIS AND
THEIR PRACTICAL IMPLICATIONS

Andreas Chrambach
Section on Macromolecular Analysis, Laboratory of Theoretical and Physical
Biology, National Institute of Child Health and Human Development, National
Institutes of Health Bethesda MD 20205, USA

Leonard M. Hjelmeland
Laboratory of Vision Research, National Eye Institute, National Institutes
of Health Bethesda MD 20205, USA

Introduction

1) Unification of the theory of electrophoresis
2) Application of moving boundary electrophoresis to charge fractionation
    a) Isotachophoresis
    b) Electrofocusing
    c) Preparative moving boundary electrophoresis on gels
3) Application of moving boundary electrophoresis to size fractionation
    a) Analytical and preparative concentrating gels
    b) Selective stacking
    c) Unstacking
    d) R_f manipulations.

During the past year, three papers have been published from this laboratory
concerning the theoretical treatment of moving boundaries. The first (1)
provided a historical perspective of moving boundary electrophoresis, and
explored the equivalence of many electrophoretic separation methods which
are commonly considered to be novel and distinct. In this paper, we concluded that in their basic theoretical treatments, moving boundary electrophoresis, isotachophoresis, multiphasic zone electrophoresis, steady state

electrophoresis and displacement electrophoresis were identical.

Our second paper (2) dealt with a reformating of the presentation of the available discontinuous buffer systems across the pH scale previously described by Jovin, Dante and Chrambach (3), as well as with the computation of several new systems which were specifically designed for use in the study of native proteins. These new systems have uniformally low trailing ion net mobilities at various trailing phase pH values for both polarities. The intent was to provide the practitioner of gel electrophoreis with a simple terminology and a restricted list of useful buffer systems.

Our third paper (4) was a theoretical investigation of the movement of boundaries when the sole available counterions are protons and hydroxyl ions. This in fact was an overt attempt to develop a theory for the establishment and the dynamics of pH gradients in buffer electrofocusing (5). This paper (4) convincingly demonstrated that not only weak acids and bases, but also biprotic ampholytes could be uniformly treated by a simple extension of the moving boundary theory, and that the prediction of both the formation and the displacement of pH gradients in these systems resulted from such treatment. It is our feeling, that although the mathematical treatment is restricted to monovalent weak acids and bases as well as biprotic ampholytes, the implications extend beyond this limited case and in fact constitute a novel approach to the understanding of the dynamics of pH gradients in electrofocusing in general.

It is the purpose of this report to spell out the practical implications of those 3 reports.

1) Moving Boundary Electrophoresis: A Unifying Concept for Contemporary Electrophoretic Separation Methods.

Electrophoretic separation methods can be divided into two simple categories, viz. electrophoresis in a buffer medium which is homogeneous as opposed to one which possesses one or more discontinuities. It should be said here that systems with multiple discontinuities of buffers are viewed simply as multiple systems of single buffer discontinuities and thus present no fundamental difference. Since practically in view of the many advantages of discontinuous buffer systems discussed below not much attention is being paid to methods involving homogeneous systems of buffers. electrophoretic theory deals nearly exclusively with an analysis of discontinuous buffer systems. Since buffer discontinuities give rise to one or several moving boundaries, we are left with the moving boundary as the principal theoretical construct of modern electrophoretic methods.

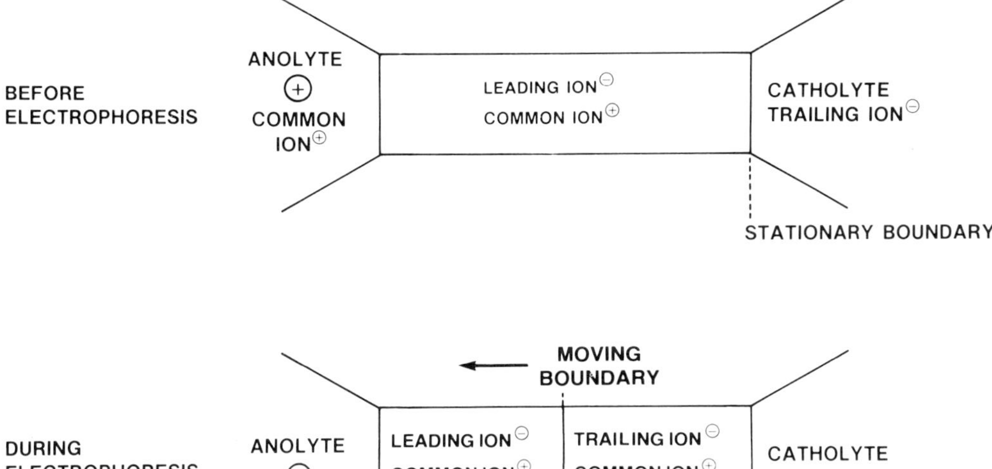

Fig.1: Schematic representation of a moving boundary and its genesis.

A typical moving boundary is shown schematically in Fig.1. The elements of this system are a leading ion, a trailing ion, which both have the same charge. and which both possess a common counterion with opposite charge. The leading ion in combination with the common counterion constitutes the leading phase. The trailing ion is separated from the leading ion by an interface (a boundary). In the electric field, the leading ion migrates out of the volume element it originally occupied, and the trailing ion migrates into that volume element. Thereby, the boundary moves in the moves in the direction of the migration of the leading and trailing ions. The moving boundary now separates the receding leading phase from an advancing new trailing phase. The volume element representing the trailing phase is the central concern of moving boundary theory since the properties of the trailing phase. i.e. the concentrations of all ions and the pH of the trailing phase, are completely determined by the properties of the leading phase (1).

When another trailing ion with an intermediate mobility is present in the volume element, this ion can create a distinct phase of its own between the leading and trailing phase by the mechanism discussed above. This phase then acts as a trailing phase for the initial leading phase of this system, and at the same time acts as a leading phase for the original trailing phase. In this way, extensions to an indefinite number of phases and boundaries is both theoretically and practically possible. In precisely the same fashion in which the buffer concentrations and pH of the initial trailing phase are determined by the leading phase in the simple system, all trailing phases in the multiple ion system are thus regulated in their phase compositions and give rise to sequentially migrating multiple moving boundaries [Fig.2 of (1)]. In such a case, the moving boundaries align in the order of the net mobilities of the constituents which define the phases. Constituents described here need not be simple buffers, but in fact may be any electrolyte, and thus include macromolecules as well.

The quantitative aspects of trailing phase "regulation" have been exhaustively treated by almost every author attempting to deal with the theory of electrophoresis. The major developments in this theory for both weak electrolytes and strong electrolytes were however essentially completed by Svensson (Rilbe) and Longsworth' school in the late 1940's (1).

More sophisticated treatments of electrophoretic phenomena than moving boundary theory have recently appeared, and are undoubtedly more rigorous in their basic assumptions (6). Such treatments, however, appear capable of dealing only with a very limited number (2-3) of ionized constituents, for reasons related to the extensive computational needs in solving systems of coupled nonlinear differential equations. Thus although the equations

of Longsworth and Rilbe are "old", their very simplicity and ease of computation makes them ideal for a general and widespread application in the
real world of electrophoretic separations.

It should be obvious at this point that the real physical basis of all of
the methods we have discussed does not depend on any particular apparatus
or technique. Thus, although multiphasic zone (disc) electrophoresis and
steady–state stacking on gels. isotachophoresis in the capillary apparatus,
displacement electrophoresis in the rotating tube apparatus, and moving
boundary electrophoresis in the Tiselius apparatus have been developed as
different electrophoretic techniques. all depend on the operation of the
same discontinuous buffer systems and share exactly the same theoretical
basis. The same holds for isoelectric focusing, although this fact becomes
practical only provided that the ionic mobilities of its constituents are
known (see below) (Fig. 2).

Moving Boundary Electrophoresis

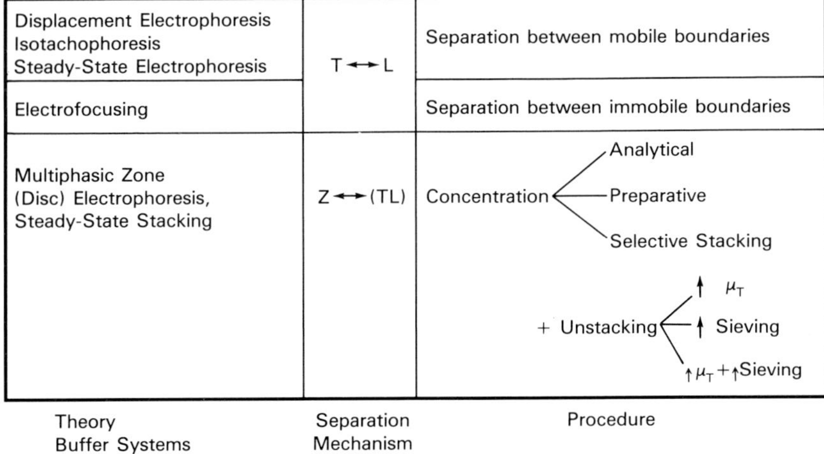

Fig.2: Modes of application of moving boundary electrophoresis.

Separation problems fall into 2 categories: Those which attempt to separate molecules differing predominantly in their net charge, by exploiting these charge differences ("charge fractionation"); and those which separate molecules differing predominantly in their size and shape, by exploiting these size differences ("size fractionation"). Commonly. isotachophoresis and electrofocusing are thought to be effective electrophoretic charge fractionation methods, and gel electrophoresis in a restrictive gel an effective size fractionation method. It seems, however, important to realize that all three methods involve different procedural forms of moving boundary electrophoresis. In isotachophoresis, separation occurs within a mobile train of sequential moving boundaries, and between the ions trailing these boundaries in order of their net mobilities at the steady state. The same is true for electrofocusing. except that here the train of sequential moving boundaries has been arrested by substituting the proton and hydroxyl ion for the common ion of the system. In the third procedure, gel electrophoresis, separation occurs between species migrating with a net mobility less than that of the trailing ion of the moving boundary or train of sequential moving boundaries; or, it may occur between species migrating within the train of moving boundaries and those migrating more slowly than the trailing ion. The deceleration of charged constituents behind the trailing ion may be effected either by accelerating the trailing ion through a change of pH; or it may be effected by decelerating a macromolecule through molecular sieving effects. In either case, we are dealing with manipulations relative to the trailing ion of a moving boundary, i.e. with moving boundary electrophoresis. Compared to gel electrophoresis in a homogeneous buffer, the main advantage of moving boundary electrophoresis in this context is the "natural" (to borrow a term from electrofocusing) concentration of the species of interest prior to separation. 2) Application of Moving Boundary Electrophoresis to Charge Fractionation rare separation problem – the absence of significant molecular size differences. They become of general importance only when coupled to a size fractionation in a second stage or dimension.

a) Isotachophoresis: Within a system of multiple sequential moving boundaries, the ions constituting the various trailing phases are separated in order of net mobility at the steady state. This separation is not readily exploited, however, since the phases are contiguous. If all sequential trailing phases are occupied by mixtures of the same chemical class, e.g. proteins, staining methods are inapplicable for their detection, and have to be replaced by relatively sophisticated and expensive electronic detection devices sensitive to the differences in conductance, or Joule heat, or absorbance among the trailing phases (7). Even with such devices, practical separations appear to depend on the availability of "spacers", i.e.species with intermediate net mobilities and chemical characteristics differing from those of the species of interest. For protein separations, synthetic carrier ampholyte mixtures (SCAMs) are conventionally used as spacers. It appears highly dubious, however, that they possess significant numbers of species with intermediate mobilities, at least in gels (8). Aminoacids also do not furnish an adequate number of spacers (8). Separation within a mobile system of sequential moving boundaries furthermore depends on ascertaining the positions of leading and trailing constituents, to ensure the isotachophoretic nature of the separation, i.e.the positioning of the constituents of interest between them, and it depends on ascertaining that the steady state has been attained. This requires a finding of "pattern constancy" as time and migration path are varied. A central problem with regard to the steady state is that the time and migration path needed to attain the steady state increases with the multiplicity of the mixture to be separated, as well as with the load. This means in practice, that complex mixtures can only be resolved in finite time when the loads are very small. Thus, extreme detection sensitivity is required. For all of these reasons - cost of detection, spacer ambiguities, difficulty in locating leading and trailing ions and in ascertaining the steady state - charge fractionation on stationary trains of sequential moving boundaries, i.e. electrofocusing, appears more popular and promising. It is possible, nonetheless, that analytical gel isotachophoresis will be of importance in 2-dimensional separations, the resolving power of which depends on even zone distribution across each of the 2 dimensions. Since systems of sequential moving boundaries provide contiguous mobility compartments the width of which can be regulated at will, they may lend themselves more readily to obtaining even protein zone distributions within a given length of migration path than an isoelectric distribution of proteins. They also avoid the insolubility problems encountered with isoelectric proteins.

b) Electrofocusing   Classically. isoelectric focusing has not been interested in the genesis of the pH gradient. Taking it for granted, classical theory has investigated the isoelectric condensation of ampholytes in the pre-existing gradient. The formation of natural pH gradients by exclusively non-amphoteric bases differing in pK (9) led however, to the attempt to account for the genesis of pH gradients, and for their dynamics, by application of moving boundary theory (4).

As pointed out above. the pH of the trailing phase is one of the parameters regulated by the leading phase in a simple moving boundary system (Fig.1). It follows that the pH in a train of sequential moving boundaries at the steady state varies from one phase to the next giving rise to a monotonic step function of pH. Due to diffusion at the phase boundaries, a smooth pH gradient arises across the train of sequential moving boundaries. Isotachophoresis can thus be viewed as separation within a mobile pH gradient. where the counterion in each phase provides a high degree of ionization of the various trailing ions and therefore electrophoretic mobility. By contrast, when the counterion is eliminated and replaced completely by protons and hydroxyl ions derived from the solvent, ionization is repressed and the pH gradient becomes near-stationary. It can then be used for electrofocusing.

Figure 3 depicts the physical properties of each phase of a system of 6 nonamphoteric bases forming sequential moving boundaries at the steady state, computed by a very simple program (10). A stationary pH gradient between pH 10.29 and 12.18 was predicted. Experimentally, without any correction for CO_2 and temperature, a pH gradient between pH 9.5 and 11.5 was found (Fig.4). Note that the dynamics of such a pH gradient qualitatively mimicks that previously determined on an isotopically labeled pH-range 3-10 Ampholine gradient [Fig.2 of (11)]. Similar natural pH gradients and dynamics were predicted and found for a moving boundary system consisting of 6 acids (3 non-amphoteric, 3 amphoteric). A mixture of the acidic and basic sequential moving boundary systems yielded experimentally the sum sum of the two pH gradients (10).

Predicted Properties of System B

	Ethanolamine	Morpholine	N-ethyl-morpholine	N-(2-hydroxyethyl)-morpholine	Lutidine	Bistris
r [1]	0.86	0.73	0.62	0.61	0.60	0.38
pK_a [1]	10.35	8.85	8.03	7.19	7.00	6.88
conc (M)	1.00	0.87	0.76	0.75	0.74	0.49
pH	12.18	11.40	10.96	10.53	10.43	10.29
ϕ	1.50×10^{-2}	2.85×10^{-3}	1.19×10^{-3}	4.55×10^{-4}	3.68×10^{-4}	3.93×10^{-4}
$\bar{r}$	1.29×10^{-2}	2.08×10^{-3}	7.37×10^{-4}	2.78×10^{-4}	2.21×10^{-4}	1.45×10^{-4}
κ (mhos/cm)	1.91×10^{-3}	3.08×10^{-4}	1.09×10^{-4}	4.11×10^{-5}	3.27×10^{-5}	2.21×10^{-5}
ν (cm³/Coulomb)	1.85×10^{-3}	1.85×10^{-3}	1.85×10^{-3}	1.85×10^{-3}	1.85×10^{-3}	1.85×10^{-3}
v [2] (cm/day)	6.85×10^{-6}	6.85×10^{6}	6.85×10^{-6}	6.85×10^{-6}	6.85×10^{-6}	6.85×10^{-6}
w (cm)		9.33×10^{-3}	4.29×10^{-3}	1.67×10^{-3}	4.08×10^{-3}	1.74×10^{-3}

[1] 0°C from Fig.4 of Ref. 27

[2] I mA/ 0.27 cm² of gel

Fig. 3: Properties of a sequential moving boundary system constituted by 6 basic non-amphoteric electrolytes with solvent counterions giving rise to a natural pH gradient (10).

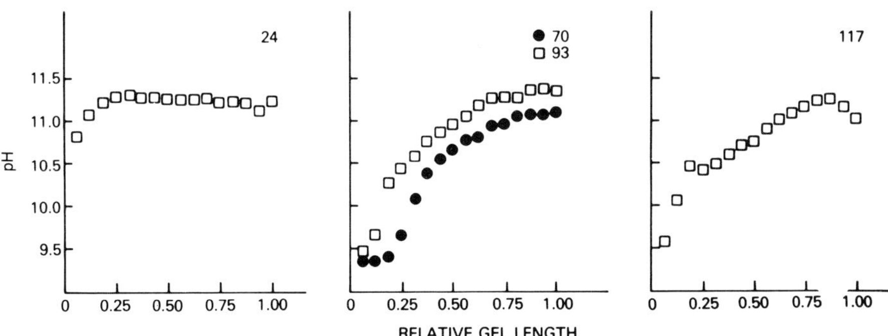

Fig. 4: pH Gradients formed by the moving boundary system define by Fig. 3.

The hypothesis advanced here that the pH gradient in electrofocusing is formed by a moving boundary mechanism is further evidenced by the following previous observations: i) Cathodic and anodic drift: Drift in any direction is qualitatively accounted for by boundary displacement. (Quantitatively, there remains a unexplained discrepancy between prediction and experiment: Experimental drift is larger by several orders of magnitude in all cases. Although, not surprisingly, electroendosmosis can be induced by incorporation of charged groups into polyacrylamide (12), it has not been measurable in either the polyacrylamide gels or even the IsoGel agarose gels under the conditions of the study.) The drift is usually cathodic since the ionic mobilities of cations exceed those of anions, and therefore the boundary displacement of a train of acids exceeds that of a sequential moving boundary system of bases being displaced in the opposite direction. In the case of the acidic and basic systems for which ionic mobility values were available, the displacement rate of the acidic sequential moving boundaries was computed to be higher by one order of magnitude than the basic ones. In the case of reported drift reversal, by changing the catholyte to a stronger base (lysine to arginine) while keeping the anolyte a very weak acid (threonine) (13), the boundary displacement rate of the cationic train is reduced below that of the anionic train by the increase in the leading hydroxyl ion concentration (pH) and increasing dissociation in the sequential phases. ii) Instability of pH gradients with strongly acidic and basic anolyte and catholyte: Strongly acidic and basic leading ions of a system of sequential moving boundaries moving either cathodically or anodically enhance the ionization of all of the trailing constituents and thereby render the system relatively isotachophoretic. The enhanced boundary displacement is perceived as an enhanced rate of pH gradient decay. By contrast, the leading ions of both acidic and basic moving boundaries, i.e. the ions with the highest net mobility, when applied as anolyte and catholyte, should stabilize the system and do so in fact (14). iii) The higher the concentration of anolyte and catholyte, the more stable the gradient (14): This is clearly in line with computations that show a decrease of boundary displacement with increasing concentration of the leading phase. iv) The dynamics of pH gradients is more dependent on the anolyte than the catholyte: This finding (15) is again in line with the much higher displacement rate of cationic moving boundary systems compared to anionic ones. v) Finally, the moving boundary theory predicts inverse pH gradients whenever net mobility differences between trailing ions are large as compared to their pK differences. This is clearly a case favored only by polymeric substances, and is experimentally substantiated by a system of immunoglobulin cations where the leading most basic species migrate at pH 5.7 and the trailing least basic ones of the train at pH 6.5 while the leading and trailing phase pHs (25#o$C) are 5.84 and 4.80 respectively (10,16). A promising corollary to the hypothesis that electrofocusing is slowed down isotachophoresis is the possibility to construct moving boundary systems with intermediate properties between the two. Thereby it might be possible to avoid the defects of both methods - isoelectric precipitation of one, and too limited a migration path in order to attain the steady-state in the other. This possibility, however, remains to be verified experimentally.

c) Preparative moving boundary electrophoresis on gels: This application of moving boundary electrophoresis is in principle the same as the procedure of isotachophoresis except that the trailing phases are greatly widened by an increase in load (at a concentration regulated by the leading phase) and a decrease of ionic strength (17). Since, as pointed out above, the time and migration path required to attain the steady state are proportional to load and sample multiplicity, the high load required for widening the trailing phases necessitates a decrease in sample multiplicity. Once the trailing phases are widened to 1 cm or more each, they can be sliced into 1 mm gel slices. Since, both theoretically and experimentally protein concentrations in regulated trailing phases at the steady state range from 20 to 100 mg/ml (17,18), a 1 cm segment of an 18 mm diameter cylindrical gel has a capacity for 50 to 250 mg of protein which is homogeneous at the steady state except in the narrow region across the leading and trailing boundaries (19). This represents a load capacity which is at least by one order of magnitude higher than that of any other preparative electrophoretic method.

3) Application of Moving Boundary Electrophoresis to Size Fractionation

a) Analytical and preparative concentrating gels: As pointed out above,

92

the key advantage in using moving boundary electrophoresis in size fractionations is "natural" concentration of the species of interest prior to
resolution. This concentration is due to, and depends on, migration of
the species of interest within a system of sequential moving boundaries.
Since protein net mobilities are low as compared with those of buffer
ions, sufficiently low trailing ion net mobilities must be found to bring
about this condition. Experience has shown that usually, proteins are able
to migrate at 1 - 1.5 pH units above or below their pIs with a mobility
(expressed relative to the mobility of Na^+) of 0.050. We have therefore
computed buffer systems yielding trailing phases at various pH intervals
with a net mobility of the trailing ion of approximately 0.05 (Fig.5) (2).
These are applied analytically to determining the optimal pH, and other
optimal conditions, by systematic experiment.

Buffer System Number 12

Trailing Phase $pH_0°$	Trailing Ion Mobility	Leading Phase				$pH_{25}°$	Trailing Phase		$pH_{25}°$	Anolyte	Catholyte
		Leading ion cacodylic acid	Common ion Bistris	4 x Leading ion cacodylic acid	Common ion Bistris		Trailing ion TES	Common ion Bistris			
		(M)		(g/100ml)			(M)				
7.05	0.045	0.1016	0.0358	5.61	3.00	5.78	0.0905	0.0247	6.59	4.18g Bistris + 10.0ml 1N HCl/1; $pH_{25}°$ = 6.50	4.58g TES + 10.0ml 1N KOH/1; $pH_{25}°$ = 7.44
7.39	0.086	0.0530	0.0477	2.93	3.99	6.29	0.0472	0.0419	6.94		
7.61	0.127	0.0360	0.0677	1.99	5.67	6.71	0.0321	0.0638	7.17		
8.15	0.250	0.0163	0.1972	1.01	16.67	7.61	0.0183	0.1992	7.74		

Fig. 5: Representative buffer system for moving boundary electrophoresis
(2).

To determine the optimal pH for separation, i.e. the one closest to the pI of the species of interest compatible with a migration rate of 0.05 relative mobility units, "non-restrictive" gels are set up in the leading phases corresponding to the various trailing phase pHs of interest, starting with a system at high or low pH. After electrophoresis, the moving boundary position on the gel is excised and assayed for the species of interest (20). The position of the boundary is located on the gel by loading a "tracking dye" with mobility higher than 0.05 and less than that of the leading ion, or by gel slicing and pH analysis, or by chemical assay for the trailing or leading ion. The buffer system with pH closest to the pI is selected. If recovery of the species of interest within the sequential moving boundary system has not been quantitative, the experiment at the optimal pH is repeated with gels varying systematically in ionic strength, detergent to protein ratio, polyol concentration, metal ions, chelating agents, degree of reduction (thioglycolate concentrations in a pre-electrophoresis), temperature, protease inhibitors, until recovery becomes quantitative (20).

Preparatively, the same gels serve for the extraction of gel slices (from resolving gels or electrofocusing gels) in the electric field and simultaneous concentration of the extracted species (21-23). Depending on the protein load, the gel surface area is increased or decreased. The train of sequential moving boundaries is allowed to migrate into a collection cup with semipermeable floor attached to the gel end (22). After buffer exchange and purification by gel filtration, the collected species is lyophilized. Yields of protein by this preparative procedure are about 70%, purity better than 90% (24).

b) Selective stacking: It is convenient at this point to recall the designation by Ornstein (25) of a system of sequential moving boundaries as a "stack", in analogy to a stack of coins, each coin representing a moving boundary which migrates as a unit and at a single speed (iso-tachophoretically). This term then allows us to define the useful verbs "stacking" and "unstacking", where the latter designates deceleration below the net mobility of the trailing ion. A concentrating gel, in which the species of interest migrates with a net mobility between that of the leading ands the trailing ion, i.e. within a train of sequential moving boundaries, is accordingly termed a stacking gel.

Let us assume that the species of interest has been stacked at an optimal pH in a buffer system with relative mobility of 0.05 (see above). Then it is advantageous to make stacking specific by increasing the trailing ion net mobility in the same buffer system (Fig.5). Again, concentrating gels are set up in leading phases corresponding to the desired values of the trailing ion net mobility (20). The highest value for that mobility is selected which is capable of stacking the species of interest quantitatively. Making stacking selective, allows one to separate the species of interest from contaminants by a charge fractionation, prior to separation of the stacked proteins by size fractionation.

c) Unstacking: To achieve size fractionation, one must transfer the species of interest from a stack, in which all components migrate at the same speed, to a trailing phase in which components migrate at different rates in gels of different restrictive concentrations. There are 3 ways to achieve the necessary unstacking (Fig.2). The first is a manipulation on the moving boundary, viz. an increae in trailing ion net mobility to unstack the species of interest. In this case, the gel concentration remains "non-restrictive". Secondly, the trailing ion net mobility is maintained at a value providing stacking of the species of interest in a "non-restrictive" gel. Assuming that this species is a macromolecule, its migration rate may be decreased by molecular sieving, i.e. by an increase in gel concentration, to a degree that this migration rate falls below the trailing ion net mobility. The third approach is a combination of the first two. It is the most practical one since it avoids very labile gels (approach 1) or extremely hard gels (approach 2). It should be noted that the unstacking gel ("resolving gel") is as much a moving boundary system as the stacking gel ("concentrating gel"), and that both assume their functions of either stacking or unstacking of the species of interest by action on the trailing ion net mobility, i.e. by a measure of moving boundary electrophoresis.

d) R_f manipulations: In size fractionations, the central problem is to optimize the gel concentration for the resolution of the species of interest from contaminants migrating in its vicinity. This optimal gel concentration can be calculated from the linear relation between the logarithm of mobility (relative to the moving boundary), R_f, and gel concentration, the so-called Ferguson plot (23). To obtain Ferguson plots, several R_f values between 0.25 and 0.85 are required. To achieve these values at convenient gel concentrations, it is useful to increase the trailing ion mobility (which decreases R_f values at any gel concentration) or to decrease it (which increases R_fs).

SDS-PAGE poses special problems in the detection of moving boundaries and in the selection of trailing ion net mobilities (26). i) Since micellar SDS migrates with a rate similar to SDS-proteins, it is concentrated within sequential moving boundaries with the same leading and trailing ion net mobilities. Since in SDS-PAGE, the catholyte forms a large reservoir of SDS, this concentration proceeds progressively with electrophoresis time, and it extends the length of the stack progressively since the concentration of micellar SDS is regulated. Therefore, the position of the trailing ion on the gel recedes progressively, allowing one to mark the leading and trailing ion positions in the "extended stack" by separate dye markers or otherwise (26). It is important in that situation to employ the leading ion to mark the reference boundary position. ii) Since mobility of SDS-proteins depends on their degree of saturation with SDS, SDS-PAGE may be restricted to saturated SDS-proteins by setting the trailing ion net mobility at a high value (0.2 to 0.3 relative to Na^+). This provides some assurance that the direct relation between log mobility and MW (which depends on saturation) holds for all unstacked zones derived from such a stack. iii) With increasing gel concentration, micellar SDS (and dyes like Pyronin-Y associated with it) unstack progressively, thereby allowing the dye marker originally trailing the stack to move to a position close to that of the leading ion.

4) Conclusions

Moving boundary electrophoresis comprises all major electrophoretic methods at this time, and is applicable equally to the various forms of apparatus and procedure within electrophoresis. It is of equal importance for charge fractionation methods such as isotachophoresis and electrofocusing in which separation proceeds within a system of sequential moving boundaries, as it is for size fractionation methods such as gel electrophoresis at restrictive gel concentrations in which separation proceeds between zones which migrate behind the trailing ion of the moving boundary system, or between such zones and components migrating within the moving boundary system. Preparatively, moving boundary electrophoresis provides a general tool for the electrophoretic extraction and concentration of resolved macromolecules from gel slices. When applied to charge fractionation of a few components, its load capacity is the highest among existing electrophoretic procedures.

References

1) Hjelmeland, L. M. and Chrambach, A.: Electrophoresis 3, 9-17 (1982).

2) Chrambach, A. and Jovin, T. M.: Electrophoresis, (1983) in press.

3) Jovin, T. M., Dante, M. L. and Chrambach. A. "Multiphasic Buffer Systems Output", National Technical Information Service, Springfield, Virginia 22151, 1970, PB 196085 to 196091, 259309 to 259312, 203016.

4) Hjelmeland, L. M. and Chrambach, A.: Electrophoresis 4, 20-26 (1983).

5) Cuono, C. B., Chapo. G. A.: Electrophoresis 3, 65-75 (1982).

6) Bier, M., Palusinski, O. A., Mosher, R. A. and Saville, D. A.: Science 219, 1281-1287 (1983).

7) Everaerts, F. M., Beckers, J. L. and Verheggen, Th. P. E. M.: Isotachophoresis. Theory, Instrumentation and Applications. Elsevier, Amsterdam-Oxford-New York 1976.

8) Nguyen, N. Y. and Chrambach, A.: Anal. Biochem. 94, 202-210 (1979).

9) Chrambach, A. and Nguyen, N. Y.: Electrofocusing and Isotachophoresis (B.J.Radola and D. Graesslin, eds.), Walter de Gruyter. Berlin-New York, 1977 pp. 51-58.

10) Buzas, Zs., Hjelmeland, L. M. and Chrambach, A.: Electrophoresis 4, 27-35 (1983).

11) Baumann, G. and Chrambach, A.: Progress in Isoelectric Focusing and Isotachophoresis (P. G. Righetti, Ed.) Elsevier, Excerpta Medica, North Holland, Assoc. Sci. Publ., Amsterdam, pp.13-23. (1975).

12) Righetti, P. G. and Macelloni, C.: J. Biochem. Biophys. Methods $\underline{6}$, 1-16 (1982).

13) Nguyen, N. Y., McCormick, A. G. and Chrambach, A.: Anal. Biochem. $\underline{88}$, 186-195 (1978).

14) Nguyen, N. Y. and Chrambach, A.: Anal. Biochem., $\underline{79}$, 462-469 (1977).

15) An der Lan, B. and Chrambach, A.: Electrophoresis $\underline{1}$, 23-27 (1980).

16) Chrambach, A., An der Lan, B., Mohrmann H. and Felgenhauer, K.: Electrophoresis $\underline{2}$, 279-286 (1981).

17) Baumann, G. and Chrambach, A.: Proc. Nat. Acad. Sci. U. S. $\underline{73}$, 732-736 (1976).

18) Chen, B., Rodbard, D. and Chrambach, A.: Anal. Biochem. $\underline{89}$, 596-608 (1978).

19) Nguyen, N. Y. and Chrambach, A.. Gel Electrophoresis of Proteins: A Practical Approach (Hames, B. D. and Rickwood D., eds.) Information Retrieval Ltd. Press, London and Washington DC, pp. 145-155 (1981).

20) Chrambach, A.: J. Molecular and Cell. Biochem. $\underline{29}$, 23-46 (1980).

21) Nguyen, N. Y. and Chrambach, A.: J. Biochem. Biophys. Methods $\underline{1}$, 171-187 (1979).

22) An der Lan, B. Horuk, R. Sullivan, J. V. and Chrambach,A.: Electrophoresis, submitted. (1983).

23) Chrambach, A. and Rodbard, D.: Gel Electrophoresis of Proteins: A Practical Approach (Hames, B. D. and Rickwood D., eds.) Information Retrieval Ltd. Press, London and Washington DC, pp. 93-143 (1981).

24) Nguyen, N. Y., Baumann G., Arbegast D. E., Grindeland, R. F. and Chrambach, A.: Preparative Biochem. $\underline{11}$, 139-157 (1980).

25) Ornstein, L.: Ann. N. Y. Acad. Sci. $\underline{121}$, 321-349 (1964).

26) Wyckoff, M., Rodbard, D. and Chrambach, A.: Anal. Biochem. $\underline{78}$, 459-482 (1977).

27) Jovin, T. M.: Ann. New York Acad. Sci. $\underline{209}$, 477-496. (1973).

ISOELECTRIC FOCUSING IN STABLE PREFORMED BUFFER pH GRADIENTS

M. Bier, R. A. Mosher, W. Thormann and A. Graham
Biophysics Laboratory, University of Arizona
Tucson, Arizona, 85721, U. S. A.

Introduction

It is generally accepted in isoelectric focusing (IEF) that
the pH gradients are best generated by the focusing process
itself. Ampholine and other such carrier ampholytes are
complex mixtures, their components migrating towards their
isoelectric points, thereby establishing a 'natural' and
relatively stable pH gradient. In many attempts by our
group (1,2) and others (3-6) to replace these preparations
with mixtures of well-defined ampholytes or buffers, the pH
gradient was still generated by the electric current.

In a recent publication of a generalized mathematical model
of electrophoretic transport processes (7), we have shown
that pH gradients are ubiquitous in all modes of electrophoresis, as any concentration gradients involving acids or
bases will generate pH changes. Most of these pH gradients
are rapidly migrating and thus useless for protein focusing.

In the present paper we wish to report the existence of
quite stable pH gradients formed using a simple system of a
weak acid and a weak base, mixed in the proportion required
to cover the desired pH range. Thus, these pH gradients are
preformed and are not generated by the electric current.
Our work extends the original suggestion of Kolin (8) for
IEF, by precising the conditions of gradient stability. As
will be shown, these gradients are stable only in two component

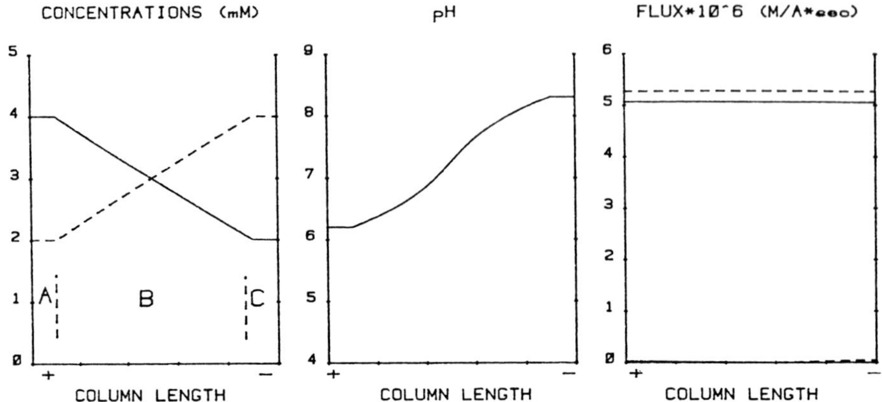

Fig. 1. Schematic presentation of TRIS-cacodylate concentration and pH gradients. Microcomputer-calculated flux values are also plotted, assuming for TRIS (dashed lines) a mobility of 2.42 cm^2/V-sec and a pK of 8.30, and for cacodylate (solid lines) a mobility of 2.31 cm^2/V-sec and a pK of 6.21.

systems. As a result, their usefulness is limited to a 4 pH units range around neutrality, beyond which H^+ and OH^- ions begin to play the role of a third component. Our experiments were all in free solution and should not be confused with the recently introduced 'Immobilines', where a precast pH gradient is actually copolymerized into the gel matrix (9). Our work bears closer similarity with the 'dilution boundaries' first described by Kohlrausch (10). Such boundaries are stationary, but exhibit too great conductivity gradients to be useful for IEF.

Theoretical Background

We wish to analyze the behavior of a system of an acid and a base, both weak electrolytes, as presented in Fig. 1. Zones A and C represent the electrolyte composition at both ends of the electrophoretic column, assumed to be connected to

virtually infinite electrode buffer volumes of identical
composition. The pH gradient is formed in the intermediate
zone B, which encompasses the boundary between the acidic
and basic portions of the column. The actual shape of the
gradients and the relative concentrations of components are
of no consequence for this discussion. The flux F_j
(mole/cm^2-sec) for component j is given by:

$$F_j = -M_j z_j \Omega_j \nabla\phi - \frac{RT}{e} \Omega_j \nabla M_j \qquad (1)$$

where $\nabla\phi$ is the field strength (V/cm), Ω the mobility
coefficient (cm^2/V-sec), M the concentration, z the valence,
R the universal gas constant, T the absolute temperature and
e the Faraday constant (eq. 4 of ref. (7)). In domains of
constant composition, A and C, net diffusional transport is
nil and the second term of above equation can be eliminated.
Further simplification is possible by equating $i = -\nabla\phi\kappa$
where i is the current density, and κ the conductivity.
This yields the much simpler expression

$$F_j = \frac{i\mu_j M_j}{\kappa} = \frac{i}{e} \cdot \frac{\mu_j M_J}{\Sigma\mu_j M_j} \qquad (2)$$

where $\mu_j = \Omega_j z_j$. This expression can be easily evaluated
on any microcomputer. The flux so calculated is also shown
in Fig. 1. It is near constant across the whole length of
the column, including the zone B, encompassing the pH gradient.
This constancy of flux is the essential requirement for
gradient stability. Disregarding again the zone B, we can
restate that the requirement for a stable gradient is that
the flux of each component be identical in the two zones A
and C. As the current must be identical in the two zones,
we can further simplify to obtain a parameter ρ:

$$\frac{M_{a-}^A \, M_{b+}^C}{M_{b+}^A \, M_{a-}^C} = \rho \qquad (3)$$

which is predictive of the stability of a system. Perfect

stability would require $\rho = 1$, but in effect it is always greater than 1 because of H^+ and OH^- ion fluxes. We have found the gradients to be quite stable below $\rho = 1.02$.

Experimental and Computer Simulation Results

The first experiments were conducted in an LKB column for IEF, modified to increase the volume capacity at both electrodes. Because of the unwieldiness of this apparatus, a simpler column was assembled. It consists of two small water jacketed distillation condensers clamped side by side and joined at the bottom by a "Y" shaped piece of rubber tubing. The pH gradient, stabilized by a sorbitol gradient, was poured in one of these condensers. The other served to balance hydrostatic pressure. The "Y" shaped rubber tubing provided a means for sample collection. Platinum wires located at the tops of both condensers served as electrodes. To assure constancy of electrolyte composition, anolyte and catholyte were recirculated around these electrodes.

The cacodylate/TRIS system shown in Fig. 1 was used to separate the human hemoglobin variants HbA and HbC, pI's 7.2 and 7.8 respectively. Fig. 2 is a photograph of this fractionation. Once established, this pattern was stable for at least 17 hours after which samples were collected and focused on a polyacrylamide gel with the appropriate standards to confirm their identity. A unique experiment was devised to further document the stability of such precast pH gradients. In place of the linear pH gradient, a sinusoidal pH gradient was constructed. This was accomplished by carefully layering alternate aliquots of catholyte and anolyte on top of each other. A total of eight 0.3 ml layers were applied, each with a successively lower sorbitol concentration to provide stability. Hemoglobin HbA was uniformly distributed throughout

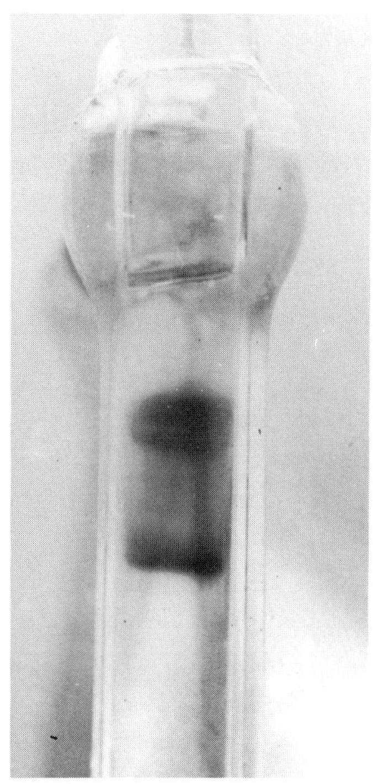

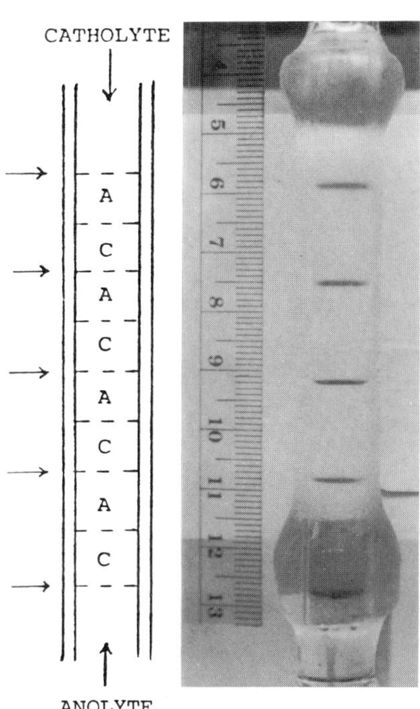

Fig. 2 (Left panel). Photograph of the separation of human hemoglobins HbA, pI = 7.2, and HbC, pI = 7.8. The precast system shown in Fig. 1 was used to form the pH gradient. The anode is at the bottom. Several intermediate minor bands were visible. This pattern was stable for at least 17 hours, in a sorbitol gradient stabilized column.

Fig. 3 (Right panels). Photograph of the focusing of hemoglobin HbA in a sinusoidal pH gradient formed by layering successive aliquots of anolyte (A) and catholyte (C) as shown in the schematic drawing. The arrows indicate the interfaces where the protein focused.

the column. Application of the electric field resulted in rapid focusing of the sample to those interfaces where the pH increased in the direction of the cathode. Figure 3 is a photograph taken after five minutes of focusing. This pattern was stable for one hour, the bottom gradient being the first to degrade by diffusion.

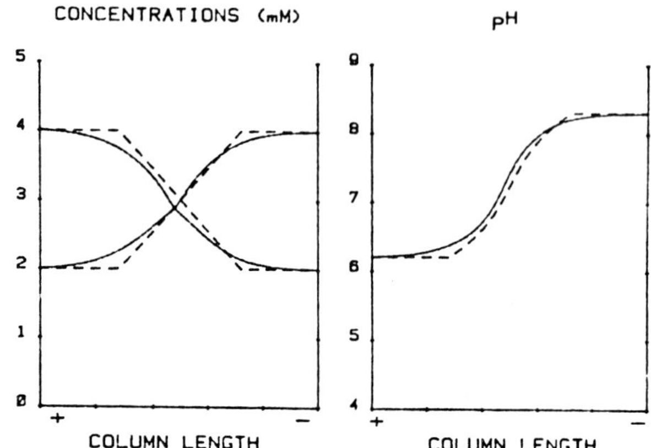

Fig. 4. The results of a computer simulation of the time-dependent evolution of the buffer system used in experiment of Fig. 2. Dashed lines represent the initial condition, solid lines representing the computer predicted gradients after 1 hour electrophoresis at 1.5 mA/cm^2. A slight anodic drift is seen.

The stability of the gradients has been further confirmed by the exercise of the computer program described in ref. (7). The initial conditions were similar to those in the experiment of Fig. 2, and are shown as dashed lines in Fig. 4. The solid lines show the predicted profiles after one hour of electrophoresis at 1.5 mA/cm^2. As expected, the actual profiles of the acid and base are a function of diffusion and current applied, the gradient becoming steeper with increased current density.

Discussion

It was not previously known that it is possible to obtain stable pH gradients using simple buffer systems, precast in the appropriate proportions. The stability of such systems were demonstrated experimentally, in density stabilized columns, and by computer simulation. The system of electrolytes described in the present paper defies ready classification in terms of conventional modes of electrophoresis. It is certainly not isotachophoresis, as the boundaries do not migrate. Though the system can be used for focusing of superimposed proteins, it is not isoelectric focusing, as none of the components of the buffer are isoelectric. At best it can be classified as a _sui generis_ moving boundary electrophoretic system with virtually immobilized boundaries.

The new process is distinguished by near stationary profiles resulting from constant mass fluxes for each component across the length of the column. It is useful for generating nonmigrating pH gradients for preparative IEF in density stabilized columns, in the pH range 5 to 9. An advantage is that it combines effective buffering with wide choice of ionic strength. This may eliminate precipitation of many proteins. The stability of such gradients increases with buffer concentration, while the pH range is only a function of the acid/base ratio. Obviously, also, such buffers are much cheaper than those produced by commercial carrier ampholytes. Moreover, contamination with ill-defined polyelectrolytes is avoided.

A fine distinction of boundary conditions in this sytem are necessary. The theoretical discussion presented here assumes the constancy of composition in zones A and C, of Fig. 1. This obviously implies large electrode reservoirs to minimize the effect of electrode reactions on buffer composition. Under these conditions, our theoretical analysis shows that a true steady state is never achieved, due to the contribution of hydrogen and hydroxyl ions to total flux.

An alternate model can also be postulated. This requires constancy of composition at the A/B and B/C boundaries only,

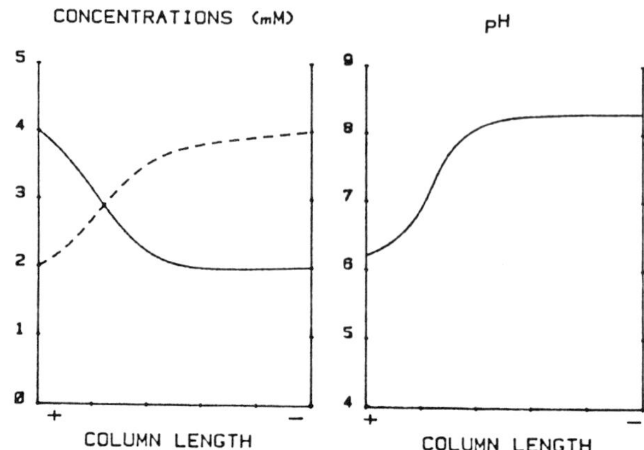

Fig. 5. Computer generated steady state concentration
profiles for TRIS (dashed line) and cacodylate (solid line)
under equivalent conditions to those presented in Fig. 4.
The only difference is in boundary conditions, as explained
in the text. These profiles are stationary and non-diffusing.
Note the anodic shift present also in Fig. 4.

but not in zones A and C. The two conditions are deceptively
similar, but have quite different consequences. Experimentally,
such a system could be formed by the introduction of electrically neutral membranes at A/B and B/C boundaries, i.e., at
both ends of zone B, and recirculation of two buffers, with
composition A and C, externally to the two membranes. Our
computer analysis shows that under such conditions a steady
state is readily achievable at all pH values, and is not
limited to neutrality. The existence of a steady state
implies that precasting of the pH gradient is not necessary,
as the final conditions will be automatically achieved by
the passage of current. A detailed analysis of this process
will be presented at a later date. At this point we only
wish to show in Fig. 5 the predicted steady state of a
system identical to that presented in Fig. 4, but with the
new boundary conditions. The similarity of the two profiles
is obvious, but the difference in the two models should not
be neglected.

References

1. Bier, M., Mosher, R.A., Palusinski, O.A., J. Chromatogr., <u>211</u>, 313 (1981).

2. Palusinski, O.A., Allgyer, T.T., Mosher, R.A., Bier, M., Saville, D.A., Biophys. Chem., <u>13</u>, 193 (1981).

3. Chrambach, A., Nguyen, N.Y., in Electrofocusing and Isotachophoresis, (Radola, B.J., and Graesslin, D., Eds.), DeGruyter, Berlin, 1977, p. 51.

4. N.Y. Nguyen and A. Chrambach, Anal. Biochem. <u>79</u>, 462, (1977).

5. Prestidge, R.L., Hearn, M.T.W., Anal. Biochem., <u>97</u>, 95 (1977).

6. Lundahl, P., Hjerten, S., Ann. N. Y. Acad. Sci., <u>209</u>, 94 (1973).

7. Bier, M., Palusinski, O.A., Mosher, R.A., Saville, D.A., Science, <u>219</u>, 1281 (1983).

8. Kolin, A., J. Chem Phys. 22, 1628 (1954).

9. Trademark of LKB, AB, Bromma, Sweden.

10. Kohlrausch, F., Ann. Phys. (Leipzig), <u>62</u>, 209 (1897).

DETERMINATION OF TRANSIENT AND STEADY STATES IN ELECTROPHORESIS

W. Thormann*, D. Arn and E. Schumacher

Institute for Inorganic Chemistry, University of Bern,
Switzerland
*Present address: Biophysics Technology Laboratory, University
of Arizona, Tucson, AZ, U.S.A.

Introduction

Electrophoretic techniques are recognized as being combinations
or just one of the four classical modes, zone electrophoresis
(ZE), moving boundary electrophoresis (MBE), isotachophoresis
(ITP) and isoelectric focusing (IEF) [1,2,3]. These modes
differ only in terms of their initial conditions which represent the distribution of sample and buffer components
along the axes of the separation space and the boundary
conditions at the ends of the column. Large electrode
volumes are used in ZE, MBE, ITP and IEF of ampholytes in
moving or precast, non-migrating pH gradients [3,4], but the
electrode buffer volume is minimized in IEF in natural pH
gradients [2]. Therefore, a universal separation trough can
be used to monitor the evolution of all electrophoretic
processes in homogeneous solution. By having many equidistant
detectors along the column [5] or by scanning the separation
trough repeatedly with a moving detector [6], transient and
steady state zone distributions can be observed, measuring a
general or specific sample property.

We constructed an apparatus for continuous registration of
the electric field which is the most general physical property
characterizing the separand pattern in electrophoresis within

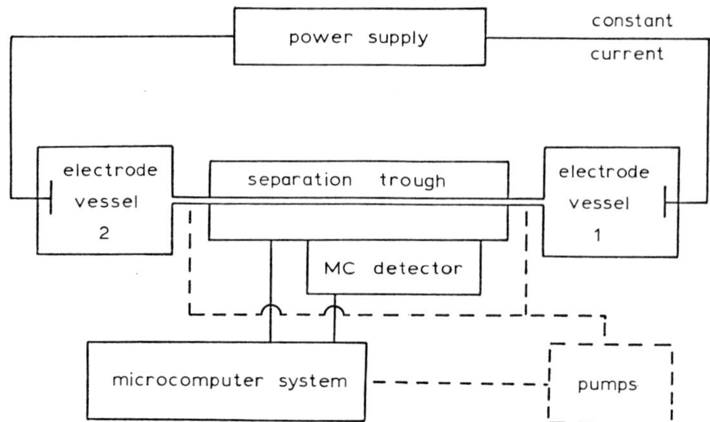

Fig 1 Schematic of an automated apparatus with a multichannel
detector to follow the dynamics of electrophoretic processes.

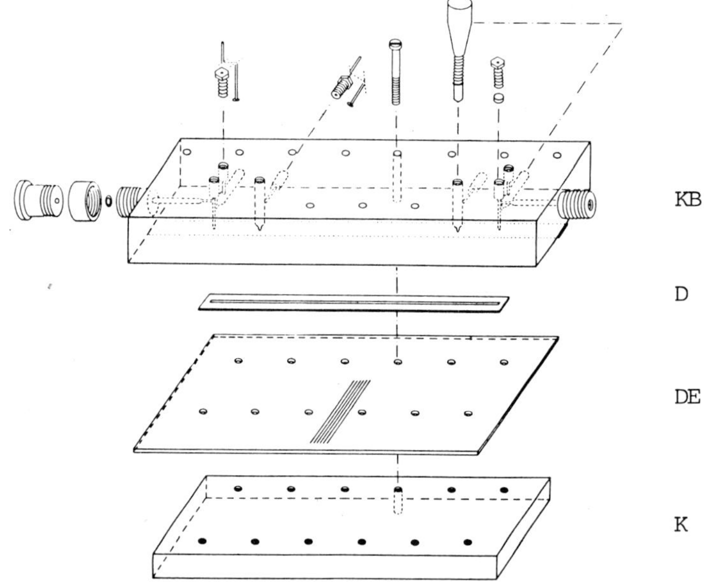

Fig 2 Exploded assembly of the separation cell, consisting
of a glass plate with the detection electrodes (DE), a
silicone gasket (D), a plexiglass block (KB) and a slab of
aluminum (K). For further explanation see text and ref. [3].

the separation column [7]. It consists of a multichannel
detection device, which enables us to make quasi simultaneous
measurements at 255 equidistant positions along the separation
tube, controlled by a microprocessor system (see Fig 1).
This permits automatic monitoring of the time development of
each zone boundary in ITP, MBE, ZE and IEF. The device is a
useful tool for validation studies of computer modeling in
electrophoresis [2,8] and of predictions made by models
based on migration (constant Kohlrausch regulating function)
only [3,9]. This leads to a better understanding of these
processes and provides a basis for the design of new types
of electrophoretic systems. In ITP the multichannel detection
method gives a computer interpretable criterion for the
approximation of the steady state at each boundary which is
lost if single channel detection at the end of the separation
trough is used [7].

The central functional unit of the system is the separation
cell, consisting of 4 main parts, as shown in Fig 2. It
comprises a rectangular separation trough (about 0.4 x
0.7mm) with an ordered array of 256 electrodes over a length
of 10cm and perpendicular to the current flow. Electrodes
are photoetched on a supporting glass plate from a vapor
deposited thin layer of In_2O_3/SnO_2 (80%/20%) or SnO_2 semi-
conductor. Each electrode has a width of 60 µm, a height of
about 0.1 µm, and they are spaced 340 µm apart. The electrode
bearing glass plate is in good contact with a cooled plate
of aluminum. The rectangular trough (10-17 cm long) is
defined by a silicone rubber gasket between the glass plate
and a block of plexiglass. This block contains valves for
the establishment of the initial electrolyte distribution,
the sample inlet system and the connections to the compartments
with the driving electrodes. A 20 kV constant current power
supply of our own design, appropriate detection electronics
for single and multichannel runs and fluid handling devices

complete the apparatus [3,7]. The electrode array is scanned
mechanically at the wide, enlarged end of the detection
electrodes [7]. It consists of two graphite brushes (medium
hard pencils) which are moved from pair to next pair by a
stepper motor device, controlled by a microprocessor. The
measured potential gradient between adjacent electrodes is
recorded and stored in the computer memory of a Heathkit H8
computer. We either can read data of a single pair of
electrodes as a function of time (single channel run) or can
scan over the whole of the detection array and get the data
of each channel nearly simultaneously (multichannel run).
With the former operational mode, information about stable
zone patterns in ITP, MBE and ZE is obtained by comparing
the detected electropherograms from at least 3 different
locations. Using the scan mode, the dynamics of electrophoretic processes (including IEF in natural pH gradients) are
monitored. This allows the detailed asymptotic approach to
the steady isotachophoretic state to be followed in a completely
automatic fashion. As soon as the criterion for stability
is fulfilled, the microcomputer takes several scans over the
total component spectrum to ensure a specified precision of
zone length. All control functions, data treatment and
printout of results (field strength, length of zones with
standard error, zone boundary width and shape factor) are
handled by the computer.

Results

The results we are showing here represent only a small
number of the performed experiments [3,7,10]. They are
selected to show the capabilities of the apparatus in detecting
transient and steady states of electrophoretic processes.
To simplify matters, all experiments presented are anionic
analyses requiring large electrode volumes, for which the

conditions are listed in Table I. The initial distribution
of the electrolytes consists of a filled separation trough
with the anolyte (from the anode to the cathodic sample
inlet compartment), and the catholyte being between the
sample inlet system and the cathode only. Samples are
introduced in small quantities (1-7 µl) into the preestablished
boundary between both electrode solutions before turning on
the constant driving current. The distance between the
sample inlet and the first detector is about 4 cm.

The 4 pherograms of Fig 3 represent transient and steady
state zone structures in ITP(a), MBE (b), combination of ITP
and MBE (c), as well as the non steady state behavior of a
moving sample zone in ZE (d). These data are recorded at
the indicated detection channels as functions of time (single
channel runs), the lower trace being the detected potential
gradient and the upper (in a-c) the first derivative of the
measured signal. In experiment a) 4 components are separated
isotachophoretically between HCl and lactic acid. Under the
given conditions, the 1.75 nmol of each component are not
fully separated until the zone structure moves across the
first detector, except maleic acid has already formed its
own zone. Between this zone and the displacing electrolyte
several small mixed zones [7] are still present. They are
resolved until the zone pattern reaches detection channel
120, where the expected 4 zone structure is monitored. That
this represents a steady state isotachopherogram is proven
by observing the moving zone pattern further in the separation
trough, e.g. in channel 240, where the same pattern is
obtained. The example b) is the simplest case of a MBE
experiment. Having two components with different mobilities
in the catholyte, one zone between leading and terminating
electrolyte can be established. The moving boundaries are
steady state concentration gradients but the length of the
zone is increasing as long as there is an infinite supply of

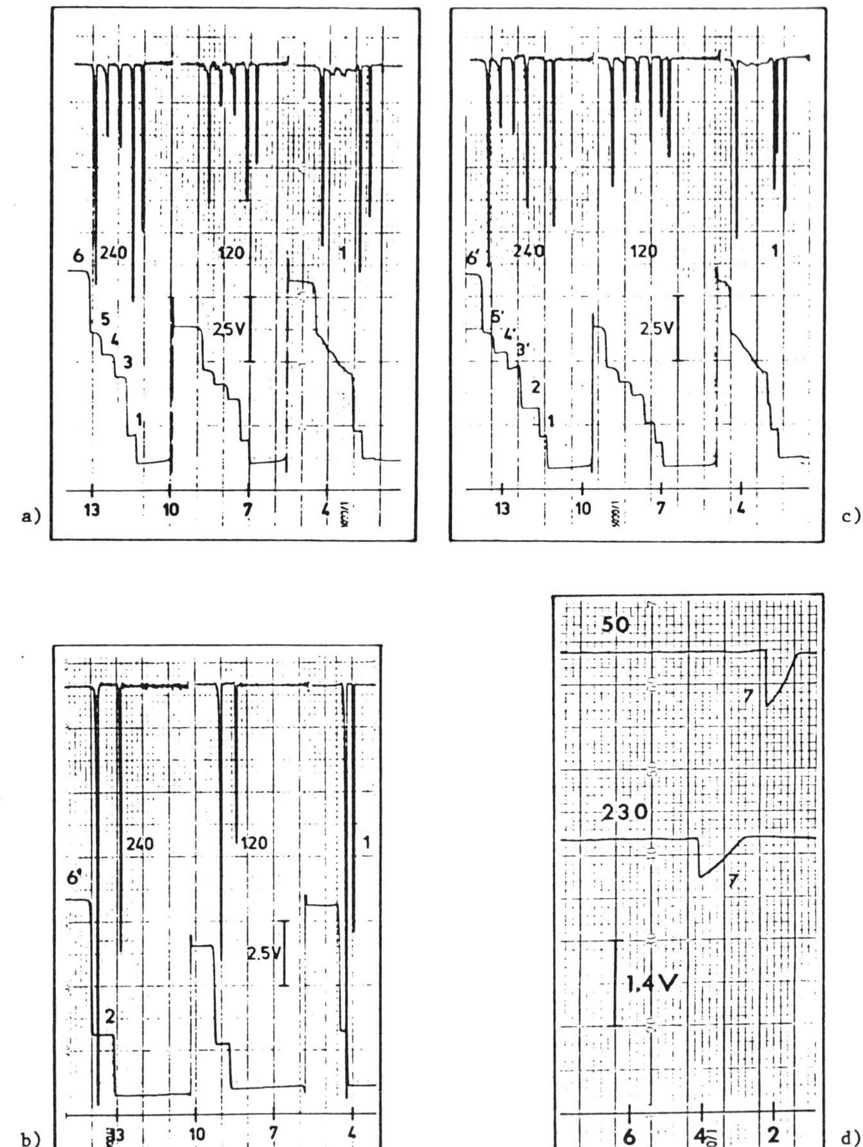

Fig 3   Single channel detection of transient and steady
state zone structures in ITP (a), MBE (b), combination of
ITP and MBE (c) and of the non steady state behavior of a
moving sample zone in ZE (d).

Table I

Experimental conditions

Experiments	3a	3b	3c	3d
anolyte*	2.5mM HCl	2.5mM HCl	2.5mM HCl	5mM HAc +Histidine (pH=6.1)
catholyte	10mM NaLactate	10mM NaLactate 0.1mM EDTA	10mM NaLactate 0.1mM EDTA	5mM HAc + Histidine (pH=6.1)
current [μA]	100	100	100	50
temperature	ambient	ambient	ambient	ambient
sample components	maleic acid tartaric acid citric acid malic acid	–	maleic acid tartaric acid citric acid malic acid	LiCl
sampled amounts	1.75 nmol each	–	1.75nmol each	500 pmol

Identification of detected zones:

  1) Maleic acid; 2) EDTA; 3) Tartaric acid; 3')Tartaric acid + EDTA
  4) Citric acid; 4') Citric acid + EDTA; 5) Malic acid; 5') Malic
  acid + EDTA; 6) Lactic acid; 6') Lactic acid + EDTA; 7) Cl^-

-----

*All anolytes contain 0.05% of polyvinyl alcohol (Mowiol 8-88,
Hoechst, Frankfurt G.F.R.) in order to minimize the influence of
electroosmosis in the separation trough.

the electrode solution. This can be recognized by comparing the detected zone lengths in channels 1, 120 and 240. Fig 4c presents a combination of experiments a and b, resulting in a steady state zone (maleic acid) between HCl and EDTA (zone 2) and in an isotachophoretic substructure (zones 3'- 5'), moving between zone 2 and the displacing electrolyte. EDTA is migrating from the terminating electrolyte through zones 3'- 5' because of having a higher mobility than the components within these zones, but is rejected by the maleic acid zone, due to a smaller mobility than this constituent. The EDTA zone (2) never reaches a constant length as can be seen in the detected pherograms. The 3 zone isotachophoretic structure approaches the steady state between detection channels 1 and 120, similar to the case in experiment a), but having an additional excess of EDTA in each zone. This structure represents a stepwise, moving pL-pH gradient, required for the ion focusing of metal complexes invented by Schumacher [11]. In all three experiments (a-c), moving pH gradients are formed, due to the non-buffering behavior of the counter component H_3O^+ [3]. Fig 3d shows the evolution of a moving Cl^- zone (500 pmol), detected in channels 50 and 230. It is a system, where the moving sample component has a higher mobility than the carrier constituent moving in the same direction. This results in a sharp, stable boundary at the rear side and a "trailing", non steady state boundary at the front side of the moving chloride zone, which corresponds nicely with the results of computer modeling [2,8] as well as with the predictions by the model based on migration only [9]. Comparing the curves, detected at two different locations, clarifies the fact that the migrating Cl^- zone never reaches a steady state. The width of the zone is smaller at detection channel 50 than at the second location and the peak height decreases substantially between the two detector positions.

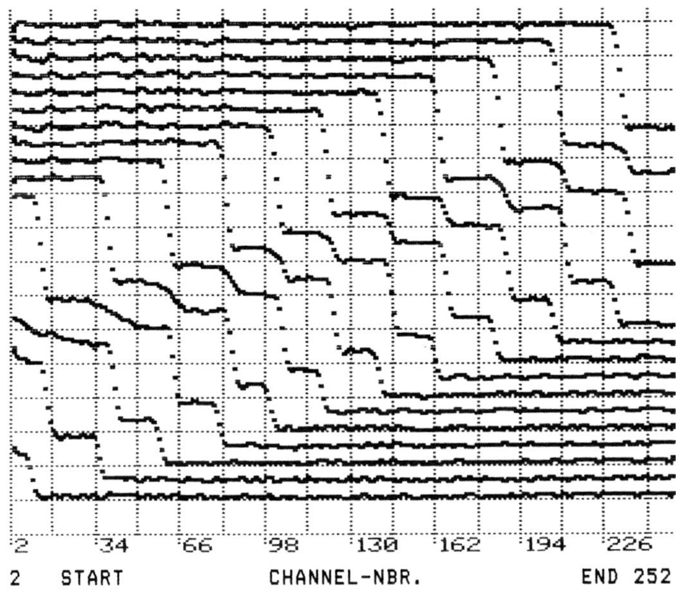

Fig. 4   Multichannel detection of a 3 zone isotachopherogram. For further explanation see text.

Fig 4 shows data points of scans 3 to 16 over nearly the
whole detection window of our equipment, representing an ITP
analysis similar to that given in Fig 3a, but without
citric acid as sample component. The injected amount of the
three acids is 3.5 nmol each. A scan represents nearly an
exact distribution of the electric field after a certain
time with current flow. In our experiment, one full scan
requires about 32.4 sec. This means that there is a small
position shift between channels 2 and 252 due to different
detection times. A new scan begins every 46.5 sec. This
experiment clearly documents that the 3 zone ITP pattern
reaches the steady state moving within the detection window.
The boundary between maleic and tartaric acid is already

fully established migrating across detector 2 (see in scan 4 and in the following ones), but the steady state moving boundary between tartaric and malic acid is not existent until scan 9. During scans 10 to 16 shown in Fig 4, the steady state zone pattern is observed. Data treatment of this experiment with a microprocessor system will be presented in another paper [12].

Acknowledgement: This work has been financially supported by CIBA-GEIGY, AG, Basel, and by "Kommission zur Foerderung der wissenschaftlichen Forschung", Bern, Grant No. 1074.

References

1. Electrophoresis, a survey of techniques and applications, Deyl, Z. (ed), Elsevier (1979).

2. Bier, M., Palusinski, O. A., Mosher, R. A., Saville, D. A.: Science, $\underline{219}$, 1281 (1983).

3. Thormann, W.: Dissertation, University of Bern, Switzerland (1981).

4. Bier, M., Mosher, R. A., Thormann, W., Graham, A.: This symposium.

5. Schumacher, E., Ryser, P., Thormann, W.: Helv. Chim. Acta $\underline{60}$, 3012, (1977).

6. Hjerten, S.: Protides Biol. Fluids, Proc. Colloq. $\underline{22}$, 669 (1975).

7. Schumacher, E., Thormann, W., Arn, D.: Analytical Isotachphoresis, Everaerts, F. M., (ed), p. 33, Elsevier 1981.

8. Vacik, J., Fidler, V: Analytical Isotachophoresis, Everaerts, F. M., (ed), p. 19, Elsevier 1981.

9. Thormann, W.: Submitted to Electrophoresis

10. Arn, D.: Dissertation, University of Bern, Switzerland, in preparation.

11. Schumacher, E.: UCRL - Report #10624, p. 208, (1963).

12. Schumacher, E., Arn, D., Thormann, W.: This symposium.

ISOTACHOPHORESIS AND SEPARATION OF PROTEINS USING CARRIER-FREE
ELECTROPHORESIS APPARATUS

Kaoru Yasukawa, Kiyotsugu Kojima, Takashi Manabe,
Tsuneo Okuyama

Dept. of chemistry, Fac. of science, Tokyo Metropolitan Univ.,
Tokyo, Japan.

Introduction

We have studied on separation of membrane proteins, and tried
to develop a carrier-free electrophoresis apparatus, without
supports such as polyacrylamide gel or agar gel (1). The new
type apparatus was designed for the carrier-free electrophoresis
which could be connected to a pumping-out device of peristaltic
pump, a UV detector and a fraction collector(2). and that could
be connected to high-performance liquid chromatography in future.
In this paper, we describe on the basic conditions for the separation of proteins by the carrier-free isotachophoresis. Condition for isoelectric focusing by the apparatus have been described previously (2).

Materials and methods

Fig-1 shows the block diagram of the apparatus for carrier-free
isotachophoresis. This apparatus was combined with a pumping-
out device composed of a peristaltic pump (C), a UV detector (D)
and a fraction collector (E).

Electrophoresis '83

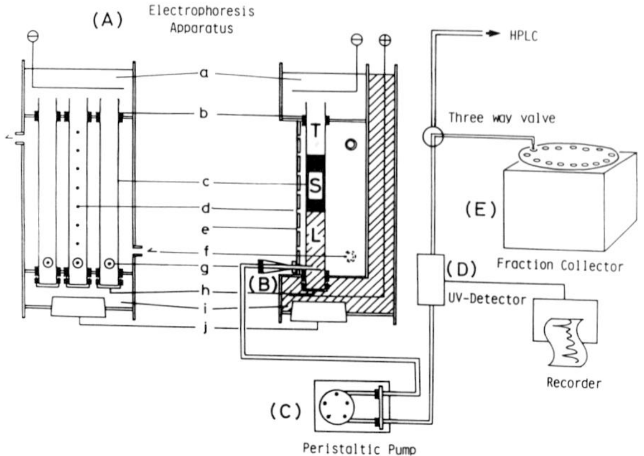

Fig-1   The system for carrier-free electrophoresis.
(A) apparatus for carrier-free electrophoresis. a) cathode
chamber, b) silicon rubber packing, c) electrophoresis
chamber, d) guide hole for manual taking out, e) sealed
tape for d), f) cooling chamber, g) guide hole for pumping
out, h) dialysis membrane, i) silicon rubber stopper.
(B) needle for pumping out the sample. (C) peristaltic pump.
(D) UV detector. (E) fraction collector.

Conditions of isotachophoresis are as follows:
Separation column : polyethylene tubing, ϕ 5mm x 145mm.
Leading solution   : 10 mM HCl, pH 9.0 (Ammediol), 0.04%
                     Hydroxypropyl methyl cellurose (HPMC),
                     0.05% Agarose.
Terminating solution : 10 mM ε-amino caproic acid, 10 mM
                     Ammediol, pH 10.8 (NaOH), 0.04% HPMC,
                     0.05% Agarose.
Current   : 1 mA constant current.
Sample solution   : bovine hemoglobin (18.5 mg/ml), bovine
                     ceruloplasmin (10.0 mg/ml), asparagine
                     (10.0 mg/ml).

Fig-2 shows the oparation steps of the system. The steps are
done in numerical order. When isotachophoresis is repeated for
another sample, the steps 3)-7) are performed.

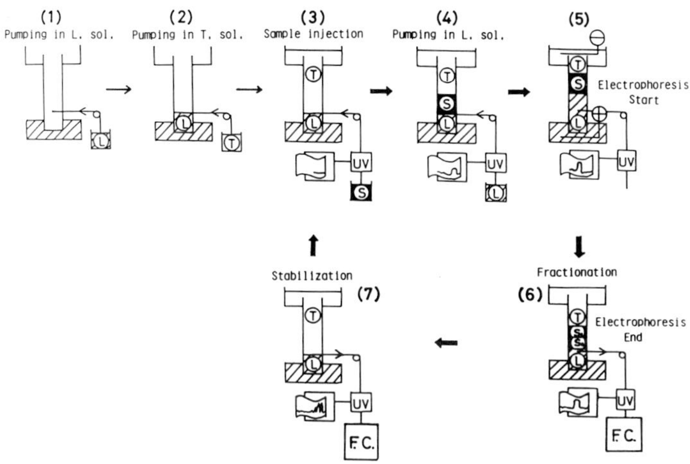

Fig-2 Oparation steps of the carrier-free isotachophoresis
       system.
   1) pumping in leading solution into the column.
   2) pumping in terminating solution untill filling anode
      chamber.
   3) pumping in a sample solution.
   4) pumping in leading solution.
   5) electrophoresis (start).
   6) electrophoresis (end) and fractionation (start).
   7) fractionation (end) and washing (pumping out one
      column volume of terminating solution).

Results and discussions

1) Carrier-free isotachophoresis of hemoglobin. Fig-3a shows
the zone width and current value for the certification of isotachophoresis. Fig-3b shows hemoglobin band of 0 min., 4 min.

122

and 10 min. The zone width of hemoglobin (1.0 ml, 1.85 mg/ml)
was concentrated 16-fold from 63 mm of band width to 4 mm for
4 min. After concentration, the zone of hemoglobin migrated
in uniform velocity.

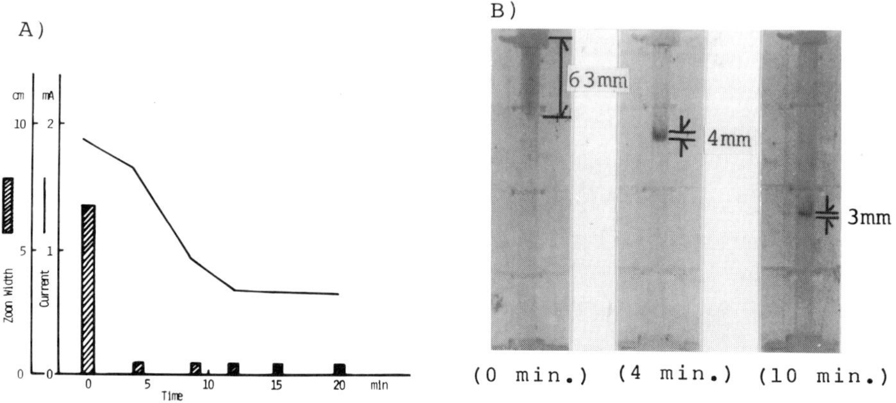

Fig-3A Carrier-free isotachophoresis of hemoglobin.
sample : 1.0 ml (1.85 mg/ml) of bovine hemoglobin soln.
▨▨▨▨   zone width, ──── current value.
Fig-3B Carrier-free isotachophoresis of hemoglobin.
photograph of hemoglobin band at 0 min., 4 min. and 10 min.

2) Separation of hemoglobin and ceruloplasmin by carrier-free
isotachophoresis. Since the polyethylene tubing was translu-
cent, the migration process of colored proteins, hemoglobin and
ceruloplasmin could be observed. During isotachophoresis, the
bans of two proteins were concentrated into adjacent two sharp
bands.(Fig-4) When asparagine was added, the separation of
hemoglobin and ceruloplasmin was improved but due to the spreading of bands. Fig-5 shows the separation of hemoglobin and
ceruloplasmin with spacer neutral amino acid (asparagine).

The amount of sample applicable could be increased by increasing the number of tubing for electrophoresis. Now, we could

applied on 10 mg proteins at least. This apparatus could be used the preparation. For the purpose of fractionation of proteins, neutral amino acids were used as spacers instead of ampholytes.

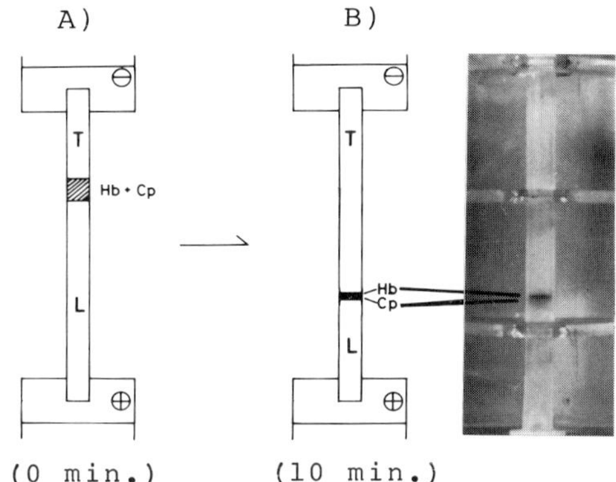

Fig-4   Separation of hemoglobin and ceruloplasmin.
sample : Hb-Cp mixture soln. 100µl (Hb: 18.5mg/ml, 50µl and
Cp: 10.0mg/ml, 50µl). A) the band of Hb and Cp mixture.
B) the separated bands of Hb and Cp.

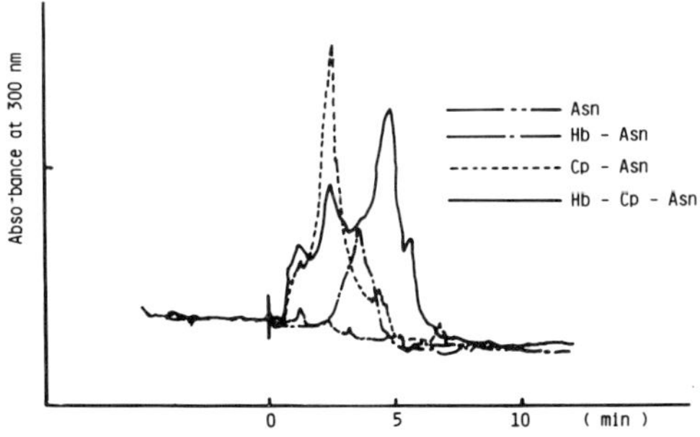

Fig-5   Separation of hemoglobin and ceruloplasmin with spacer
        neutral amino acid (asparagine).
sample : — - - —   Asn soln.,   — - —   Hb-Asn soln.,
- - - - -   Cp-Asn soln. and ————————   Hb-Cp-Asn soln.

Reference

1. Kojima, K., et al.; J. chromatography, <u>239</u>: 565, 1982.
2. Yasukawa, K., et al.; Seibutsu Butsuri Kagaku(The physico-
chemical biology) in press.

APPLICATION OF A RULE TO ILLUSTRATION OF ELECTROPHORETIC
RESULTS AND STANDARDIZATION OF ELECTROPHORESIS APPARATUS

Tsutomu Inoue
Department of Biology, Tokyo Gakugei University
Tokyo Japan

Introduction

In an attempt to facilitate uniformity in publication of
electrophoretic results. In 1971 the IUPAC-IUB Commission(1)
recommended that standardized rules for the systematic
orientation of electrophoretic photographs and diagrams be
adopted. Since that time, however, many publications have
appeared ignoring these recommendations(2). Recently, we
also proposed general rules for the orientation of
electrophoretic photographs and diagrams and, in addition,
standarization of electrophoresis apparatus to avoid the risk
of mishandling.(3)

Recommendations

1). Electrophoretic photographs and diagrams
    A. Horizontal type: Anode should be oriented to the right-
hand side of the page, but acid side for isoelectric focusing
should be oriented to the left-hand side. (Fig.1)
    B. Vertical type: Anode should be oriented below but acid
side for isoelectric focusing should be oriented (Fig.2)

C. Two-dimensional electrophoresis
a) O'Farrell method (Fig.3)
First dimension: Anodic side
should be oriented to the
left-hand side of the page
(Fig.3)
Second dimension: Anode
should be oriented to the
down.
b) Crossed immunoelectro-
phoresis (Fig.4)
First dimension: Anode
should be oriented to the
right-hand side of the page.
Second dimension: Anode
should be oriented to the
top side.
D. Densitograms(Fig.5)
Anode should be oriented to
the right-hand side of the
page.

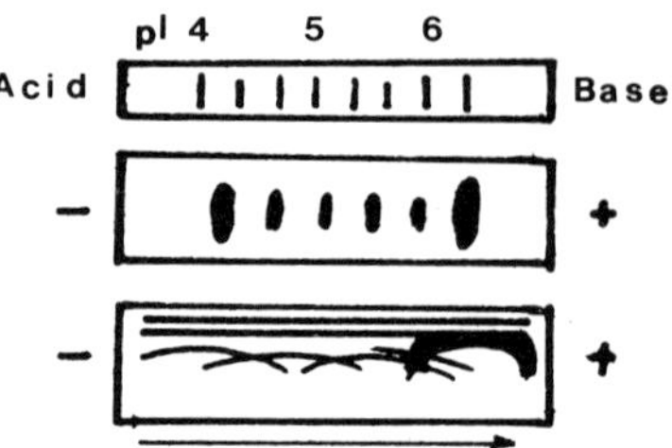

Fig.1 Horizontal electrophoresis
upper: isoelectric focusing

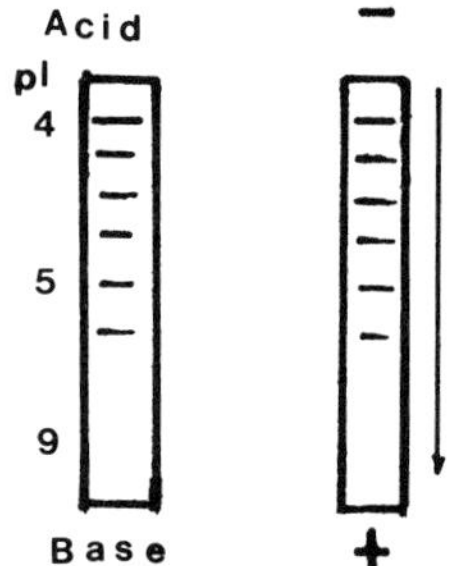

Fig.2 Vertical electrophoresis
left: isoelectric focusing

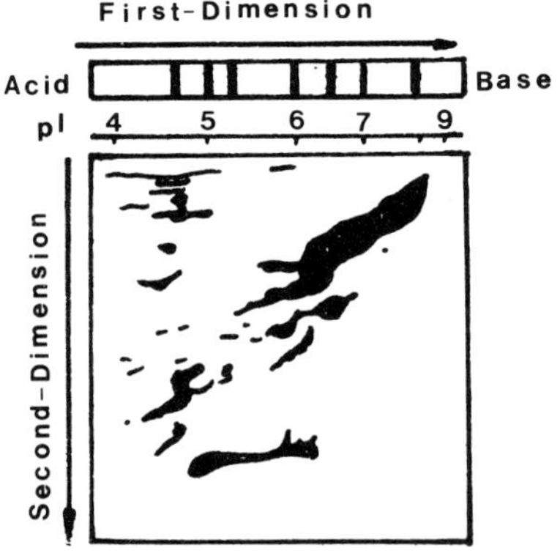

Fig.3 O'Farrell two-dimensional
electrophoresis

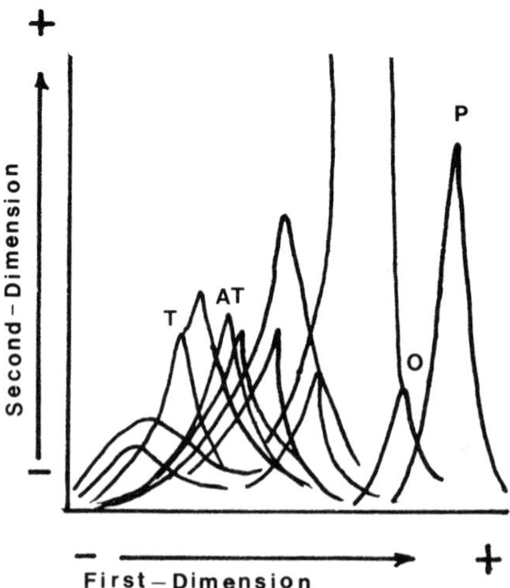

Fig.4 Crossed immunoelectrophoresis

2) Apparatus

A. Power supply

a). Position of the ampere and volt meter

Ampere-meter should be located to the left-hand of the meters arranged horizontally or up side of the meters arranged vertically. (Fig. 6 and 7)

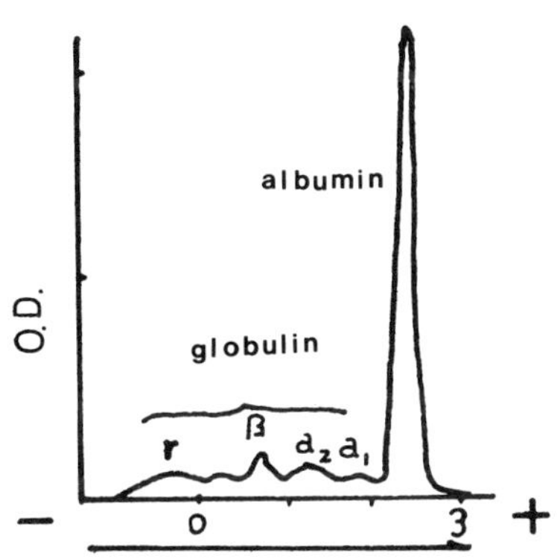

Fig.5 Densitometer
Arow indicates a direction of electrophoresis

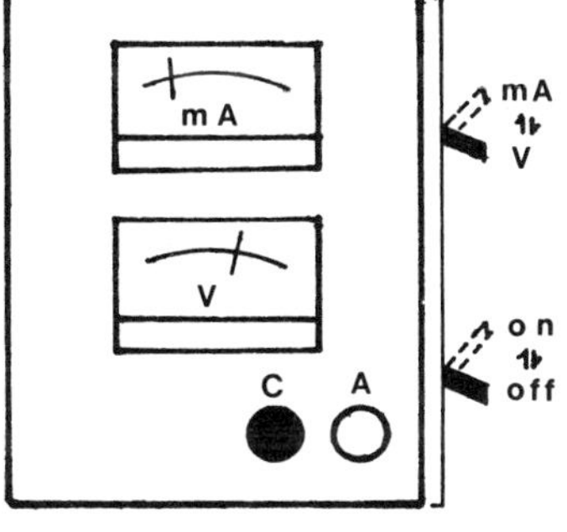

Fig.7 Position of Ampere and volt meter arranged vertically.

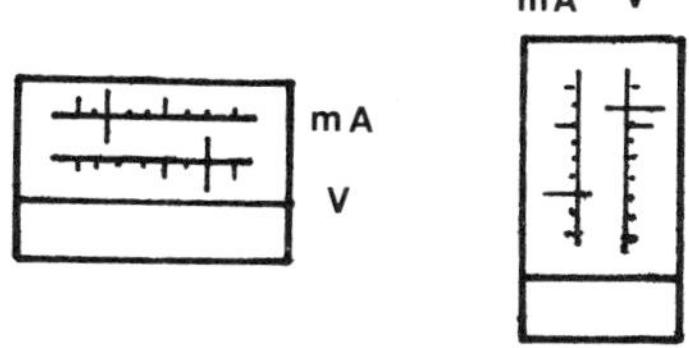

Fig.8 Position of Ampere and volt meter in A-V meter.

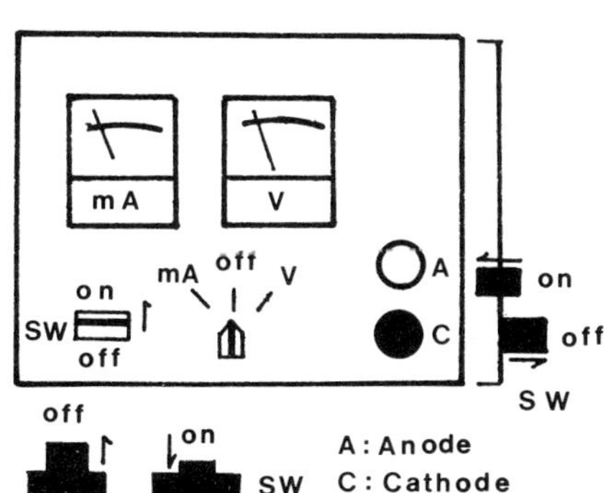

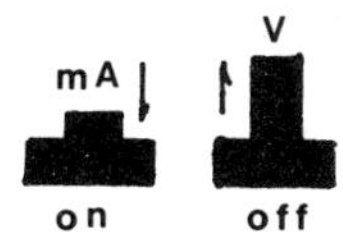

Fig.9 Functional switch

Fig.6 Position of Ampere meter, volt meter and power switch (SW) arrenged horizontally.

In the A-V meter, ampere-meter should be located to the
up side or left side. (Fig.8)
b). Power switch
Knob switch: Up or left side of the Knob switch should be
indicated as "ON".

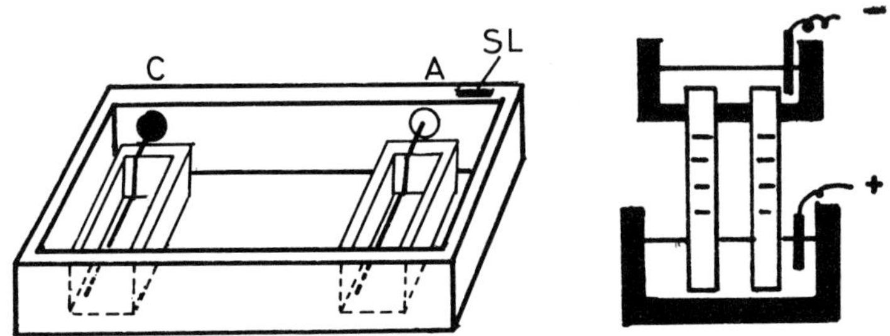

Fig.10 Horizontal and vertical electrophoresis chamber

Push buttom switch: Push side should be indicated as
"ON".
c). Functional switch of the current and voltage (Fig.9)
Knob switch: Up or left-hand side should be indicated as
"Current". (Fig.7)
Push buttom switch: Push side should be indicated as
"Current".
d). Position of the terminal electrodes placed in a pair
on the frontal panel.(fig.10)
Anodic terminal arranged horizontally should be located
to the right-hand side, or up side arranged vertically.
B. Safty device
Electrophoresis chamber should be set up a regulatory
switch synchronized with the power supply for the
prevention of risk by the mishandling.(Fig.10 SL)

References

1.   IUPAC-IUB Commission on Biological Nomenclature.: J.
     Biol. Chem. 246, 6127-6128 (1971).
2.   Electrophoresis '79 symbol mark.: 1979.
3.   Inoue, T., Physico-Chem. Socie., 26, 39 (1982)

Two-Dimensional Electrophoresis

ELECTROPHORETIC METHODS IN HORIZONTAL ULTRATHIN-LAYER
POLYACRYLAMIDE GELS - A VERSATILE TIME AND COST SAVING SYSTEM

Reiner Westermeier, Wilhelm Postel and Angelika Görg
Lehrstuhl für Allgemeine Lebensmitteltechnologie der
Technischen Universität München,
D-8o5o Freising-Weihenstephan, F.R.G.

Introduction

Electrophoretic separation techniques in nonrestrictive media
(agarose gels, cellulose acetate films) as well as isoelectric
focusing in polyacrylamide or agarose gels are mostly performed
in the horizontal plane. Electrophoretic techniques which are
utilizing the molecular sieving effect of the anticonvective
medium (starch gels, polyacrylamide gels) are almost exclusively carried out in vertical cylindrical gel rods or vertical
gels slabs. The methodical procedure of the vertical technique
is more tedious and troublesome compared to horizontal separation systems. Thus, particularly in routine laboratories,
where many samples are to be analysed, those methods which are
feasible in the horizontal technique are preferred to those
performed vertically.

The reason for running electrophoresis within molecule -
restrictive media in the vertical direction is that the use of
a horizontal system introduces some problems such as electrodecantation of proteins during the entry into the gel and
unequal heat dissipation within the gel layer, because it is
cooled only from one side. Smithies was the first who observed
these effects and reversed starch gel electrophoresis from a
horizontal (1) to a vertical (2) system. On these grounds, the
fundamental works in polyacrylamide electrophoresis, as for
instance disc electrophoresis by Ornstein (3) and Davis (4),

SDS electrophoresis by Weber and Osborn (5), gradient gel electrophoresis by Margolis and Kenrick (6), and high-resolution two-dimensional electrophoresis by O'Farrell (7) were all based on the vertical electrophoresis technique.

With the introduction of ultrathin-layer polyacrylamide gels for isoelectric focusing (8) and electrophoresis (9) the uneven cooling of the gel layer and electrodecantation of the proteins in the only o.1 mm deep sample slots of the o.24 mm thick gel layers were abolished. Thus, if ultrathin polyacrylamide gel layers are used, also electrophoretic techniques utilizing the molecular sieving effect of the gel can be carried out with the horizontal technique, resulting in a reduction of time, costs and methodical efforts.

Because ultrathin-layer gels are polymerized on a plastic film as a protection against mechanical stress, handling is facilitated and the gels do not swell or shrink during the various steps of the visualization procedures. Due to the short diffusion distances for ultrathin layers, the staining is very fast and effective. Compared to the methods using conventionally thick gels (1 - 3 mm), the ultrathin-layer methods are therefore more sensitive (o,1 µg protein per band is easily detected with the Coomassie stain), and the separation results are obtained in a fraction of time needed for staining of conventional gels.

In this report we describe the practicabilities of the horizontal ultrathin-layer gel electrophoresis techniques and some methodical modifications that must be applied, when special electrophoretic methods are adapted from the vertical to the horizontal system.

Material and Methods

Ampholine pH 3.5 - 1o, acrylamide, NN'-methylenebisacrylamide

(Bis), ammonium persulphate, N,N,N',N'-tetramethylethylene-
diamine (TEMED), tris (hydroxymethyl) aminomethane (Tris),
Sodium dodecyl sulphate (SDS), urea, dithiothreitol, GelBond
PAG-film, Repeal-silane, Coomassie Brilliant Blue G-25o and
R-25o were from LKB (Bromma, Sweden). All other chemicals were
of analytical grade. Equipment used: LKB 2197 power supply,
LKB 2117 Multiphor, IEF and electrophoresis kit, 2D and gra-
dient gel kit, LKB 2oo1 vertical electrophoresis unit.

Ultrathin-layer gels with homogeneous polyacrylamide composition were cast according to refs. 8 and 9. Polyacrylamide
gradient gels were cast as described in ref. 1o; for casting
linear gradients the gradient mixer was used with both chambers open, and the light solution contained acrylamide/Bis
corresponding to 1o % T, 2.5 % C. Further hints for specific
gel and buffer compositions and running conditions are given
in "results and discussion".

Results and Discussion

At first, an important aspect has to be considered, when
ultrathin-layer polyacrylamide gels serve as matrices for
molecular sieving: Conventionally thick gels and ultrathin
gels, which are polymerized from the identical acrylamide and
Bis concentration, have different retardation effects on
electrophoretically migrating macromolecules. Ultrathin gels
are higher porous than conventional gels, due to a greater
influence of the marginal layers on the polymerization. Thus,
when a vertical electrophoresis technique is reversed to the
horizontal ultrathin-layer system, the polymerization solution must contain a higher concentration of acrylamide. A
general correction formula or table is not yet available, but
a few examples are given in this report.

Another point is that ultrathin-layer horizontal gels are run

with an open surface, whereas the vertical separation techniques are performed in a closed system. If a gel buffer
contains a volatile substance, it would gradually leave the
gel and, thus, the pH buffering conditions would change during
electrophoresis. In order to overcome this effect, the gel surface is simply covered with a chemically and electrically
inert sheet of plastic, for instance a mylar polyester film.

In figure 1 an example is demonstrated for an electrophoretic
separation of prolamins in an o.24 mm gel layer with a homogeneous acrylamide composition and an o.188 M acetic acid/
glycine buffer system pH = 3.1. This method was adapted from
the vertical technique according to Maier and Wagner (11) to
the ultrathin-layer horizontal technique. Whereas in the
original method the separations are run in 3 mm thick gels

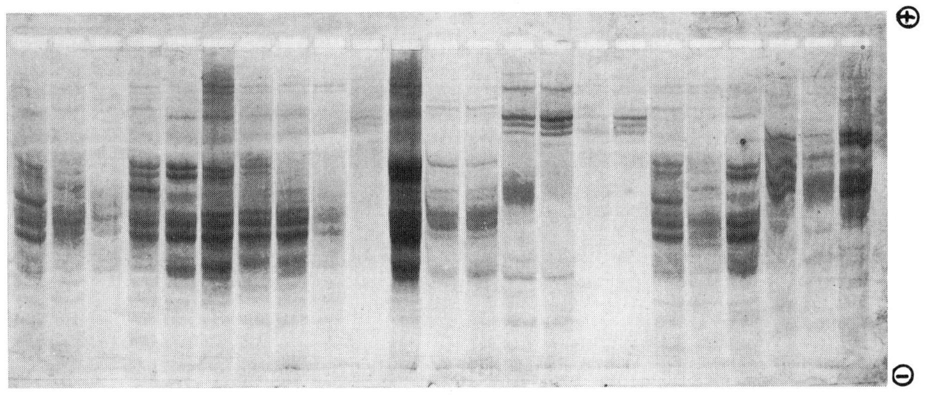

Fig. 1: Polyacrylamide gel electrophoresis of prolamins in a
horizontal ultrathin layer (o.24 mm). Buffer system: o.188 M
acetic acid/glycine, pH = 3.1. PAG: 1o % T, 2.5 % C. Separation conditions: 1o °C, 1 h at max 8oo V, max 2o mA, max 2o W.
Gel surface was covered with mylar polyester film. Samples:
Gliadins of (1) - (5) wheat varieties (1) Arkas, (2) Aureum,
(3) Durum, (4) Horizont, (5) Kolibri, (6) wheat flour type
1o5o, (7) wheat flour type 4o5, (8) wheat pollard, (9) Graham
flour type 17oo (1o) Graham flour B, (11) wheat gluten, (12)
and (13) pasta, (14) rye flour A, (15) rye flour type 18oo,
(16) white bread, (17) black bread. Staining with o.1 %
Coomassie Brilliant Blue R-25o in 12 % TCA.

with an acrylamide concentration corresponding to 6 % T and
a cross-linking of C = 4 %, the ultrathin-layer gel contained
acrylamide and Bis corresponding to T = 1o % and C = 2.5 %.
The gel surface was covered with a sheet of Mylar polyester
film in order to prevent the acetic acid from evaporating out
of the gel. For a comparison, the surface area of three sample
tracks was left open as indicated in figure 1. The negative
effect of the changing pH and buffering conditions due to
loss of acetic acid during the separation run is obviously
visible. Because of the effective heat dissipation within the
ultrathin gel layer, higher field strength can be applied
than in conventional gels. Thus, under the separation conditions given in figure 1, the electrophoresis run is complete
within one hour. The results are available after another hour
of staining. The separation in conventional gels lasts for
two hours and staining is performed overnight.

When buffer systems are adapted to the horizontal technique,
which do not contain volatile compounds, covering of the surface is not necessary. The positions of the slots for sample
application can be chosen within the separation plane, so that
cathodally as well as anodally migrating sample fractions are
separated within a single matrix (9). In the vertical technique the samples are applied either at the cathodal or at
the anodal extreme of the gel, resulting in a loss of the
cathodally respectively the anodally migrating fractions in
the buffer reservoir.

Very high resolution and very sharp bands are obtained, when
SDS pore gradient electrophoresis is performed in horizontal
ultrathin gel layers (12). In figure 2 an example is shown
for a separation of pathological urine proteins in a o.24 mm
gel, which contains a linear polyacrylamide gradient from
8 - 1o % T, 2.5 % C. The molecular weight fractionation of the
proteins present in an urine sample provides informations
about the proteinurie type the patient suffers from. 24

samples are analyzed in a gel matrix of the size 25 x 12 cm. The urine samples are mixed with 1 % SDS, 4 µl of each are pipetted into the small slots in the gel surface. After 2 hours of electrophoresis under the separation conditions, as indicated in fig 2, and 1 hour of fixing, staining, and destaining, the pherogram is ready for evaluation. Sensitivity can be increased by applying the silver stain method on the gels.

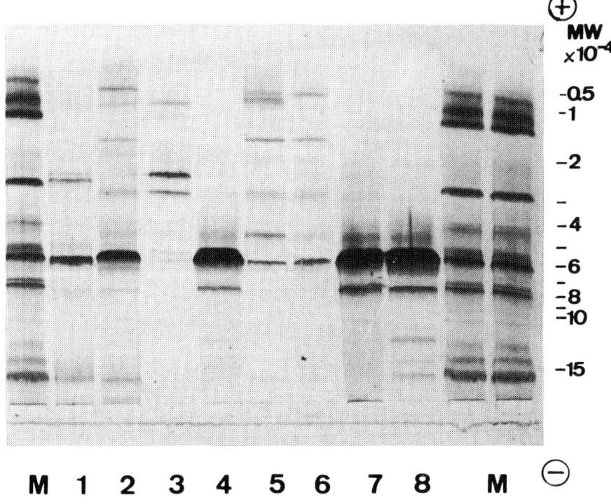

Fig. 2: SDS pore gradient electrophoresis of pathological urine proteins in a o.24 mm thick horizontal gel. PAG: linear gradient 8-18 % T, 2.5 % C. 375 mM Tris-HCl pH 8.8, o.1 % SDS. Urine samples contain 1 % SDS. Separation conditions: 5 °C, 2 h at max 5o mA, max 3o W, max 1ooo V. Staining with o.1 % Coomassie Brilliant Blue R-25o. (M) molecular weight markers.

The method of two-dimensional (2D) electrophoresis is substantially facilitated when carried out in the horizontal ultra-thin-layer technique (9, 1o, 13). After the first dimension run, the strip containing the sample track to be analyzed with 2D electrophoresis is cut off the gel and is simply laid into a preformed gel trench in the surface of the second dimension gel. Handling of the gel strip is no problem, because it firmly adheres to the plastic support film. No overlay with agarose, like in the conventional methods, is necessary.

The quality of the separation results are comparable with

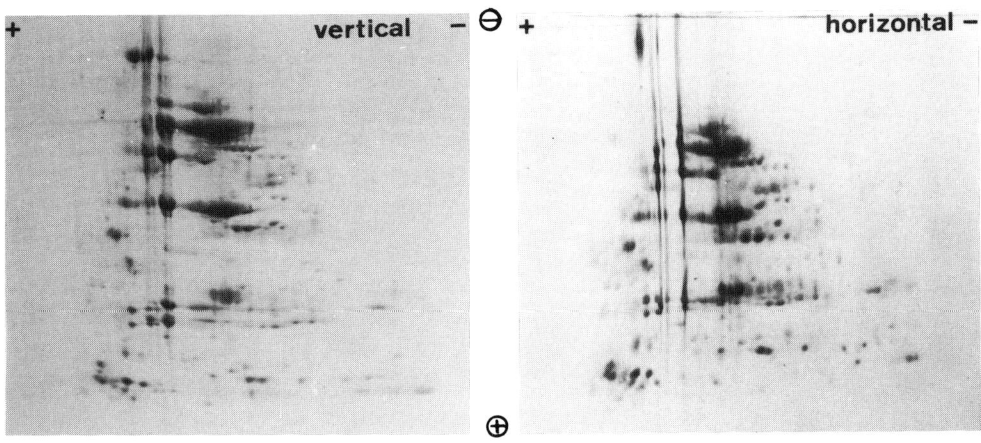

<u>Fig. 3</u>: High-resolution 2D electrophoresis of broad bean seed
proteins. First dimension: IEF in o.24 mm gel containing 8 M
urea. PAG: 4 % T, 6 % C. Separation distance: 1o cm. 1o ^{o}C,
9o min at max 6 W, 2ooo V. Second dimension: SDS pore gradient
electrophoresis. Vertical: o.75 mm gel, linear gradient
1o − 22.5 % T along 16 cm, 2.5 % C. 5 ^{o}C, 5 h at max 2o mA,
max 2o W, max 5oo V. Horizontal: o.36 mm gel, linear gradient
14 − 22.5 % T along 12 cm, 2.5 % C. 5 ^{o}C, 2 h at max 5o mA,
max 3o W, max 1ooo V. Staining with Coomassie Brilliant Blue
R-25o.

those obtained with the vertical technique, as it is demonstrated in figure 3 for the example of a high-resolution 2D
electrophoresis of broad bean (<u>Vicia</u> <u>faba</u>) seed proteins. The
first dimension was isoelectric focusing in an o.24 mm gel
layer containing 8 M urea. Two cut-off gel strips are equilibrated for two minutes in 15o mM Tris-HCl pH 8.8, 1 % SDS, and
o.1 % dithiothreitol. As reducing reagent dithiothreitol is
preferred to ß-mercaptoethanole, because it is nonvolatile. One
strip was applied on a vertical o.75 mm thick SDS gel, containing a linear polyacrylamide gradient from 1o − 22.5 % T,
the other one was applied into the trench of a o.36 mm
horizontal SDS gel, containing a linear gradient from 14 −
22.5 % T. Both gels were crosslinked corresponding to 2.5 % C.

Ultrathin-layer gels are not only advantageous because of the
enormous reductions in costs for isoelectric focusing, they

also provide the basis for a versatile separation system, which allows to perform all gel electrophoresis methods in the horizontal plane instead of in the more complicated vertical technique. The separations and the staining procedures are for ultrathin gels complete within a fraction of time compared to conventional gels. Another benefit of ultrathin-layer poly-acrylamide gradient gels is that they are easily dried without the risk of cracking, which is an important factor, when autoradiography is employed for protein visualization.

References

1. Smithies, O., Biochem. J. 61, 629-641 (1955).

2. Smithies, O., Biochem. J. 71, 585-587 (1959).

3. Ornstein, L., Ann. N.Y. Acad. Sci. 121, 321-349 (1964).

4. Davis, B.J., Ann. N.Y. Acad. Sci. 121, 4o4-427 (1964).

5. Weber, K., Osborn, M., J. Biol. Chem. 244, 44o6-4412 (1969).

6. Margolis, J., Kenrick, K.G., Nature (London) 214, 1334-1336 (1967).

7. O'Farrell, P.H., J. Biol. Chem. 25o, 4oo7-4o21 (1975).

8. Görg, A., Postel, W., Westermeier, R., Anal. Biochem. 89, 6o-7o (1978).

9. Görg, A., Postel, W., Westermeier, R. in Electrophoresis '79 (B.J. Radola, ed.) W. de Gruyter, Berlin, New York 198o, pp. 67-78.

1o. Görg, A., Postel, W., Westermeier, R., Gianazza, E., Righetti, P.G., J. Biochem. Biophys. Methods 3, 273-284 (198o).

11. Maier, G., Wagner, K., Z. Lebensm. Unters.-Forsch. 17o, 343-345 (198o).

12. Görg, A., Postel, W., Westermeier, R., Z. Lebensm. Unters. Forsch. 174, 282-285 (1982)

13. Görg, A., Postel, W., Westermeier, R., Gianazza, E., Righetti, P.G. in Electrophoresis '81 (R.C. Allen, P. Arnaud, eds.) W. de Gruyter, Berlin, New York 198o, pp. 257-27o.

MICROCOMPUTER-AIDED TWO-DIMENSIONAL DENSITOMETRY

Tosifusa Toda, Toshiko Fujita and Mochihiko Ohashi

Department of Biochemistry, Tokyo Metropolitan Institute of
Gerontology, 35-2 Sakae-cho, Itabashi-ku, Tokyo 173, Japan

Introduction

Two-dimensional electrophoresis has become a powerful tool
in various fields of the life sciences. The high resolution
of two-dimensional electrophoresis provides the capability
of detecting small changes of protein, which appear with
aging, carcinogenesis, disease states and therapy. We have
developed a new type of two-dimensional electrophoresis on
cellulose acetate membrane (1) to analyze age-related
changes of protein. The method has many practical properties: among others, only a short time is required, the
supporting membrane is easy to handle, enzyme activities are
detectable after electrophoresis, several replicas of a
protein map are available. However, the most serious
problem of two-dimensional electrophoresis is the difficulty
in estimating quantities of protein in the spots. Several
groups have developed methods of two-dimensional densitometry (2,3,4) to overcome the problem using a minicomputer
system. For the same purpose, we have deviced two kinds of
microcomputer-equipped systems, a scanning-densitometer-
equipped system, and a CCD-camera-equipped system, because a
minicomputer is not accessible. Details of the scanning-
densitometer-equipped system were reported at the 33rd
Annual Meeting of the Society of Electrophoresis (Japan),
and will be published in a separate paper (5). In this

paper, the method and the performance of the CCD-camera-equipped system is described and discussed.

Materials and Methods

Scanning-densitometer-equipped system.   The system consists of a Joyce Loebl Microdensitometer 6 (Gateshead, UK), a DG Minicomputer NOVA4 (Westboro, USA) and a SORD Microcomputer M223 (Tokyo, Japan). The minicomputer is used only to control the scanning densitometer. All the data processing is performed on the SORD microcomputer.

CCD-camera-equipped system.   The system consists of an image processor unit and a SORD Microcomputer M223. The image processor unit is equipped with a CCD camera, 4 pages of frame memory, a page of display memory, a CRT monitor screen, and a digitizer (Fig.1). The intelligence for basic functions of the image processor unit is built into the CPU board as the firmware. The unit is connected to the SORD microcomputer through 2 ports of parallel interfaces, and its functions are discharged by the system commands issued from the host computer. The image processor unit is available from ADS Co. Ltd (Nara, Japan) as Two-Dimensional Densitometer Model E.

Fig. 1.   A schematic view of the hardware of CCD-camera-equipped system. The system has two ports of parallel interfaces to connect with a host computer.

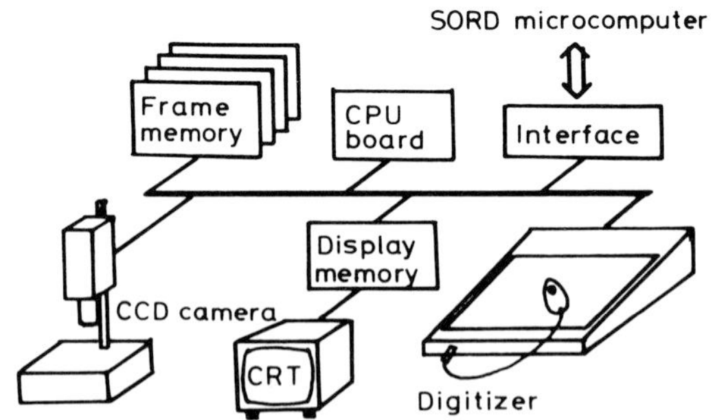

Software.   The core structure of the software is shown
in Fig. 2.   FIMC is a subroutine package, which we wrote in
Z-80 assembly language for two-dimensional densitometry.
FIMC COMMANDs listed on Fig. 2 can be used in BASIC programming.   The utility programs, which direct our actual
procedures of two-dimensional densitometry and the subsequent analyses, are written in BASIC language.   A skeletal
program for two-dimensional densitometry is shown in Fig. 3.
We used sevaral modifications of this program in our
studies.

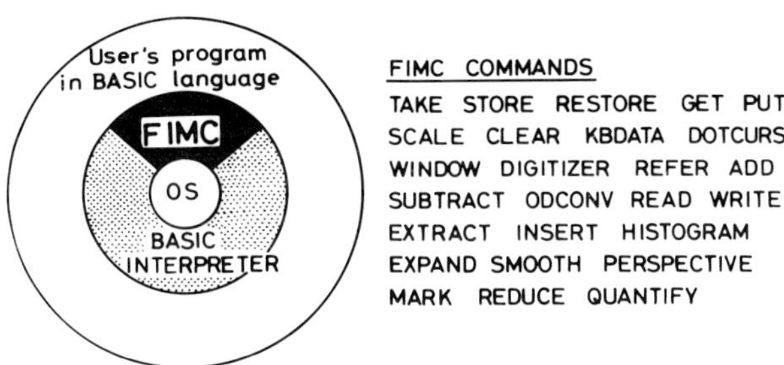

Fig. 2.   FIMC is a subroutine package for two-dimensional
densitometry.   Users are allowed to write their utility
programs in BASIC language using the FIMC COMMANDs.

Two-dimensional electrophoresis.   Two-dimensional
electrophoresis of rat tissue extracts was performed as
described previously (1) with the following modifications.
10%(w/v) homogenates in distilled water were centrifuged at
10,000 x g for 20 min, and 15 µl aliquots of the supernatants were used for two-dimensional cellulose acetete
electrophoresis.   In the first-dimensional direction,
concentrating electrophoresis was carried out in two steps
as follows: first, isotachophoretic concentration was
performed on a strip of cellulose acetate membrane at 200 V
for 10 min, followed by electrophoretic separation at 0.4 mA
per strip for 50 min.   Second-dimensional isoelectric
focusing was performed on a sheet of cellulose acetate

membrane absorbing 8%(v/v) LKB Ampholines mixed in the following proportions: 1 part at pH range 4-6, 4 parts pH 5-7, 1 part pH 7-9, and 2 parts pH 8-9.5. Proteins on the membrane were stained with Coomassie Blue G-250 as described previously (1).

```
10   INPUT "TAKE A BLANK",A$              200 INPUT "Next spot? (Y/N)",A$
20   CALL #0,"TAKE"                           :IF A$="Y" THEN GOTO 150
30   CALL #0,"STORE","2"                  210 INPUT "Another pattern?",A$
40   CALL #0,"RESTORE","2"                220 GOTO
50   CALL #0,"STORE","0"                  230 /SUBROUTINE
60   INPUT "TAKE A SAMPLE",A$             240 LET X1=150 :LET Y1=120 :D1=0
70   CALL #0,"TAKE"                       250 CALL #0,"DOTCURSOR",X1,Y1,D1
80   CALL #0,"REFER"                      260 LET F=0
90   INPUT "Is that OK?",A$                   :CALL #0,"WINDOW",X1,Y1,F
     :IF A$="N" THEN GOTO 60             270 IF F=0 THEN GOTO 280
100  CALL #0,"STORE","0"                      :ELSE GOTO 250
110  PRINT "MEASURE A STANDARD SPOT"     280 LET B=0
120  GOSUB 230                               :CALL #2,"EXPAND",X1,Y1,B
130  LET K1=X1 :LET K2=Y1-(X1*B/8)       290 CALL #2,"SMOOTH",0
140  LET S1=8 :LET S2=5.96               300 CALL #2,"PERSPECTIVE"
150  PRINT "MEASURE A PROTEIN SPOT"      310 LET X2=0 :LET Y2=0
160  CALL #0,"RESTORE","0"                   :CALL #0,"DIGITIZER",X2,Y2
170  GOSUB 230                           320 CALL #2,"PAINT",1
180  LET Y1=Y1-(X1*B/8)                  330 CALL #2,"REDUCE",X1,Y1,"0"
     :LET M1=S1*X1/K1                    340 CALL #2,"QUANTIFY",X1,Y1,"0"
     :LET M2+S2*Y1/k2                    350 RETURN
190  PRINT "Size =";M1;"(mm.mm)"         360 END
     :PRINT "IOD =";M2;"(OD.mm.mm)"
```

Fig. 3.   A skeletal program for two-dimensional densitometry. All of the basal procedures of two-dimensional densitometry can be excuted in this short program.

Results and Discussion

Reliability of IOD.   Any light source other than an ideal point source is subjected to some unevenness. To help correct for nonuniformities in the camera's view, two steps of subtractions were introduced into our procedure of two-dimensional densitometry. Fig. 4 shows the complete (III) and two abridged procedures (I and II). In the complete procedure, CCD camera takes a "BLANK" image before taking a "SAMPLE" image. The differential image between them is obtained by "BLANK-SUBTRACTION". The internal scale of optical density is calibrated using the standard chip of exposed photofilms. Then, IOD of the protein spot is calculated. Here, the "LOCAL-BACKGROUND" is subtracted from the IOD. To evaluate the efficiency of the two "SUBTRAC-

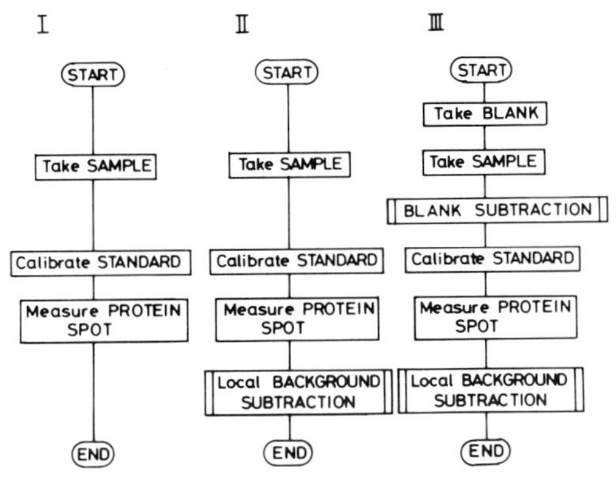

Fig. 4. Procedures of two-dimensional densitometry. III is the complete procedure with two "SUBTRACION"s. I and II are its abridged procedures. "BLANK SUBTRACTION" is omitted in I and II, and "Local BACKGROUND SUBTRACTION" is omitted in I.

TION"s, position dependency of IOD was examined in these three procedures. The camera's field of view was devided into 9 portions, and a standard chip was measured 5 times in every portion. The mean IOD and the SD obtained in each portion are shown in Table 1. Procedure I with no "SUBTRACTION" showed quite a serious position dependency. "LOCAL-BACKGROUND SUBTRACTION" in procedure II reduced the

Table 1. Position dependency of IOD measurement in the three procedures. The camera's field of view was divided into 9 portions. A standard chip of exposed photofilm was measured 5 times in each portion. The outside values are the mean and SD of all 45 measurements. The scanning-densitometer-equipped system produced 3.69 ± 0.08 from 10 measurements.

I NO CORRECTION (Mean± SD)

6.91±0.25	5.87±0.52	6.88±0.42
5.81±0.49	3.71±0.28	5.96±0.22
7.14±0.47	6.18±0.37	7.09±0.46

6.17 ± 1.08 (N=45)

II "LOCAL-BACKGROUND SUBTRACTION"

4.63±0.15	4.23±0.17	4.57±0.14
4.60±0.14	3.63±0.06	4.54±0.18
4.75±0.18	4.09±0.14	4.68±0.14

4.41 ± 0.37 (N=45)

III "BLANK SUBTRACTION" and "LOCAL-BACKGROUND SUBTRACTION"

3.75±0.02	3.76±0.02	3.75±0.03
3.76±0.02	3.77±0.02	3.78±0.02
3.78±0.02	3.76±0.01	3.76±0.02

3.76 ± 0.02 (N=45)

Joyce Loebl Microdensitometer 6
3.69 ± 0.08 (N=10)

dependency partially.   The combination of the two "SUB-
TRACTION"s in procedure III cancelled the dependency almost
completely.   In the following experiments, we used procedure
III.

Linearity of IOD.   The linearity of IOD was examined
using the test pattern, which was made of photofilms with
various densities.   The absolute optical densities of the
films were determined using a Joyce Loebl Microdensitometer
6.   The results, shown in Fig. 5, demonstrate that the
linearity of the CCD-camera-equipped system is as high as
that of the scanning-densitometer-equipped system.

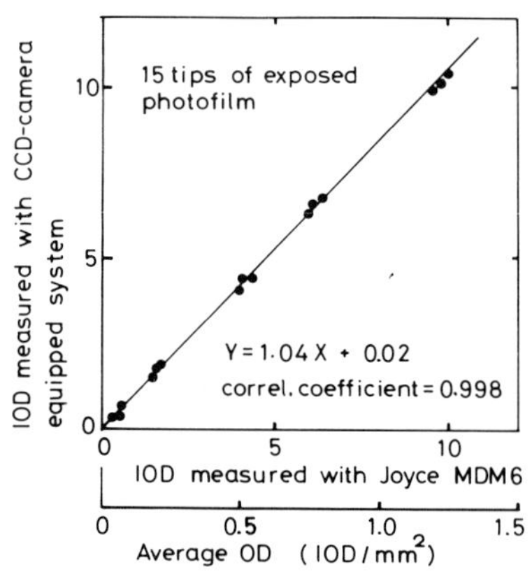

Fig. 5.   Linearity of IOD.
Various chips of differently
exposed photofilms were measured by both the scanning-densitometer-equipped system and
the CCD-camera-equipped system.
The average OD was obtained by
division of the IOD by the
size (8 mm^2).

Quantitative comparison of two-dimensional electrophoretograms.   As an example of quantitative analysis, two
electrophoretograms of rat liver protein were compared.
Procedures of sample preparation and electro- phoresis are
described in Materials and Methods.   Fig. 6 shows   the
scheme of the electrophoretogram.   More than 90 spots were
observed, and 26 spots, named A to Z, were compared for
%total IOD as shown in Fig. 7. Quantitative differences of
protein in spots, corresponding to each other on the two
patterns, are expressed in divergence from the diagonal
line.   In this example, spots Y, Z, M and G were increased,
spots E, D, R, U and I were decreased with aging. In prac-

tice, many samples should be compared each other to eliminate individual variation. We are making such comparisons systematically using the SORD microcomputer as follows: when a sample pattern is analyzed by two-dimensional densitometry, the IOD data are stored in the data bank on a floopy diskette; when the "pattern-to-pattern" comparison is per-

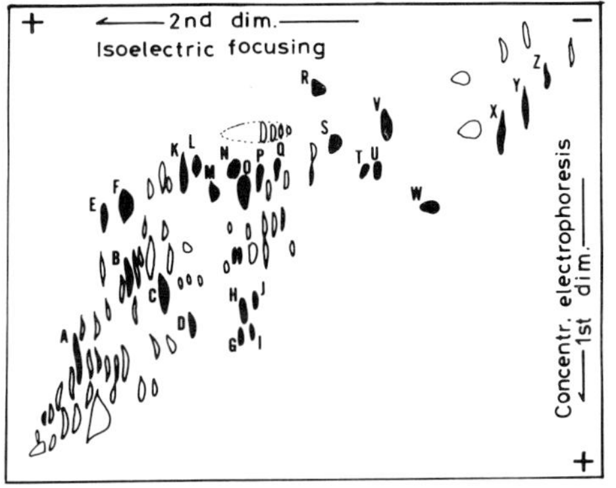

Fig. 6. Scheme of two-dimensional electrophoretogram of rat liver protein. Electrophoresis and Coomassie staining were performed as described in Materials and Methods. The size of the membrane was 6 x 11 cm. More then 90 spots can be observed.

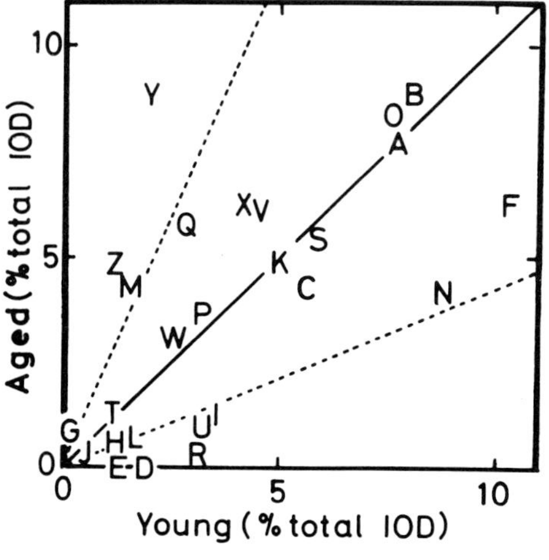

Fig. 7. An example of "pattern-to-pattern" comparison. In this case, liver protein of a 30-months-old rat was compared with that of a 3-month-old rat. Spots Y, Z, M and G were increase, spots E, D, R, U and I were decreased with aging.

146

formed, the computer recalls the data files of the two patterns from the bank, and calculate the divergence of all the corresponding spots automatically. This program enables us to compare all cases among a large number of patterns stored in the data bank. We named this method "Round-Robin Comparison".

Conclusion

The novel method of two-dimensional densitometry has many special features, that is, the hardware is handy, the software is compact, the excution is rapid. And the most distinctive feature is the simplicity of the programming system. FIMC is a convenient subroutine package containing many tasks for image processing. It enables us to write our utility program for two-dimensional densitometry in BASIC, which is the most familiar language for users of microcomputers. The novel method is useful to quantitative analysis of protein on two-dimensional electrophoretograms.

References

1. Toda, T., Fujita, T., Ohashi, M.: Anal. Biochem. $\underline{119}$, 167-176 (1982).

2. Lemkin, P. F., Lipkin, L. E.: Comput. Biomed. Res. $\underline{14}$, 272-297 (1981).

3. Aycock, B. F., Weil, D. E., Sinicropi, D. V., McIlwain, D. L.: Comput. Biomed. Res. $\underline{14}$, 314-326 (1981).

4. Vo, K. -P., Miller, M. J., Geiduschek, E. P., Nielsen, C., Olson, O., Xuong, N. -H.: Anal. Biochem. $\underline{112}$, 258-271 (1981).

5. Toda, T., Fujita, T., Ohashi, M.: Physico-Chem. Biol. in press.

MICRO TWO-DIMENSIONAL ELECTROPHORESIS OF SOLUBLE PROTEINS OF
ALBUMEN GLAND

Yoko Suda, Masatoshi Fujishiro and Tsutomu Inoue
Department of Biology, Tokyo Gakugei University
4-1-1 Nukui-kitamachi, Koganei-shi, Tokyo 184, Japan

Introduction

The albumen gland in Pulmonates is a secretory organ
providing nutritive substances to the egg. During the
feeding period, the secretory cells in the albumen gland of
Helix pomatia produce secretory substances consisting of
galactogen, proteins, glycoproteins, amino acids and calcium
(1,2,3). The biochemical properties of galactogen have been
studied by many authors (4,5,6), whereas those of the
protein constituents are not well known. Morrill et al.(7)
reported that the soluble proteins isolated from the egg of
Lymnaea palustris are derived from the albumen gland and
their electrophoretic patterns show species specificity in
nature. Recently, we analyzed the soluble proteins in the
albumen gland and the egg of Japanese land snails, _Euhadra
peliomphala_, _Euhadra hickonis_ and _Euhadra quaesita_ by disc
electrophoresis and immunoelectrophoresis (8,9).
In this paper, we report the protein constituents of the
albumen gland and the relationship of the soluble proteins
in the egg with the albumen gland studied by the micro
two-dimensional electrophoresis which was recently developed
in our laboratory.

Electrophoresis '83
© 1984 Walter de Gruyter & Co., Berlin · New York

148

Materials and Methods

1) Japanese land snails

Euhadra peliomphala, *Euhadra hickonis* and *Euhadra quaesita* were used in all experiments. The albumen gland was taken out, homogenized in 0.15M NaCl solution and centrifuged at 12,000 xg for 30 min at 4 °C. The extract of the egg within 1 day after oviposition was prepared by the same procedure.

2) Micro two-dimensional electrophoresis

Our recently developed two-dimensional electrophoresis on polyacrylamide gel without SDS was used for the analysis of proteins having biological activities and takes only 5h to perform. Apparatus were made by Kayagaki Company (Tokyo, Japan) (10). The first dimensional separation was performed with micro gel isoelectrofocusing containing equal volumes of Ampholine pH3.5-10 and pH3.5-5 using apparatus C in Fig.1. The second dimensional separation was performed with micro plate disc electrophoresis without SDS using apparatus B in Fig.1. Regulated d.c. power supply shown as apparatus A in Fig.1 stabilizes the current (μA-mA). The electrolytes for micro two-dimensional electrophoresis are summarized in Table 1.

3) Macro two-dimensional electrophoresis

The method of O'Farrell was used (11). Sample materials were dissolved in a solution containing 1% SDS and 8M urea. In the first dimension, isoelectrofocusing was carried out at a constant voltage (200 V) for 12h. In the second dimension, polyacrylamide slab gel electrophoresis with SDS (7.5-30% polyacrylamide gradient gel) was carried out at a constant current (25 mA) for 7h.

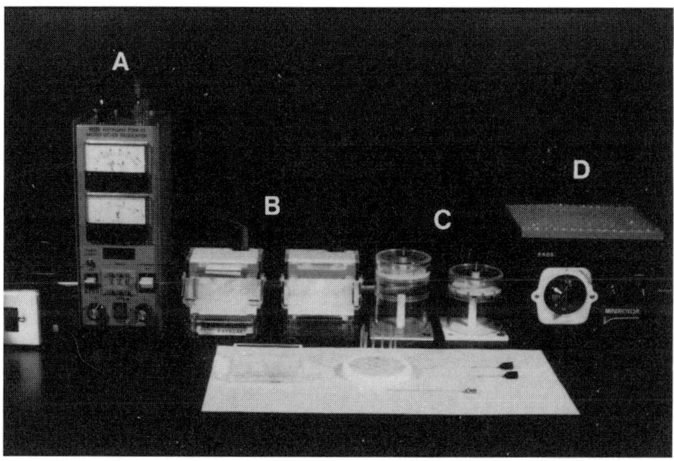

Fig. 1 Total view of micro two-dimensional electrophoretic
apparatus
A: Regulated d.c. power supply
B: Apparatus for second dimensional electrophoresis
(micro plate disc electrophoresis)
C: Apparatus for first dimensional electrophoresis
(micro gel isoelectrofocusing)
D: Minirotor for staining and destaining

Table 1. Stock solution for two-dimensional micro electrophoresis

First dimensional E.F. (micro gel isoelectrophoresis)		Second dimensional E.F. (micro plate disc electrophoresis)	
A[#]	Acrylamide 20 g Bis 0.6 g D.W. to 100 ml	A	Tris 8.6 g TEMED 0.63 ml D.W. to 100 ml 3.6N H_2SO_4 to adj. pH 8.8
B[#]	40 % Ampholine pH 3.5 - 5 pH 3.5 - 10	C	Acrylamide 40 g D.W. to 100 ml Bis 1.07 g
C[#]	TEMED 0.23ml D.W. to 100 ml	[#]2	APS 0.14 g D.W. to 100 ml
D[#]	APS 0.1 g D.W. to 100 ml	B	Tris 5.95 g TEMED 0.45 ml D.W. to 100 ml 1N HCl to adj. pH 6.7
Electrode solutions anode 0.01 M H_3PO_4 cathode 0.05 M NaOH		D	Acrylamide 10 g D.W. to 100 ml Bis 2.5 g
		E	Riboflavin 4 mg D.W. to 100 ml
		Electrode buffer Tris 6 g Glycine 28.8 g	

4) Detection of proteins and glycoproteins on the gel
    Detection of proteins or glycoproteins was carried out
    by the silver staining method of Oakley (12) or by the
    periodic acid-silver method of Dubray (13), respective-
    ly.

Results and Discussion

1) Standard protein map
    To prepare a standarized protein map of the micro
    two-dimensional electrophoresis, each soluble protein
    isolated by disc gel was used for analysis by micro
    two-dimensional electrophoresis. Separated proteins
    were stained by the silver-staining method. The results
    are summarized in Fig.2. About 60-70 distinct protein
    spots from the albumen gland of E. peliomphala in the
    feeding period were found. The code number of the
    protein in the figure was tentatively named as the
    compound code combined from the values of pI and Rm on
    the two-dimensional gel. Each protein is described
    according to its location on the standard map. This
    protein map will be used as standard for comparative
    biochemical studies of the albumen gland and the egg in
    the land snail Euhadra and others.
2) Glycoproteins of albumen gland
    All protein bands of the albumen gland separated by disc
    electrophoresis were found to be PAS sensitive.
    However, these spots separated by micro two-dimensional
    electrophoresis were stained weakly with PAS reagents.
    Therefore, the sensitive periodic acid-silver staining
    reported by Dubray et al.(13) was used for micro
    detection of glycoproteins obtained from the albumen
    gland and the egg of E. peliomphala. Consequently, not
    only the major protein spots but also the minor protein

spots were readily detected and furthermore the minor spots using Coomassie brilliant blue were more strongly evident than those of the major protein spots.

3) Comparison of the protein maps

The seasonal variation of the wet weight of the albumen gland was tentatively divided into four periods named as hibernation, feeding, estivation and egg laying. The albumen glands in each period were examined according to the protein maps. Certain changes in the protein spots were observed. The protein map of the albumen gland in hibernation was a quite different from that of the albumen gland in feeding, as seen for spots A742, A841 and A831. For the albumen gland in the juvenile and after oviposition, the large major protein spots A742, A841 and A831 were altered.

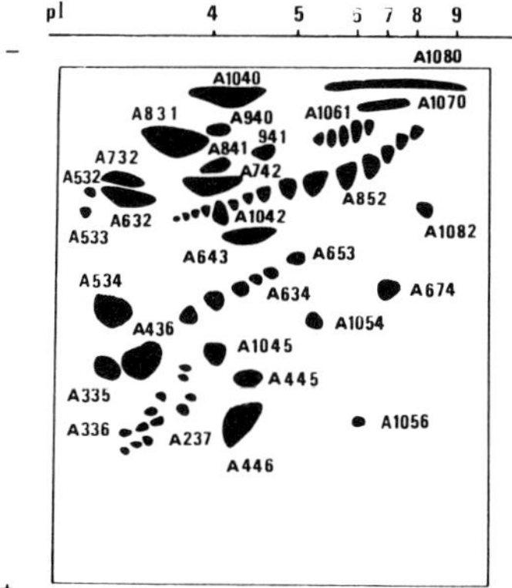

Fig. 2 Standard protein map of albumen gland in *E. peliomphala* during feeding period obtained by micro two-dimensional electrophoresis. Experimental conditions are described in the text.

4) Comparison of proteins from egg and albumen gland
    The albumen gland is located between the hermaphroditic
    duct and the oviduct, and provides nutrient for the
    development of the fertilized egg. Therefore, to
    determine what proteins in the albumen gland are
    transferred into the egg, we compared the protein maps
    obtained from the egg and the albumen gland. Although
    these maps were quite similar, one major spot at pI6.5
    and some of minor spots were found only in the egg
    protein map. It is considered that these proteins may
    be included in the zygote or the shell layers.

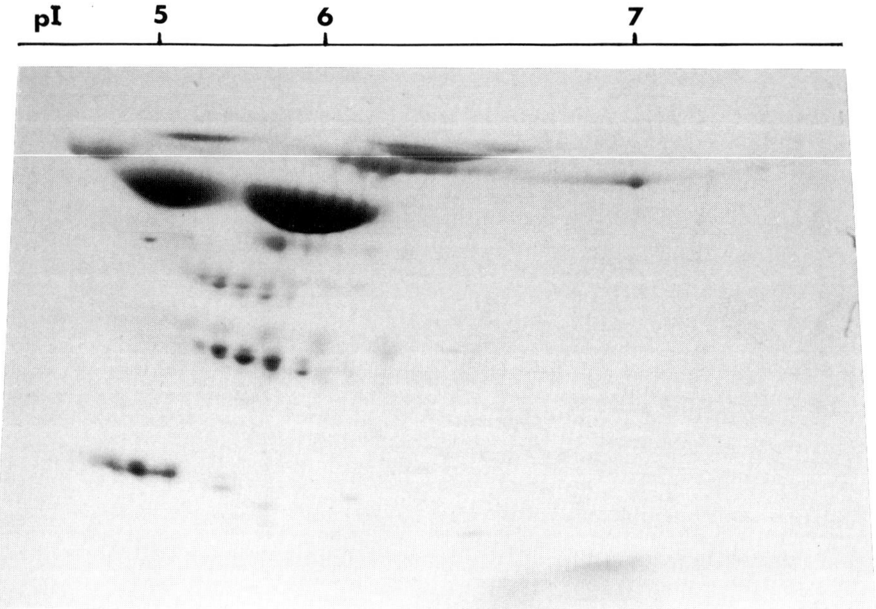

Fig. 3 Peptide maps of albumen gland in E. peliomphala
        during feeding period. Experimental conditions are
        described in the text.

5) Peptide map of albumen gland and egg

The peptide maps of the albumen gland were observed by the method of O'Farrell using Coomassie brilliant blue R-250. The specific peptide map of the albumen gland in E. _peliomphala_ was shown in Fig.3. The peptide maps of the albumen gland were expressed essentially according to the principle described for that of the egg proteins and the albumen gland proteins. The large spots of pI4.5 with M.W.56,000 and pI3.6 with M.W.63,000 were the major proteins in both the albumen gland and the egg.

References

1. Nieland, M.L., Goudsmit, E.M.: J. Ultrastruct. Res. 29, 119-140 (1969).

2. Bayne, C.J.: Comp. Biochem. Physiol. 23, 761-773 (1967).

3. Correa, J.B.C., Dmytraczenko, A., Duarte, J.H.: Carbohydrate Res. 3, 445-452 (1967).

4. Goudsmit, E.M., Ashwell, G.: Biochem. Biophys. Res. Commun. 19, 417-422 (1965).

5. Uhlenbruck, G., Steinhausen, G., Geserick, G., Prokop, O.: Comp. Biochem. Physiol. 59 B, 285-288 (1978).

6. Chatterjee, B.P., Chatterjee, S., Prokop, O., Uhlenbruck, G.: Biol. Zbl. 98, 85-90 (1979)

7. Morill, J.B., Norris, E., Smith, S.D.: Acta Embryol. Morph. Exp. 7, 155-166 (1964).

8. Inoue, T., Nakada, I.: Physico-Chemical Biol. 25, 155-160 (1981).

9. Inoue, T., Suda, Y.: Physico-Chemical Biol. 25, 208 (1981).

10. Inoue, T., Iwase, M.: Physico-Chemical Biol. 26, 128 (1982).

11. O'Farrell, P.H.: J. Biol. Chem. 250, 4007-4021 (1975)

12. Oakley, B.R., Kirch, D.R., Morris, N. R.: Anal. Biochem. 105, 361-363 (1980).

13. Dubray, G., Bezard, G.: Anal. Biochem. 119, 325-329 (1982).

14. Laemnli, U.K.: Nature (London) 227, 680-685 (1970).
15. Bayne, C.J.: Proc. Malac. Soc. Lond. 38, 199-211 (1968).

HIGH PERFORMANCE HORIZONTAL TWO-DIMENSIONAL ELECTROPHORESIS OF SERUM PROTEINS

M.Yoshida, K.Okano and M.Itoh
Central Research Laboratory, Hitachi, Ltd., Kokubunji Tokyo
185, Japan

Abstract

Simple, rapid and reliable method of two-demensional electrophoresis suitable for automation is presented. Separation is
performed in both dimensions on a horizontal polyacrylamide
gel backed with a silanized glass plate. A narrow isofocusing
gel plate is mounted in a groove of a gradient gel plate to
perform the second electrophoresis. The separated proteins
are detected by a new rapid method whereby they are converted
into artificial peroxidase by means of adsorption of hemin.
They are stained by the oxidative coupling reaction of dia-
minobenzidine and xylenol in the presence of H_2O_2.

Introduction

Since O'Farrell (1), Anderson and Anderson (2), and Manabe et
al (3) developed high resolution two-dimensional electrophoresis, attempts have been made to apply this technique to
clinical diagnosis. The present paper deals with simplification and acceleration of this technique, which would enable it
to be used in clinical laboratories. The electrophoretic
separation of serum proteins without denaturating agents (3)
was performed on a horizontally placed polyacrylamide gel (PAG)
chemically bound to a silanized glass. The separated proteins
were stained by a new, rapid and sensitive method, named the

hemin method. The proteins were converted into an artificial peroxidase by adsorption of hemin, and were stained by oxidative coupling of 3.3'-diamino benzidine (DAB) with xylenol in the presence of H_2O_2. Oxidation of DAB in the presence of H_2O_2 has been previously used for the detection of cytochromes in PAG (4).

Materials and Methods

1) Preparation of polyacrylamide gel plates.
   Polyacrylamide gel for isoelectric focusing. The mixture for the preparation of isofocusing gel was poured into a gel casting vessel with narrow grooves (4×80×0.5mm), and the gel mixture was covered with silanized glass plates (4×90×1.0mm). After gelling, a polyacrylamide gel strip (T=4%, C=4%, 2% LKB Ampholine, size 4×80mm, thickness 0.5mm) bound to silanized glass plates was obtained.

   Polyacrylamide gradient gel. The gel was formed between silanized and non-silanized glass plates (110×110×1.5mm with a spacer (4×110×0.5mm). Several sets of the two glass plates were vertically placed in a gradient gel casting chamber with a water jacket. The mixture for the gradient gel was pumped into the bottom of the chamber at 6°C through a gradient forming device (Gradient mixer, Pharmacia; Peristalic pump, Atto, Japan). After pouring the mixture into the gel casting chamber, the temperature of the water jacket was increased to 25°C. Polymerization of the acrylamide mixture was then rapidly completed, and gradient gels (T=4.2~18%, size 100×100mm, thickness 0.5mm) bound to glass were obtained. The silanization of the glass plates was performed according to Radola (5) with slight modifications.

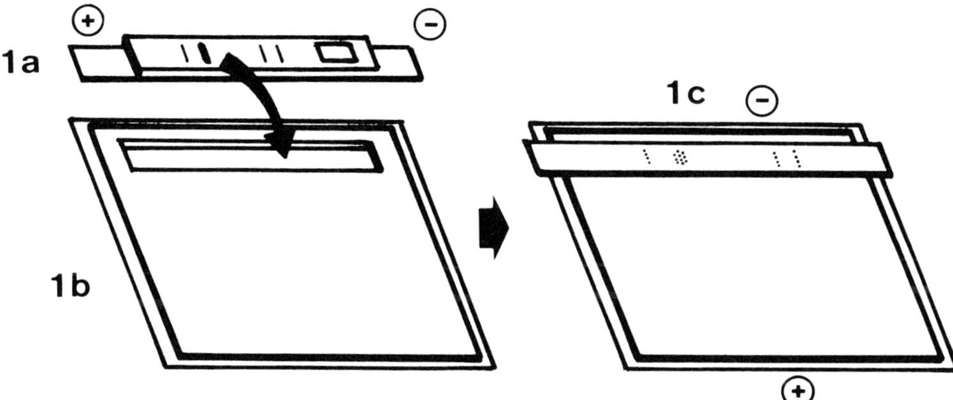

Fig.1 Two-dimensional electrophoresis with two open type gels
bound to silanized glass plates. Serum is applied to a slot
on an isofocusing gel (1a). After focusing, the narrow gel
strip is placed in a groove of a gradient gel (1b), and they
are attached to each other with a low melting temperature
agarose (1c). The bands in Fig.1a and 1c are drawn arbitrarily
for explaining the first dimensional separation. The electric
field is shown by plus and minus signs.

2) Electrophoresis

An isofocusing gel bound to silanized glass plates (Fig.
1a) was placed on the electrofocusing cooling plate (set
at 4°C) and was connected with electrolytes (0.04M NaOH,
catholyte, and 0.01M H_3PO_4, anolyte) by paper bridges (No.
800, Toyo Scientific Industry, Japan). 3 µl of serum was
applied to a slot cut in the gel 5mm from the cathodic
side. After focusing at 1500V×h (Ca 3.5h), the gel
surface was finely undulated, and serum proteins with
differences in pI value of 0.02 pH were separated. The
strip of isofocusing gel was placed in a groove of the
gradient gel (Fig.1b). The two gel plates were attached
to each other with a low melting temperature agarose, 1%
Seaplaque (Marine Colloids) and 0.038M Tris-HCl buffer
solution (pH 8.9). (Fig.1c) The electrolyte for the
second electrophoresis was 0.05M Tris Glycine buffer solution (pH 8.3). After the second electrophoresis at 1000V
×h (Ca 2.5h), separated proteins were fixed and stained.

3) Staining. Hemin and 2.6-xylenol were obtained from Tokyo Kasei Co.; DAB from Aldrich Co.; catalase from Boehringer Mannheim Co.; and, angiotensins I, II, angiotensin III inhibitor, bradykinin and δ-sleep-inducing peptide from Peptide Institute, Inc. Following electrophoretic separation, the gel plate was treated with hemin (1×10^{-4}M) in a phosphate buffer (0.07M, pH 7.5) containing 50% methanol for 10 min., and was then rinsed with the same phosphate buffer without hemin to remove background hemin. It was then treated with a mixture of 2mM DAB and 4mM xylenol in the buffer for 5 min. Finally, H_2O_2 (0.1M) was added to the mixture, and blue spots appeared within 10 min. To stop color development, the gel was treated with catalase (1.7×10^{-8}M) in a phosphate buffer (0.07M, pH 7.5) without methanol. The separated proteins were thus stained within 30 min.

Results and Discussion

1) Isoelectric focusing
The result (Fig.2a) obtained by the present method of iso-focusing was superior to that obtained by the conventional method which uses a rod-like gel in a glass capillary (Fig.2c). The band of albumin in the open gel (see ref.6 for identification of isolated spots) is not diffused to a direction of the pI axis. The diffusion of the albumin band is thought to be caused by mechanical compression of the rod-like gel, which is inevitable when the gel is forced from the capillary.

2) Separation in the gradient gel
Improvement in the method of forming open type gradient gels was made in the following two respects: i) the thickness of the gel was reduced to 0.5mm or less, ii)

strong silanization of the glass was performed at a high
concentration, 2% silan coupling agent and 50% ethanol.
We performed the second electrophoresis, using this
improved gel plate, at 10W. A sufficient separation was
attained after 2.5 hrs (Fig 2a and 2b). Even greater
acceleration of the electrophoretic process will be
possible, increasing the cooling efficiency of the apparatus.

3)   Two-dimensional electrophoretograms obtained by hemin and
Coomassie method. The phoretogram obtained after a 30
min. hemin method staining procedure (Fig.2b) is similar
to that obtained after 15 hours of a conventional staining-
destaining procedure with use of Coomassie Blue (Fig.2a).
The staining speed was greatly (approx. 30 times) enhanced
by the present hemin method. Detailed examination of the
two patterns revealed that the staining intensity of the
protein at acidic region is higher when obtained by the
hemin method. However staining intensity of IgG was,
somewhat lower when obtained by the hemin method.

4)   Application of hemin method to detection of model peptides
The active center of natural peroxidase is generally
thought to be heme coordinated with a histidine residue.
Therefore, it may be deduced that the present hemin method
is effective especially for staining proteins with
histidine residues. To test this outcome, a peptide with
or without histidine residue was electrofocused, and the
peptide in the gel was stained by hemin method. 10µg each
of angiotensin I with 2 histidine residues, angiotensin II
with one residue, and angiotensin III inhibitor with one
residue were stained within 30 min. by this method. In
contrast, peptides without histidine residue (bradykinin
and δ-sleep-inducing peptide, 10µg each) were not detected
by this method.

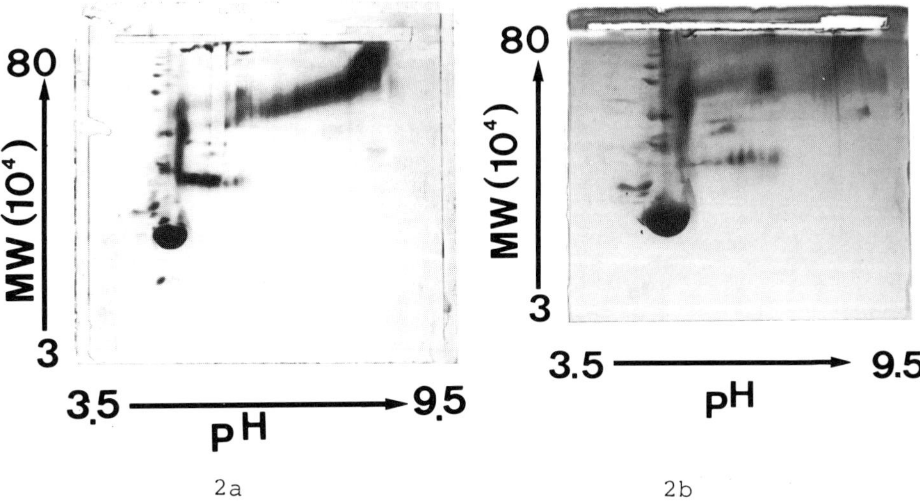

Fig.2 Comparison of two-dimensional electrophoretogram of serum proteins obtained by a conventional staining-destaining procedure with use of Coomassie Brilliant Blue R-250 for 15 hours in 7% acetate 50% methanol (2a) with that obtained by the hemin method for 30 min.(2b) Electrophoresis of 2a and 2b was performed as in Fig.1. The phoretogram obtained by combination of rod-like gel in a capillary glass tube at the first dimension and the open-type horizontal gradient gel at the second dimension was also given (2c). For 2c, the gel was stained by Coomassie Blue.

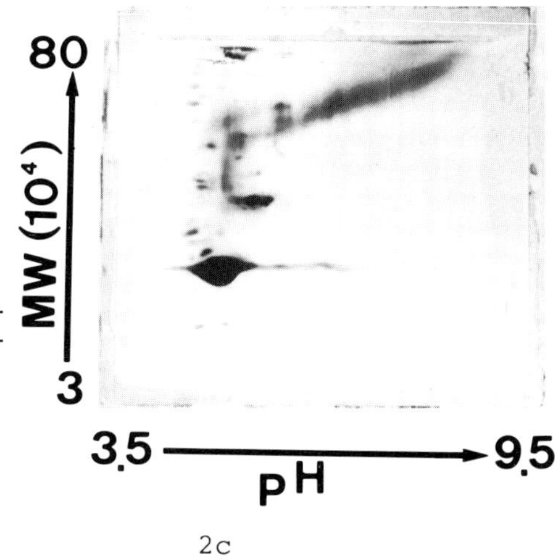

In conclusion, the present method of electrophoresis employing open-type horizontal gel plates is extremely useful when applied in combination with the hemin method in the following respects: i) simplicity and reliability; in conventional two-dimensional electrophoresis, thin rod-like gel made in a glass tube is forced from the tube and firmly placed on the top of a vertical gel sandwiched between glass plates. Handling of the rod-like gel often causes destruction of the gel, and the use of the vertical type gel often results in water leakage between upper and lower electrode buffer tanks. These troubles were avoided by the present technique of electrophoresis: ii) speedy staining; the hemin method gave approx. 30-fold speed up of staining as compared with the conventional method with Coomassie Blue. Since the gel was firmly adhered to the glass plate, it was not damaged during staining procedure.

A method of horizontal electrophoresis for isoelectric focusing and protein mapping was developed previously (7) using a thin layer of PAG on a cellophane foil. In the cellophane method, the second electrophoresis was performed on a gel foil of a constant concentration of acrylamide with no gel concentration gradient. It remains for a future work to determine whether or not the gradient of gel concentration, a fundamental element in sharp separation of proteins, can be made without disorder on a cellophane sheet. Mechanical handling of the gel foil by automatic machine may be difficult. The present electrophoretic technique using a gel on a glass plate may be more suitable for automation of the analytical process.

References

1. O'Farrell, P.H.: J.Biol.Chem. 250, 4007-4021 (1975)
2. Anderson.L., Anderson.N.G.: Proc. Natl. Acad. Sci. 74, 5421-5425 (1977)
3. Manabe, T., Tachi, K., Kojima, K., Okuyama, T.: J.Biochem. 85 649-659 (1979)
4. Mc Donnel, A., Staehelin, L.A.: Anal. Biochem. 117, 40-44 (1981)
5. Radola, B. J.: Electrophoresis 1, 43-56 (1980)
6. Manabe, T., Kojima, K., Jitsukawa, S., Hoshino, T., Okuyama, T.: J. Biochem. 89, 841-853 (1981)
7. Gorg, A., Postel, W., Westermeier, P.: Electrophoresis, 79, 67-78 (1980)

AGE-RELATED CHANGES OF TWO-DIMENSIONAL ELECTROPHORETIC PROTEIN PATTERN OF RAT LIVER

Toshiko Fujita, Tosifusa Toda and Mochihiko Ohashi
Department of Biochemistry, Tokyo Metropolitan Institute of Gerontology

35-2 Sakae-cho, Itabashi-ku, Tokyo 173, Japan

Summary

Age-related changes of proteins in rat liver were studied using our new method of two-dimensional cellulose acetate electrophoresis(1). Liver specimens of both male and female rats 1.5, 6, 12, 18, 24, 30 months old were analyzed. Coomassie-stained protein pattern showed more than ninety spots. For the first step of the analysis, several spots, which appeared to change with aging, were selected, and possible quantitative changes were confirmed by micro-computer-aided two-dimensional densitometry. From this study, it was obvious that six spots changed with aging. Moreover, the change was common to both Wistar and Fischer strains of rats. However, there was a sexual difference in the change. Subsequently, we studied the effect of castration on these age-related proteins. Two of the spots were clearly affected by castration.

Abbreviations: 2DCAE, two-dimensional cellulose acetate electrophoresis; 2DD, two-dimensional densitometry; 2DE, two-dimensional electrophoresis; 1DE, one-dimensional electrophoresis; IEF, isoelectric focusing; IOD, integrated optical density; CAIEF, cellulose acetate isoelectric focusing.

Introduction

Two-dimensional electrophoresis is a very useful technique for analysis of protein mixtures. We have developed a new type of electrophoresis, 2DCAE (1). This method offer a number of practical advantages : Cellulose acetate membrane is an easy-to-handle supporting medium ; It is possible to analyze a large number of samples ; Several replicas of a protein pattern can be obtained ; Many kinds of detection methods, such as, protein staining, enzyme staining and immunochemical assay, can be carried out at one time. Enzyme staining is possible, because no denaturing reagent is used. We applied this new method to the analysis of age-related changes of proteins in rat liver.

Materials and Methods

Animals.   Male and female SPF rats of Wistar/Slc and Fischer 344/DuCrj strains 1.5, 6, 12, 18, 24, and 30 months old were used. In the castration study, Wistar/Slc rats 3 to 4 months were used.

Preparation of samples. Tissues were homogenized with 9 vol. of distilled water in a glass homogenizer. The homogenates were centrifuged at 10,000xg for 20 min. The supernatants were stored at -70° C until use. Supernatants containing 70 µg of protein were applied to the 2DCAE.

The amount of protein was determined by Lowry's method (2).

Two-dimensional electrophoresis. First-dimensional electrophoresis consists of two steps, i.e. isotachophoretic concentration and electrophoretic separation. Both steps were carried out on a cellulose acetate strip (Titan III,

76x10mm). In the second dimension, isoelectric focusing was carried out· on a cellulose acetate membrane (Separax EF, 60x110mm) impregnated with 5%(v/v) LKB Ampholines mixed in the following proportions, 5 parts of pH4-6, 5 parts of pH5-7, 5 parts of pH7-9 and 6 parts of pH8-9.5. The focusing was started with a constant current of 1 mA per sheet, and when the voltage reached 1200 V, the voltage was automatically maintained at a constant level. The whole process of the focusing was completed within 3h. The temperature of circulating coolant was regulated by a refrigerated circulating bath (RTE-9B, NESRAB INSTRUMENTS, INC VOLTAS, Portsmouth, U.S.A) at -3°C.

Two-dimensional densitometry.   Quantitative analysis of protein spots was performed by two kinds of 2DD apparatuses. One was a scanning-densitometer-equipped system, and another a CCD-camera-equipped system. Details of the 2DD are reported in a separate paper in these Proceedings.

Chemicals and proteins.   Titan III cellulose acetate plate was purchased from kabushiki-kaisha Helena Kenkyujyo (Urawa--shi, Japan) ; Separax EF cellulose acetate membrane from Fuji Photo Film Company (Tokyo, Japan). Tris-(hydroxymethyl-)aminomethane was from Merck (Darmstadt, F.R.G). Ampholine carrier-ampholytes were from LKB (Stockholm, Sweden). Other chemicals of analytical reagent grade were purchased from Wako Pure Chem. Ind., Ltd. (Osaka, Japan).

Results and Discussion

1. Resolution of protein in CAIEF and 2DCAE.
The optimum ratio of Ampholine mixing was determined to obtain the best resolution for our studies. Fig.1A shows a

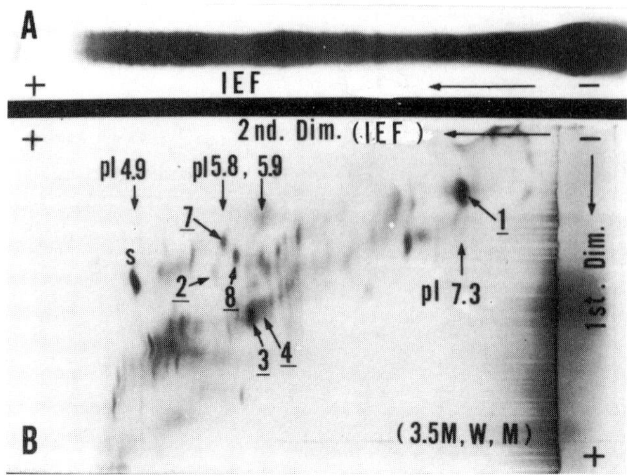

Fig. 1.   A. CAIEF pattern of rat liver extracts.   5 µl of
the extract containing 63 µg protein was applied. CAIEF was
performed on a cellulose acetate membrane impregnated with
8%(v/v) LKB Ampholines mixed in the following proportions
(pH4-6:pH5-7:pH7-9:pH8-9.5=1:4:1:2). IEF was started with a
constant current of 1 mA per sheet, and when the voltage
reached 1200 V, the power supply was changed to a constant
voltage. IEF was completed within 4 h.
B. 2DCAE pattern of rat liver extracts. 5.5 µl of the same
extract as above was applied. The second dimension was
carried out under the same conditions as above. Other
experimental conditions are as described in Materials and
Methods.

CAIEF pattern of protein extracted from male rat liver.
Fig.1B shows a 2DCAE pattern of the same sample. On the
CAIEF pattern, about 50 bands, on the 2DCAE pattern, about
90 spots were observed. In the CAIEF pattern, all bands have
almost the same shape, but in the 2DCAE pattern, spots
appear in various shapes, and the two-dimensional distribution helps to identify unique spots on the pattern. 2DCAE
is superior to CAIEF in resolution.

2. Age-related change of proteins in rat liver.   Fig.2
shows parts of 2DCAE patterns of young, adult and aged male

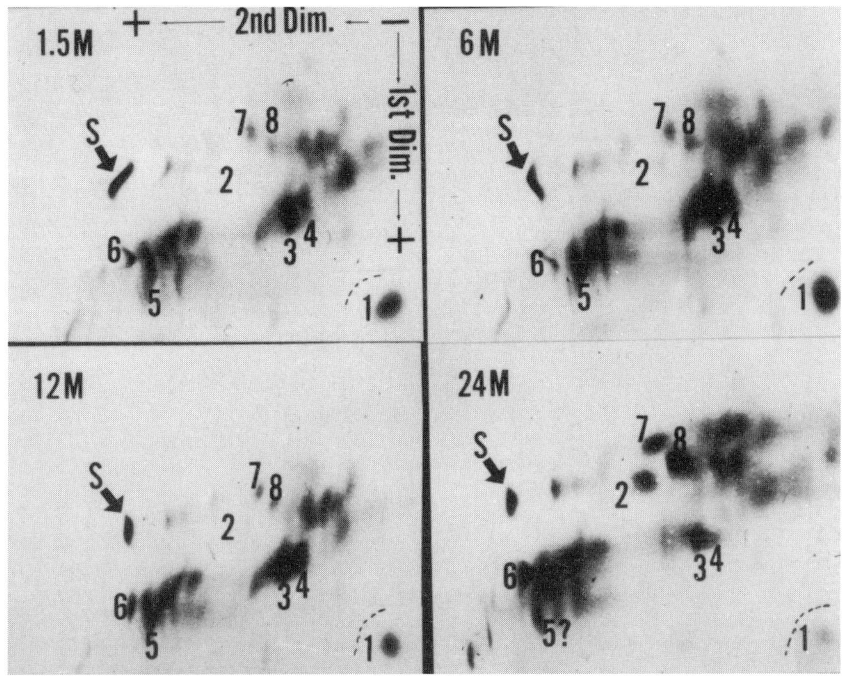

Fig. 2   Parts of 2DCAE pattern of liver extracts from male
Wistar rats. The ages of the rats are 1.5, 6, 12 and 24
months. Spot 1 shown here is transferred from the basic
side. Electrophoresis and Coomassie staining were performed
as described in Materials and Methods. The spot S indicated
with an arrow was used as internal standard for the
quantitative studies in Fig.4 and Fig.5.

Wister rats. Spot 1 clearly decreased at 24 months. Three

spots, numbered 2, 7, and 8, increased at 24 months. Two

spots, 3 and 4, increased by 6 and 12 months and then

decreased at 24 months. Same changes were also observed in

male Fischer rats. The 2DCAE patterns from female rats are

shown in Fig.3. The level of Spot 1 was high in young and

adult male rats, but low in aged male rats and in female

rats at any age. Spots 3 and 4 increased by 6 and 12 months,

and then decreased at 24 months. These changes were common

in both sexes. Fig.4 shows the quantified data of

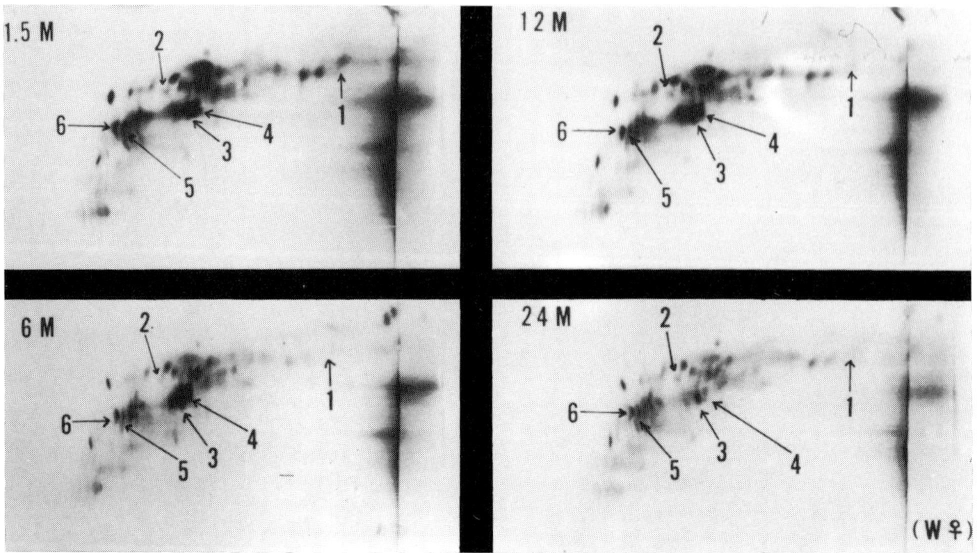

Fig.3.   2DCAE pattern of liver extracts from young, adult and aged female Wistar rats.   The ages of the rats were 1.5, 6, 12 and 24 months. Electrophoresis and Coomassie staining were performed as described in Materials and Methods.

age-related protein spots. Quantitative analysis was per — formed with scanning-densitometer-equipped system. Spot 1 decreased markedly with aging in the male rats of both strains. On the other hand, it existed in a small amount in female rats of both strains at any age. Spots 2, 7 and 8 in the male rats of both strains, increased at 24 and 30 months, but such changes were not observed in the female.

3. Effect of castration on age-related protein spots. Sexual differences were also observed in these protein spots. We therefore determined the effect of castration on the amounts of proteins in these protein spots. Fig.5 shows the quantified IODs of the protein spots of livers from castrated and control rat, spot 1 clearly decreased, while spot 2 increased after castration. Spots 7 and 8 increased slightly. Spot 1, the level of which was low in female and

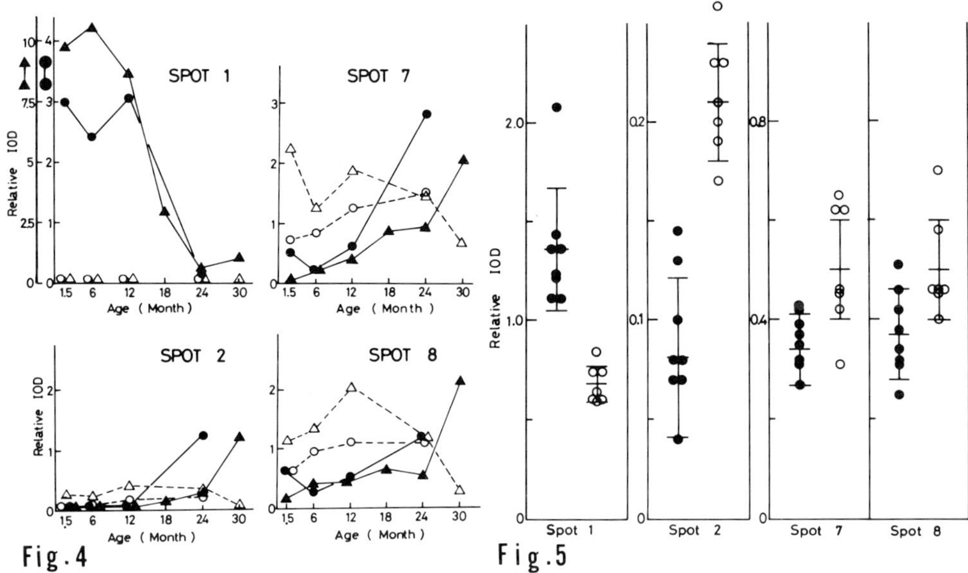

Fig.4

Fig.5

Fig.4.   Age-related changes of protein ratio on 2DCAE of rat liver extracts. Relative IOD was obtained from the calculation in which the determined IOD was divided by IOD of the spot S as shown in Fig.2. Spot numbers shown here correspond to the numbers shown in the 2DE. (●—●) male, Wistar, (O··O) female, Wistar, (▲—▲) male, Fischer, (△··△) female, Fischer.

Fig.5.   Effect of castration on age-related protein. Rats were killed eight weeks after the operations. Seven castrated (O) and eight control (●) rats were analyzed by 2DCAE and 2DD. Mean±Standard Deviation was shown in the figure.

decreased in male with aging, also decreased after castration. This result indicates that the decline of spot 1 might be a marker of sexual aging of male rat. Spot 2, which increased with aging, was also increased by castration. However, the spot quite small in females of both strains at any age. Spot 2 may be affected by not only aging but also by sexual glands. At present, we are examining the effect of testosterone. The changes of spots 3 and 4 were common in both sexes, therefore, they might be affected by aging only.

Conclusion

Aging reserch is still complecated, and there are many approaches to aging mechanisms. Genetic studies and metabolic studies are equally important aspects in aging research. Because proteins are end products of gene coding, and also act as catalysts in substance metabolism, analysis of protein may give fruitful informations in both genetic and metabolic aspects.

Thus, we introduced 2DCAE and 2DD to our studies on aging. In this paper, age-related and/or sex-related changes of protein were analyzed by the methods. And, we screened out six candidates of age-related proteins. Two of them changed commonly in both sexes. However, other four spots had sexual difference and changed only in male rats. And, the two of them also changed by castration. This observation and the fact that functions of sexsual glands decline with aging may indicate that we found the markers of sexual aging. Sexual aging is one of phenomena which show the steps of animal aging.

To survey other aging markers, we need much more samples, and require a method to analyze small changes. In this situation, we are also making an automatic comparator which allows all cases of comparison among samples in a large data bank.

References

1. Toda, T., Fujita, T., Ohashi, M.: Anal. Biochem. 119, 167-176 (1982).
2. Lowry., O.H., Rosebrough, N.J., Farr, A.L., Rnadoll, R.J.: J. Biol. Chem. 193, 265 (1951).

SLAB GEL ELECTROPHORESIS AT ANY THICKNESS WITH EFFECTIVE
SAMPLE MOVEMENT & GRADIENT FLATTENING OF SDS AND IEF GELS

Yasuyuki Yamada
Department of Obstetrics and Gynecology, Yamanashi Medical
University, Tamaho, Nakakoma-gun, Yamanashi-ken 409-38 Japan

New methods for two-dimensional electrophoresis has been
developed using my reported methods for microelectrophoresis
(1, 2, 3, 4). The methods of microelectrophoresis are important methods to study effects of hormones and chemicals on an
intact single cell in vivo and in vitro, and disclosed variety effects (1, 2, 3, 4). Two-dimensional electrophoresis
employing isoelectric focusing (IEF) in the first dimension
and slab gel electrophoresis in sodium dodecyl sulfate (SDS)
in the second is a powerful method for separation and analysis of proteins and peptides (5, 6, 7). The new slab-gel apparatus has been developed to facilitate a good, effective
movement of protein from a first-dimension gel into a second-
dimension gel. This new simple apparatus makes running time
of the slab gel short and is convenient for the simultaneous
processing of many polyacrylamide gel slabs and ultra-thin
(100μm) gels. Many gels can be run at the same time in a small space and be readily inserted or removed. Multiple gels
can easily be prepared. The plates can have any desired shape and dimension, and their thickness can easily be changed.
The method is very simple and inexpensive, has a wide application and many other advantages. An apparatus for IEF has
also been developed to reduce anodic and cathodic drifts in
vertical gel rod IEF. This apparatus is inexpensive and can
be easily used for many more gels as compared with commercially available apparatuses. The length, diameter and shape of
the gel can be arbitrarily changed with the apparatus in con-

trast with commercially available apparatuses. High concentrations of detergent can be used to dissolve protein samples.
Removal of gel cylinders from glass tubes is easy. Using the
same system, molecular weight gradient flattening of SDS polyacrylamide gel and pH gradient flattening of IEF gel has be-
en achieved at any gel segment. Any crowded gel segment where congregated components are not separated well can easily
be widened for good separation and any dispersed gel segment
where components are separated too much can easily be narrow-
ed. Therefore every gel segment can be used effectively and
meaningfully because the gradient curve can be adjusted to
any distribution of the components. In the crowded area any
small spots of components, which could not be detected previously because of nearby heavy staining or strong radioactivity of an abundant component, can be sufficiently separated
from the nearby spots in a small gel without sacrificing other areas.

I. Microelectrophoresis in vivo and in vitro

It is now widely accepted that the hormone secretion is regulated by feedback mechanism. Some hormones, especially sex
hormones, are strongly regulated by the brain with releasing
and inhibiting factors. For the regulation mechanism, some
brain cells or neurons must be sensitive to the hormone in
the circulating blood, and I found prolactin-, estrogen-,
betamethasone- and testosterone-sensitive neurons (1, 2, 3,
4). The method of microelectrophoresis must be used to study
the hormone and chemical sensitivity of neurons and intrveno-
us or systemic administration of hormones and chemicals has
limited value for the cellular sensitivity, because the nervous system is composed of a network of many neurons and some
insensitive cells can be immediately affected indirectly through this network. But with this in vivo method, it is impo-

ssible to change the medium around the recorded cell, to observe the recorded living cell, and to discriminate the sensitive site on the cell, i.e. a point on the cell body, proximal dendrite, periferal dendrite, axon hillock, axon, etc. So that instead of the microelectrophoretic method of stereotaxic measurement the similar microelectrophoretic method under microscopic observation of a single neuron can be used. With the in vitro method the single cell responses to the blood components or humoral composition around the neurons can be studied. The in vivo microelectrophoretic method is schematically illustrated in Fig. 1. Seven-barreled pipette contains hormones and chemicals. Recording electrode is glued to it. An anesthetized animal is placed in a stereotaxic instrument. Hormones and other chemicals were applied in the immediate vicinity of a single cell by passing a constant current through each of the multibarreled glass micropipette with an overall tip diameter of 2μm. Extracellular unit activity was recorded through a glass microelectrode cemented to the multibarreled micropipette and filled with 4M NaCl. The effects of applied chemicals were evaluated from the change in frequency of spike discharges which were counted for each second by a pulse counter coupled with the preamplifier and depicted with a pen recorder. The in vitro microelectrophoretic method is shown in Fig. 2. In the schematic cross section of the observation chamber, the brain tissue (T), medium (M), stimulating electrode which contains a chemical (S), drainage (D), and objective lens of long working distance (O) are shown. The medium around the thin-sectioned brain tissue was changed by replacing the flowing medium by a new one. With the multibarreled method the chemicals can be applied through the different micropipettes of the same electrode either at the same time or in succession. Ionized hormones and chemicals can be applied electrophoretically and nonionized ones electro-osmotically.

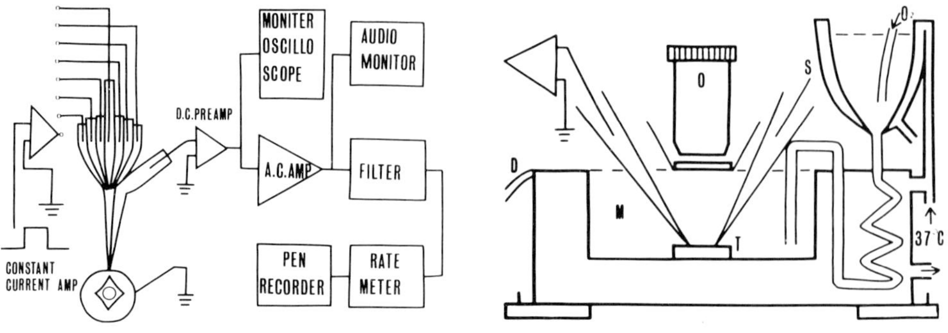

Fig.1. in vivo method

Fig.2. in vitro method

II. Slab gel electrophoresis

This method does not need any special expensive apparatus,
since all the necessary equipment can be assembled in the
laboratory and is inexpensive. A number of gels are simply
piled up between two ordinary glass dishes for electrophoresis. All the procedures from acrylamide preparation to slab
gel drying are simplified. This method facilitates good movement of protein sample from the first-dimension gel into the
second-dimension gel. The running time of the slab gel electrophoresis is reduced and many slab gels can very easily be
run simultaneously. Ultra-thin (100μm thick) gels can easily
be made, and the size and shape of the slab gel can be easily
changed. Acrylamide polymerization can be adjusted by cooling with ice. Plate-sealing layers are straight and narrow.
The glass plates used do not require notches. The samples
adsorbed and dried on filter paper pieces can be applied directly. During the run the plates are freely ventilated on
both faces. Rapid heat loss makes it possible to use more

power (more than 50 W per gel). The method of sealing the
edges provides a narrow seal and straight margin even with
very thin gels. The method of sealing does not use agarose
and does not require heating, eliminating the risk of denaturing protein or cracking the glass. The dye band is resolved
into a sharp and straight line, and the separated protein positions may be measured accurately in relation to the dye band. Many gels can be prepared simultaneously and stored on
an ordinary test-tube rack. Disposable syringes can be used
to make either a linear or a logarithmic gradient. All the
necessary apparatuses, from the gradient maker to the gel
dryer, are laboratory-build and inexpensive. For sealing of
bottom and sides of thicker slab gels (more than 1mm thickness), the bottom is covered by Parafilm for a width of 1cm
using rubber bands. A small portion of the heavy (higher
concentration) acrylamide solution (15% is used at present)
is chilled on ice and 20% ammonium persulfate (1.5% vol.) and
Temed (3% vol.) are added. This solution is applied on both
sides from the inside to cover the Parafilm spacer and onto
the bottom, where it polymerizes in a few minutes, providing
a complete seal with staight margins. This sealing layer is
narrow and sharper than one prepared with hot agarose. For
the sealing of thinner slab gels (100µm to 1mm thickness),
especially for the 100µm gel, the plates are rinsed with 5%
SDS and placed in the test-tube rack, oriented such that the
acrylamide pouring direction is parallel to the gravity line,
to drain and half-dry, a step that eliminates bubbles when
the gels are poured. Even if any bubbles appear during pouring, they will float up to the top because of the trace amount of SDS used to coat the plate surface. For 100µm gels,
Parafilm strips of 5mm width are used for spacers. For 200µm
gels, Parafilm spacers are prepared from strips 1 cm wide
folded over once and set in place at the sides. Similarly,
any thickness can be made according to the number of folds.
These Parafilm spacers can be used repeatedly. An outside

cover of Parafilm is not necessary since at a thickness of
$100\mu m$ to 1mm capillary action readily retains the sealing
layer. For a thickness of $100-200\mu m$ the cold sealing acrylamide is applied at the outer edges, i.e. the bottom and both
sides. For gel preparation disposable polyethylene syringes
can be used to prepare the gradient as shown in Fig. 3. Two
syringes are connected with a short section of small-bore latex tubing. The second syringe (A) has an additional tip
made by making an opening at the circumference of the barrel
with a hot wire and fusing onto it a tip broken off from a
third syringe which is discarded. A glass bead or a small
stone to help in mixing is placed in this syringe, which is
then filled with the heavy acrylamide solution, while the
other syringe (B) is filled with the light acrylamide solution. The syringes are bound together with transparent adhesive tape such that the flange of the piston of syringe A over-
laps that of syringe B and the two advance synchronously when
the piston of syringe B is pushed with the thumb. Mixing in
syringe A is provided by manual shaking aided by the presence
of the glass bead or small stone. Exponential gradients can
be made by using disposable syringes of different sizes.
Only for $100\mu m$ gels, before pouring, a bundle of a few plastic filaments or a metal filament, either of which has been
rinsed with 5% SDS beforehand like the glass plates, is placed on the center line between the paired glass plates in parallel to the pouring direction in order to lead the poured
acrylamide solution directly onto the bottom. This is the
same principle as I described previously for leading solutions into microelectrophoretic glass capillaries of only $1\mu m$
diameter (1, 2, 3, 4). For the run the first-dimension IEF
gel is placed and a strip of filter paper 1mm wide, which has
previously been soaked in 0.03% (w/v) bromophenol blue in methanol and dried, is placed on the first-dimension gel and a
coiled filter paper is pushed onto it. Alternatively, the
tracking-dye strip can be placed in the coiled filter paper

beforehand. The sample can be run in the same way as the tracking dye, i.e. the sample is absorbed on a strip of filter paper and applied directly to the top of the slab gel without the intervening first dimension gel.

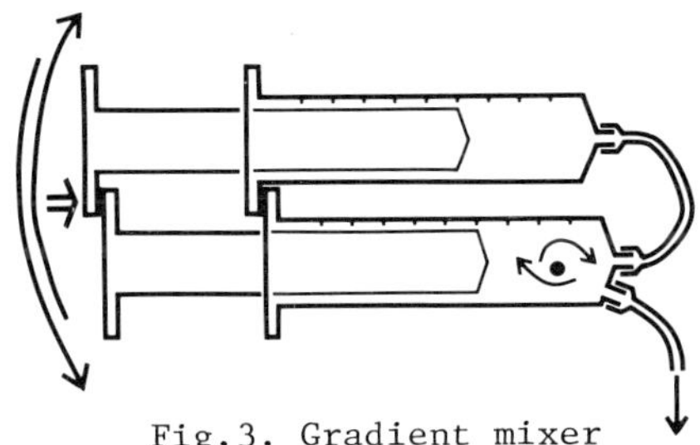

Fig.3. Gradient mixer

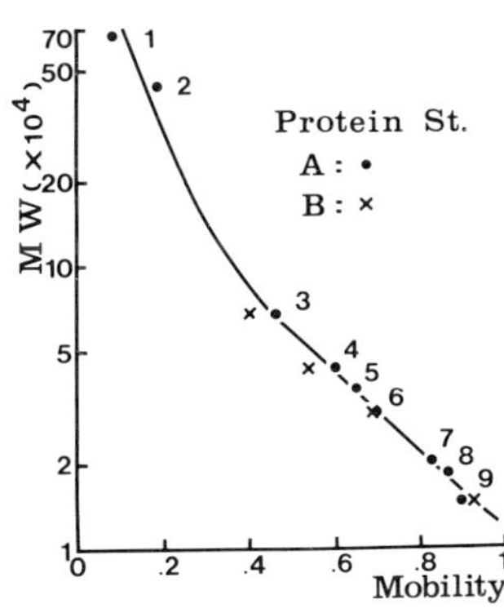

Fig.5. Mobility in 750μm(A) and 100μm(B) thickness

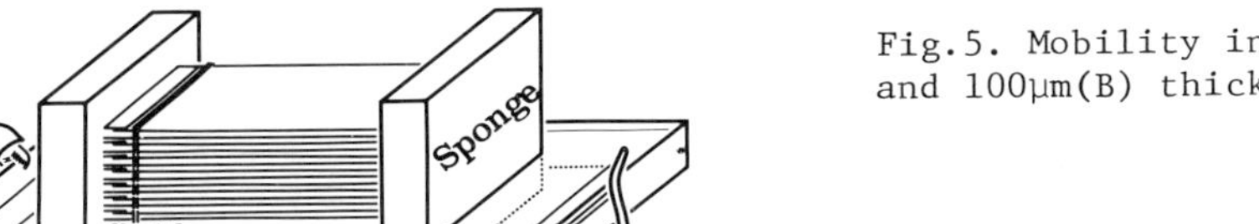

Fig.4. The setup of the system

III. Isoelectric focusing with reduced anodic and cathodic

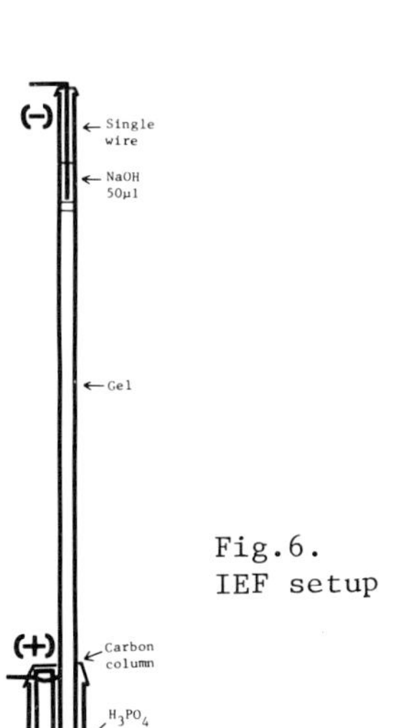

Fig.6. IEF setup

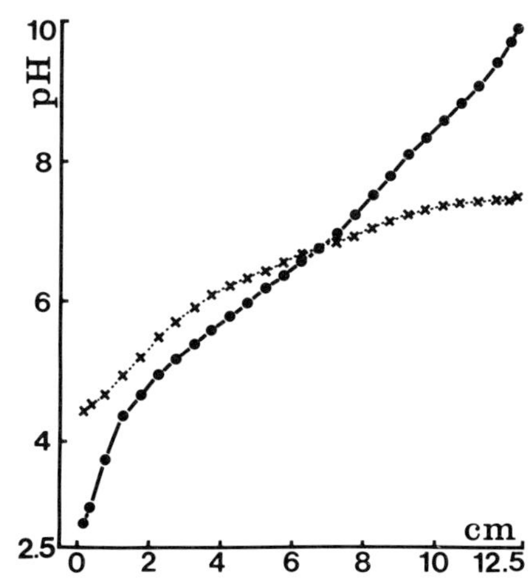

Fig.7. pH gradient by the present method (O) and the commercial apparatus (X)

178

VI. Gradient flattening of SDS and IEF gels

A thin plate is used to change depth of a part of the slab in
order to change electromotive force in any selected specific
area of the gel. The calculation of electromotive force in
a slab gel is done as follows. The whole current at any point of the slab gel is same anywhere from the gel inlet to outlet, even if any kind of electric current or voltage is del-
ivered to the gel. Therefore, I = V/R, i = v/r, I = i, where
V: voltage, R: resistance, I: electric current, and the capital and small letters are used for the values in the thick
(non-plated) and the thin areas. Accordingly v/V = r/R and
using depth (D), v/V = D/d. For IEF gel a fine plastic bar
is used to change the thickness.

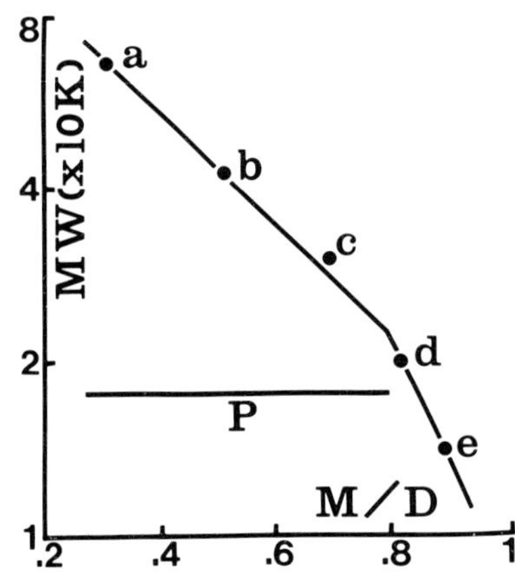

Fig.8. Mobility vs the dye

Fig.9. pH gradient with
the thin segment

References

1. Yamada, Y.: Neuroendocrinology 18, 263-271 (1975)
2. Yamada, Y., Nishida, E.: Brain Res. 142, 187-190 (1978)
3. Yamada, Y., Ueshima, H.: Endocrinol. Japon. 25, 397-401
   (1978).
4. Yamada, Y.: Brain Res. 172, 165-168 (1979).
5. to 7. Yamada, Y.: J. Biochem. Biophys. Methods (in press)

TWO-DIMENSIONAL ELECTROPHORESIS OF CEREBROSPINAL FLUID PROTEINS IN THE
ABSENCE OF DENATURING AGENT AND IMMUNOCHEMICAL IDENTIFICATION AFTER
PARALLEL NITROCELLULOSE BLOTTING

Takashi Manabe, Yuko Takahashi, Tsuneo Okuyama
Department of Chemistry, Faculty of Science, Tokyo Metropolitan University,
Setagaya-ku, Tokyo, Japan

Atsushi Hiraoka and Isao Miura
School of Health Science, Kyorin University, Hachioji City, Japan

Osamu Murao
Shimofusa National Sanatorium, Chiba City, Japan

Introduction

In a previous report, we described a two-dimensional electrophoresis technique, which does not employ denaturing agents (1). Using the technique,
we prepared a 'normalized map' of human plasma proteins which showed
standard positions (pI and molecular weight) of native plasma proteins and
identified about 80 protein spots out of 128 spots on the map (2). Micro-
scale-multisample version of the technique (3) combined with silver staining
enabled us to analyze body fluids of low protein concentration, such as
cerebrospinal fluid, urine, abdominal fluid, etc. In this report, we
present a standard distribution pattern of human cerebrospinal fluid (CSF)
proteins obtained by two-dimensional electrophoresis in the absence of
denaturing agent. A technique of parallel nitrocellulose blotting followed
by immunochemical identification was developed to identify CSF proteins.

Materials and Methods

<u>Materials</u>—Antisera against human plasma proteins were kindly provided by

180

Dr. Y. Hasebe (Hoechst Japan). Nitrocellulose sheets (0.45 μm pore size)
were obtained from Schleicher and Schüll (Dassel, W. Germany). Peroxidase-
conjugated goat anti-rabbit IgG was purchased from Miles Laboratories.
Cerebrospinal fluid was obtained by lumbar puncture with a fine needle and
after total protein determination, 3-7 ml was concentrated by about 30-fold
in a Minikon-CS15 concentrator (Amicon, Lexington, MA). Sucrose was added
to the concentrated CSF to give a concentration of 40% (w/v) and the mixture was kept at -20°C.

<u>Two-dimensional electrophoresis</u>—Two-dimensional electrophoresis in the
absence of denaturing agent was performed as described previously (4).
Isoelectric focusing was run in 4% polyacrylamide (0.2% methylenebisacryl-
amide) gel columns (ϕ 4 mm) containing 2% Ampholine pH 3.5-10 and 0.5%
Ampholine pH 3.5-5, followed by electrophoresis in 4-17% polyacrylamide
gradient slab gel (0.2-0.85% bisacrylamide gradient), 160 mm wide x 120 mm
high x 3 mm thick in size. Micro two-dimensional electrophoresis was performed as described previously (3), employing ϕ 1.3 mm capillary gel columns
in isoelectric focusing and gradient slab gels of 38 mm wide x 35 mm high x
1 mm thick in size.

<u>Staining of slab gels</u>—Macro slab gels were stained in 0.025% Coomassie
Brilliant blue R-250-50% (v/v) methanol-7% (v/v) acetic acid overnight and
destained in 7% (v/v) acetic acid for 2-3 days. Micro slab gels were
stained in 0.1% Coomassie blue R-250-50% (v/v) methanol-7% (v/v) acetic
acid for 15 min and destained in 30% (v/v) methanol-7% (v/v) acetic acid
for 2 h. Silver staining of micro slab gels were performed as described
by Oakley et al. (5) with some modifications for staining of micro slab
gels.

<u>Electrophoretic transfer of proteins from multiple micro 2-D gels to nitro-
cellulose sheets</u>—An apparatus for parallel electrophoretic transfer of
proteins from four 2-D gels to four nitrocellulose sheets was devised.
2-D gels and nitrocellulose sheets were set horizontally, then the sheets
were easily replaced to new ones. Electrophoretic transfer was performed
at 20V/cm (0.2A) in 0.025M Tris-0.19M glycine (pH 8.3), replacing new
nitrocellulose sheets (40 x 40 mm, 0.45 μm pore size) every 10 min, for

50 min. Therefore, five copies on nitrocellulose sheets (blots) could be obtained from each micro 2-D gel, thus 20 blots from 4 slab gels. The blot may be stained with 0.1% Coomassie Brilliant blue R-250-50% (v/v) methanol-7% (v/v) acetic acid for 5 min and destained in three changes of 80% (v/v) methanol-7% (v/v) acetic acid for 10 min.

<u>Immunochemical Identification of CSF proteins on nitrocellulose sheets after blotting</u>—After blotting, nitrocellulose sheets were treated as described by Towbin et al. (6) with some modifications: 1) Each sheet was put in a plastic container with a lid (64 x 58 x 21 mm H), soaked in 5 ml of 3% bovine serum albumin (BSA) in saline (0.9% NaCl, 0.01% NaN_3, 10 mM Tris-HCl, pH 7.4) for 2 h at room temperature; 2) Specific rabbit antiserum (10-25 µl, 1:500-1:200 dilution) was added to the BSA solution and kept at room temperature for 30 min; 3) Washed in saline (five changes during 30 min); 4) Soaked in peroxidase-conjugated goat anti-rabbit IgG (5 µl) + 5 ml of 3% BSA in saline for 60 min at room temperature; 5) Washed in saline; 6) Stained in 10 ml of 2 mM diaminobenzidine-saline + 50 µl 3% H_2O_2 for 15-30 min.

Results and Discussion

One example of the two-dimensional patterns of human cerebrospinal fluid proteins was shown in Fig. 1. A cerebrospinal fluid sample, 10 mg protein /dl, was concentrated 30-fold, added sucrose to a concentration of 40% (w/v), and one microliter of the mixture (about 3 µg protein) was subjected to micro 2-D electrophoresis and the slab gel was silver stained. The pattern resembled to the patterns of plasma proteins, although the following points were characteristic to the CSF pattern. 1) High molecular weight plasma proteins, IgM, LDL, and haptoglobin polymers, which are present in human plasma at a high concentration of more than 100 mg/dl, were not observed. 2) A dense spot was observed at prealbumin position. 3) Arrowed spots which are not present in plasma patterns were observed: a) The spots at pI 6 and of slightly higher apparent molecular weight than transferrin (tau-proteins); b) Broad spots at pI 8-6 and of apparent molecular weight higher than albumin. These characteristics were also

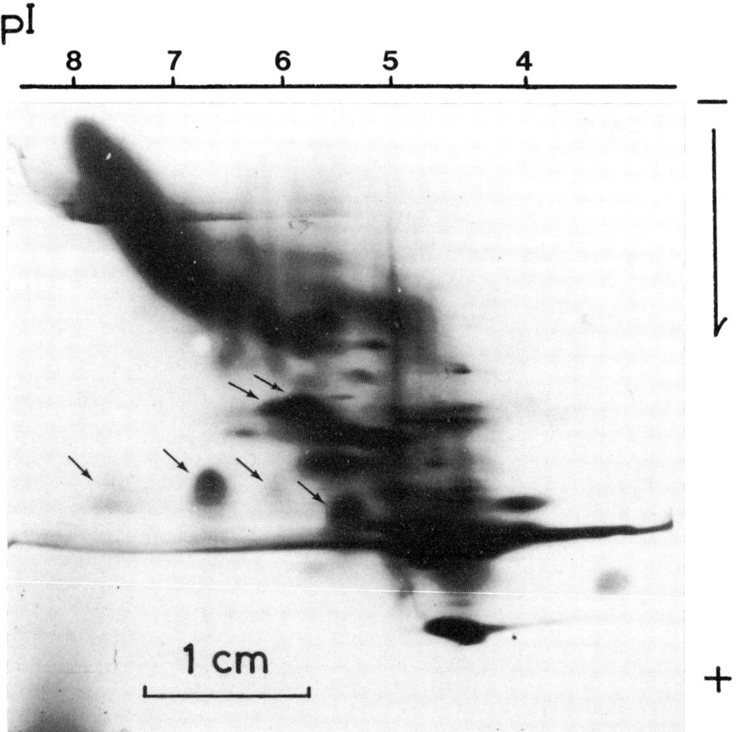

Figure 1. A micro 2-D electrophoretic pattern of a human cerebrospinal fluid sample. One microliter of 30-fold concentrated CSF (about 3 μg protein) was applied and the gel was silver stained. Arrows show the positions of the spots which are not observed in plasma patterns.

ascertained for 20 CSF samples of normal protein level (less than 20 mg protein/dl CSF). Thus most of the constituent proteins in CSF seemed to be filtrates of plasma proteins, although there are some exceptions.

In order to know the locations of plasma proteins in 2-D patterns of CSF, we tried blotting of CSF proteins to nitrocellulose sheets, followed by specific antiserum-peroxidase conjugated anti-IgG staining. An apparatus

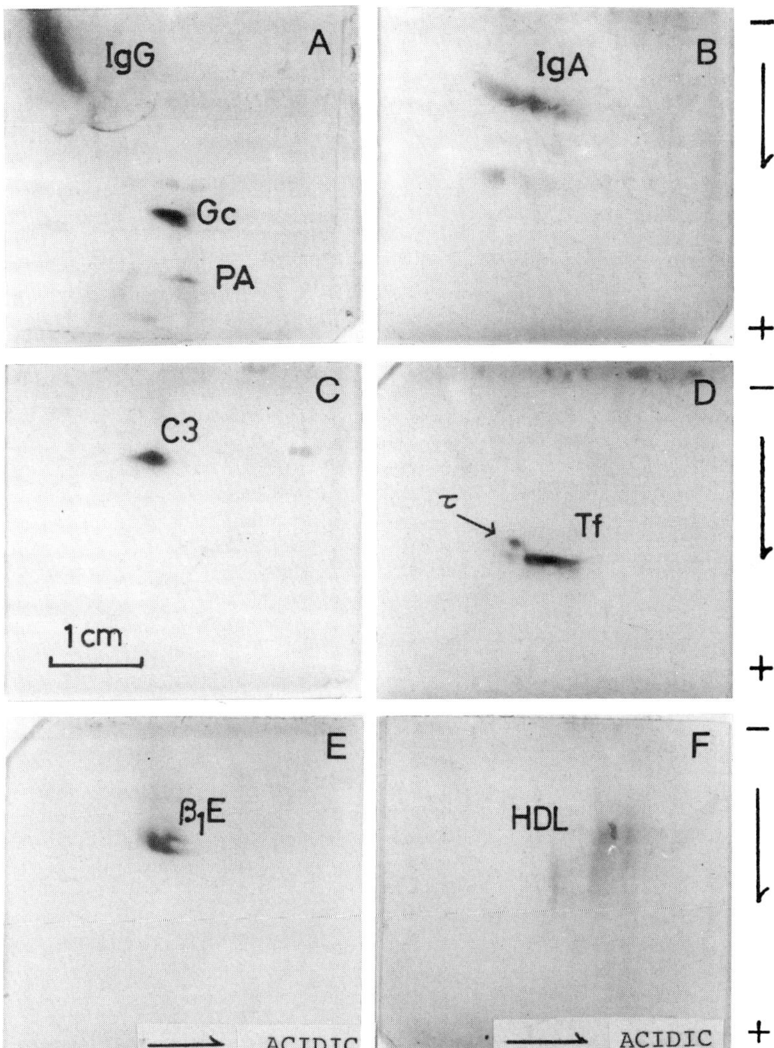

Fig. 2. Some of the nitrocellulose sheets after parallel blotting-immunochemical identification of CSF proteins. For abbreviations, see text.

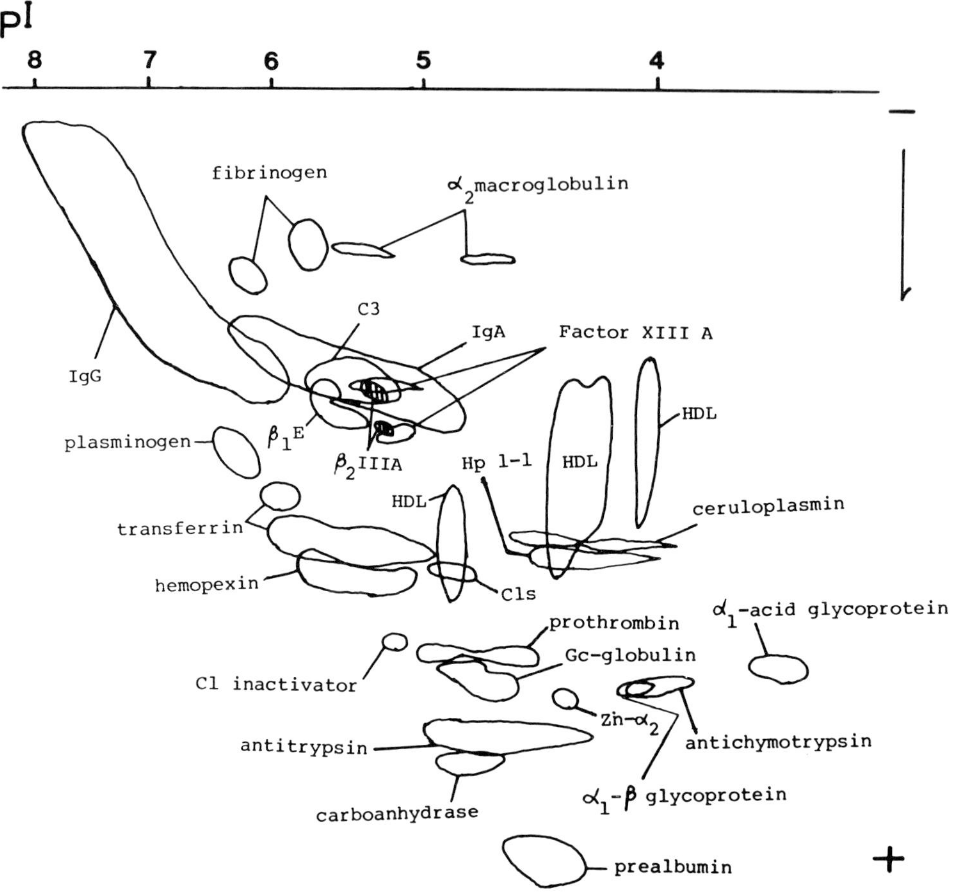

Fig. 3. The results of blotting-immunochemical identification of CSF proteins. Four micro slab gels were set horizontally on a blotting apparatus and 20 blots were obtained within 50 min electrophoresis. Abbreviations used are: C3, complement C3; β_1E, β_1E-glycoprotein (C4); β_2IIIA, β_2IIIA-glycoprotein; Hp 1-1, haptoglobin phenotype 1-1; HDL, high-density lipoprotein; Cls, complement Cls; Zn-α_2, Zn-α_2 glycoprotein. Antisera to low-density lipoprotein (LDL), IgM, retinol binding protein, α_2PA-glycoprotein, C-reactive protein, and β_2-glycoprotein did not give clear spots.

for parallel electrophoretic transfer was devised, which enabled to obtain 20 blots from 4 micro 2-D gels simultaneously. A CSF sample was subjected to micro-multi-2-D electrophoresis, and then four micro 2-D gels were set on the blotting apparatus. The first blot from each 2-D gel was stained for IgG + Gc-globulin + prealbumin and used as a mobility standard for the

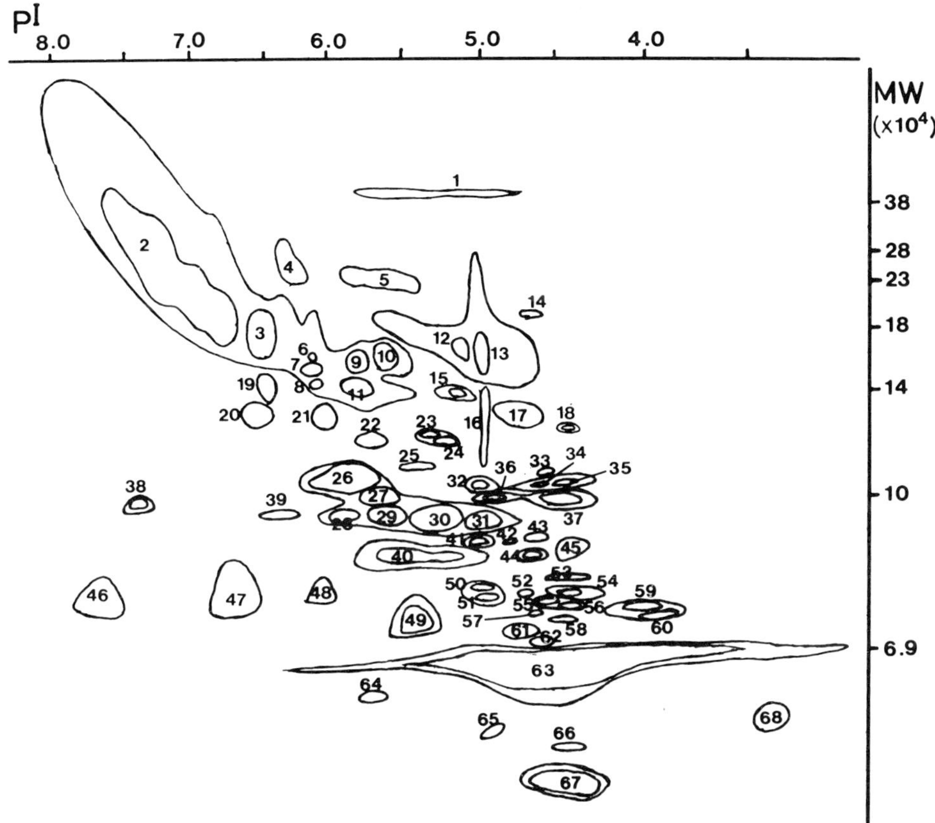

Fig. 4. A standard distribution map of CSF proteins obtained by 2-D electrophoresis in the absence of denaturing agent. Proteins were identified or tentatively identified (*) to be: 1, α_2-macroglobulin; 2-3, IgG; 4, fibrinogen; 5, IgA dimer; 6-8,11, β_1E-glycoprotein (C4)*; 9-10, complement C3; 12-13, IgA (also includes HDL); 16, HDL; 17, inter-α-trypsin inhibitor; 26-31, transferrin (26, tau-protein); 32,34-35, ceruloplasmin; 36-37, haptoglobin phenotype 1-1; 40, hemopexin; 41, complement Cls ; 50-51, Cl inactivator*; 58, prothrombin*; 59, antichymotrypsin*; 60, α_2HS-glycoprotein; 61-62, Gc-globulin phenotype 1-1; 63, albumin (also includes antitrypsin at pI 5-4.5); 67, prealbumin. Spots 46-49, which were characteristic to CSF patterns, were not identified.

second to the fifth blots, each of which was stained for a specific plasma protein to be located. Figure 2 shows some examples of the nitrocellulose sheets, after staining. Each nitrocellulose sheet was treated with: A, anti-IgG + anti-Gc-globulin (Gc) + anti-prealbumin; B, anti-IgA; C, anti-C3a complement (C3); D, anti-transferrin; E, anti-β_1 E glycoprotein (β_1E); F, anti-high-density lipoprotein (HDL). As shown in Fig. 2 D, tau-proteins, arrowed in Fig. 1, reacted with anti-transferrin. Since the 2-D electrophoresis technique does not use dissociating agents, HDL was detected as three broad spots at pI 5-4 with streaking in vertical direction (Fig. 2 F). When SDS was added to the second dimension electrophoresis buffers, HDL dissociated to its apoprotein and lipids.

Figure 3 summarizes the results of the immunochemical identification of CSF proteins. Only the positions of densely stained protein species were shown. The location of identified CSF proteins was quite similar with that of plasma proteins (2), except IgM and low-density lipoprotein (LDL) were not detected in the CSF pattern.

We compared twenty micro 2-D patterns of CSF samples of normal protein level (less than 20 mg protein/dl CSF) with that shown in Fig. 1 and prepared a contour map which represents a standard distribution pattern of CSF proteins (Fig. 4). CSF proteins were located on the map by immunochemical identification (Figs. 2 and 3) and by comparing the map with a 'normalized map' of plasma proteins (2). The locations of major CSF proteins are in consistent with those reported by Felgenhauer and Hagedorn (7), although the first dimension mobilities of CSF proteins in their pattern are considerably different from ours, since they employed agar gel electrophoresis in the first dimension.

When the 2-D patterns of CSF samples of abnormal protein level (above 50 mg protein/dl CSF, an example was shown in Fig. 5) were compared with the map shown in Fig. 4, the presence of high-molecular weight plasma proteins such as LDL, IgM, and haptoglobin polymers were clearly observed. Attempts to apply the technique of micro 2-D electrophoresis to diagnosis of multiple sclerosis and infectious disorders of the central nervous system are in progress.

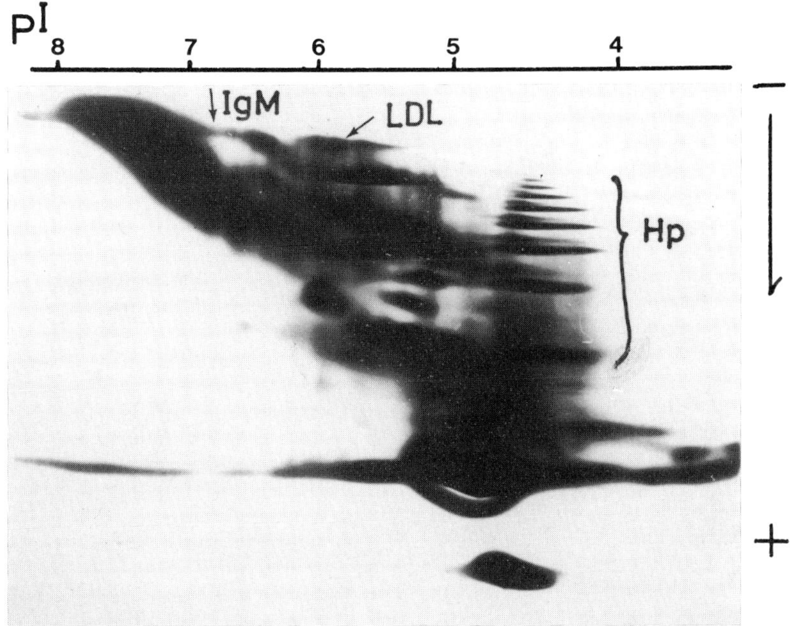

Fig. 5. One of the 2-D patterns of CSF samples of abnormal protein level.
A CSF sample (50 mg protein/dl) was concentrated by 30-fold and 1 μl was
subjected to micro 2-D electrophoresis and the gel was silver stained.
Note the presence of IgM, LDL, and haptoglobin polymers.

References

1. Manabe, T., Tachi, K., Kojima, K., Okuyama, T.: J. Biochem. 85, 649-
    659 (1979).

2. Manabe, T., Kojima, K., Jitzukawa, S., Hoshino, T., Okuyama, T.: J.
    Biochem. 89, 841-853 (1981).

3. Manabe, T., Hayama, E., Okuyama, T.: Clin. Chem. 28, 824-827 (1982).

4. Manabe, T., Kojima, K., Jitzukawa, S., Hoshino, T., Okuyama, T.:
    Clin. Chem. 28, 819-823 (1982).

5. Oakley, B. R., Kirsch, D. R., Morris, N. R.: Anal. Biochem. 105, 361-
    363 (1980).

6. Towbin, H., Staehelin, T., Gordon, J.: Proc. Natl. Acad. Sci. USA 76,
    4350-4354 (1979).

7. Felgenhauer K., Hagedorn, D.: Clin. Chim. Acta 100, 121-132 (1980).

MOUSE LIVER PROTEIN VARIANTS DETECTED BY TWO-DIMENSIONAL ELECTROPHORESIS

Leslie J. Baier, Samir M. Hanash, and Robert P. Erickson

Department of Pediatrics and Human Genetics, University of Michigan
Medical School
Ann Arbor, Michigan 48109

Introduction

High-resolution two-dimensional polyacrylamide gel electrophoresis allows
the simultaneous detection of a large number of protein gene products
(1). One potential application of the technique is the detection of
genetic damage induced by radiation or other mutagens (2). Radiation-
induced deletions in the mouse albino region (3) provide a useful model
for the ability of 2-DE to detect genetic damage. In preparation for
such a study we have undertaken a 2-DE analysis of the amount of genetic
variation in liver proteins from several mouse strains.

Homozygous mice from B1.10A, 129 Sv/Sn and A/J stocks were analyzed.
Additionally, we analyzed $\underline{c}^{ch}/c^{3H}$ mice as representatives of the $\underline{c}^{3H}$
strain. $\underline{c}^{3H}$ homozygotes do not survive beyond the newborn period because
of the albino deletion.

To obtain interstrain F_1 generations, we crossed $\underline{c}^{ch}/\underline{c}^{3H}$ mice with homozygotes from the other 3 strains and analyzed the offspring.

Materials and Methods

Livers from 9 newborn mice were analyzed for electrophoretic variants.
Three mice were from the $\underline{c}^{3H}$ stock, two were from the B1.10A stock, two
were from the 129 Sv/Sn stock and two were from the A/J stock. Each
stock was maintained on its own inbred background. We also crossed the
$\underline{c}^{3H}$ strain with each of the other 3 strains. $\underline{c}^{3H}$xB1.10A, $\underline{c}^{3H}$x129 Sv/Sn
and $\underline{c}^{3H}$xA/J genotypes were obtained. Three livers from each of these
heterozygous genotypes were analyzed.

Electrophoresis '83
© 1984 Walter de Gruyter & Co., Berlin · New York

Two-dimensional electrophoresis

Excised livers were stored at -80°C for subsequent analysis. Preparation
of the sample for electrophoresis involved slicing a small section of
approximately 5 mg. from the frozen liver and immediately solubilizing
the fragment in 9 M urea. The mixture was gently agitated in a 1.5 cc
centrifuge tube until the liver sample was no longer visible. Solubilized
samples were centrifuged for 3 minutes. For each liver sample, 20 and 40
µl of supernatant were applied into the first dimension gel. Therefore
duplicate gels for each sample were obtained.

Electrophoresis was carried out using the Iso-Dalt system (4). Twenty
gels were electrophoresed simultaneously. First dimension gels contained
.75 ml of pH 3.5-10 Ampholines and 30 ml of NP40 supernatant. Isofocusing
was done at 700 V for 16 hours and 1200 V for an additional 2-hour
period. SDS electrophoresis was performed as previously described. Two-
dimensional gels were stained by the silver technique of Merril et al (5).

Results and Discussion

The two-dimensional liver protein pattern of spots for the $\underline{c}^{3H}$ strain is
shown in Figure 1. All of the gels from the four different strains were
remarkably similar. Approximately 700 spots were routinely visualized
in each 2-DE gel. Most spots assumed the same relative intensity in
different strains.

The majority of spots were uniformly distributed throughout the central
surface region of the gel. This region, between pH 4 to 7.5 and M.W. of
80,000 to 20,000 was selected for comparison of 2-DE patterns. Four
hundred and fifty spots were selected for analysis on the basis of
intensity, such that if the spot were to split in two, both spots could
be visualized. These spots were scored for their presence or absence in
all of the gel patterns.

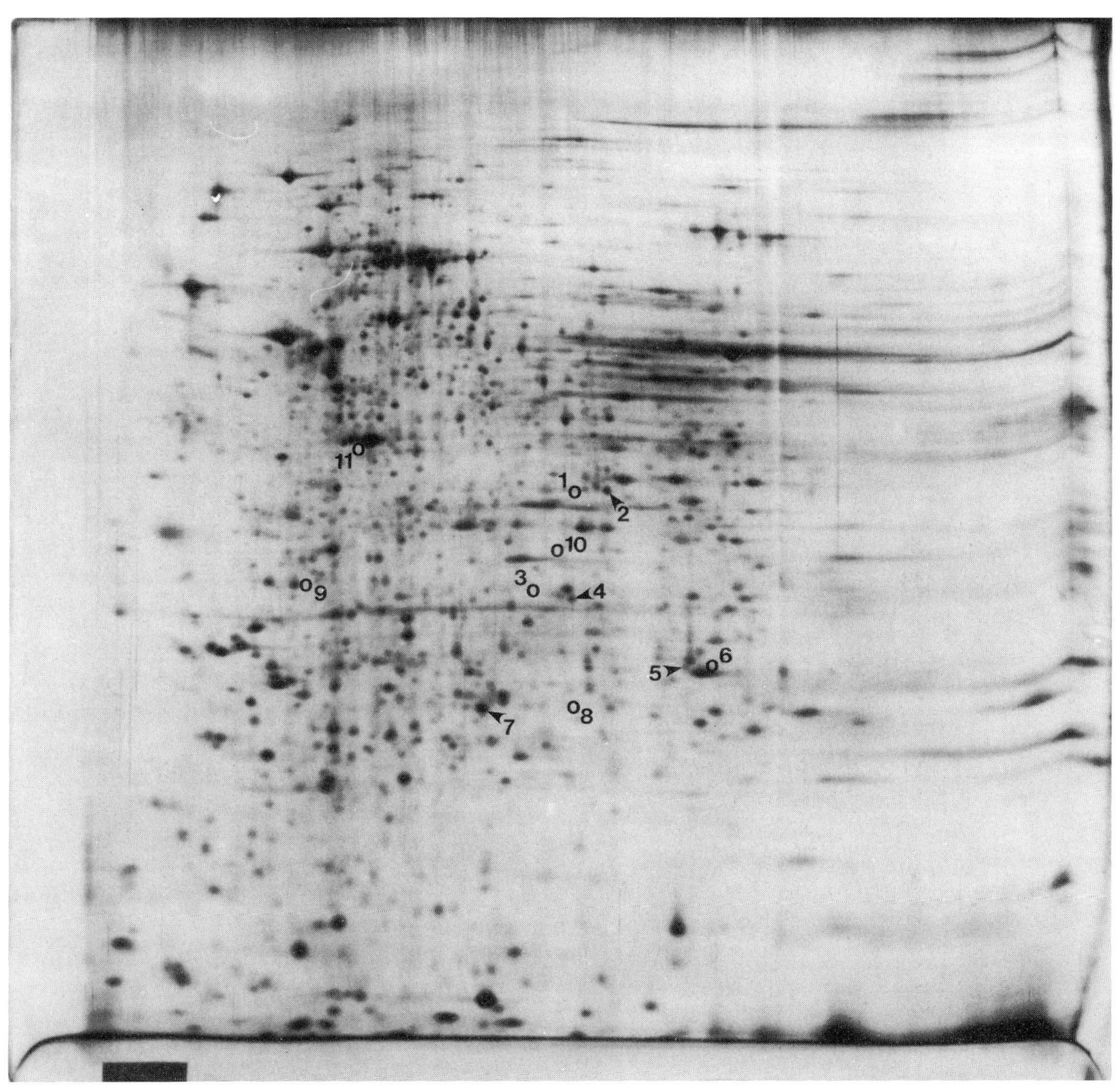

Figure 1.
Two-dimensional liver protein pattern for $\underline{c}^{3H}$ strain.

Spots that were missing in the gel pattern from one or more strains, were
identified. Eleven spot differences were observed (Table 1).

Table 1

	$\underline{c}^{ch}/\underline{c}^{3H}$	A/J	129 Sv/Sn	B1.10A
1	0	+	0	0
2	+	0	+	+
3	0	0	0	+
4	+	+	+	0
5	+	0	+	0
6	0	+	0	+
7	+	0	0	0
8	0	+	+	+
9	0	+	+	+
10	0	0	+	+
11	0	0	+	0

Comparison of the location of these spots on the gels indicated that
eight of the spot differences resulted from the presence of four electrophoretic variants. Identical molecular weights, but slightly different
pI's were observed for four pairs of spots. Only one spot from each pair
was present in the patterns of homozygote. The remaining three spots
are missing from some of the strains, without a variant spot being
identified.

The eleven spot differences were then analyzed in the $\underline{c}^{3H}$x B1.10A,
$\underline{c}^{3H}$x129 Sv/Sn and $\underline{c}^{3H}$xA/J heterozygotes. In instances when one parent
strain was homozygous for a spot and the other parent strain was homo-
zygous for the variant spot, the F_1 heterozygotes showed both spots at
half the intensity of the respective polypeptide in the parental strain
(Figure 2). F_1 heterozygotes, in instances where one of the two parent
strains was missing a spot, showed the spot at lower intensity.

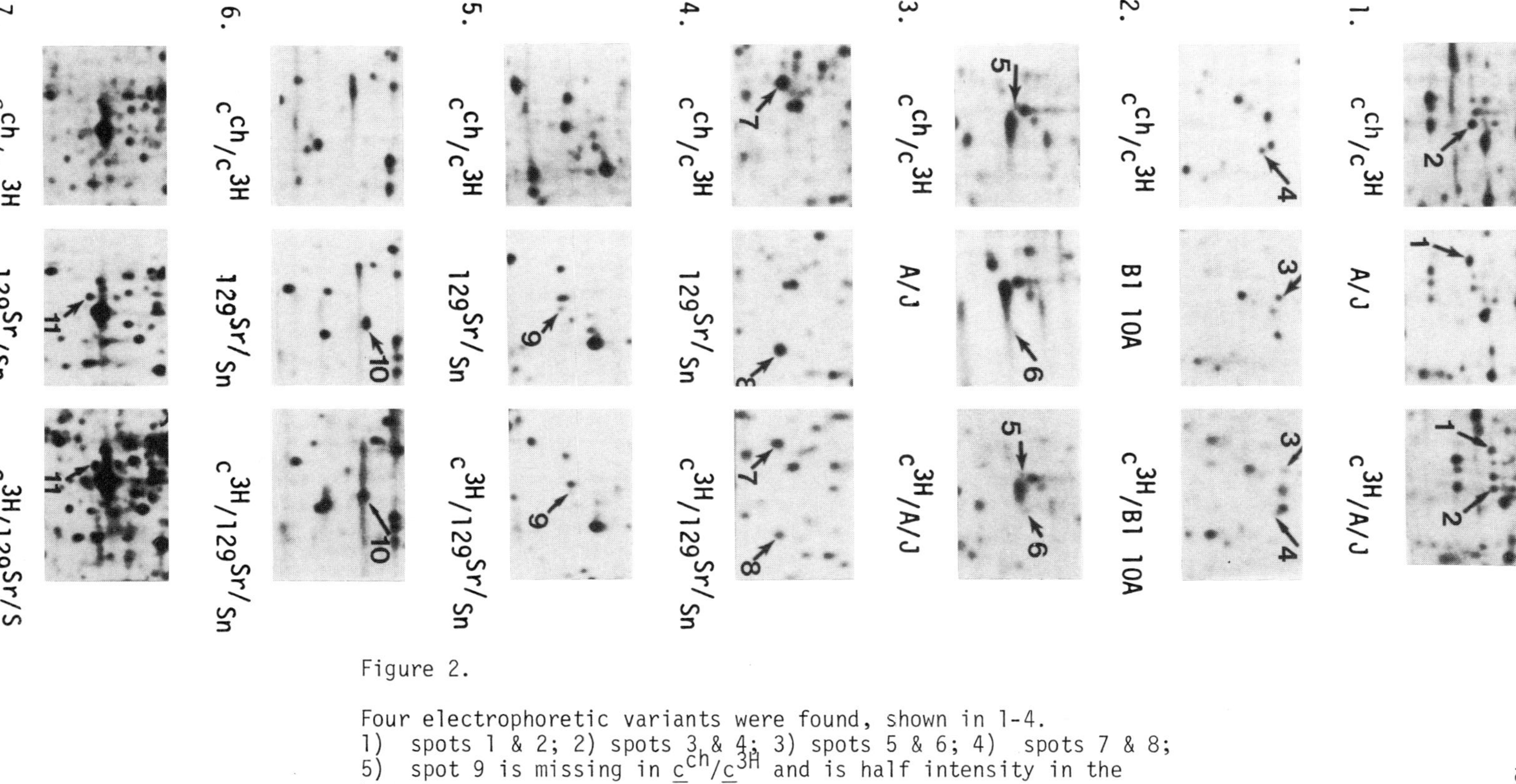

Figure 2.

Four electrophoretic variants were found, shown in 1-4. 1) spots 1 & 2; 2) spots 3 & 4; 3) spots 5 & 6; 4) spots 7 & 8; 5) spot 9 is missing in c^{ch}/c^{3H} and is half intensity in the heterozygote.

The analysis presented here was undertaken as part of a study aimed at the identification of protein deficiencies associated with mouse albino deletions (data to be published). The results presented here pertaining to variant proteins are in agreement with previously published data. Elliott identified an average of eight genetically-determined differences between adult mouse livers from various strains (6). Her analysis included about 250 spots which were visualized with Coomassie blue. In this study of newborn livers, we analyzed about 450 spots which were visualized by silver staining. While it is not possible to derive a heterozygosity index based on the data presented, recent data would indicate that the amount of genetic variation detectable by 2-DE is in agreement with earlier estimates based on one-dimensional electrophoresis of single enzymes or proteins (7,8,9).

The ability to detect genetic variants among multiple gene products suggests that two-dimensional electrophoresis could be a useful approach for genetic monitoring (2).

Acknowledgements

Supported by Program Project NCI-CA-26803 HD 11738

References

1. O'Farrell, P.H.: J. Biol. Chem. $\underline{250}$, 4007-4021 (1975).

2. Neel, J.V., Rosenblum, B.B., Sing, C.F., Skolnick, M.M., Hanash, S.M., and Sternberg, S.: In, Methods and Applications of Two-Dimensional Gel Electrophoresis of Proteins (in press).

3. Gluecksohn-Waelsch, S.: Cell $\underline{16}$, 225-237 (1979).

4. Anderson, N.L., and Anderson, N.G.: Anal. Biochem. $\underline{85}$, 341-354 (1978).

5. Merril, C.R., and Goldman, D., Sedman, S.A., and Ebert, H.H.: Science $\underline{211}$, 1437-1438 (1981).

6. Elliott, R.: Genetics $\underline{91}$, 295-308 (1978).

7. Hamaguchi, H., Yamada, M., Noguchi, A., Fuji, K., Shibasaki, M., Mukai, R., Yabe, T., and Kondo, I.: Hum. Genet. $\underline{60}$, 176-180 (1982).

8. Rosenblum, B., Neel, J., and Hanash, S.: Proc. Natl. Acad. Sci. USA (in press).

9. Harris, H. The Principles of Human Biochemical Genetics, 3rd ed, Elsevier/North Holland Biomedical Press, Amsterdam, pp. XV and 554 (1980).

TWO-DIMENSIONAL GEL ANALYSIS OF PROTEINS IN MATURE ERYTHROID BURSTS

Barnett B. Rosenblum and Samir M. Hanash

Departments of Pediatrics and Human Genetics, University of Michigan Medical School
Ann Arbor, Michigan 48109

Introduction

Erythroid stem cells in the bone marrow and in the peripheral blood can be stimulated in culture to differentiate and mature into hemoglobinized erythroblasts. Previous biochemical analysis of the human peripheral blood early erythroid progenitors (BFUe; burst forming unit-erythroid) has been limited primarily to the study of hemoglobin synthesis (for review see 1). Hemoglobin, as the major red cell protein, accumulates in the maturing red cell colony. The relative proportion of the various globin chains has been quantitated by several techniques (2, 3, 4, 5). However, studies of non-hemoglobin proteins in maturing BFUe have been limited by the inability to detect low quantities of individual polypeptides. In this report we demonstrate the usefulness of 2D-PAGE coupled with the ultrasensitive silver stains to study cultured erythroid cells. Changes in the polypeptide pattern associated with erythroid maturation are described.

Materials and Methods

BFUe cultures were established using nonadherent mononuclear cells (primarily lymphocytes) isolated from normal adult peripheral blood which was obtained following informed consent. The cells were isolated on Ficoll-Paque (Pharmacia Inc.) and macrophages and monocytes removed by adherence to plastic (6). Cultures were established in 35 mm dishes. The medium contained methylcellulose, α MEM, 30% FCS, 10% BSA and 2 IU erythropoietin/ml of medium as previously described (7). Cultures were maintained at 37°C in a humidified incubator with 5% CO_2 for 14 days. Mature erythroid bursts were washed from the culture medium with normal saline followed by centrifugation.

Two-dimensional electrophoresis of peripheral blood nonadherent mononuclear cells and red cells and of mature BFUe was performed essentially as previously described (8, 9). Gels were initially stained with Coomassie brilliant blue R-250 in ethanol:acetic acid:water (10:1:9) and destained in the same solution with decreasing concentrations of ethanol. Following the final destaining, gels were allowed to equilibrate in distilled water. All gels were subsequently stained with silver by the technique of Merril _et al_ (10).

Results and Discussion

These studies were designed to identify polypeptide changes which correlated with the differentiation and maturation of erythroid progenitors (BFUe). The 2D PAGE pattern of mature peripheral blood BFUe (day 14 of culture) was compared to the 2D PAGE pattern of the nonadherent mononuclear cells used to establish the erythroid cultures. Mature BFUe showed over 500 polypeptides on 2D gels (Fig. 1). The general pattern of spots and the apparent relative concentrations of most of them was similar to the pattern for nonadherent mononuclear cells (Fig. 2) used to culture the erythroid stem cells. In contrast, the pattern of polypeptides observed in the mature BFUe (Fig. 1) showed little resemblance to the overall pattern of mature red cells obtained from the peripheral blood of a normal adult (Fig. 3). The polypeptide pattern for the mature red cells included both cytosol and membrane polypeptides.

The major red cell cytosol proteins, hemoglobin and carbonic anhydrase, were present in the mature BFUe. Hemoglobinization is the major recognizable feature of mature BFUe in culture. The colonies acquire a bright red color as hemoglobin accumulates within individual cells. Globin chains (primarily γ and β) and carbonic anhydrase are identified in the mature BFUe 2-D pattern

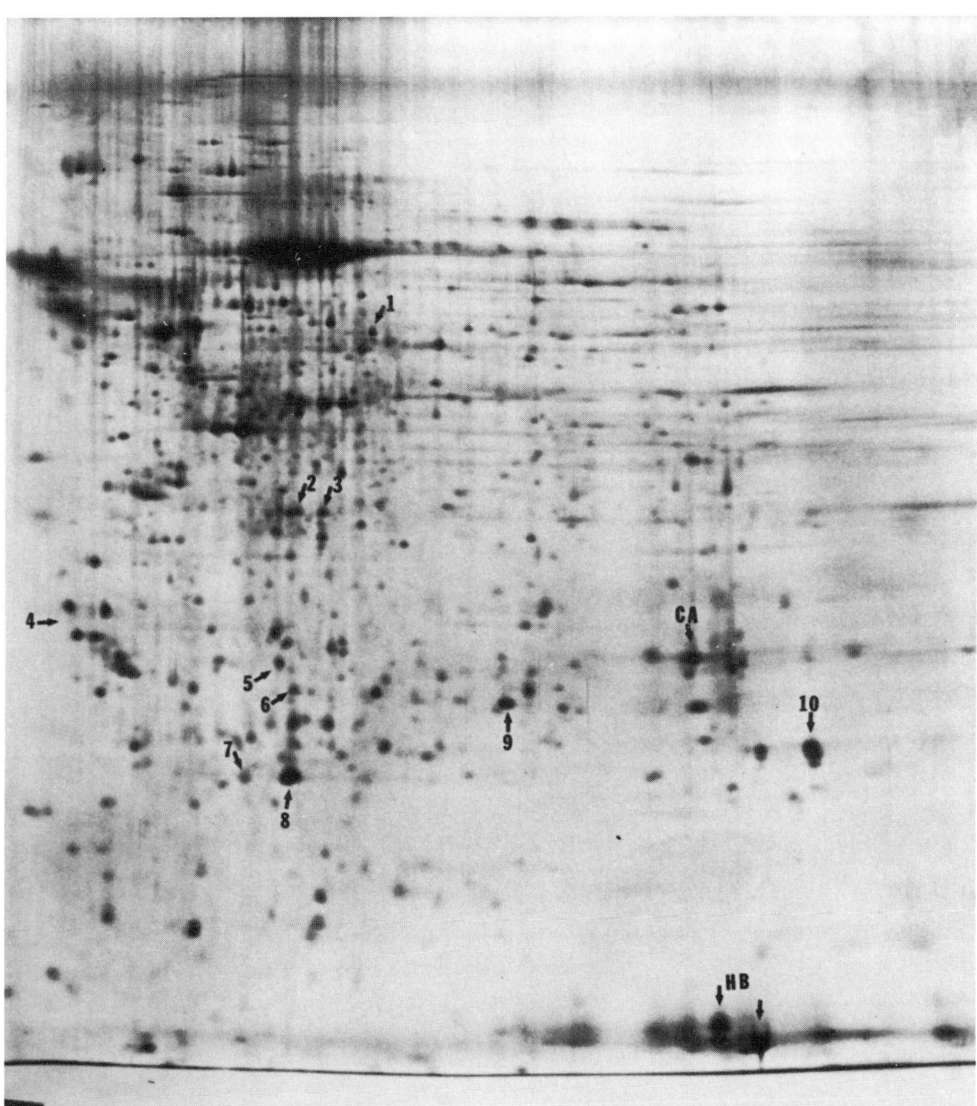

Figure 1. Two-dimensional gel pattern of mature BFUe stained with an ultrasensitive silver stain (10). Pattern is similar to the pattern for nonadherent mononuclear cells (Fig. 2). Hemoglobin and carbonic anhydrase are labeled HB and CA, respectively. Other polypeptides which showed either qualitative or quantitative changes are indicated by numbers. All patterns are oriented with the acidic side to the left and the basic side to the right and with decreasing molecular weights from the top to the bottom of the photograph.

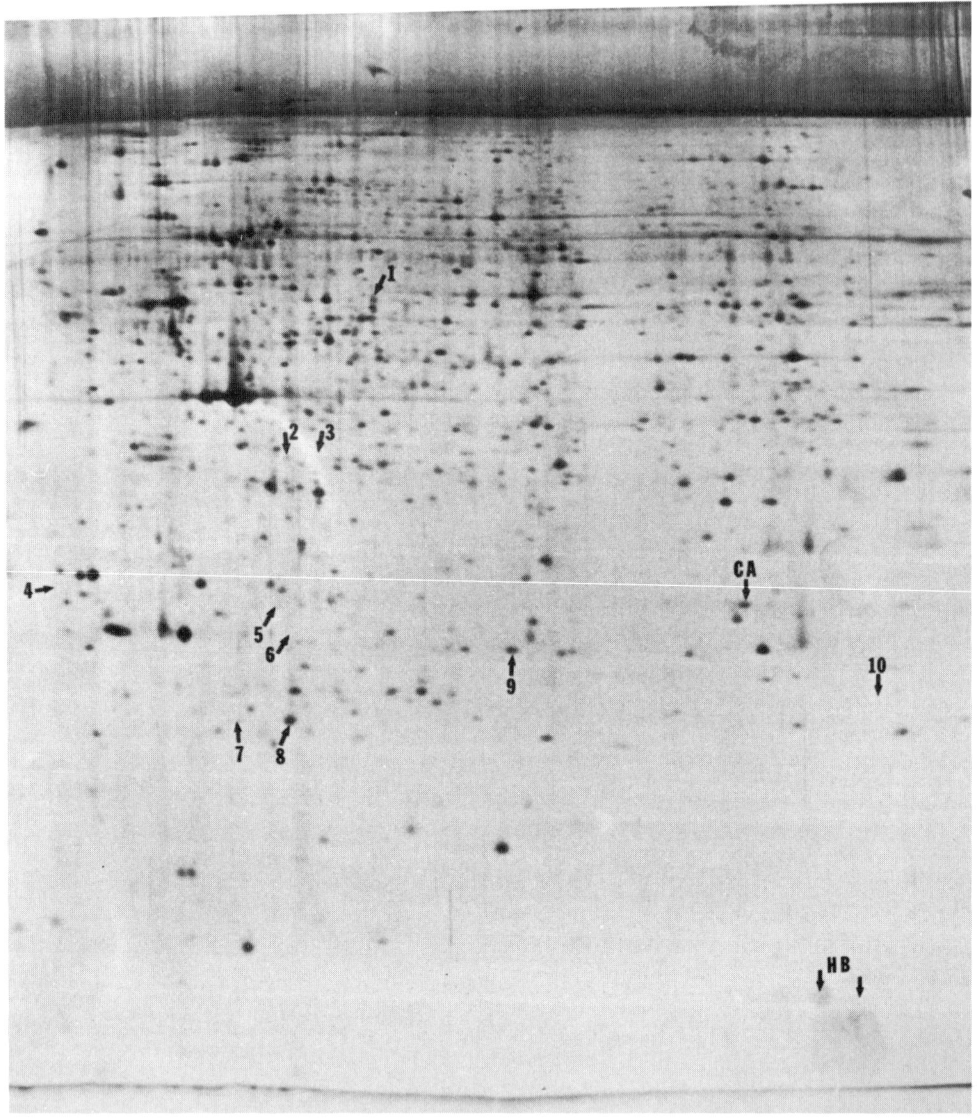

Figure 2. Nonadherent mononuclear cells separated by 2D PAGE and stained with silver as in Fig. 1. Cells isolated by Ficoll-Paque density centrifugation and by adherence to plastic were used to culture BFUe and for electrophoretic analysis. Spots or locations indicated by numbers or HB or CA correspond to Fig. 1.

Figure 3. Two-dimensional PAGE pattern of total red cell polypeptides.
Whole red cells were solubilized prior to electrophoresis. Hemoglobin (HB),
because of its large amount appears as a diffuse band along the buffer
front. Carbonic anhydrase (CA) appeared as discreet spots after Coomassie
blue staining but as a diffuse band following silver staining. Spots 7 and
8 are also seen in mature BFUe (Fig. 1).

(Fig 1) as HB and CA, respectively. The nonadherent mononuclear cells prepared on Ficoll-Paque, prior to culture, were usually contaminated with red cells and thus showed small amounts of globin chains and carbonic anhydrase on the initial 2D PAGE pattern (Fig. 2). The contaminating mature red cells lyse after two days in culture as evidenced by the absence of globin chains in the 2D pattern (data not shown). At least two other spots of similar molecular weight in the BFUe pattern (Fig. 1, spots 7 and 8) were also clearly observed in the red cell 2D PAGE pattern (Fig. 3). The heaviest of these two spots (spot 8) was observed in the nonadherent mononuclear cell pattern (Fig. 2) and increased in density with the maturation of the erythroid progenitors. The more acidic spot (spot 7) was absent from the mononuclear cell pattern.

In addition to the 4 polypeptides which were observed in both mature erythroid progenitors (BFUe) and red cells, 8 other spot differences between nonadherent mononuclear cells and mature BFUe were observed. Four of these consisted of the appearance of new polypeptides in the mature erythroid progenitors (spots 3, 5, 6 and 10). The other four differences in the polypeptide pattern represented quantitative changes in spot intensity. One spot, a high molecular weight polypeptide (spot 1), increased in density relative to adjacent polypeptides. Spot 4 involved a change in two acidic polypeptides including a decrease in the density of the more basic polypeptide with a corresponding increase in the density of the more acidic polypeptide. Spots 2 and 9 showed quantitative increases in the density of the respective spots relative to adjacent spots in mature BFUe (Fig. 1) as compared to nonadherent cells (Fig. 2).

Because the red cell lysate (Fig. 3) and BFUe (Fig. 1) patterns are quite dissimilar accurate spot comparison between the two patterns is difficult. Further studies with antibodies to specific red cell proteins should facilitate the analysis.

The use of 2D PAGE and the sensitive silver stains has enabled us to study polypeptides in mature erythroid progenitors other than globin chains. Further studies of the polypeptides identified in this study as well as of other specific red cell enzymes and proteins, should provide further data concerning the maturation and differentiation of erythroid progenitors in culture.

Acknowledgements

This study was supported in part by NIH Grant HL-25541.

References

1. Stamatoyannopoulos, G. and Nienhuis, A.W. eds.: Cellular and Molecular Regulation of Hemoglobin Switching, Grune & Stratton, New York 1979.

2. Clegg, J.B., Naughton, M.A., Weatherall, D.J.: J. Mol. Biol. 19, 91-108 (1966).

3. Comi, P., Giglioni, B., Ottolenghi, S., Gianni, A.M., Ricco, G., Mazzo, U., Saglio, G.Camaschella, C., Pich, P.G., Gianazza, E., Righetti, P.G.: Biochem. Biophys. Res. Commun. 87, 1-8 (1979).

4. Strahler, J., Rosenblum, B., Hanash, S., Zuckerman, K., Bukunus, R.: Blood 60, 58a (1982).

5. Leary, A.G., Porter, P.N., Ogawa, M.: Am. J. Hematol. 14, 67-73 (1983).

6. Mangan, K.F., Desforges, J.F.: Exp. Hematol. 8, 717-727 (1980).

7. Stamatoyannopoulos, G., Rosenblum, B.B., Papayannopoulou, Th., Brice, M., Nakamoto, B., Shepard, T.H.: Blood 54, 440-450 (1979).

8. Rosenblum, B.B., Neel, J.V., Hanash, S.M.: Proc. Natl. Acad. Sci. USA, In Press.

9. Hanash, S.M., Tubergen, D.G., Heyn, R.M., Neel, J.V., Sandy, L., Stevens, G.S., Rosenblum, B.B., Krzesicki, R.F.: Clin. Chem. 28, 1026-1030 (1982).

10. Merril, C.R., Goldman, D., Sedman, S.A., Ebert, M.H.: Science 211 1437-1438 (1981).

POLYPEPTIDE DIFFERENCES BETWEEN T-LYMPHOCYTES AND T-LYMPHOBLASTS DETECTED BY TWO-DIMENSIONAL ELECTROPHORESIS

S. Hanash, S. Schwartz, L. Baier, B. Rosenblum and K. Springstead

Department of Pediatrics
University of Michigan Medical School
Ann Arbor, Michigan 48109

Introduction

Some of the abnormal features of cancer cells could be related to protein alterations including the synthesis of polypeptide products of genes that are generally not expressed in normal differentiated cells. Consequently there exists considerable interest in identifying protein differences among and between normal and neoplastic cells. A variety of approaches such as histochemical staining and reactivity with antisera or more recently with monoclonal antibodies have been utilized for that purpose. Most of these approaches target a specific protein for analysis. Two dimensional electrophoresis (2-DE) however, because of its resolving power, allows a systematic search for differences involving multiple protein subsets simultaneously (1). Prior studies of abnormally proliferating cells, using 2-DE have focused on established cell lines or autoradiographic analysis of polypeptide patterns (2-4). In this study we have investigated the use of 2-DE coupled with ultrasensitive silver staining to identify polypeptide differences between normal peripheral blood T-lymphocytes and T-lymphoblasts obtained from leukemic individuals.

Materials and Methods

Lymphoblasts were isolated at the time of diagnosis from peripheral blood of three children with T-cell acute lymphoblastic leukemia (ALL). The diagnosis was based on morphological features, negative histochemical staining for peroxidase, esterase and Sudan black, lack of surface or cytoplasmic immunoglobins; and the occurrence of rosetting with sheep

Electrophoresis '83
© 1984 Walter de Gruyter & Co., Berlin · New York

erythrocytes. The patients were selected on the basis of a high blast count in their peripheral blood to maximize the yield of leukemic cells and minimize contamination. Blast cells were isolated from the buffy-coat by gradient centrifugation on Ficoll-Hypaque. More than 98% of the cells analyzed consisted of lymphoblasts.

Several cell populations were prepared from anticoagulated peripheral blood of various healthy subjects. The whole blood was centrifuged for 20 minutes at 125 g to obtain a platelet rich plasma (PRP) fraction. The upper two-thirds of the PRP was recentrifuged at 400 g for two minutes to remove contaminating red cells or white cells. Platelets were then centrifuged at 3400 g. The resultant platelet pellet was washed twice with phosphate buffered saline (PBS) prior to solubilization. Normal lymphocytes were isolated from peripheral blood by Ficoll-Hypaque centrifugation followed by removal of monocytes on a Sephadex G-10 column. T-cells were separated from non-T lymphocytes following incubation of neuraminidase treated sheep red cells and isolation of resultant rosettes on Ficoll-Hypaque. The neutrophil rich red cell sediment, obtained following the Ficoll-Hypaque separation, was diluted with two volumes of 3% Dextran in PBS. The upper fraction containing neutrophils was aspirated and centrifuged at 400 g. The neutrophil pellet was washed three times with PBS. Contaminating red cells were lysed by osmotic shock.

Some of the T-lymphocytes isolated from three healthy subjects were stimulated to undergo blast transformation by treatment with 20 µg/ml phytohemogglutinin-P (Difco, Detroit, MI) for 96 h. at 37° in a humidified 5% CO_2 incubator. At the end of incubation the cells were washed with PBS prior to pelleting.

Cell pellets were solubilized by addition of 40-100 µl of a solubilization mixture containing per liter; 9 M urea, 40 ml of NP40 surfactant, 20 ml of pH 3.5-10 ampholytes, 20 ml of β-mercaptoethanol and 0.2 mM of phenyl methyl sulfonyl fluoride in distilled, deionized water. In most cases 20 µl containing 3×10^6 solubilized cells were applied immediately after solubilization onto isofocusing gels. First dimension gels contained 20 ml of pH 3.5-10 ampholytes per liter. Isofocusing was done at 700 volts for 16 hours and 1200 volts for the last two hours. Twenty gels were run simultaneously. For the second dimension separation, an 11.0 to 13.5 g/L acrylamide gradient was used. Protein spots in gels were visualized by the silver staining technique of Merrill (5). Unsolubilized aliquots were frozen as pellets at -80°C.

Results and Discussion

The 2-D patterns for unstimulated and PHA-stimulated T-lymphocytes as well as lymphoblasts from children with T-cell leukemia were remarkably similar (Figure 1). A total of 350 major spots that were common to all three preparations were identified. These spots were invariably present in all 2-D gels and did not show marked quantitative variation between the three types of T-cells studied.

We searched for consistent spot differences between the three types that would represent either substantial quantitative differences in spot intensity or the appearance or disappearance of a spot. Eight such spots were identified (Figure 1). Four spots (1, 2, 5, 6) were observed in leukemic 2-D patterns but not in patterns of stimulated or unstimulated T-cells. Two spots (4 and 7) were present in normal and stimulated T-cells but missing from leukemic T-cell patterms. The intensity of spot 4 decreased with PHA stimulation. Two spots showed quantitative differences. Spot 3 was markedly increased in intensity in leukemic T-cells and intermediate in intensity in stimulated T-cells. Spot 8 was decreased in intensity in leukemic cells relative to the other two preparations. Three of the eight spots clustered together focusing around a pH of 6.0 with an approximate molecular weight of 22,000. The remaining spots were scattered over the focusing range and between 20,000 and 40,000 in molecular weight.

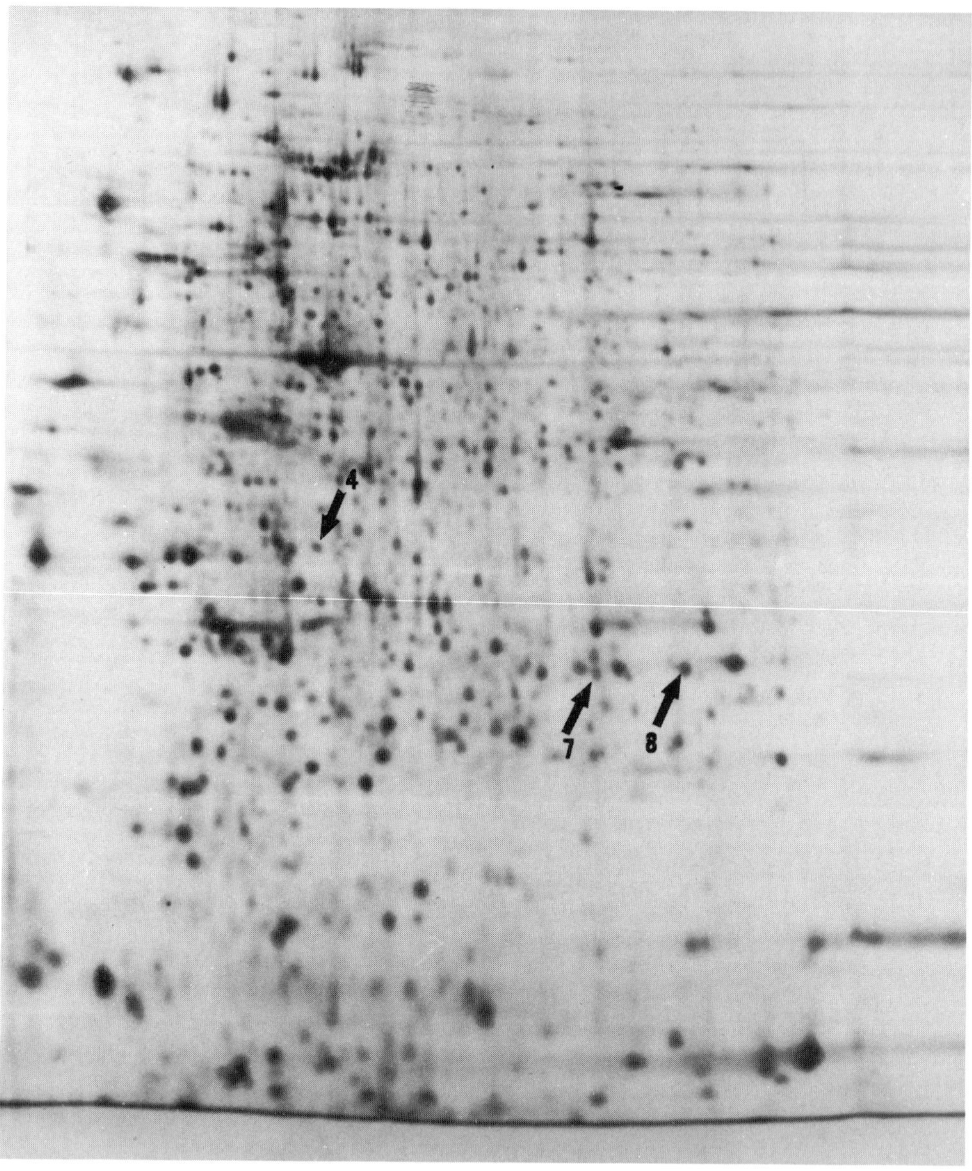

Figure 1. Two-dimensional patterns of (a) normal T-lymphocytes, (b) stimulated T-Lymphocytes and (c) leukemic T-cells. Arrows point to spots that showed differences between patterns.

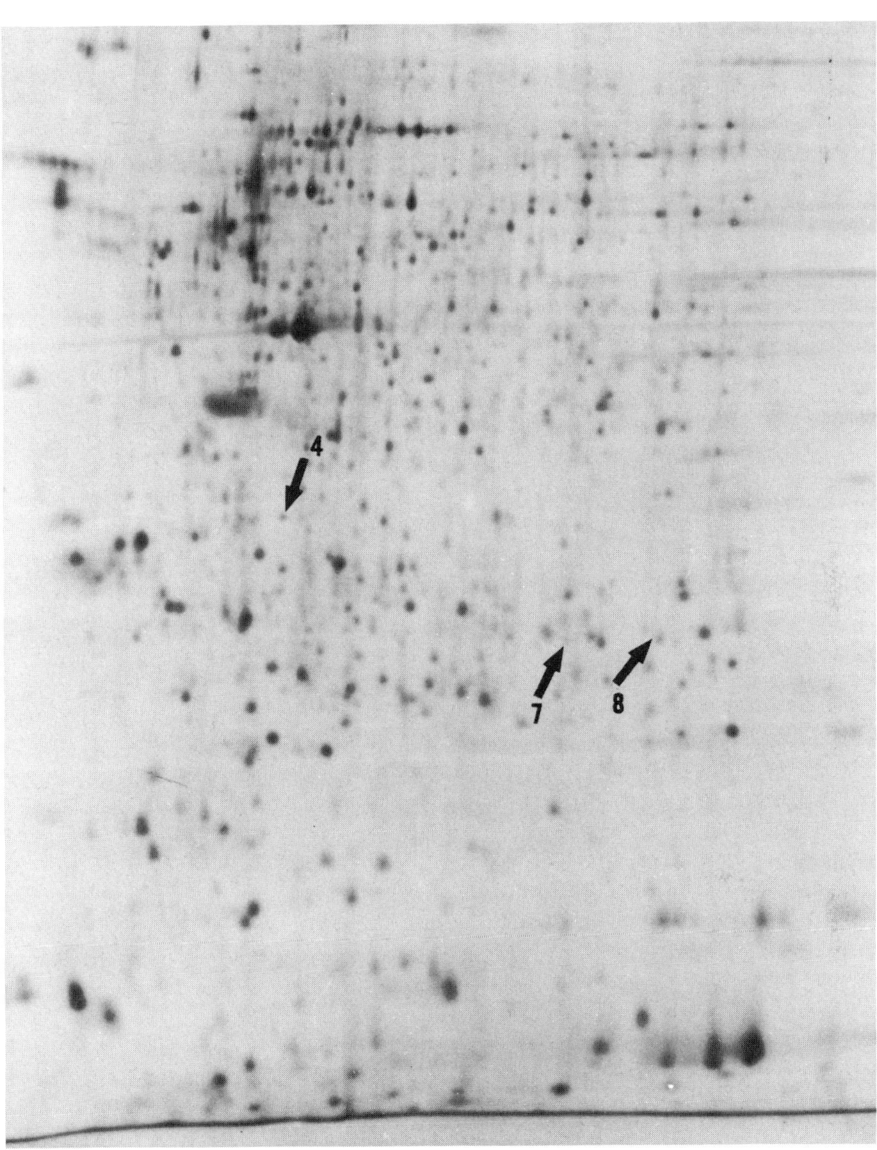

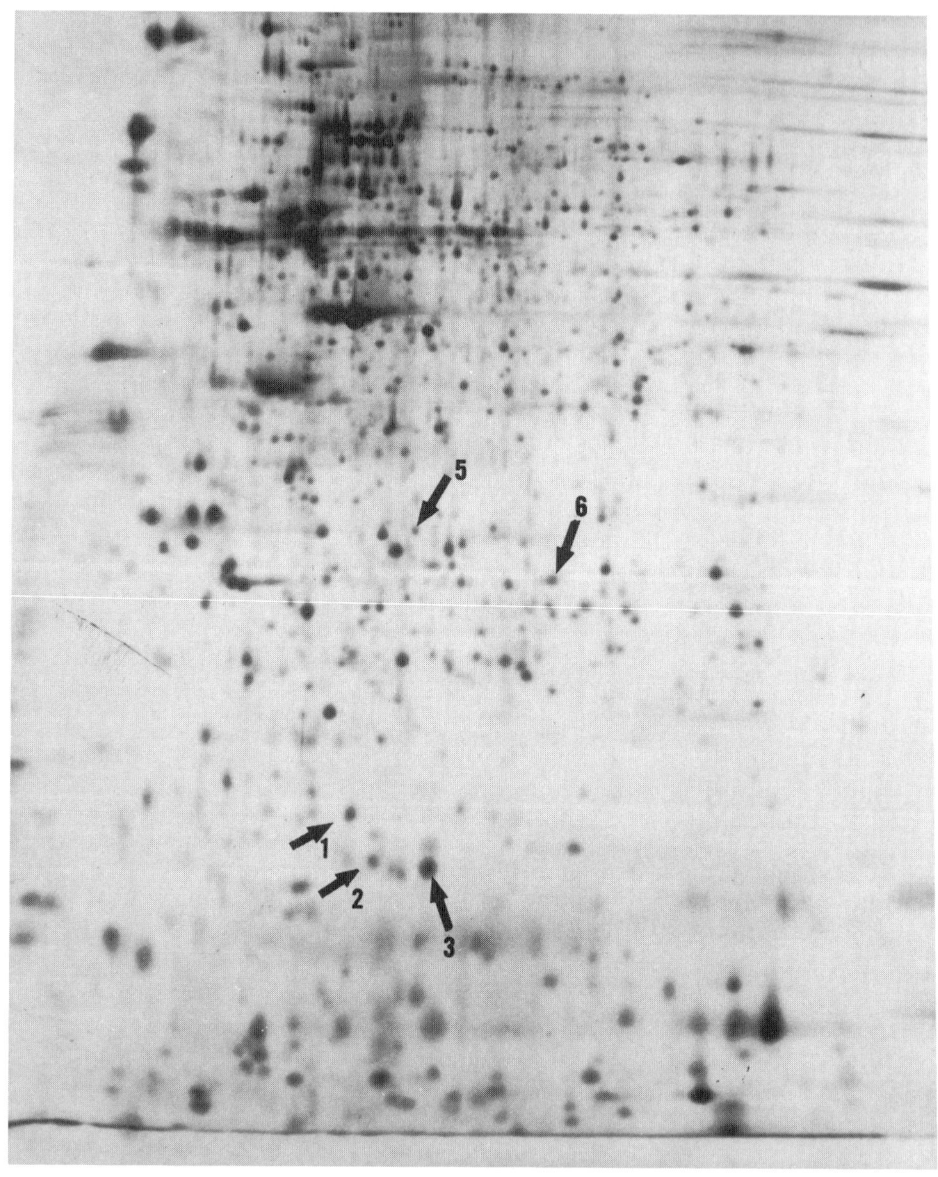
5
6
1
2
3

None of the spots that were distinct to the leukemic cells in this analysis could be observed in platelet, neutrophil or monocyte preparations. Spot 3 which was faint in normal T-cells was also observed as a faint spot in platelets, neutrophils and monocytes.

These studies represent our initial attempt at identifying polypeptide differences between a leukemic cell population and its normal counterpart in the peripheral blood. It is clear that most of the major lymphocyte polypeptides are present in leukemic cells. Indeed, it appears that a substantial number of polypeptide spots in 2-DE gels of lymphoid cells are also present in patterns of other peripheral blood mononuclear cell populations. Consequently, the identification of discrete differences between related cell populations requires rigorous standardization of the 2-DE technique and meticulous attention to detail.

T-cell leukemia accounts for a small proportion of cases of childhood leukemia (6). However because normal peripheral blood provides an easy source of related mature T-lymphocytes for comparison we undertook the study of this type of leukemia first. The significance of the spot differences we have identified remains to be determined. It is interesting that no qualitative changes were observed following phytohemagglutinin stimulation of lymphocytes as was previously observed with autoradiographic methods (3). This suggests that the leukemia distinct spots we have observed are not due to cell cycle differences. These spots could be related to the differentiation stage of the leukemic cells. Alternatively, they might represent oncogene products (7) or foreign viral proteins (8).

Purification of polypeptides distinct to leukemic cells from 2-DE gels will permit their structural characterization. Furthermore the development of monoclonal antibodies to these proteins will allow detailed analysis of their distribution in lymphoid tissues and bone marrow. Other studies in progress, including the analysis of a much larger number of samples from patients with common ALL should also be of relevance.

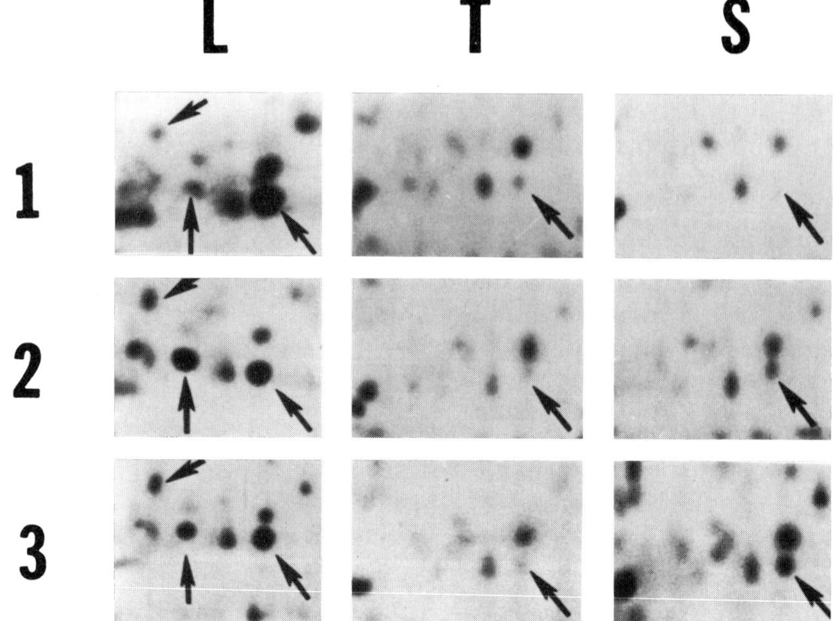

Figure 2. Close up of a cluster of spots in leukemic (L) T-cells from three individuals and normal T (T) and stimulated (S) T-lymphocytes from three normal subjects. Arrows point to spot differences.

Acknowledgements

This study was supported by NCI grant ROI CA 32146.

References

1.   O'Farrell, P.H.: J. Biol. Chem. 250, 4007-4021 (1975).

2.   Bravo, R., Alis, J.E.: Clin. Chem. 28, 949-954 (1982).

3.   Lester, E., Lemkin, P., Lipkin, L., Cooper, H.L.: J. Immunol. 126, 1428-1434 (1981).

4.   Strand, H., August, J.T.: Proc. Natl. Acad. Sci. USA 79, 2729-2734 (1977).

5.   Merril, C.R., Goldman, D., Sedman, S.A.: Science 211, 1437-1438 (1981).

6.   Foon, K.A., Schroff, R.W., Gale, R.P.: Blood 60, 1-10 (1982).

7.   Weinberg, R.A.: Cell 30, 3-7 (1982).

8.   Gallo, R.C., Wong-Staal, F.: Blood 60, 545-557 (1982).

GENETIC ANALYSIS OF OVARY SPECIFIC PROTEINS IN DROSOPHILA MELANOGASTER

Yasuhiko Sakoyama and Sumiko Nakai

Department of Genetics Osaka University Medical School Kita-ku Osaka 530 Japan

Introduction

To investigate from a molecular and a genetic standpoint the development of <u>Drosophila,</u> we have begun to search for stage or tissue specific proteins in developmental processes. We detected a few stage-specific proteins during oogenesis and embryogenesis. Previous reports had shown that a protein named OVA-1A is present only in the ovary and not in embryos and somatic tissues (head or thorax)(1, 2). We searched for OVA-1A variants in different stocks of <u>D. melanogaster.</u> By screening studies, we identified molecular weight variants and charge

variants of OVA-1A. It is not clear whether the variants are the result of an altered allelic structural gene or post-translational modification of the OVA-1A protein by a modifying enzyme (s) (3,4). To answer these questions, we must, first of all, determine the genetic locus and whether the variants are codominantly expressed or not. The present work describes the detection of molecular weight variants and the location of the gene for OVA-1A and then we discuss the possible biological function of OVA-1A protein during oogenesis.

Materials and Methods

1) <u>Strains:</u> The maternal effect mutants used were already

described in previous reports (2). Normal females examined were those of Oregon R **wild type** and of a virginator stock of Wright and Green, T(Y;2) CyO DTS-513 (5).

2) <u>Sample preparation and electrophoresis</u>: Newly eclosed females were separated from males and kept on fresh medium for three to five days at 25°C.

For the preparation of ovary extract, females (3-5 days old) were dissected in <u>Drosophila</u> Ringer's solution kept on ice. The ovaries were rinsed once with the Ringer and transfered immediately into an all glass micro homogenier containing 20-25 μl of the **lysis buffer**. About 15 pairs of mature ovaries and 50 pairs of immature ovaries were used. Two-dimensional polyacrylamide gel electrophoresis used was that of O'Farrell with slight modifications (1,2,6).

Results and Discussion

The protein patterns of mature ovary were analysed by two-dimensional polyacrylamide gel electrophoresis technique of O'Farrell (6). Approximately 270-300 protein spots were identified reproducibly. The ovary specific protein (OVA-1) was found. It was detected only in the mature ovary but not in the immature ovary (Fig. 1), the embryo or female somatic tissues (head or thorax, 1,2).

<u>Description of electrophoretic variants of OVA-1 protein</u>:
The OVA-1A protein is usually detected in most of the strains studied. By screening studies of the 2 normal and 38 maternal effect mutants, we found natural variants of the OVA-1A protein. One of them, named OVA-2A, has an altered isoelectric point (pI), and migrates to the acidic side of OVA-1A protein (Fig. 2, b2). The other type of variant forms displayed changes in the mobility on SDS-PAGE. The OVA-1A protein spot was missing in the ovaries of tuh/tuh, fs(2)75K41/SM1, fs(3)

a mature ovary

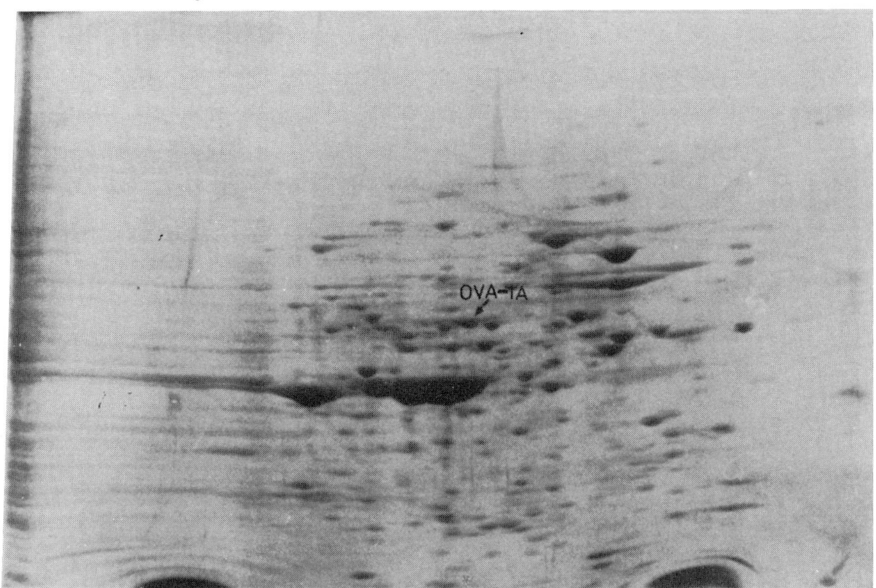

b immature ovary

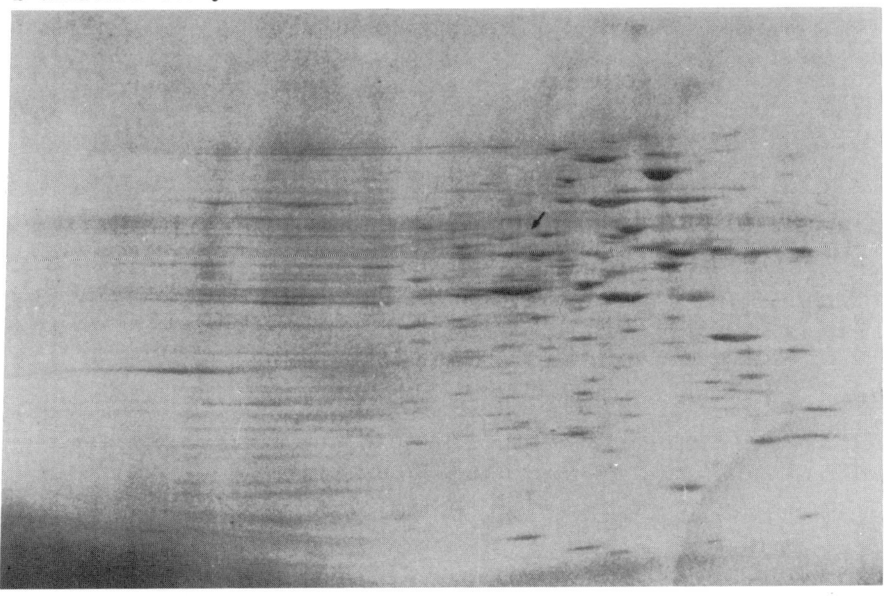

Fig. 1. Gel patterns of mature ovary(a) and immature ovary(b) proteins of T(Y;2) CyO DTS-513 strain. Ovary homogenate (15 pairs for mature ovaries and 50 pairs for immature ovaries) was loaded on a 130mm long isoelectric focusing gel for first dimension electrophoresis. The second dimension was on a sodium dodecyl sulfate(SDS) linear gradient polyacrylamide slab gel (8-22.5% acrylamide). OVA-1A indicates the ovary specific protein, which is not detected in immature ovaries.

3049/TM6 and fs(3)L2/TM3 and new protein spots named ONA-1B
and -1C with altered mobility on SDS-PAGE appeared instead
of OVA-1A (Fig. 2a). The OVA-1B protein is detected in the
ovaries of the tuh and fs(3)3049 strains. The OVA-1C with a
lower molecular weight is found in the ovaries of the fs(2)75
K41 and fs(3)L2 strains. Eggs laid by the females of fs(2)75K
41 and fs(3)L2 strains did not show this newly observed spot,
indicating that this new protein spot is ovary specific
although we have not tested on eggs of the tuh and fs(3)3049
strains. Are the molecular weight variants the product of an
altered structural gene or the modified products by modifying
enzyme(s)?. Genetic change at structural locus would be
expected to codominant expression of two variant forms,
whereas a mutation of a post-translational modifying enzyme
would be either recessive or dominant. Figure 3 shows that
both OVA-1A and -1C or OVA-1A and -1B proteins are expressed
in the ovaries of heterozygous females from mating Oregon R (
OVA-1A)Xfs(2)75K/SM1 (OVA-1C) or r^{39k}/FM6 (OVA-1A)Xtuh/tuh (
OVA-1B). These molecular weight variants are codominantly
expressed in heterozygotes, implying that the variants are
due to a structural gene alteration of OVA-1A protein.
Postlethwait also isolated a molecular weight variant of yolk
polypeptide-2 from <u>Drosophila</u> which alters the mobility in SDS
-PAGE and these variants are expressed codominantly (3,4).
However, final resolution of these argument must be a wait
amino acids sequencing of peptides or DNA sequencing of the
gene.

<u>Chromosomal localization of the OVA-1 protein</u>
As shown in figure 2b, ovaries of homozygous fu/fu fly
contained only a OVA-1A protein and heterozygous fu/FM3 fly
have both OVA-1A and -2A proteins. The other heterozygous state
of the FM3 chromosome, m^D/FM3 and lz/FM3, also have both OVA-
1A and -2A proteins. The OVA-2A variant spot is closely linked
to FM3 chromosome. This result indicates that the <u>Ova</u> gene
resides on the X chromosome, since FM3 is the balancer

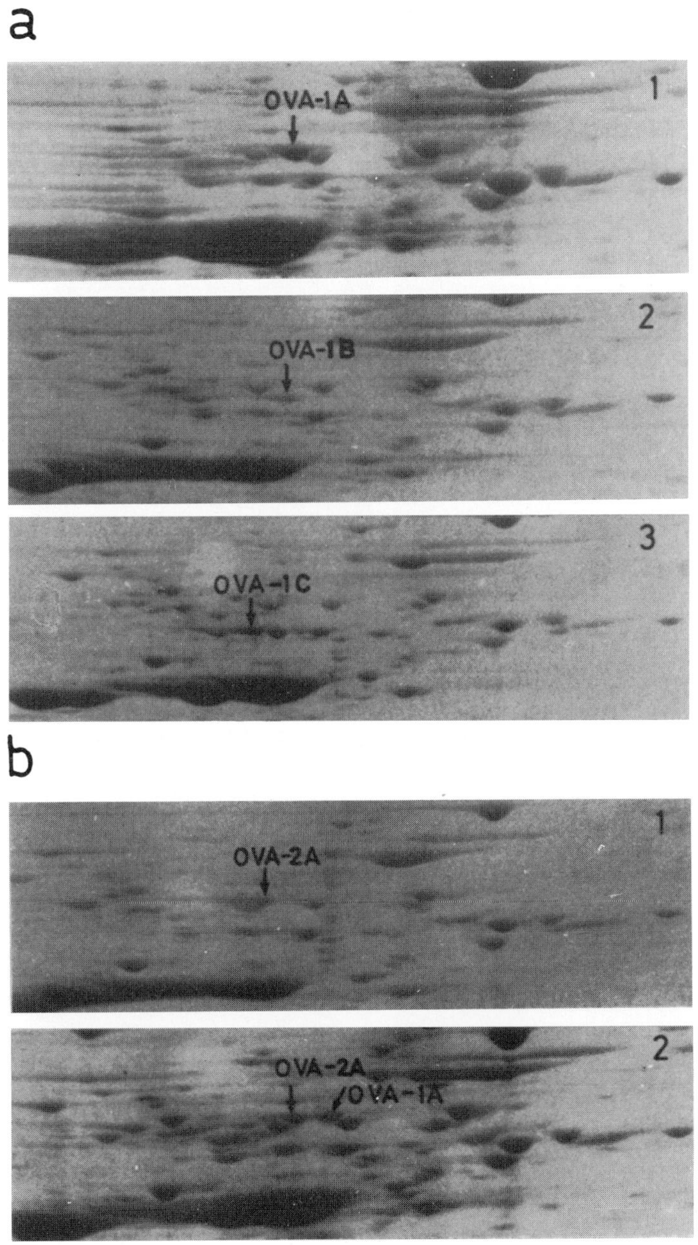

Fig. 2 Gel pattern of genetic variants of ovary specific protein. Alteration of molecular weight (a) and changes in net charge of peptides (b) are observed. a1: T(Y;2) CyO DTS-513(OVA-1A usually detected in most strains), a2: tuh/tuh (OVA-1B instead of OVA-1A), a3: fs(2)75K41/SM1 (OVA-1C instead of OVA -1A and -1B). Figures show selected portions of the gels.

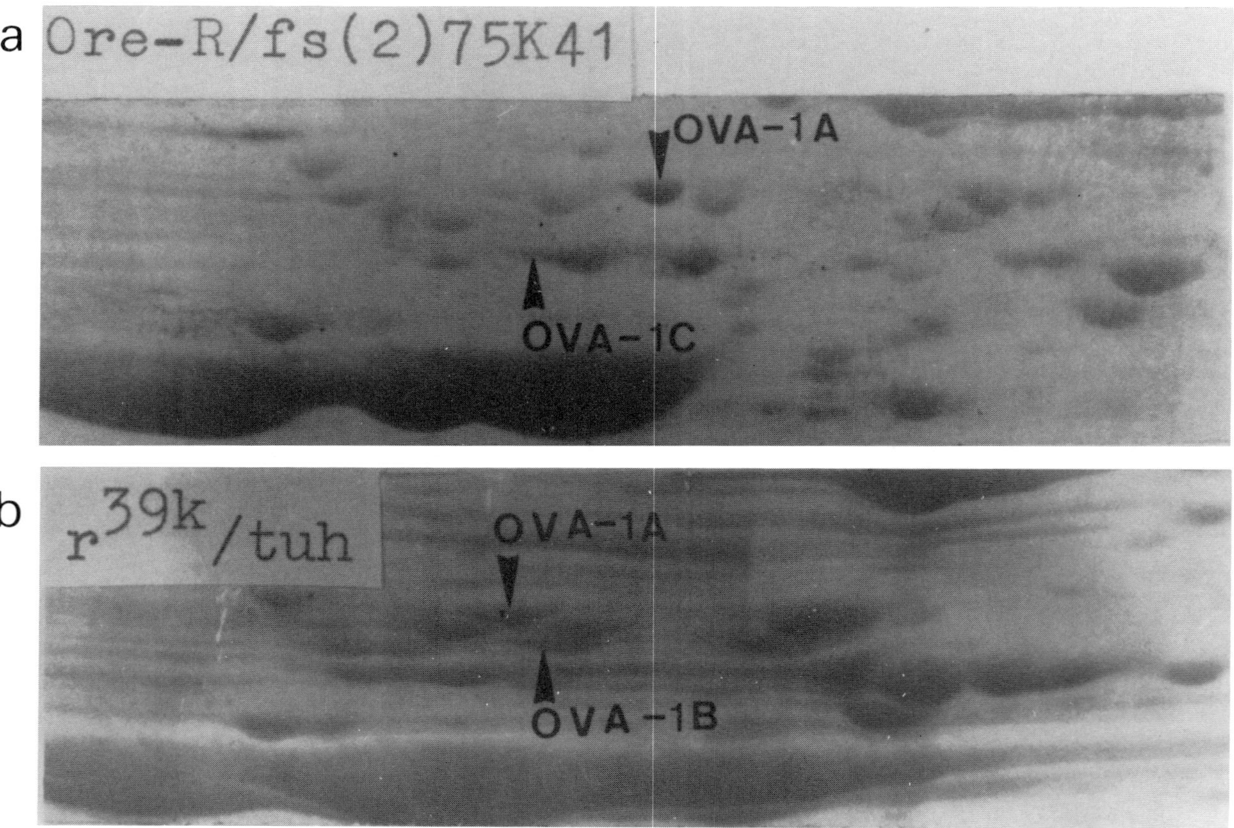

Fig. 3 Codominant expression of molecular weight variant forms in heterozygotes. a: Oregon R/fs(2)75K41, b: r39K/tuh. Figure shows selected portions of the gels.

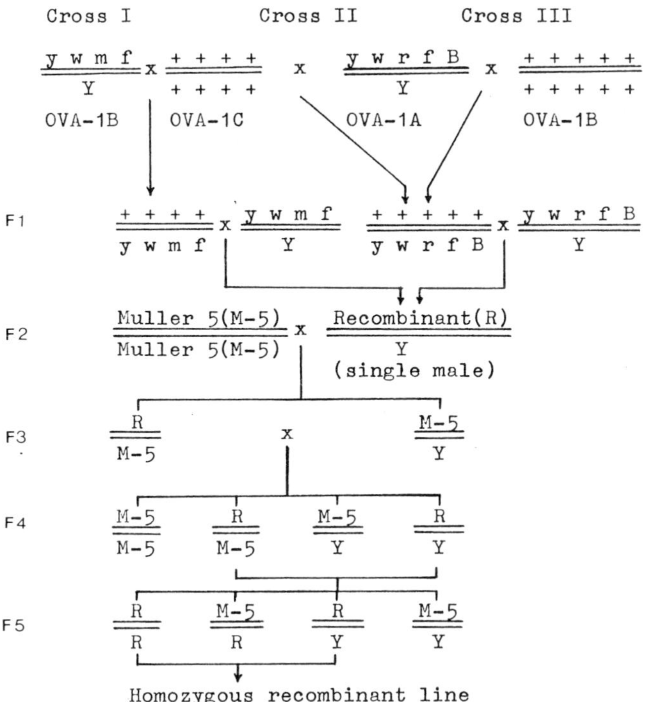

Fig. 4 Mating scheme for mapping of ovary specific protein(
OVA-1). We picked up the homozygous recombinant chromosome by
Muller 5 and constructed a homozygous recombinant line from
the mating of F3, F4 and F5. Ovaries from homozygous re-
combinants were tested for their <u>Ova</u> genotypes. OVA-1C in Cross
I and II: fs(2)75K41 strain, OVA-1B in Cross III: tuh strain.

chromosome for X. The position of the <u>Ova</u> loci on X chromosome
was found by mapping experiments using the multiple marked
chromosome y w m f or y w r f B as follows: the chromosome
which carries the OVA-1 protein variant was determined by
mating y w m f/Y (OVA-1B) or y w r f B/Y (OVA-1A) to female
flies which have variant forms of OVA-1 protein, OVA-1C: fs(2)
75K41, OVA-1B: tuh/tuh. F1 females were backcrossed to y w m f
or y w r f B. A single recombinant male of the F2 was mated to
Muller 5 (M-5) and we picked up the homozygous recombinant
chromosome from the mating of F3, F4 and F5 as shown in Fig. 4.
Ovaries from individual homozygous recombinant lines were

218

Table 1. Mapping of Ova locus

Cross I			Cross II			Cross III		
Recombinant chromosomes*	Ova type 1B	1C	Recombinant chromosomes	Ova type 1A	1C	Recombinant chromosomes	Ova type 1A	1B
y w + +	4	10	y w + + +	0	2	y w + + +	0	6
+ + m f	6	8	y w + f B	0	1			
y w + f	3	5	+ + r f B	1	4	+ + r f B	2	1
+ + m +	4	4	+ + r f +	1	1	+ + r + +	0	1
y w m +	7	0						
+ + + f	0	6						

* To determine Ova genotype, homozygous recombinant chromosomes were constructed and then subjected to electrophoretic analysis (see also Fig. 4).

On the basis of recombination data obtained in cross I, the map position of Ova was determined: 1.5+(36.1-1.5) x [(6+10+4+5)/(6+10+4+5+4+3+8+4)] x 100 = 21.4

tested for their Ova genotypes by two-dimensional PAGE. Table 1 summerizes the results of mating analysis. The results from Cross II and III show that Ova loci are located on the region between the w and r locus. By further mapping experiments of Cross I, the Ova loci are located on the region between w and m (at about 21 on X chromosome).

Biological function

An interesting question concerns the biological functions of the protein. The 38 maternal effect mutants of D. melanogaster studied were classified into oviposition-positive and -negative strains. Histological studies of developmental processes in these mutants have already been reported (2). All strains in oviposition-positive strains develop to stage 14 (final stage of ovarian development) and their ovaries contained almost normal levels of the OVA-1 protein and vitellogenine (Fig.5b). Whereas oviposition-negative strains included a variety of defects ranging from stage 4 to 14 (i.e. immature ovary to mature ovary). However, OVA-1 protein was not detected although vitellogenines were deposited in some mutant strains (Fig.5a).

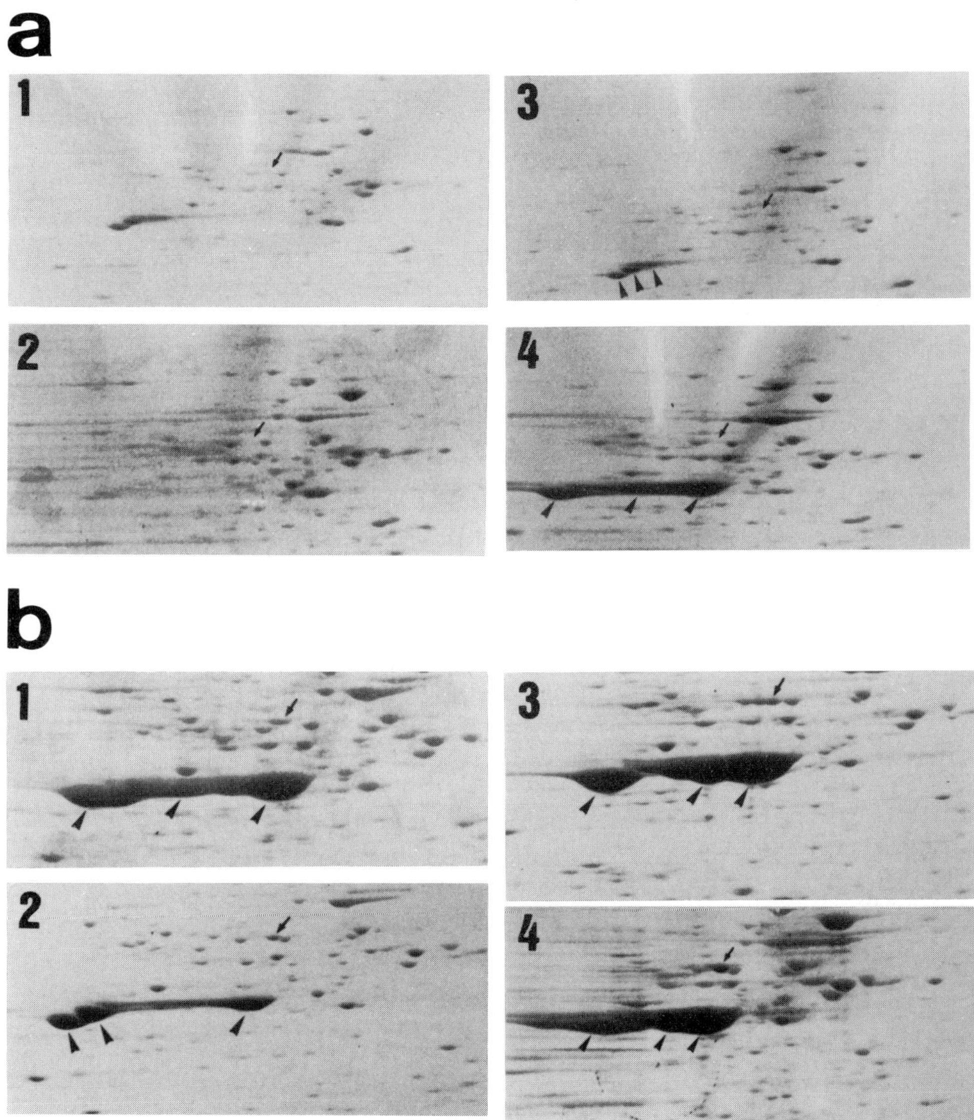

Fig.5 Ovary specific protein pattern of maternal effect
mutants. Oviposition-negative strains (a) and -positive strains
(b). A few examples are shown in figures. Arrow heads indicates
major yolk proteins. Solid arrows indicates OVA-1 protein.
a1:fs(2)34, 2:fs(2)811, 3:str(2)350, 4:fs(2)32; b1:r, 2:fs(2)
75K161, 3:fs(3)mat3, 4:Oregon R (wild strain)

Table 2. Ovarian development of maternal effect mutant strains

Oviposition-negative				Oviposition-positive			
mutants	ovarian development	YP[a]	OVA[b]	mutants	ovarian development	YP	OVA
fs(2)1	stage 4	+	−	34 strains of maternal effect mutants*	stage 14	+++	+ (OVA-1A)
fs(2)34	stage 4	+	−				
ty	stage 9	+	−				
fs(2)2201	stage 9	++	−	tuh fs(3)155 fs(3)3049	stage 14	+++	+ (OVA-1B)
fs(2)811	stage 9-10	+	−				
str(2)350	stage 8-10	+	−	fs(2)75K41 fs(3)L2	stage 14	+++	+ (OVA-1C)
fs(2)16	stage 13	+++	−				
fs(2)32	stage 14	+++	−				

*amx, dor, fu, fs(1)Y^{a-2}, mel(1)R1^{0}, mel(1)R1^{r}, r^{39k}, fs(2)39, fs(2)182, fs(2)c1, fs(2)c2, fs(2)75K161, da, fs(2)A2-297, fs(2)301, fs(2)D45, fs(2)9-1, mat(3)2, mat(3)3, mat(3)4, mat(3)5, mat(3)6, mat(3)7, mat(3)8, mat(3)9, mat(3)10, fs(3)609, fs(3)707, fs(3)1506, fs(3)1868, fs(3)1936, fs(3)2361, fs(3)2796, fs(3)3076.
a:vitellogenine, b:ovary specific protein

Table 2 summerizes the present results. The OVA-1 protein is found in the mature ovary (stage 14) but not in the immature ovary (immediately after eclosion). The OVA-1 protein appeared during maturation of normal ovary. Therefore, the oviposition-positive strains resemble the normal mature ovary, whereas the negative strains resemble the immature ovary. From these results, we infer that the OVA-1 protein is related to oviposition and maturation of the ovary.

References

1. Sakoyama, Y. and Okubo, S.: Develop. Biol. 81, 361-365(1981)
2. Sakoyama, Y.: SBBKA 4 (The Physico-chemical Biology), 26, 463-468 (1982).
3. Pothlethwait, J. H.: Mol. Gen. Genet. 253-260 (1980)
4. Pothlethwait, J. H. and Jowett, T.: Cell, 20, 671-678 (1980).
5. Wright, T. R. F. and Green, M. M.: DIS, 51, 108-109 (1974).
6. O'Farrell, H. O.: J. Biol. Chem. 250, 4007-4021 (1975).

ELECTROPHORETIC ANALYSIS OF PANCREATIC PROTEASES AND ZYMOGEN-ACTIVATING
FACTORS IN MICE

Masaharu Isobe and Zen-ichi Ogita

Department of Biochemical Pathology, Research Institute for Oriental
Medicines, Toyama Medical and Pharmaceutical University, 930-01 Toyama,
Japan

Introduction

The mechanisms of conversion into active proteases have been studied
mainly in the bovine animal. In the case of bovine proteases, at first
trypsinogen is activated by enterokinase (enteropeptidase, EC 3,4,4,8)
[1] or autoactivation in the intestine [2]. Following activation of
trypsin, other pancreatic zymogens 'chymotrypsinogen, proelastase and
procarboxypeptidase' are activated by trypsin. It is generally believed
that trypsin is the key enzyme of this cascade activation. It is not
known whether other enzymes can convert chymotrypsinogen into chymotrypsin directly. Analyzing the mechanisms of the activation other than
in bovines is difficult, because of problems encountered in purifying
zymogens and zymogen-activating factors at the same time. To avoid these
difficulties the method of visualization of non-specific protcase and of
specific protease was developed in conjunction with the use of two-
dimensional electrophoresis to the analysis of pancreatic zymogens and
intestinal-activating factors. In this paper we report the application
of these techniques to characterizing the mouse pancreatic protease and
new types of activating factors which exist in the intestine.

Materials and Methods

<u>Samples</u>   The duodenum was used for preparation of luminal fluid. The
intestine was excised from a mouse and rinsed gently with ice-cold 0.9%

NaCl in a test tube. The luminal fluid together with residual mucus was collected in 1 ml of ice-cold saline. The precipitate was removed by centrifugation at 12,000 x g for 30 min. The resulting supernatant was analyzed for protease activities. Pancreas from mouse or rat was homogenized in 4 volumes of sodium-maleate buffer, pH 5.6, at 4°C with a glass homogenizer. Intestine from mouse or rat was homogenized in 4 volumes of deionized water at 4°C using a glass homogenizer. The resulting homogenates were centrifuged at 12,000 x g for 30 min, and the supernatants were used as samples.

<u>Electrophoresis</u>   Thin layer electrophoresis was carried out essentially in the manner previously described by Ogita [3]. Electrophoresis was performed in 8% polyacrylamide gel. The gels were made up in the buffer containing 38 mM Tris, 21 mM boric acid, and 2.6 mM NaCl, pH 8.8. The electrode buffer(pH 8.2) contained 50 mM NaOH and 300 mM boric acid. One-dimensional electrophoresis was carried out at 4°C for 150 min. at a constant voltage of 150 V.

Two dimensional electrophoresis was carried out for estimation of zymogen, enterokinase and intestinal activating factors. First dimensional electrophoresis for separation of zymogens or intestinal activating factors was carried out at 4°C for 90 min. at a constant voltage of 150 V in an 8% polyacrylamide gel under conditions similar to those described earlier. Following electrophoresis, a 3 mm wide strip cut by razor blade from the first-dimensional gel was placed on an 8% second-dimensional gel, followed by a 3 mm wide strip of filter paper containing intestinal extract or 0.1% bovine trypsinogen or pancreatic extract placed on the gel strip. This was then incubated at 37°C for 60-120 min. in a moisturized container. After incubation, the paper strip was removed and underlying gel subjected to a second electrophoretic separation in the perpendicular dimension at 4°C for 120 min. at a constant voltage of 150V. The electrode buffer used was the same as for one-dimensional electrophoresis. Following electrophoresis, the gel strip was removed and protease activity was observed.

<u>Non-specific protease staining</u> Non-specific protease was stained as a slight modification of Poort's [4] method. Following electrophoresis, a cellulose acetate membrane containing 1% casein in 0.1 M phosphate

buffer, pH 7.6 was placed on the polyacrylamide gel and incubated for 30-60 min. at 37°C. After incubation, the membrane was stripped from the gel, stained for 1 hour in a solution containing 0.01% Coomassie brilliant blue R-250, 12.5% trichloroacetic acid and destained in 10% acetic acid. Protease activities were apparent as white bands on a blue background.

Specific protease staining   Staining of specific protease was accomplished by improvement of Dijkhof's method [5]. Benzoyl-DL-arginine-p-nitroanilide (BANA) and Glutaryl-L-phenylalanine-p-nitroanilide (GPNA) were used as substrate for trypsin and chymotrypsin respectively: 10 mg of BANA or 15 mg of GPNA were dissolved in 0.5 ml of dimethylsulfoxide, 0.1 M, phosphte buffer, pH 7.6, was added to make a final volume of 2.5 ml. The cellulose acetate membrane was soaked in the substrate. Following non-specific protease staining, the membrane was placed on the same gel, and incubated until yellow bands of p-nitroaniline appeared (30-120 min.). The membrane was then stripped from the gel and soaked for 15 min. in 40 ml of 0.1 M, Tris-HCl buffer, pH 8.8, containing 50 mg of N-1-naphthylethylenediamine dihydrochloride and 100 mg of sodium nitrite. The membrane was briefly rinsed in 40 ml of deionized water containing 500 mg of ammonium sulfamate, and then soaked in 2 N HCl. Protease activities were seen as clear violet bands.

Results

Protease in the mouse luminal fluid and in mouse tissues   Mouse luminal fluid and mouse tissue extracts were separated by thin layer polyacrylamide gel electrophoresis. Zymograms of mouse protease are shown in Fig.1. It can be seen that mouse luminal fluid has 7 bands in the anodal area, and one band in the cathodal area. Casein digestive activity is not present in the mouse intestine. Also, a weak band was visualized on the channel for the pancreas. When samples of pancreas and intestine are mixed and incubated for 60 min. at 37°C, electrophoresis reveals 7 bands in the anodal area, a pattern nearly identical to that of luminal fluid except that the band in the cathodal area is missing. These proteases are staind with synthetic substrates, and named for mobility and

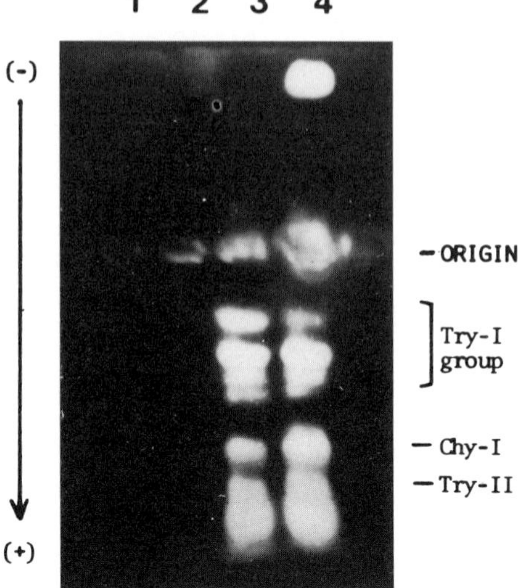

FIG. 1 PROTEASES IN THE MOUSE LUMINAL FLUID AND
IN MOUSE TISSUES. CHANNELS:1, INTESTINE;
2, PANCREAS; 3, MIXTURE OF THE SAMPLES OF
PANCREAS AND INTESTINE; 4, LUMINAL FLUID.

substrate specificity. There are two types of trypsin (Try-I group and Try-II) and one type of chymotrypsin (Chy-I) in mouse luminal fluid and in the mixture of samples of mouse pancreas and intestine.

<u>Electrophoretic analysis of the origin of zymogens and activating factors</u> To determine whether the zymogens occur in pancreas or intestine, pancreas and intestine samples of mouse and rat were reciprocally combined. When mouse pancreas is combined with either mouse intestine or rat intestine, the resulting protease zymograms are mouse type. When rat pancreas is combined with either rat intestine or mouse intestine, the resulting protease zymograms are rat type. In both cases, all zymogens are pancreatic and all zymogen activating factors are intestinal.

<u>The analysis of mouse pancreatic zymogens</u> Two-dimensional electro-phoresis was used for the estimation of mouse pancreatic zymogens. Pancreatic zymogens contained in mouse pancreas were separated by one-dimensional gel. After the first run, these zymogens were activated by intestinal extract, and then second dimensional electrophoresis was carried out. Following electrophoresis, protease activities were made

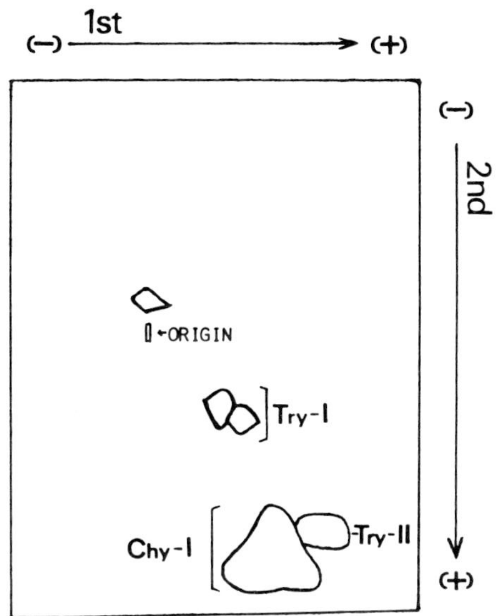

FIG. 2 TWO-DIMENSIONAL ELECTROPHORESIS FOR THE ESTIMATION
OF PANCREATIC ZYMOGENS.

apparent by the use of casein and synthetic substrate. The zymogram is shown diagrammatically in Fig.2. The result shows that Try-I group and Try-II are derived from different trypsinogens, Try G-I group and Try G-II, respectively. Chy-I is derived from a chymotrypsinogen, Chy G, and moreover, not only trypsinogen but also chymotrypsinogen are activated directly by mouse intestinal extract.

<u>Mouse enterokinase</u>   The most probable candidate for the intestinal-activating factor is enterokinase. To estimate mouse enterokinase, we used two-dimensional electrophoresis with bovine trypsinogen as substrate. Mouse intestinal extract containing enterokinase was separated by first-dimensional electrophoresis. A strip of filter paper containing bovine trypsinogen was placed on the gel strip of the first run. Following activation of bovine trypsinogen by mouse enterokinase, a second electrophoresis was carried out. A single band of bovine trypsin activated by mouse enterokinase was found in the cathodal area near the origin of the one dimensional gel. This result shows that mouse enterokinase has a slow mobility.

<u>Two-dimensional electrophoresis for the estimation of the mouse</u>

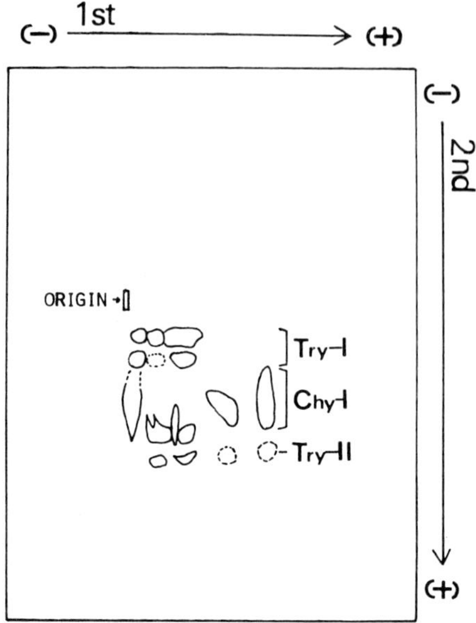

FIG. 3 TWO-DIMENSIONAL ELECTROPHORESIS FOR THE ESTIMATION
OF INTESTINAL ACTIVATING FACTORS.

intestinal-activating factors   Two-dimensional electrophoresis was used
for the analysis of mouse intestinal-activating factors. First, mouse
intestine was separated by one-dimensional gel. A filter paper containing mouse pancreas was placed on the gel strip of the first run.
Following activation of mouse pancreatic zymogen, a second electrophoresis was carried out. After electrophoresis, protease activities
were demonstrated by staining methods of non-specific and specific
protease. Bands identified by their substrate specificities and
mobilities of secondary direction are illustrated diagrammatically in
Fig.3. It can be seen that pancreatic zymogens are activated not only
near the origin but also in more anodal area of a one-dimensional gel.
This result indicates the presence of several kinds of new activating
factors other than mouse enterokinase.

Discussion

Mouse pancreatic proteases and zymogens   By the use of thin layer polyacrylamide gel electrophoresis, we can separate into at least 8 bands, the active proteases that exist in the intestine under physiological conditions. Furthermore, we can produce nearly identical bands in vitro. Thus the proteases observed in the mixture of intestinal and pancreatic extracts play an important role as digestive enzymes. Digestive proteases are produced as inactive forms in the pancreas, and are converted into active proteases in the intestine. In the mouse, zymogens are also produced and stored in the pancreas, while zymogen-activating factors occur in the intestine. Judging from mobilities and substrate specificities, bands corresponding to our Try-I, Try-II and Chy-I are reported by Watanabe et al. [6]. Since we separated Try-I into 4 bands, we call these bands the Try-I group.

Two dimensional electrophoretic analysis of mouse zymogens clearly shows that there are at least two different trypsinogens (Try G-I group and Try G-II) and one chymotrypsinogen (Chy G), supporting the report of Watanabe et al. [7]. It is known that there are multiple activation forms in bovine chymotrypsin [8]. In the mouse, we believe that there are multiple activation forms as well. Supporting this view is the fact that the change of Chy mobility was observed depending on the extent of activation. In spite of the presence of multiple activation forms, we think that the enzyme is derived from a single zymogen, because we found only one chymotrypsinogen, though we tried two-dimensional electrophoretic analysis under various conditions of electrophoresis.

<u>Mouse enterokinase and intestinal activating factors</u> Enterokinase has been studied mainly using bovine trypsinogen as substrate, and has been reported for porcine animals [9], rat [10] and bovine animals [11]. The molecular weight of porcine enterokinase is 195,000 consisting of two molecules [12]. Two dimensional electrophoresis showed that mouse enterokinase has a slow mobility, thus suggesting it has a high molecular weight.

Two-dimensional electrophoretic analysis of intestinal activating factors indicates there are several other activating factors in the intestine besides enterokinase, the existence of factor(s) directly activating Chy G is indicated from the results of two-dimensional electrophoretic analysis of zymogens. A considerable amount of research on enterokinase has been reported, [9,10,11,12], but other intestinal activating factors have not been studied, because, bovine trypsinogen has been used for the substrate to detect the enterokinase. Moreover, zymogens from the same species or the same individual have rarely been used to investigate the intestinal activating factors. These new activating factors are being characterized by the use of purified mouse zymogens. Precise data will be reported elsewhere [Isobe et al. in preparation].

References

1. Kunitz,M.: J. Gen. Physiol. $\underline{22}$, 429-446 (1939)

2. Kay, J., Kassell, B.: J. Biol. Chem. $\underline{246}$, 6661-6665 (1971)

3. Ogita, Z.: Isozymes, Vol.II, pp.289-314, Academic Press, New York 1975 (Markert, C.L., ed.)

4. Poort, C., Venrooy, W.J.W.: Nature $\underline{204}$, 684 (1964)

5. Dijkhof, J., Poort, C.: Anal. Biochem. $\underline{62}$, 598-600 (1974)

6. Watanabe, T., Ogasawara, N., Goto, H.: Biochem. Genet. $\underline{14}$, 697-707 (1976)

7. Watanabe, T., Ogasawara, N.: Biochim. Biophys. Acta $\underline{717}$, 439-444 (1982)

8. Wilcox, P.E.: Methods in Enzymology, Vol. XIX, pp. 64-108, Academic Press, New York 1970 (Perlmann, G.E., ed.).

9. Louvard, D., Maroux, S., Baratti, J., Desnuelle, P.: Biochim. Biophys. Acta $\underline{309}$, 127-137 (1973)

10. Nordström, C., Dahlqvist, A.: Biochim. Biophys. Acta $\underline{242}$, 209-225 (1971)

11. Liepnieks, J., Light, A.: J. Biol. Chem. $\underline{254}$, 1677-1683 (1979)

12. Baratti, J., Maroux, S., Louvard, D., Desnuelle, P.: Biochim. Biophys. Acta $\underline{315}$, 147-161 (1973)

IDENTIFICATION OF FETAL POLYPEPTIDES IN AMNIOTIC FLUID USING
TWO-DIMENSIONAL GEL ELECTROPHORESIS

Kathryn E. Kronquist, Barbara F. Crandall and Lina G. Cosico
Mental Retardation Research Center
Department of Psychiatry
University of California Medical Center
Los Angeles, California 90024

Introduction

Soluble proteins in amniotic fluid (AF) have been largely
ignored for the diagnosis of fetal disorders at mid-gestation
because the majority were shown to be derived from maternal
serum (1). However, fetal proteins originating from a variety
of sources, such as lung, skin, the gastrointestinal and
genitourinary tracts and the fetal membranes may be released
into AF and qualitative or quantitative changes in these
polypeptides may reflect normal as well as abnormal
developmental processes. We undertook a detailed analysis of
the polypeptides in AF and in matched samples of maternal
plasma (MP). The objects of this study were to establish the
normal two-dimensional profile for the major AF polypeptides
present at mid-gestation and, also, to compare the polypeptide
pattern of AF with that of MP in order to discover polypeptides
which were present at a higher relative concentration in AF.
This group of polypeptides would be expected to include
proteins originating from the fetus and its membranes and
maternal proteins derived from the uterus.

In this paper we report the identification of twenty-one serum
proteins on two-dimensional gels of mid-gestational AF. In
addition, we have found ten AF polypeptides which were not
detected in MP. Two of these ten species have been identified

230

as known fetal proteins and we have begun the characterization
of the the remainder by searching for macromolecules with
identical electrophoretic mobilities in cord serum (CS) and
fetal urine (FU).

Specimens and Reagents

Fifty-eight matched samples of AF and MP were obtained at 15-21
weeks of gestation from women with uncomplicated pregnancies
undergoing amniocentesis because of advanced maternal age (>35
years). Thirteen samples of term amniotic fluid (TAF) were
obtained at 38-42 weeks of gestation from patients with normal
pregnancies requiring amniocentesis to determine fetal lung
maturity prior to delivery by repeat cesarian section.
Gestational age was determined by ultrasonographic measurement
of the fetal biparietal diameter (16). Cord serum (CS) and
samples of the first voided urine, which was used as an
approximation of fetal urine (FU), were obtained at 22 - 25
weeks of gestation from 5 healthy, premature infants. All
samples were centrifuged at 400 x g for 5 min, followed by a
second spin at 12,800 x g for 3 min to remove cellular elements
and were then stored at $-70\,^{\circ}$C.

Protein A-Sepharose CL-4B was from Pharmacia. Broad-range
(Isodalt grade) ampholytes, pH 3-10, were from Serva and narrow
range ampholytes, pH 2.5-4 and pH 5-7, were from LKB. Other
reagents for electrophoresis were from sources previously
described (3). Rabbit anti-human albumin and anti-transferrin,
purified human IgM and serum IgA were from Cappel Laboratories.
Rabbit antiserum to human kappa light chains, ceruloplasmin,
α-fetoprotein α_2-macroglobulin, α_1-antitrypsin, hemopexin and
haptoglobin were supplied by Dako Corporation. Human plasma
fibronectin and antiserum to plasma fibronectin were from
Bethesda Research Laboratories. Rabbit antiserum to human
apolipoprotein AI was the kind gift of Dr. Paul Roheim,

University of Louisiana, New Orleans, Louisiana. Purified apoVLDL containing apolipoprotein E was prepared by the heparin precipitation method of Utermann (14).

Sample Preparation and Two-Dimensional Electrophoresis

50-60% of the albumin was immunoadsorbed from AF, TAF, MP and CS samples using anti-human albumin bound to protein A-Sepharose. Selected serum proteins in AF and MP were identified by immunoadsorption using small amounts of protein A-Sepharose containing the appropriate specific antiserum. Details of these procedures are described in (3). FU was desalted using a modification of the method of Tuszynski (5) described in (3). Samples were dissolved at room temperature in dissociation buffer and were subjected to isoelectric focusing on broad range gels (pH 4-9) followed by electrophoresis on 5-15% linear polyacrylamide gradient gels in the presence of sodium dodecyl sulfate (SDS) using the method of O'Farrell (6) modified as described in (3). Gels were stained using the silver stain method of Merril et al. (7) and were photographed on Kodak Contrast Process Pan film.

Results and Discussion

The treatment of samples with anti-albumin bound to protein A-Sepharose selectively removed most immunoglobulins and 50-60% of the total albumin. Only traces of other proteins were bound (3). Over 200 polypeptides were resolved on gels of both AF (Figure 1 A) and MP (Figure 1 B) following the removal of albumin. The polypeptides corresponding to 28 MP proteins were identified on electropherograms either by comparing the distinctive mobility and conformation of the observed protein complexes with those in published electrophoretic patterns

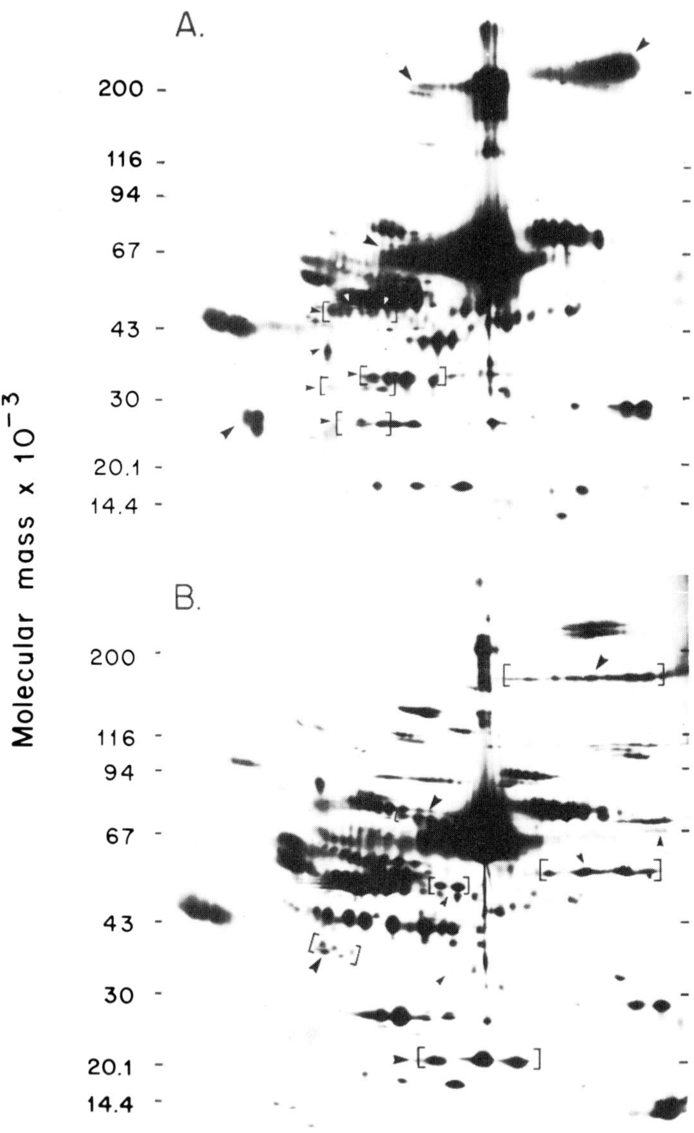

Figure 1. Two-dimensional pattern of AF and MP polypeptides. Aliquots containing 100 μg of protein from albumin-depleted AF (Gel A) or MP (Gel B) obtained at 19 weeks of gestation were analyzed by two-dimensional gel electrophoresis. Arrows on Gel A mark the position of polypeptides or polypeptide groups detected only in AF. Arrows on Gel B indicate MP polypeptides not detected in AF (Taken from reference 3).

(8-10), or by two-dimensional analysis of purified or immunologically identified proteins. Of these 28 identified serum proteins, 21 were also present in AF.

A composite diagram (Figure 2) depicts the typical pattern of AF polypeptides present at 15-21 weeks of gestation superimposed on the polypeptides detected in MP. This drawing shows the general similarities and differences observed from inspection of 58 individual sample pairs but does not display all observed polymorphisms. The identities of numbered AF polypeptides are listed in the legend to Figure 2.

In addition, ten major polypeptides or polypeptide groups, designated AF_1 to AF_{10} and indicated in Figure 2, were consistently present in all 58 mid-gestational AF samples examined and were not detected in matched MP samples containing equal amounts of total protein (3). The characteristics of AF_1-AF_{10}, such as their apparent M_r and the number of polypeptides per group are shown in Table 1, together with the results of studies of TAF, CS and FU for the presence of components with similar mobilities.

Two of these AF components have been identified immuno-logically. AF_1 is a form of fibronectin distinct in mass and appearance from the species observed in maternal plasma or in preparations of purified plasma fibronectin. AF_3 is α-fetoprotein, a well known fetal protein. AF_4 has been tentatively identified as a discrete form of α_1-antitrypsin which was not detected in MP (11). Seven polypeptides listed in Table 1 are still under investigation. Five of these were not found in either FU of CS, suggesting that their source may be fetal membranes (amnion and chorion), integument, or the respiratory or gastrointestinal tracts. It is possible that AF_7 may be a previously described fetal lung surfactant protein with a reported mass of 34 Kd (12), and AF_{10} a 25-Kd uterine protein which is present at mid-gestation but markedly

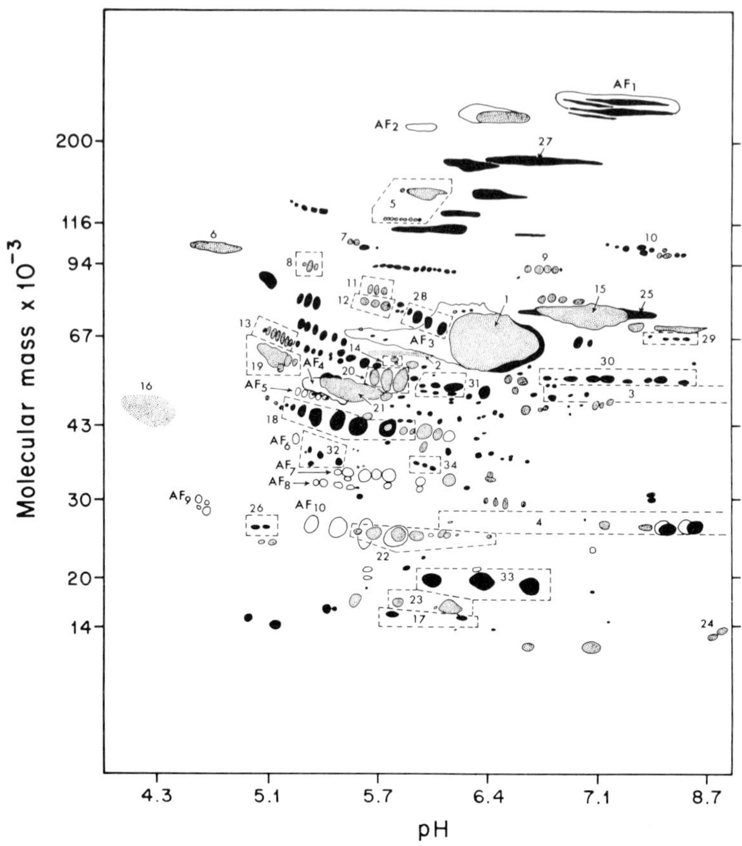

Figure 2. Composite diagram of polypeptides in mid-gestational
AF and MP showing two-dimensional pattern of AF polypeptides
(drawn in white) superimposed on those detected in MP (drawn in
black). Constituents detected in both samples are shaded in
gray. The polypeptides identified in AF are: (1) albumin, (2)
immunoglobulin A α chain, (3) immunoglobulon G γ chain, (4)
immunoglobulin κ and λ chains, (5) ceruloplasmin, (6) Cls
inhibitor, (7) α_1-antitrypsin dimer, (8) Cls, (9) C_3
activator, (10) plasminogen, (11) prothrombin, (12) α_1B-
glycoprotein, (13) α_1-antichymotrypsin, (14) anti-thrombin
III, (15) transferrin, (16) α_1-acid-glycoprotein, (17)
haptoglobin α' chain of 1-1 and 2-1 forms, (18) haptoglobin β
chain, (19) α_2HS glycoprotein, (20) GC globulin, (21)
α_1-antitrypsin, (22) apolipoprotein AI, (23) prealbumin
and (24) hemoglobin. Polypeptides numbered 25-34 were
detected only in maternal plasma (Taken from reference 3).

TABLE 1: AMNIOTIC FLUID POLYPEPTIDES NOT DETECTED IN MATERNAL PLASMA

Polypeptide	M_r (Kd)	Polypeptides Per Group	Also Detected In		
			Cord Serum	Fetal Urine	Term Amniotic Fluid
AF_1 (fibronectin)	221–225	1	–	–	+
AF_2	208	1	–	–	–
AF_3 (α-fetoprotein)	70	6–7	+	+	+
AF_4	55	4–5	N.D.	N.D.	N.D.
AF_5	51	5–7	–	–	–
AF_6	43	1	–	–	–
AF_7	33.5	4–6	–	–	+
AF_8	31	4–5	–	+	+
AF_9	27	1–5	trace	+	trace
AF_{10}	25	2–4	–	–	–

N.D. = Not determined
(Taken from Reference 11)

diminished at term (13). One AF polypeptide, AF_9, was present in CS and FU, suggesting it may be a fetal serum protein like α-fetoprotein, and one other, AF_8, was present in FU but not detected in CS, pointing to the urinary tract as a possible source. We believe that these data justify continued study of the soluble proteins in AF, both for better understanding of normal fetal development and for the detection of fetal abnormalities.

Acknowledgements

We wish to thank Dr. Khalil M. Tabsh of the Department of

236

Obstetrics and Gynecology, University of California Medical Center, Los Angeles, who contributed term amniotic fluid, cord serum and neonatal urine samples. This work was supported in part by grant HD 5615 from the National Institutes of Child Health and Human Development and by U. S. P. H. S. grant RR 05756.

References

1. Sutcliffe, R. G.: Biol. Rev. Camb. Philos. Soc. 50, 1-33 (1975).

2. Campbell, S.:J. Obstet. Gynecol. Br. Commonw. 75, 568-576 (1968)

3. Kronquist, K. E., Crandall, B. F. and Cosico, L. G. Manuscript submitted for publication (1983).

4. Utermann, G., Albrecht, G. and Steinmetz, A.: Clin. Genetics 14, 351-358 (1978).

5. Tuszynski, G. P., Knight, L., Piperino, J. R. and Walsh, P. N.: Anal. Biochem. 106,118-122 (1980).

6. O'Farrell, P. H.: J. Biol. Chem. 256, 4007-4021 (1975).

7. Merril, C., Goldman, D. and Van Keuren, M. L.: Electrophoresis 3, 17-23 (1982).

8. Anderson, N. L. and Anderson, N. G.: Proc. Natl. Acad. Sci. USA 74, 5421-5425 (1977).

9. Tracy, R. P., Carrie, R. M. and Young, D. S.: Clin. Chem. 28, 890-899 (1982).

10. Tracy, R. P., Carrie, R. M., Kyle, R. A. and Young, D. S.: Clin. Chem. 28, 900-907 (1982).

11. Kronquist, K. E., Crandall, B. F. and Cosico, L. G. Manuscript submitted for publication (1983).

12. King, R. J., Ruch, J., Gikas, E. G., Platzer, A. C. and Creasy, R. K.: J. Appl. Physiol. 39, 735-741 (1975).

13. Sutcliffe, R. G., Bolton, A. E., Sharp, F., Nicholson, L. B. and MacKinnon, R.: J. Reprod. Fert. 58, 435-444 (1980).

CHARACTERIZATION OF SERUM PROTEINS INDUCED BY PARTIAL HEPATECTOMY

Tsuyoki Kadofuku, Takeru Iijima, and Tsuneo Sato
Central Chemical Laboratory, School of Medicine,
Showa University, 1-5-8 Hatanodai, Shinagawa-ku,
Tokyo 142, Japan

Introduction

The factors that control the proliferation of liver cells
after partial hepatectomy in rat are not yet known with
certainty. A series of specific substances are said to effect
the proliferation of liver cells, but metabolic changes
possibly also act as a contributory causative factors in the
stimulation of those substances. To define these relationships, the changes in the serum concentration of various
metabolites after partial hepatectomy have been studied (1-4).
However, as for the changes in serum protein distribution,
little is known except for some proteins (2-5). We investigated changes in protein distribution in rat serum after
partial hepatectomy by means of polyacrylamide gradient gel
and two-dimensional electrophoresis, and detected newly
appearing as well as a greatly increase in others. In this
report, we describe the time-dependent changes and some
molecular properties of these proteins.

Materials and Methods

Hepatectomy and serum samples

Male rats (Wistar strain, 150g) were used. Partial
hepatectomy was performed under ether narcosis according to

the method of Higgins and Anderson (6). After hepatectomy the animals were maintained for 0-240 h. Food and water were provided _ad_ _libitum_. At scheduled time intervals blood was taken from the descending aorta of each rat. The blood was left standing at 4°C for 2 h, and then centrifuged at 3,000 x g for 10 min. Sucrose was added to the supernatant serum to attain a concentration of 10%, and the serum samples were stored at -20°C until use.

Electrophoresis

Electrophoresis was performed on a 4-10% polyacrylamide linear gradient (containing a 0-5% sucrose linear gradient) slab gel. Two-dimensional electrophoresis was carried out according to the method of Manabe _et_ _al._ (7).

Preparation of antiserum and rocket immunoelectrophoresis

The protein induced by partial hepatectomy was purified from rat serum taken 24 h after operation. The purification procedure will be presented elsewhere. A rabbit (Japanese White, 2 Kg) was injected with 2 mg of the purified protein in an equal volume of Freund's complete adjuvant. Two more injections (0.5 mg each) were given at one-week intervals. Two weeks after the last injection the rabbit was bled from the carotide artery. Rocket immunoelectrophoresis was performed according to the method of Laurell (8).

Results and Discussion

Fig. 1 shows a pattern of the polyacrylamide slab gel electro-phoresis of rat serum proteins before and after partial hepatectomy. The positions showing a large increase in protein after hepatectomy are shown by arrows in the figure. Apparently this protein increases greatly within 24-48 h after

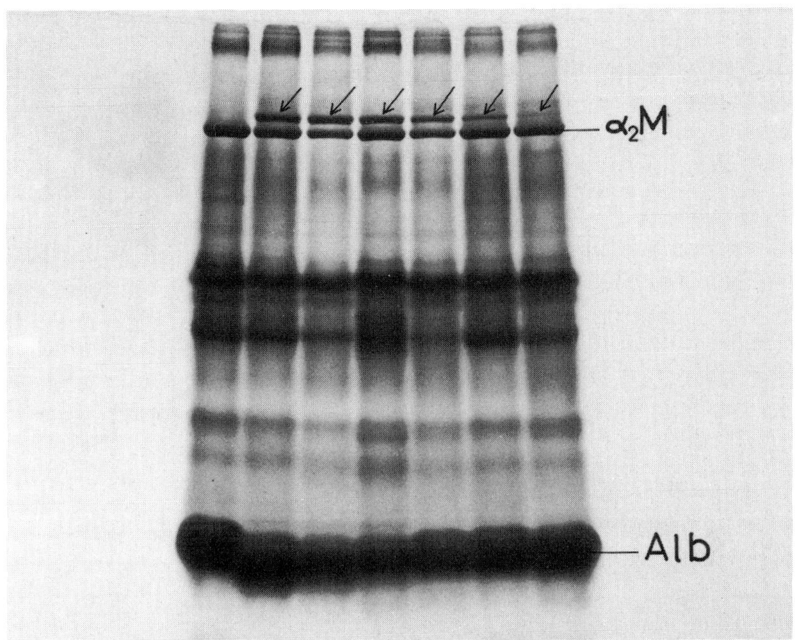

Fig. 1. Polyacrylamide slab gel electrophoresis of serum
proteins before and after partial hepatectomy. Serum samples;
from left to right, 0, 24, 48, 72, 96, 168, and 240 h after
hepatectomy. The protein which increased after hepatectomy
is shown by arrows. $\alpha_2 M = \alpha_2$-macroglobulin, Alb = albumin.

hepatectomy, gradually diminishing, and almost disappearing at
240 h. In order to study the time-dependent changes of this
protein in detail, its serum concentration was quantitated by
means of immunoelectrophoresis. Results are shown in Fig. 2.
This protein increased within 3-6 h after hepatectomy, reached
a maximun within 24-48 h (its concentration was about 1 mg/ml),
and gradually disappeared. The change of regenerative rate of
the remnant liver cells after partial hepatectomy have been
reported by Tanaka et al. (9), and it seemed that the regenerative rate parallels the concentration of this protein,

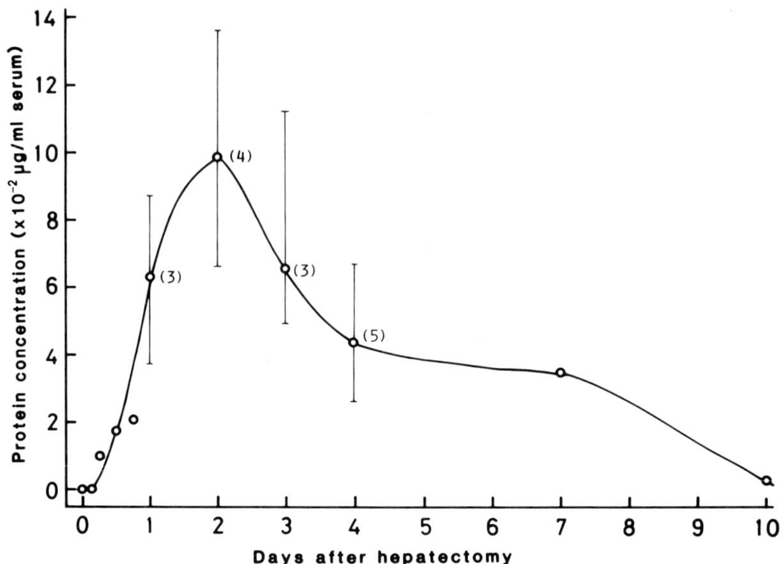

Fig. 2.   Time-dependent change of the protein induced by
partial hepatectomy.   Quantitation was carried out by means of
rocket immunoelectrophoresis.   Numerals in parentheses show
the number of animals.

suggesting that this protein is closely related to the
proliferation of the remnant liver cells after partial
hepatectomy.

To investigate whether this protein is a newly synthesized
substance after hepatectomy, the influence of cycloheximide
administration was investigated.   Cycloheximide (dissolved in
0.9% NaCl) was injected into rats immediately after partial
hepatectomy and serum concentration of the protein was
measured 24 h later by means of rocket immunoelectrophoresis.
Results are summarized in Table 1.   The amount of protein in
serum of rats induced by administration of cycloheximide was
significantly lower than in sera of control rats, suggesting

Table 1. Changes of induced protein level in rat serum by cycloheximide administration.

	Protein concentration (µg/ml serum)	
Control	628 $\pm$ 184	(n = 3)
Cycloheximide		
250 µg/Kg	382	(n = 1)
500 µg/Kg	275	(n = 1)

that this protein was newly synthesized after partial hepatectomy.

Next, by means of two-dimensional electrophoretic technique, we analyzed the serum proteins after partial hepatectomy. Fig. 3 shows one of the two-dimensional electrophoretic patterns of rat serum proteins taken at 24 h after hepatectomy. Three proteins either greatly increased in quantity or newly appeared were detected, indicated by arrows in the figure. We named those proteins Protein A, B, and C, in order of their isoelectric points. Judging from the electrophoretic mobility, it seemed that both Protein A and Protein B corresponded to the protein marked in Fig. 1 by arrows. Indeed, Protein A and Protein B reacted with our antiserum, and no antigenic differences were apparent. The time courses of the changes in the two-dimensional patterns of Proteins A, B, and C were examined. Serum samples taken at 24, 48, 72, and 96 h after hepatectomy were subjected to two-dimensional electrophoresis. In Fig. 4, the time-dependent changes at two-dimensional gel sections of area 1 and area 2 (indicated in Fig. 3) are shown. It can be seen that Protein A and B rapidly increased after hepatectomy but had almost returned to their original level at 96 h. Protein C which did not exist in normal rat serum, appeared within 24 h after hepatectomy, gradually decreased, and had almost disappeared at 96 h.

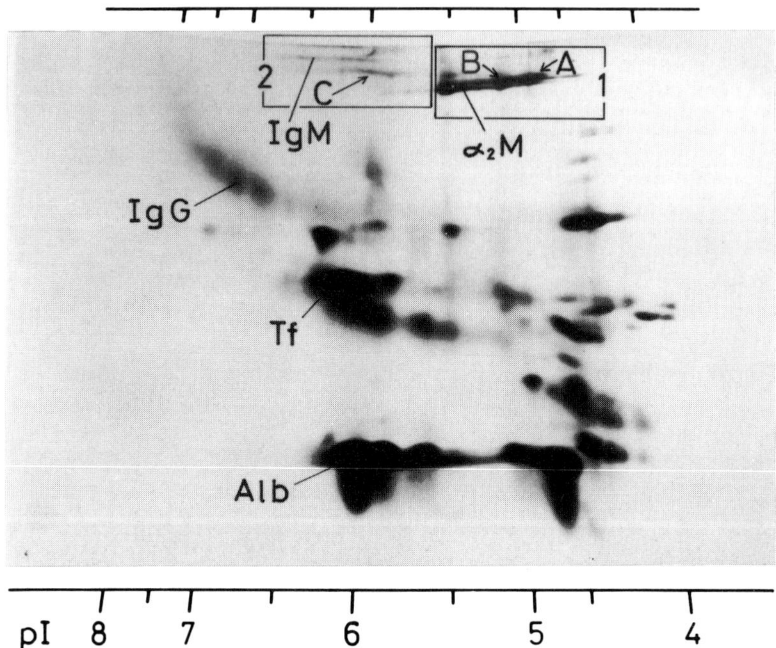

Fig. 3. Two-dimensional electrophoretic pattern of rat serum proteins taken 24 h after partial hepatectomy. Proteins greatly increased in quantity or newly appeared are shown by arrows (Proteins A, B, and C). IgG = immunoglobulin G, IgM = immunoglobulin M, α_2M = α_2-macroglobulin, Tf = transferrin, Alb = albumin.

We investigated some characteristics of these proteins. The molecular weights of Proteins A and B were estimated to be about 800,000, and their isoelectric points were 4.9 and 5.1, respectively. Proteins A and B were stained with periodic acid-Schiff reagent but not with Sudan black. The carbohydrate content of Proteins A and B were estimated to be 5% (weight % of glucose equivalent to the total weight) by the phenol-sulfuric acid method. These proteins are also found in AH-66 and AH-130 tumor-bearing rat serum. Therefore, we

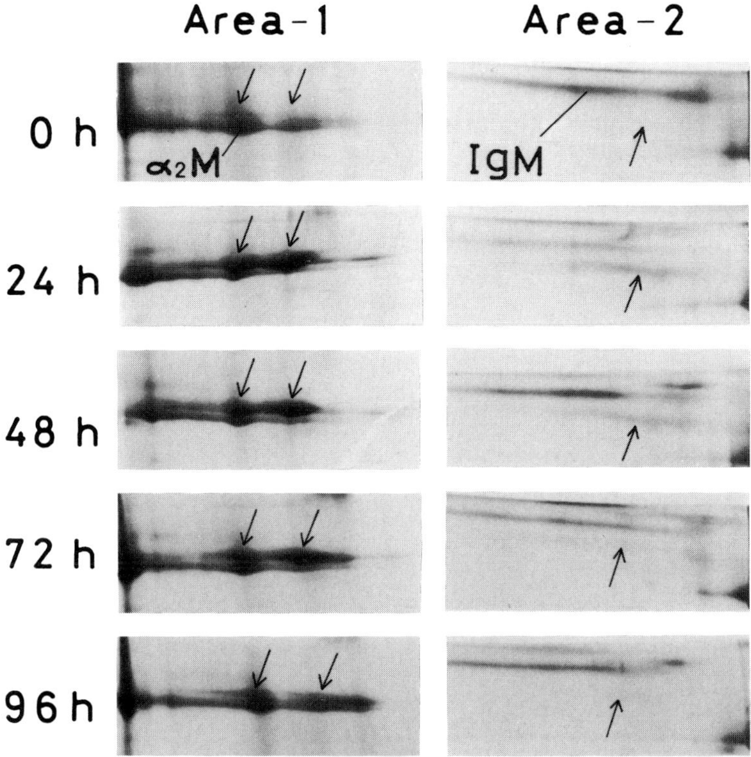

Fig. 4. Time courses of the changes in the two-dimensional patterns of Proteins A, B, and C after partial hepatectomy. Serum samples (50 µl each) taken at 0, 24, 48, 72, and 96 h after partial hepatectomy were analyzed by two-dimensional electrophoresis. Time-dependent changes at gel sections of areas 1 and areas 2 (indicated in Fig. 3) are shown. Protein positions are indicated by arrows. α_2M = α_2-macroglobulin, IgM = immunoglobulin M.

244

suggest that these proteins are closely related to cell
proliferation. Protein C showed a relatively wide pI range
(pI 5.5-6.5) and had an apparent molecular weight of about
900,000. Protein C was stained neither with periodic-Schiff
reagent nor Sudan black. We could not detect Protein C either
in AH-66 or in AH-130 tumor-bearing rat serum. This protein
may play an important role only the proliferation of the
normal remnant liver cells after partial hepatectomy.

References

1. Morley, C.G.D. and Kingdon, H.S.: Biochim. Biophys. Acta
   308, 260-275 (1973).

2. Strecker, W., Silz, S., Salem, A., and Ruhenstroth-Bauer,
   G.: Horm. Metab. Res. 12, 604-608 (1980).

3. Nagasue, N., Kobayashi, M., Iwai, A., Yukawa, H.,
   Kanashima, R., and Inokuchi, K.: Cancer 41, 435-443 (1978).

4. Coetzee, N.L., Short, J., Klein, K., and Ove, P.: Cancer
   Res. 42, 155-160 (1982).

5. Weesner, R.E., Mendenhall, C.L., Morgan, D.D., Kessler, V.,
   and Kromme, C.: J. Lab. Clin. Med. 95, 725-736 (1980).

6. Higgins, G.M. and Anderson, R.M.: Arch. Pathol. 12, 186-
   202 (1931).

7. Manabe, T., Tachi, K., Kojima, K., and Okuyama, T.:
   J. Biochem. 85, 649-659 (1979).

8. Laurell, C.B.: Anal. Biochem. 15, 45-52 (1966).

9. Tanaka, Y., Nagasue, N., Kanashima, R., Inokuchi, K., and
   Shirota, A.: Cancer 49, 19-24 (1982).

AUTOMATIC EVALUATION OF ELECTROPHEROGRAMS AT HIGH RESOLUTION

Harald Kronberg, Hans-Georg Zimmer and Volker Neuhoff
Max-Planck-Institut für experimentelle Medizin
Forschungsstelle Neurochemie, D-3400 Göttingen, Germany

Introduction

A gel scanning system for more accurate and precise automatic quantitative
evaluations of one- and two-dimensional electropherograms is being designed
in co-operation with the Carl Zeiss Comp. with the support of the German
Ministry of Research and Technology. The system consists of a scanning
microphotometer, controlled by a laboratory computer, and computer programs
for the analysis of the measuring values. The photometric performance has
been optimized by minimizing the number of optical components and by
adapting the optical, electronical and digital processing to the physical
nature of the measuring signal (1). Thus the scanner is qualified as a
reference instrument, but it is expected that the test of prototypes will
reveal criteria for a simplified design of a routine scanner e.g. for
clinical applications.

System Specifications

Fig. 1 is a sketch of the scanning microphotometer hardware including
analog and digital electronic components. The electropherogram is put on
a computer-controlled scanning stage, whose starting position can be
interactively readjusted with the aid of a tracker-ball (TRB).

Original gel-electropherograms up to 5 mm thick having a maximum size of
250 mm x 250 mm are scanned meandering-like at a step size of 0.1 mm in
both x- and y-direction and at a frequency of 2000 Hz equivalent to a

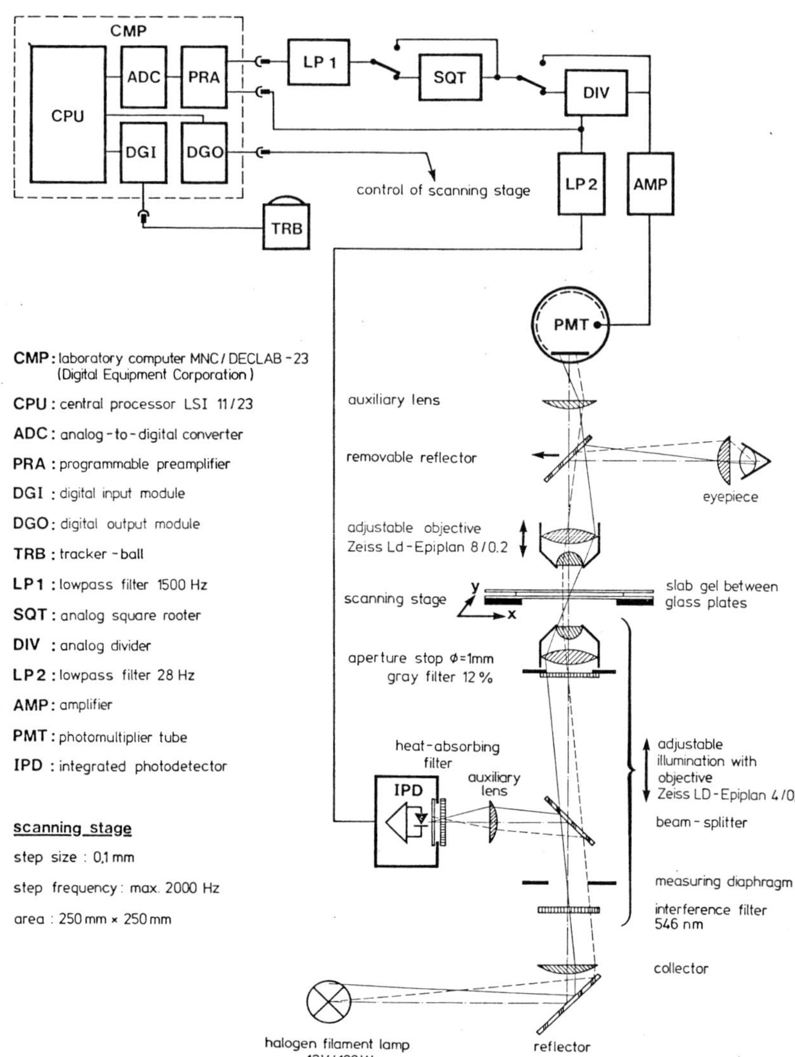

Fig. 1: Set-up of the computer-controlled scanning microphotometer

scanning speed of 200 mm/s. The sampling grid can be refined along the
lines down to 0.02 mm for the acquisition of highly resolved one-
dimensional electropherograms. Fig. 2 illustrates how overlapping circular
measuring spots about 2.5 times as large in diameter as the step size are
imaged into the gel, thus optimizing the reduction of contrast-impairing
glare and the reproducibility of quantitative results (2). Filtered
incoherent light is used to allow photometric quantification according to
Lambert-Beer's law.

The irradiation of the light source is continuously monitored by a
photodiode (IPD) in order to compensate illumination-induced fluctuations
of the photomultiplier (PMT) output current (3). Companding the electrical
measuring signal by an analog square root transformation (SQT) and
subsequent smoothing by a lowpass filter (LP1) is the optimal means of
noise reduction (4). Therefore all 4096 possible gray levels produced by
the 12 bit analog-to-digital converter (ADC) are photometrically
distinguishable, which is the basis of precise absorbance measurements in
the range of 0.0001 to 3.6.

The system software allows for a quantitative evaluation of two-dimensional
electropherograms based on implicit modeling of protein spots (1). It will
be extended to the semi-automatic comparison of different electropherograms
by specifying a few landmarks which facilitate spot matching based on

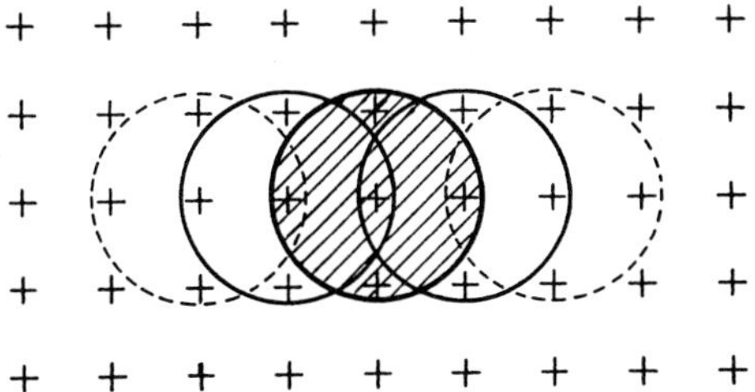

Fig. 2: Scanning grid and overlapping apertures

interspot distance relations (5). During the next developmental stage a
procedure for two-dimensional evaluations of one-dimensional electropherograms will be implemented. As an advantage over the usual one-
dimensional evaluation by slit-scanners and simple integrators even
transversally expanded highly loaded (6) or heavily distorted protein bands
are adequately taken into account. This avoids the distribution error
resulting from inhomogeneous absorbance within the slit aperture of one-
dimensional scanners. Hence the scanning microphotometer system also meets
clinical requirements of a more accurate protein analysis.

References

1. Kronberg, H., Zimmer, H.-G., Neuhoff, V.: Electrophoresis 1, 27-32 (1980)
2. Linge, H., Zimmer, H.-G., Neuhoff, V.: Proc. 6th ICPR, IEEE-Nr. 82CH1801-0,
        594-596, Munich 1982
3. Kronberg, H., Zimmer, H.-G., Neuhoff, V.: Microscopica Acta 82, 223-228
        (1979)
4. Zimmer, H.-G., Kronberg, H., Bernstein, R., Neuhoff, V.: Pattern Recog-
        nition 13, 79-82 (1981)
5. Pardowitz, I., Zimmer, H.-G., Neuhoff, V.: A least-squares method for
        spot identification in two-dimensional chromatograms.
        Submitted to Anal. Biochem. (1983)
6. Tamura, S., Tanaka, K., Miki, S., Okamoto, Y., Satake, A., Okuda, M.:
        Proc. 6th ICPR, IEEE-Nr. 82CH1801-0, 1176-1178, Munich 1982

ANALYSIS OF CULTURED FIBROBLASTS FROM PATIENTS WITH TRISOMY
18 USING ELECTROPHORETIC TECHNIQUES

Surjit Singh, Ingrid Willers, H. Werner Goedde
Institute of Human Genetics. University of Hamburg
Butenfeld 32, D-2000 Hamburg 54. F.R.G.

Introduction

Cultured human fibroblasts offer excellent material for
prenatal and postnatal detection of cytogenetic and
inherited metabolic defects. Until now such analysis has
been performed mostly by examining the different loci for
protein synthesis in seperate experiments, whereby the
majority of the proteins studied has been enzymes detected
by specific histochemical means or direct activity assay.
In order to get more knowledge about a particular genetic
defect the analysis of a large spectrum of cellular structural and functional proteins would be a worthwhile approach.
In recent years it has been realised by employing the techniques of high resolution protein mapping (Klose,1975;
O'Farrell,1975) and their recent extensions (Anderson and
Anderson,1977; McConkey ,1979; Singh et al. 1981). We have
been engaged in these studies of protein mapping of cultured
human cells in health and disease by dye staining technique
(Singh et al. 1978,1980,1981; Willers et al.1981; Klose et
al.1983). In continuation of our studies, in this report
we present further analysis of protein patterns from normal
individuals and patients with free trisomy 18. This particular trisomy, also known as Edwards Syndrom, has an incidence
of about 1:8000 with 80% of them being females. The postnatal
survival is rather poor, mean being two months.As in other
trisomies the maternal age is usually advanced.In the present
study additional methods of radiolabelling the cells and

monitoring the gels by autoradiography have also been incorporated to assess the degree of variation.

Methods and Material

1) Cell Cultures. Apart from 22 normal individuals reported
before,fibroblast cell lines have been set up from 5 patients
with trisomy 18 in Hams F-10 culture medium using upper arm
biopsy material and the cultures were handled according to
standard procedure reported before (Singh et al.1981).

2) Electrophoresis. The standardised procedures employed for
isoelectric focussing, SDS electrophoresis and two dimensional mapping have been published before (Singh et al.1981).
Only the gel dimensions have been changed to 12 x 12 cm. and
the focussing time has been increased by one hour.

3) L-^{35}S Methionine labelling. The growing cultures have
been labelled by adding 50 uci/ml of the radiolabel in
each culture well containing F-10 medium without methionine
for 24 hours. After washing the cells three times with the
physiological saline, they were lysed in buffer A of O'Farrel
(1975). 250,000 cpm were applied to each IEF gel of Ampholine
pH 4-6. The approprite regions were cut out and laid onto
SDS gel of gradient 10% to 20% to seperate the spots only
in this region. The autoradiography was performed by using
Enlightening(NEN) and Kodak Xomat film.Prior to application
the lysate was centrifuged 60000 RPM in Airfuge.
4) Evaluation. The abscence or presence of a new spot or
change in the mobility of a spot distinguished on a light
box has been referred as qualitative diferrence, while changes in the density has been referred to as quantitative
variation. The spots showing suspected differences were
quantitated manually by measuring their density in two
perpendicular directions under LKB laser beam densitometer.

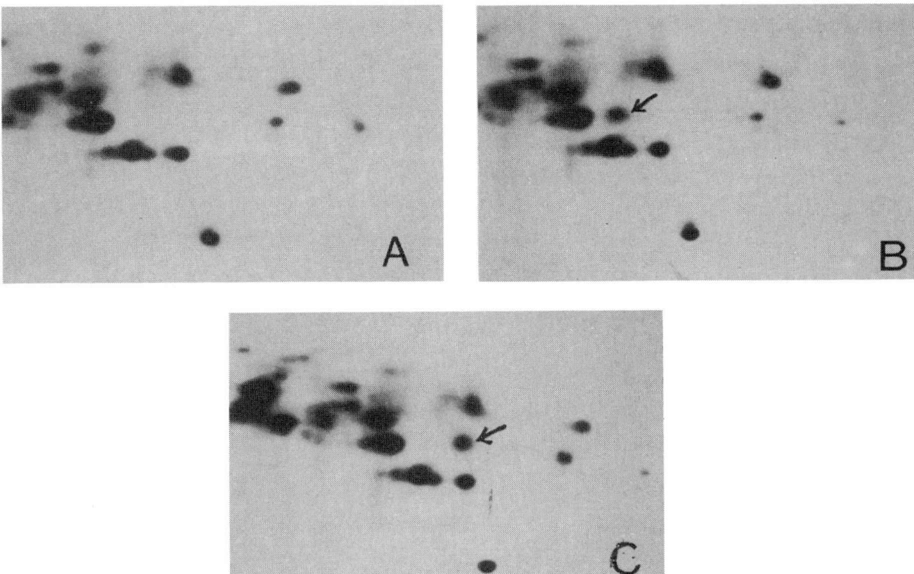

Figure 1. Autoradiographs of radioactive fibroblast extracts
from two patients with trisomy 18 each showing a variant
spot (C&B) in comparison to normal pattern (A).The relevant
region is enlarged for a better comparison.The arrows indicate the variant spots.

Results and Discussion

With these techniques very well seperated protein patterns
have been obtained from the cell lysates.560 spots have
been evaluated on the 2D gels. The interindividual variation
lies in the range of the values(2%to3%) reported earlier
by us for the qualitative variants.The lower value corresponds to autoradiographic analysis and may be due to smaller
sample examined so far(12).No differences specific to trisomy 18 have been found in single dimension IEF or SDS electrophoresis as apparently the variant spots found by 2D technique are masked by the other proteins.By the 2D techniques
in one case of trisomy 18 individual, the presence of

an extra spot(B) reported before by us (Willers et al.
1981) has been again confirmed in the same patient(fig.1).
In another patient an additional spot(C) in the same region
but not at the same place has been observed.The molecular
weight corresponds to the spot(B) and differs from it slightly
in the isoelecric point,that being more on the basic side.
In other 3 patients no such spot has been observed and the
pattern is like in healthy controls(A).Apart from normal
controls, the patients with trisomy 21 and trisomy 13 and
some other genetic defects have not shown such or other
qualitatively changed spots (Singh et al.1981a).It may be
possible that these extra spots found in trisomy 18 are
examples of very rare polymorphism. Therefore, the interpretation has been reserved pending the family studies with
respect to them. Regarding the quantitative differences,no
variants pertaining specifcally to this trisomy, and present
in all the patients, has been found,as considerable unspeci-
fic interindividual variation in the intensity of the spots
has been noted.In this respect the results in this trisomy
are in conformance to those reported for trisomy 21 by weil
and Epstein(1979) and Klose et al(1982).

References

1. Anderson,L., Anderson,N.G.: Proc.Nat.Acad.Sci.74,5421(1977)
2. Klose,J.: Hum.Genet.26, 211-234 (1975).
3. Klose,J.,Zeindel,E.,Sperling,K.: Clin.Chem.28,987-992(1982)
4. Klose,J.,Willers,I.,Singh,S.,Goedde,H.:Hum.Genet.inpress
5. McConkey,E.H.: Anal.Biochem. 96, 39-44 (1979).
6. O'Farrell, P.H.: J. Biol. Chem. 250, 4007-4021 (1975).
7. Singh,S.,Klose,J.,Willers,I.,Goedde,H.W.:Electrophoresis
8. Singh,S.,Willers,I.,Klose,J.,Goedde:Biochem.Genet.19,871-78:
9. Singh,S.,Willers,I.Klose,J.,Goedde,H.:J.Inh.Met.Dis,4,77
10. Weil,J.,Epstein,C.J.: Am.J.Hum.Genet.31,478-488(1979)(1981)
11. Willers,I.,Singh,S.,Goedde,H.W.,Klose,J.:Clin.Genet.20,
                                            217-221(1981)

DIFFERENCE IN ANDROGEN-DEPENDENT CHANGES OF SUBCELLULAR
PROTEINS BETWEEN VENTRAL AND DORSOLATERAL LOBES OF RAT
PROSTATE AS DETECTED BY POLYACRYLAMIDE GEL ELECTROPHORESES

Yuhsi Matuo, Nozomu Nishi, Yukio Tanaka, Yasuyoshi Muguruma
and Fumio Wada
Department of Endocrinology, Kagawa Medical School
1750, Miki-cho, Kita-gun, Kagawa 761-07, Japan

Introduction

The rat prostate is composed of a paired ventral lobe and a
dorsolateral lobe. The two lobes depend on the level of
androgen for differentiation and maintenance of their
structures and functional properties. The present work was
undertaken to clarify the difference in androgen-dependent
changes between the ventral lobe (VL) and the dorsolateral
lobe (DL) of adult rats (14-15 weeks of age), using polyacrylamide gel electrophoreses.

Results

1. The tissue weights were decreased by castration, and
   restored to control level by replacement of androgen, in
   the two lobes. The extent of the decrease was greater,
   and that of restoration lower, in the VL than in the
   DL (Table 1).
2. Cytosol and nuclear fractions were prepared in the
   presence of 1 mM PMSF (a potent inhibitor of serine
   protease) (1), and subjected to SDS-polyacrylamide slab
   gel electrophoresis. i) Cytosol proteins (Fig. 1): A
   glycoprotein having a molecular weight of about 16,000

Table 1. Androgen-dependent changes in tissue weights and the most abundant protein species in the cytosol and nuclear fractions from the VL and DL of adult rats.

| Rat | Tissue weight(%) [*2] | | Most abundant protein (%) [*1] | | | |
| | | | Cytosol fraction | | Nuclear fraction | |
group [*3]	VL	DL	16K/5.6 in VL	67K/6.3 in DL	20K-NHP in VL	20K-NHP in DL
Control	100	100	100	100	100	100
C-2	91	87	57	78	98	72
C-5	36	61	20	60	90	48
C-8	21	44	6	41	56	44
C-8/T-3	41	80	14	60	92	72
C-8/T-11	100	140	46	128	160	92

[*1] Relative content/100 g body wt. for the cytosol and relative content/DNA for the nuclei, respectively.
[*2] Tissue weights of control rats were 114±17 mg/100 body wt. for the ventral lobe (VL) and 66±11 mg/100 body wt. for the dorsolateral lobe (DL), respectively.
[*3] One group was composed of 4 rats for control, C-2 (castrated for 2 days), C-5 and C-8/T-11 (C-8 rats killed at 11th day after daily injection of 2 mg testosterone propionate in 0.2 ml of sesame oil).

and pI of 5.6 (16K/5.6) and a protein of 67K/6.3 were most abundant in the VL and in the DL, respectively (Fig. 1). The content of 16K/5.6 in the VL was more rapidly decreased by castration than was 67K/6.3 in the DL, whereas the content of 67K/6.3 in the DL was more rapidly restored by replacement of androgen than was 16K/5.6 in the VP. These changes were in accord with those in the tissue weights. ii) Nuclear proteins: The histone content was fundamentally alike in the two lobes. Of non-histone proteins, a species having a molecular weight of about 20,000 and pI of about 11.5 (20K-NHP) was most abundant in both prostates. The 20K-NHP content was more rapidly decreased by castration, but more slowly restored by replacement of androgen, in the DL than in the VL (Table 1). The 20K-NHP content was highest in the DL among various organs including accessory sex organs (Table 2);

Fig. 1. Two-dimensional electrophoretic separation of cytosol proteins from the ventral and dorsolateral lobes of adult rats.

Ventral lobe:

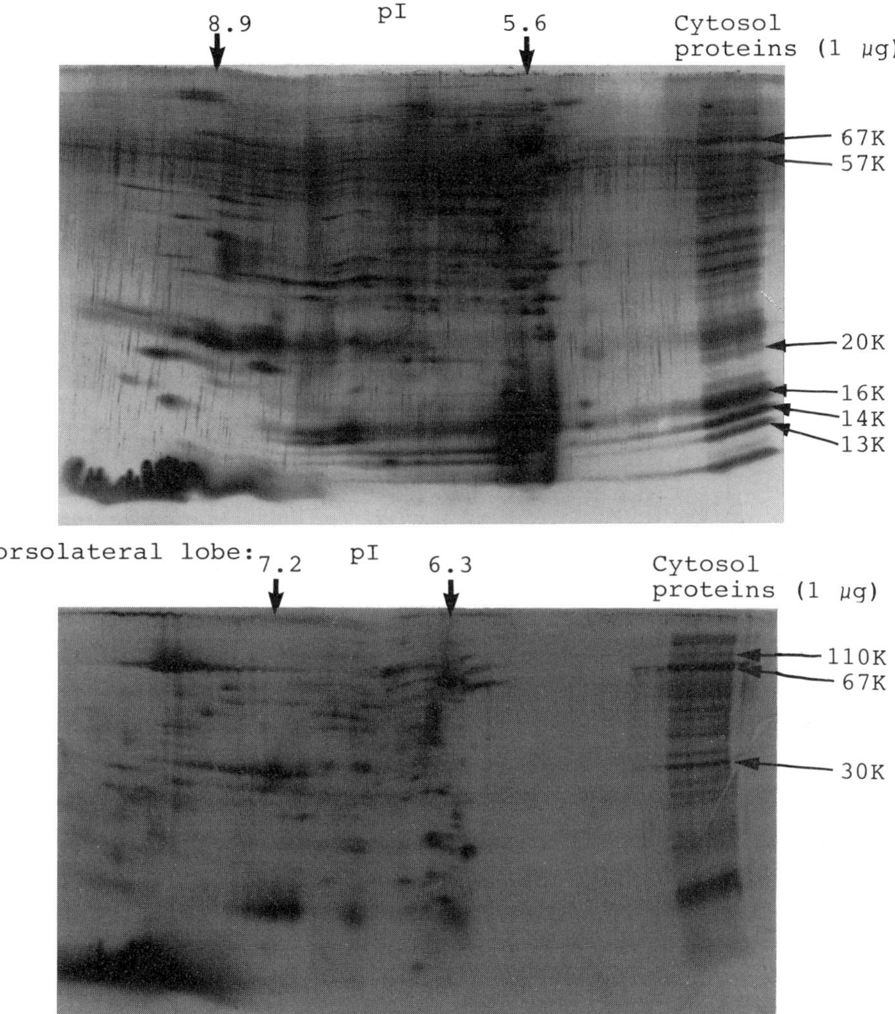

Cytosol proteins (6.6 µg) were subjected to micro disc gel (0.8mm x 6mm) isoelectric focusing and subsequent micro SDS-polyacrylamide slab gel (85mm x 50mm x 1mm) electrophoresis. The gels were stained by silver stain method. pI values indicated were estimated by O'Farell's method, using Coomassie Brilliant Blue R-250 stain.

256

Table 2. Relative content of 20K-NHP in nuclei from various
organs of adult rats.

Organ	Relative content of 20K-NHP/DNA (%)
Dorsolateral lobe	100
Coagulating gland	16
Ventral lobe	5.5
Seminal vesicle	2.1
Other organs*	not detected

* Brain, liver, kidney, spleen and thymus of adult rats.

the 20K-NHP content in the DL was at least 17 times higher
than that in the VP. The 20K-NHP in nuclei of both lobes
was effectively extracted with 0.35 M NaCl, and isolated in
homogeneous form by chromatography on a CM-Sepharose column
(2). The amino acid composition of the purified 20K-NHP was
distinctly different from those of various kinds of histones
including H1 histone, its variants and high mobility group
non-histone proteins (2). These results indicate that the 20K-
NHP, abundantly localized in the dorsolateral lobe, is a novel
species of non-histone protein with androgen dependency.

Conclusion

The most abundant species in the cytosol fractions was
16K/5.6 in the ventral lobe and 67K/6.3 in the dorsolateral
lobe. The most abundant non-histone protein in the nuclear
fractions was 20K-NHP (pI=11.5) for both lobes. Androgen
dependencies of these proteins were different for the two
lobes. The 20K-NHP content/DNA was about 17 times higher in
the dorsolateral lobe than in the ventral lobe. The amino acid
composition of purified 20K-NHP indicated that 20K-NHP is a
novel species of non-histone protein.

References

1.Matuo,Y. et al.:Electrophoresis 3,293-299 (1982).
2.Matuo,Y. et al.:B.B.R.C.109,334-340 (1982).

ELECTROSTAINING OF TWO-DIMENSIONAL POLYACRYLAMIDE GEL ELECTROPHORESIS

Setsuko Jitsukawa and Hyoichiro Sakurai
Mitsubishi Yuka Laboratory of Medical Science
1-2-10, Narimasu, Itabashi-ku, Tokyo 175, Japan

Tadao Hoshino
Div. of Chemotherapy, Parmaceutical Inst., School of
Medicine, Keio Univ.
35, Shinano-machi, Shinjuku-ku, Tokyo 165, Japan

Introduction

Electrostaining is a new and simple method for the
post-staining of proteins on plyacrylamide density gradient
gel electrophoresis. In a previous paper (1), we examined 36
dyes for this method on density gradient column gels. These
dyes were considered capable of staining proteins, but the
dyes which showed cationic behavior in electrophoretical pH
(8.3) were neglected. Among the dyes that showed anionic
mobilities in pH 8.3, 5 successfully revealed 8-11 bands on
the 4-45% polyacrylamide density gradient gel column, while
the conventional acid staining method showed 10 bands in the
same system, using human serum proteins. Two-dimensional
electrophoresis (2D-PAGE) has an extremely high resolving
power for separating proteins in many biological samples (2,
3) and is widely used. Electrophoresis is separated into two
types, i.e. the system with SDS and other denaturating
agents, and that without them. We prefer the system without
denaturating agents (3) because it may provide us with
results reflecting the true biological state of proteins
more directly than the system using SDS. Here, we report the

application of electrostaining to micro-scale 2D-PAGE (4) and the results of establishing suitable conditions.

Materials and Methods

<u>Preparation of plasma</u>: Ten-mililiters of normal human blood was drawn into a vacuum tube containing 90 units of sodium hepariante, and centrifuged at 1,000 x g for 10 min. Sucrose was added to the supernatant plasma to a 40% concentration (w/v), the sample was divided into small portions and stored at -80°C until used.

<u>2D-PAGE</u>: Micro-scale 2D-PAGE was constructed with isoelectric focusing (pH 3.5-10) and density gradient gel (4-20%) electro-phoresis without denaturing agents. Conditions of electrophoresis were according to the method of Manabe <u>et al.</u> (4) with slight modifications.

<u>Electrostaining</u>: Electrostaining was performed as described previously (1), but the electric current was 5 mA/gel and the time of run was 3 hr. Special devices were not used.

<u>Conventional acid staining of the gel</u>: After electrostaining, the gel was also stained in 0.1% dye-50% methanol-10% acetic acid (w/v/v).

Destaining of the gel was done in 25% methanol-10% acetic acid (v/v), with several changes of the staining solution.

Results

1)   Preliminary examination showed that 5 dyes were available for electrostaining, i.e. Coomassie Brilliant Blue R250 (CB-R), Coomassie Brilliant Blue G250 (CB-G), Lanasol Violet 3B (LV), Remazole Turgoise Blue G (RB) and Direct Deep Black E (DB) (1). Among these, we chose CB-G

because of its high sensitivity for proteins, high
electrophoretic velocity, and clear background. Figure 1
shows a comparison of human plasma protein maps obtained
by electrostaining (a) and by the conventional acid
staining method (b). By electrostaining, most of the
protein species that could be detected by the
conventional method, were stained. However, the binding
force between the dye and protein seemed to vary
according to the characteristics of the proteins, so that
immunoglobulin G (IgG) and haptoglobin polymers (Hp) were
not detected. The shade at the bottom of the gel (Fig.
1-a) is excess dye which has not bound to proteins. The
clear area under the albumin spot implies that the dye
should come to this position has been trapped by proteins
during the electrostaining.

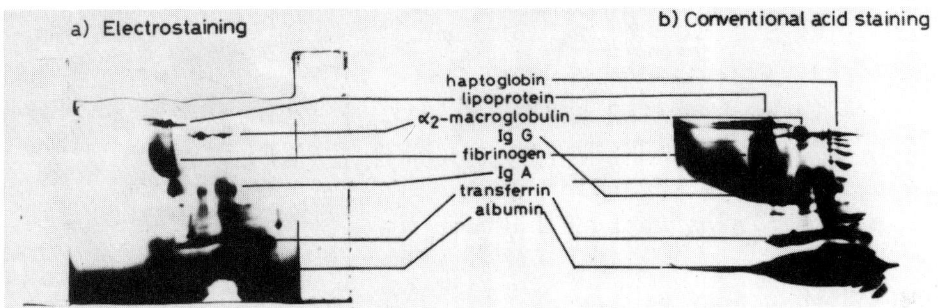

Fig. 1 Comparison of human plasma protein maps obtained
by electrostaining and by conventional acid staining.

2) Excess dye remaining at the bottom, was difficult to
    leak out from the gel. CB-G (like other dyes) has some
    hydrophobic groups in the molecule, so it may form large
    molecular weight aggregates under of the electrostaining
    conditions. To perform electrostaining more rapidly and
    to obtain a clearer map, we tested some dispersing agents
    are considered to prevent the inter-molecular hydrophobic
    interaction of the dye. Of these SDS was the most
    effective.

As shown in Fig. 2, excess dye oozed out from the gel more rapidly, when 0.5% SDS was added to the dye solution. Moreover, IgG and Hp were stained, but α_2-macroglobulin was separated into two spots differing in molecular weights. When the concentration of SDS was raised to 1%, the dye ran faster but the pattern of the map was disturbed. We suppose that SDS added to the dye solution run through the gel may result in changes in protein conformation. Thus if hydrophobic groups of proteins on the surface, the dye can contact them more easily. Because SDS is attracted by the anode, while the proteins do not move because of the pore limit effect, the period of contact of protein and SDS is relatively short. Thus, the dissociation of protein complexes are limited. SDS at such concentrations (0.5-1%) did not change the pH in the gel during electrostaining.

3) Adding 0.5% SDS to the dye solution, other dyes were also examined. Figure 3 shows the results of LV (a), RB (b), DB (c), and CB-R (d). LV and RB were less sensitive than CB-G. DB had such a high background that the protein spots were hard to distinguished from it. CB-R was as sensitive as CB-G, but an insoluble portion remained at the top of the gel. Although the dyes which have cationic mobilities originally at pH 8.3 were able to migrate anodally by the addition of SDS, they failed to bind to proteins and no protein spots were visualized.

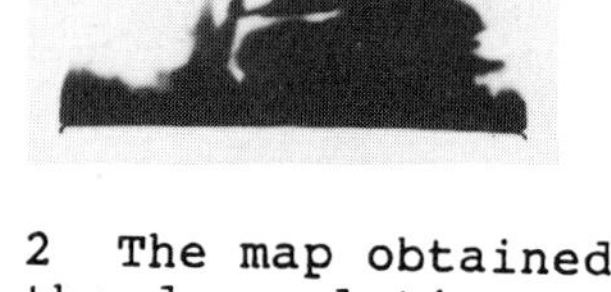

Fig. 2 The map obtained with the dye solution containing 0.5% SDS.

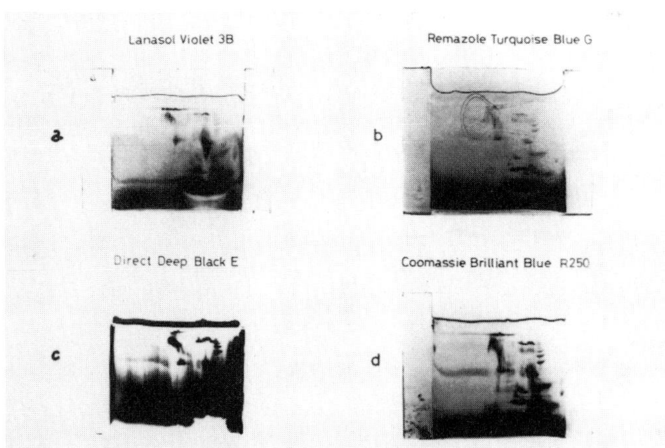

Fig. 3   Protein maps obtained by electrostaining with four kinds of dye solutions containing 0.5% SDS.

References

1. Hoshino, T., Jitsukawa, S., Tahara, M., Yamada, G., and Sakakibara, K.: Electrophoresis '81 (Eds. Allen, R.C. and Arnaud, P., Walter de Gruyter, Berlin, New York) pp. 117-125 (1981)

2. O'Farrell, P.H.: J. Biol. Chem., 250, 4007-4021 (1975)

3. Manabe, T., Kojima, K., Jitzukawa, S., Hoshino, T., and Okuyama, T.: J. Biochem., 89, 841-853 (1981)

4. Manabe, T., Kojima, K., Hayama, E., and Okuyama, T.: Clin. Chem., 28, 824-827 (1982)

TWO-DIMENSIONAL ELECTROPHORESIS OF BOVINE BRAIN PROTEINS--SOLUBLE AND
INSOLUBLE FRACTIONS--

Yuko Takahashi, Takashi Manabe, Toshihiko Kadoya, Noriaki Ishioka,
Toshiaki Isobe, Tsuneo Okuyama

Department of Chemistry, Faculty of Science, Tokyo Metropolitan
University, Setagaya-ku, Tokyo, Japan

Introduction

We have attempted the analysis of all proteins in bovine brain.
For the soluble protein analysis, we employed two types of 2-D electrophoresis technique, one was run without denaturing agents (1-4) and the
other was run using SDS only in the second dimension (5). In these 2-D
thechniques, proteins are not denatured in the focusing gels, so they are
different from that of O'Farrell's, in which 9M urea, 2% NP-40, and 5%
mercaptoethanol were used in the focusing gels (6). These techniques were
applied further for the analysis of eight fractions obtained by DEAE-
Sephadex chromatography of 'Soluble Fraction' and also for brain proteins
purified in our laboratory (7-13). Comparative analysis of brain acidic
proteins employing 2-D electrophoresis in the absence of denaturing agents
has been reported previously (4). For the insoluble protein analysis, we
studied only the 2M urea extract from Insoluble Fraction so far.

Materials and Methods

<u>Systematic purification of bovine brain proteins</u>--For total analysis of
Brain proteins, we developed a system of protein fractionation as shown
in Fig. 1. whole brain was homogenized and centrifuged as 'Soluble Fraction'
and 'Insoluble Fraction', respectively. Soluble Fraction was further
fractionated by ammonium sulfate precipitation. Three fractions were

Electrophoresis '83

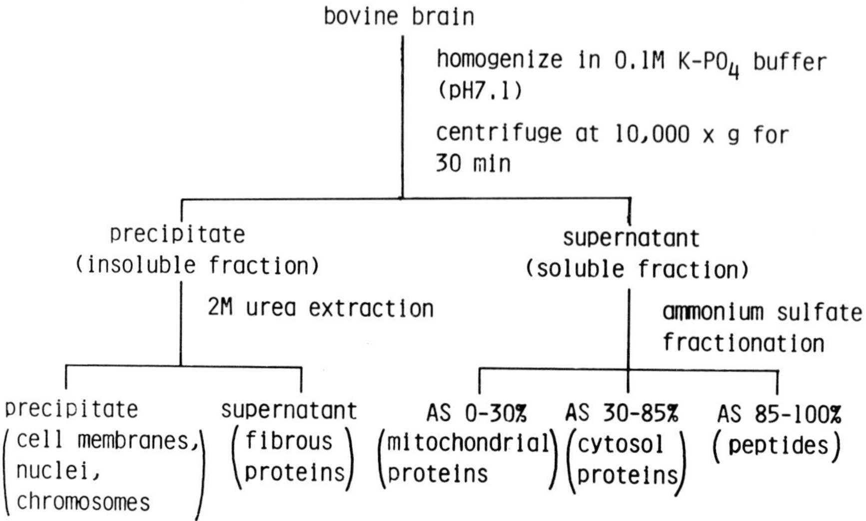

Fig. 1. Fractionation of brain proteins

obtained; the precipitate at 30% saturated ammonium sulfate (A.S.), the
precipitate at 30-85% A.S. which contained most of the cytosol proteins,
and the supernatant at 85% A.S. which is composed of polypeptides.
The precipitate of the brain homogenate, 'Insoluble Fraction', was subjected
to 2M urea extraction. The extract was rich in cellular fibrous proteins
such as glial fibrillary acidic protein (GFA), and the residue contained
nuclei, cell membrans, and chromosomes.
Each fraction was further subjected to DEAD-Sephadex A-50 column chromatography for the purification of proteins.

<u>Samples for 2-D electrophoresis</u>--An aliquot of each fraction was added sucrose
to give a concentration of 40% (w/v) and subjected to 2-D electrophoresis.
Highly concentrated soluble protein fraction for 2-D electrophoresis was
prepared as follows; Fresh bovine brain (100 mg) was mixed with 100 μl
of distilled water and homogenized with a glass homogenizer and centrifuged
at 25,000 x g for 60 min at 4 °C.

<u>Two-dimensional electrophoresis</u>--2-D electrophoresis using macro (160 x

120 x 3 mm) (1) and micro (38 x 35 x 1 mm) (3) slab gels were performed
as described previously, except 4-17% acrylamide liner gradient slab gels
(0.2-0.85% bisacrylamide gradient) were used. We also used the following
two buffer system for 2-D electrophoresis; 1) ISO-GRAD system which
employed no denaturing agents throughout the run, 2) ISO-SDS system which
employed 1% SDS in the gradient gels.

Staining density maps—Staining density maps of brain proteins were prepared as described previously (1). Six differently exposed photographs
were prepared for one slab gel and the contours of staining density were
drawn. All the density peaks were numbered.

Identification of brain proteins—Brain proteins were identified by the
following two methods; 1) Purified brain proteins were co-electrophoresed
with each brain protein fraction. 2) After 2-D electrophoresis, proteins
were electrophoretically transferred from 2-D gels to nitrocellulose
membranes and stained with specific antiserum-peroxidase staining as
described by Towbin et al. (14) with some modifications.

Quantitation of major spots on 2-D gels—The quantity of major spots on
stained 2-D gels was mesured by a TV camera-microcomputer system. Details
of the system will be described eleswhere (15).

Results and Discussion

Protein map of soluble proteins—Figure 2 shows a protein map of bovine
brain soluble proteins obtained by macro 2-D electrophoresis in ISO-GRAD
system. About 1 mg protein/ 50 μl was applied. A staining density map
of the slab gel was drawn and about 270 spots were counted on the map.
Most of the proteins (about 230 out of 270) showed apparent molecualr
weight higher than 50,000. The positions of purified proteins were shown
in the figure. Some of the proteins remained in the focusing gel at pI
5-6, and the major component was identified to be tubulin, by 2-D electrophoresis in ISO-SDS system. Streaking in the vertical direction might
show the position of lipoproteins.

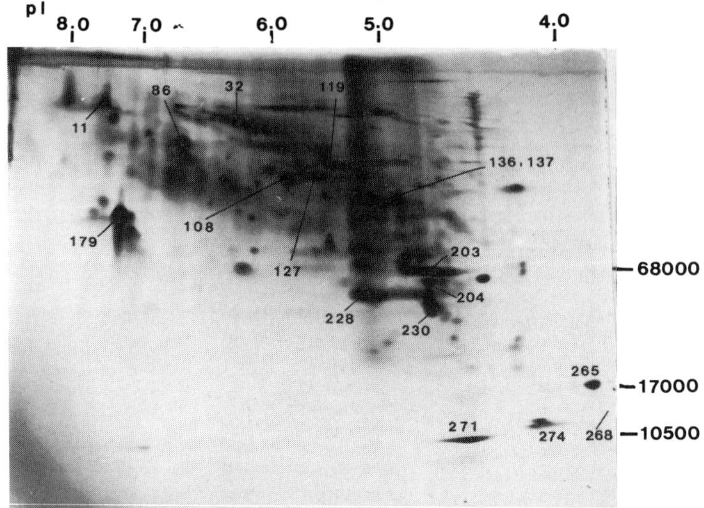

Fig. 2. A protein map of bovine brain soluble proteins. Spot numbers of
some major proteins were shown. Positions of the purified proteins were;
hemoglobin (179), albumin (203), NSE (204), calmodulin (265), micro Glu-P
(268), S-100 b (271), S-100 a (274). The mobilities of maker proteins,
bovine serum albumin (MW 68,000), calmodulin (MW 17,000), S-100 b protein
(MW 10,500) are shown.

Polypeptide map of brain soluble proteins—A 2-D pattern of bovine brain
soluble proteins, obtained in ISO-SDS system is shown in Fig. 3. Since
1% SDS was employed in gradient gel electrophoresis, the pattern represents
the map of polypeptides, not of proteins. A density map was prepared for
the gel and 401 spots were counted. Although the only difference between
the protein map (Fig. 2) and the polypeptide map (Fig. 3) was the absence
or the presence of 1% SDS in the gradient gel, about 100 spots which showed
molecular weight less than 20,000 dalton were newly appeared on the polypeptide map. Twelve purified proteins including tubulin and S-100 proteins
were located on the map. Major spots on the protein map and the polypeptide
map could be correlated according to their pI, size, and quantity.

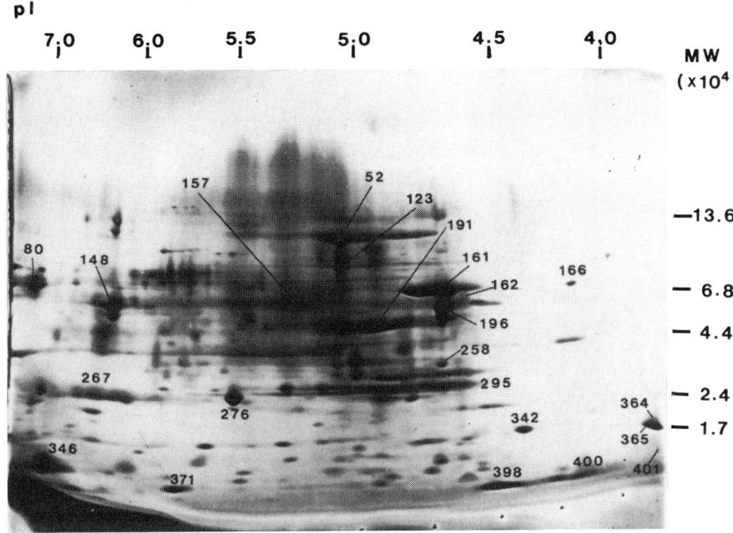

Fig. 3. A polypeptide map of bovine brain soluble proteins. Spot numbers of some major proteins were shown. Positions of purified proteins were; X-70 (spot 123), tubulin (157), albumin (161), X-56 (162), DEK protein (166), P-46 (191), NSE (196), hemoglobin (346), calmodulin (365), des-calmodulin (364), S-100 b (398), S-100 a (400), micro Glu-P (401).

Insoluble proteins--2M urea extract--Insoluble Fraction was subjected to 2M urea extraction and the extract was analyzed by 2-D electrophoresis in ISO-SDS system using 2M urea in focusing gel and 1% SDS in a gradient gel. Figure 4 shows a polypeptide map obtained by micro 2-D electrophoresis of 2M urea extract from Insoluble Fraction. About 12 µg protein/ 4 µl was applied and the gel was stained with Commassie blue. About 100 spots were detected on the gel. Glial fibrillary acidic protein (GFA) and neuro-filament tripret proteins (NFP, molecular weight 200K, 160K, and 68K) were identified on the map by the electrophoretic transfer-immunochemical detection technique. Two major spots at 16K and 45K were also indicated in Fig. 4.

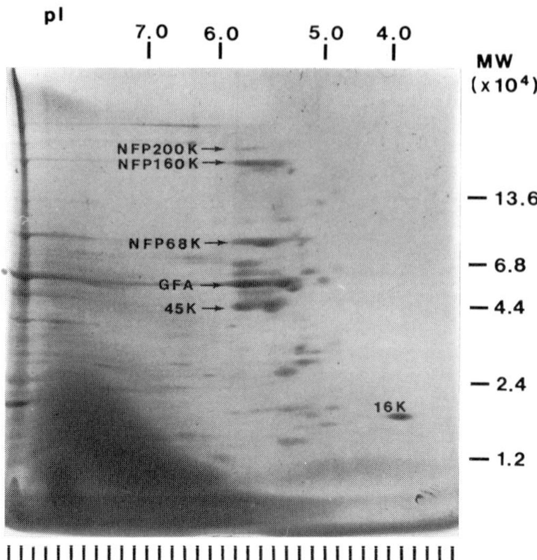

Fig. 4. A polypeptide map of 2M urea extract from Insoluble Fraction. Positions of glial fibrillary acidic protein (GFA), neurofilament tripret proteins (NFP, 200K, 160K, and 68K) and two major proteins (16K and 45K) were shown.

Comings (16) and Jackson and Thompson (17) reported the analysis of human brain proteins employing 2-D electrophoresis technique of O'Farrell. However, our purpose was on the analysis of brain   proteins maintaining their native physicochemical properties. Thus 2-D electrophoresis systems, ISO-GRAD system which used no denaturing agent and ISO-SDS system which used denaturing agent (SDS) only in the second dimension, were developed. The analysis of the ammonium sulfate fractions and their DEAD-Sephadex A-50 chromatography subfractions employing the 2-D electrophoresis systems is in progress. For the analysis of insoluble proteins, not extracted by 2M urea, the technique of O'Farrell or its improved versions should be examined.

References

1. Manabe, T., Tachi, K., Kojima, K., Okuyama, T.: J. Biochem. 85, 649-659 (1979).

2. Manabe, T., Kojima, K., Jitzukawa, S., Hoshino, T., Okuyama, T.: J. Biochem. 89, 841-853 (1981).

3. Manabe, T., Hayama, E., Okuyama, T.: Clin. Chem. 28, 824-827 (1982).

4. Manabe, T., Jitzukawa, S., Ishioka, N., Isobe, T., Okuyama, T.: J. Biochem. 91, 1009-1015 (1982).

5. Kadoya, T., Takahashi, Y., Ishioka, N., Manabe, T., Isobe, T., Okuyama, T.: Protides of the Biological Fluids 30, 591-594 (1982).

6. O'Farrell, P. H.: J. Biol. Chem. 250, 4007-4021 (1975).

7. Isobe, T., Nakajima, T., Okuyama, T.: Biochim. Biophys. Acta 494, 222-232 (1977).

8. Isobe, T., Ishioka, N., Okuyama, T.: Eur. J. Biochem. 115, 469-474 (1981).

9. Isobe, T., Ishioka, N., Okuyama, T.: Biochem. Biophys. Res. Commun. 102, 279-286 (1981).

10. Isobe, T., Ishioka, N., Kadoya, T., Okuyama, T.: Biochem. Biophys. Res. Commun. 105, 997-1004 (1982).

11. Kasai, H., Kato, Y., Isobe, T., Kawasaki, H., Okuyama, T.: Biomed. Res. 1, 248-264 (1980).

12. Isobe, T., Okuyama, T.: Eur. J. Biochem. 116, 79-86 (1981).

13. Isobe, T., Okuyama, T.: Eur. J. Biochem. 89, 379-381 (1978).

14. Towbin, H., Staehelin, T., Gordon, J.: Proc. Natl. Acad. Sci. USA 76, 4350-4354 (1979).

15. Manabe, T., Okuyama, T.: J. Chromatogr. (in press).

16. Comings, D. E., Carraway, N. G., Pekkula-Flagan, A.: Clin. Chem. 28, 790-797 (1982).

17. Jackson, P., Thompson, R. J.: J. Neurol. Sci. 49, 429-438 (1981).

ANALYSIS OF GENETIC VARIATION IN AMERINDIAN SERA BY 2-D PAGE

Jun-ichi Asakawa, Norio Takahashi
Radiation Effects Research Foundation, Hiroshima 730, Japan

Barnett B. Rosenblum and James V. Neel
Dept. of Human Genetics, Univ. of Michigan Med. Sch., Ann Arbor, MI 48109

Introduction

The advent of two dimensional polyacrylamide gel electrophoresis (2-D PAGE)
added a powerful new approach to the study of genetic variation in human
and animal populations (1,2,3). Using 2-D PAGE combined with Coomassie Blue
staining and the silver stain developed by Sammons et al. (4). We found
genetic variation in plasma samples from a predominantly Caucasoid population (5). Here, we extend our observations on variation of plasma proteins
to two Mongoloid populations: Amerindian from Central and South America,
and Japanese from Hiroshima and Nagasaki.

Results

2-D PAGE of plasma was as described by Neel et al. (5). A typical gel pattern of plasma proteins after silver stain is shown in Fig. 1. The gel was
divided into 5 areas (A-E): each spot examined was assigned a number, shown
in Table 1. A total of 23 plasma proteins (12 from Coomassie Blue stained
gel and 11 from silver) were evaluated for genetic variation, the data
obtained from 107 Amerindian and 50 Japanese trios (mother-father-child),
allowing quick confirmation of the inheritance of any variant. We observed
variation in 9 plasma proteins: 6 are known from 1-D and/or 2-D electrophoresis. Fig. 2 shows the variants. All, found in one or the other parent,
are inherited traits. Hemopexin (A-02) is polymorphic in Japanese. The
variant has the same M.W., but anodal migration results in doublet spots.

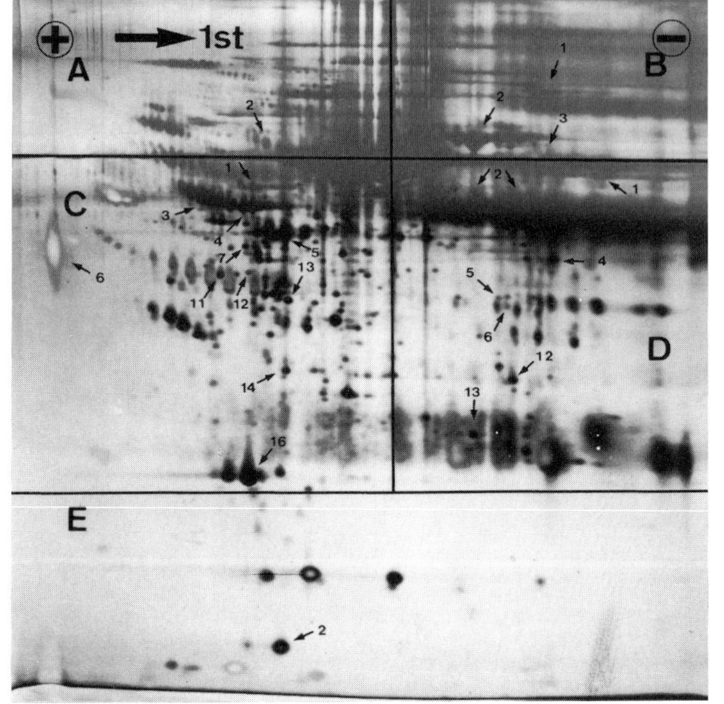

Fig. 1. Silver stained 2-D PAGE pattern of plasma proteins.
Spot designations given in Table 1.

Table 1. Spot designation of plasma proteins for genetic variation.

Spot	Protein	Spot	Protein
A-02*	Hemopexin	C-13	
B-01*	Plasminogen	C-14	Apolipoprotein E
B-02*	Transferrin	C-16*	Apo A-I lipoprotein
B-03*		D-01*	Fibrinogen α-chain
C-01		D-02*	Fibrinogen β-chain
C-03*	α_1-Antitrypsin	D-04	
C-04*	Gc-globulin	D-05	
C-05*	Fibrinogen γ-chain	D-06	
C-06	α_1-Acid glycoprotein	D-12	
C-07		D-13	
C-11*		E-02*	Prealbumin
C-12			

* : Protein was scored from Coomassie Blue stained gel.

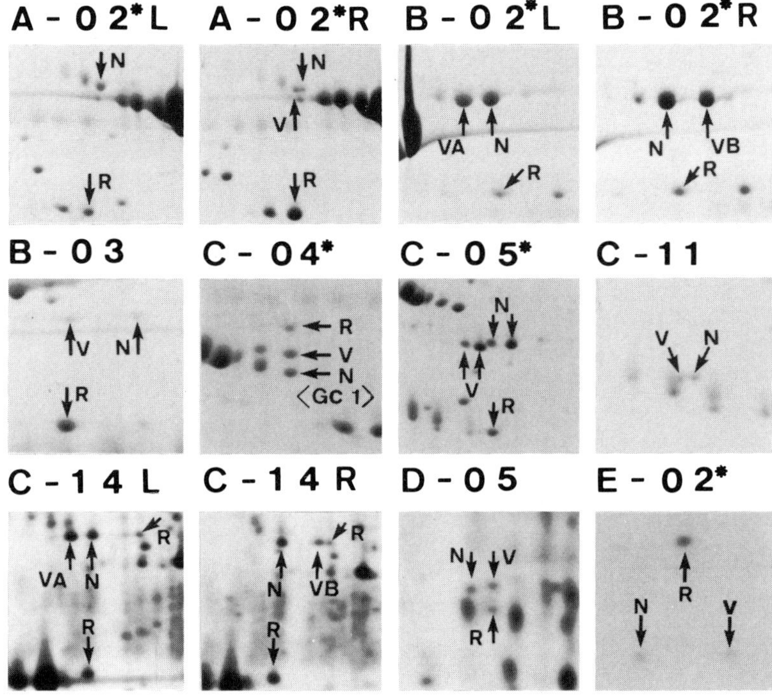

Fig.2. Genetic variants of plasma proteins. A–02 L & R: Hemopexin N & V; B–02 L & R: Tf B & D; C–04: Gc; C–05: Fibrinogen γ–chain; C–14 L & R: apo E II & IV; E–02: Prealbumin; B–03, C–11, D–05: unidentified proteins; N: Normal, V: Variant, R: Reference spot. : Identified by antisera.

Transferrin (Tf) is polymorphic in Mongoloids, but we found no variants in Amerindians. Genetic variants of unidentified protein B–03 occur in lower frequency than found earlier in Caucasoids (5). Two Gc (C–04) variants found in Japanese have higher M.W. than Gc 1, but are not separable from it by focusing. Fibrinogen γ–chain, major and minor (C–05), variant found in Amerindians, migrated anodally. C–11 (unidentified protein) exhibits genetic variation (2,5). We found 7 genetic variants in Amerindians and 1 in Japanese. Apolipoprotein E (apo E: C–14) occurs in at least 3 different isopeptides on 2-D PAGE: apo E II,III,IV (6). In Japanese all 3 forms were seen, but apo E II was not found in Amerindians. D–05 (unidentified protein) is polymorphic in our two groups, as also reported in Caucasoids (5).

We found 1 prealbumin (E-02) variant with cathodal migration in Japanese.

Table 2.   Genetic variation in plasma proteins.

	A-02	B-02	B-03	C-04[a]	C-05	C-11	C-14[b]	D-05	E-02	OTHERS[c]
Amerindian[d]										
N	106	102	105	80	41	98	66	71	107	1275
N/V	0	0	2	23	1	7	23	26	0	0
V	0	0	0	4	0	0	6	1	0	0
Japanese										
N	47	45	49	25	50	49	34	18	49	689
N/V	3	5	1	21	0	1	14	20	1	0
V	0	0	0	4	0	0	0	11	0	0
Caucasian[e]										
N	51	47	36	29	54	48	35	15	55	750
N/V	0	3	18	20	1	8	20	26	1	8
V	0	0	0	6	0	0	1	7	0	0

a): Gc 1 and Gc 2 were scored as N and V respectivly, in the Japanese two heterozygous variants (Fig. 2) were included in N/V. b): Both apo E II & IV were scored as V, apo E III as N. c): Another 14 proteins; no variants were found in Amerindian or Japanese. d): 65 samples of 107 were sera. e): Data from Neel et al. (5).

Table 2 summarizes the genetic variation of plasma proteins. The index of heterozygosity obtained in this study is: Amerindian 3.8%, Japanese 5.8% . These findings are similar to those in Caucasoids regarding heterozygosity but not unexpectedly, ethinic differences begin to emerge.

This work was partially supported by program projects NCI-CA-26803 and AC-02-82-ER-60089.

References

1. Anderson, L. et al.: Proc. Natl. Acad. Sci. USA 74, 5421 - 5425 (1977)

2. Tracy, L. A. et al.: Clin. Chem. 28, 890 - 899 (1982)

3. Wanner, R. P. et al.: Am. J. Hum. Genet. 34, 209 - 215 (1982)

4. Sammons, D. W. et al.: Electrophoresis 2, 135 - 141 (1981)

5. Neel, J. V. et al.: Methods and Applications of two Dimensional Gel Electrophoresis of Proteins (J. E. Cellis, ed.). Academic Press, New York. (in press)

6. Børresen, A.-L. et al.: Clin. Genet. 20, 438 - 448 (1981)

AN ISO-DALT ELECTROPHORESIS SYSTEM WITH REDUCED BUFFER AND CURRENT REQUIREMENTS

N. Cho
Electro Nucleonics, Inc., Oak Ridge, TN 37830, USA

S. L. Tollaksen, N. G. Anderson, and N. L. Anderson
Molecular Anatomy Program, Division of Biological and Medical Research
Argonne National Laboratory
Argonne, IL 60439, USA

Introduction

Electrophoretic techniques for use in biological systems have been under development over the past 30 years, as reviewed in reference 1, but it was not until 1970 that Stegemann (2) combined isoelectric focusing in one dimension and electrophoresis in the presence of sodium dodecyl sulfate (SDS) in the second dimension. O'Farrell optimized the two-dimensional system in 1975 (3) using small, radiolabeled samples that enabled very high resolution of approximately 1200 protein subunits of $\underline{E.\ coli}$ by autoradiography.

In order to use two-dimensional electrophoresis effectively in a clinical or research setting, development of methods for running many analyses in parallel was necessary to achieve high throughput and good reproducibility. In 1976, Anderson and Anderson (4-5) at Argonne National Laboratory began to develop the hardware and technology necessary to achieve the goal of automation of two-dimensional electrophoresis, including automation of sample preparation, pouring and electrophoresis of gels, staining, photography, and data analysis. The term "ISO-DALT" was given to the Argonne two-dimensional electrophoresis system, combining isoelectric focusing in the first dimension (ISO) with a molecular weight (expressed as daltons) separation using SDS in the second dimension (DALT). The original ISO-DALT system was designed to prepare and run twenty 1.5 mm x 7 inch (16 cm) gel rods simultaneously in the first dimension and twenty 7 x 7 inch gel slabs (10%-20% acrylamide gradient) in parallel in the second dimension.

Based on the prototype developed at Argonne, modifications and improvements have been made in the equipment used for gel preparation and electrophoresis for two-dimensional electrophoresis These modifications, described below, provide greater ease in performance and increased economy.

276

<u>ISO System</u>:

In the modified ISO system, shown in Figure 1, the ISO tubes can be removed from the setup and can be discarded or reused after cleaning in
chromic acid solution at 90 C for 2 hours. This is the only significant
modification made in the original Anderson design. In the original model,
the ISO tubes were permanently bonded to the holders, having the advantage
of holding many tubes close together, but requiring a special Teflon acid
tank to clean the fixed tubes. The modified tank system thus reduces
cost and increases operator safety.

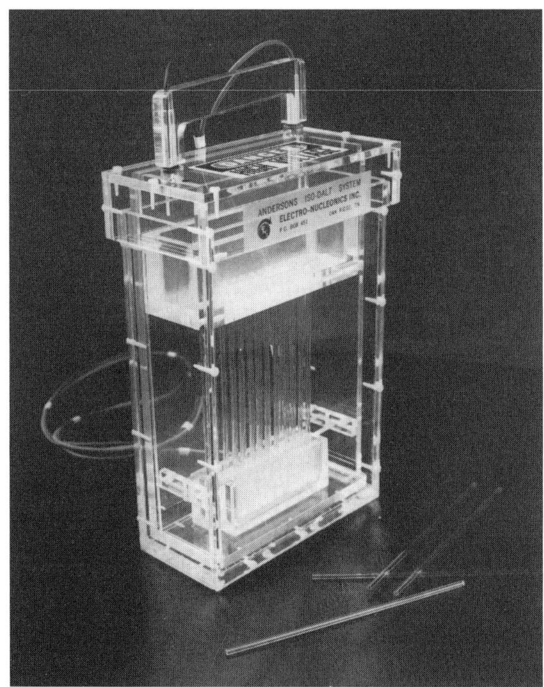

Fig. 1. The removable ISO tubes for ISO electrofocusing in the first
dimension.

<u>DALT System</u>:

The original DALT tank of 28" x 12 3/4" x 11" has been reduced to 14 1/2" x 10 1/2" x 10" (Figure 2) and the rubber seals have been improved. Figures 3A and 3B compare a representative segment of gels of a human platelet sample run in the original prototype model and the newer modified version; the results are identical.

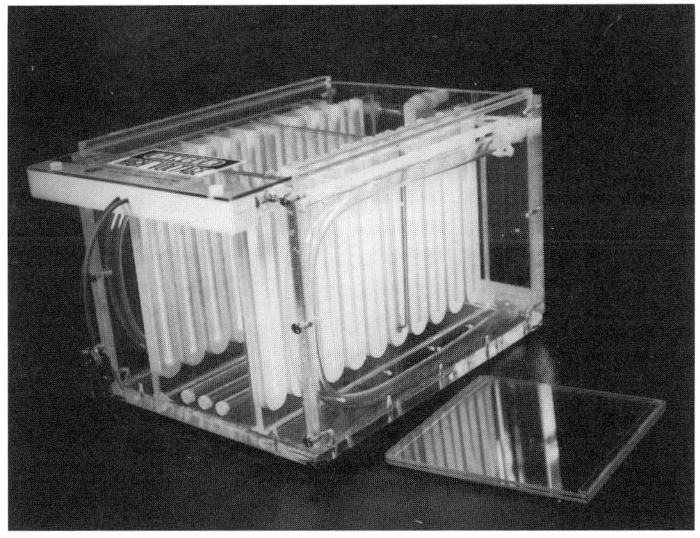

Fig. 2. DALT System. Improved seals are shown in reduced tank size for separations by molecular mass (in daltons) in the second dimension.

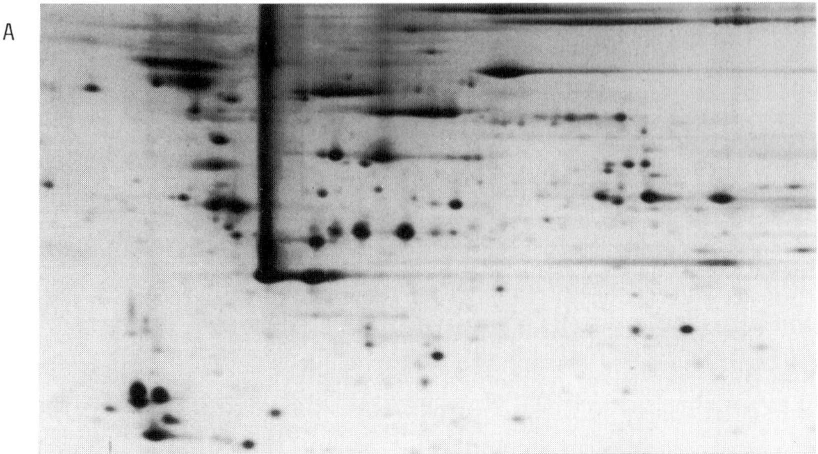

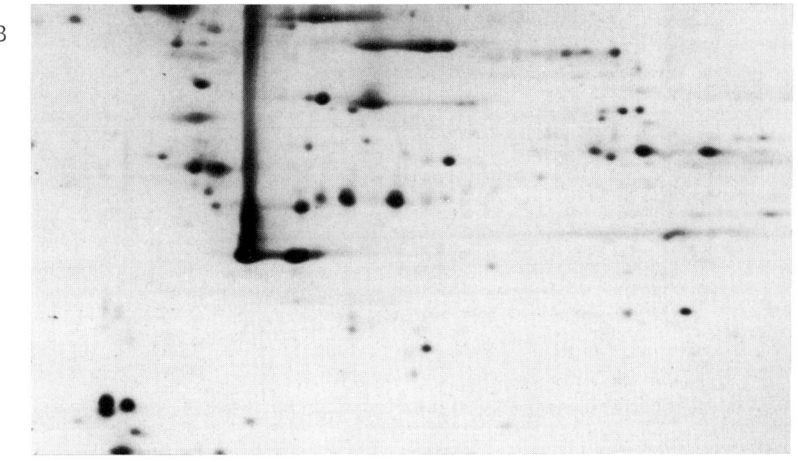

Fig. 3. Comparison of representative portions of a human platelet sample run in the original prototype model (Figure 3A) and in the newer modified version (Figure 3B). Gels are oriented with acid end to the left and basic end to the right.

This system can successfully handle ten gel plates at once. The smaller
amount of buffer (13 liters compared with 37 liters) obviously saves expensive buffer chemicals. Improvement in the rubber gel plate seals reduces the power required for one run by 120 watts, and the reduced current
leakage around plates running in parallel reduces the total run time by
about 20%. No change was needed in the buffer cooling capacity or in the
buffer pump.

In addition, the DALT system is now capable of handling four blot transfer
gels (6) at a time by simply inserting two partitions inside the center of
the tank next to the fixed partitions, as shown in Figure 4. This added
versatility of the system saves the purchase of two separate tank systems.

Fig. 4. Blot Transfer. The inserts shown contain four gels for blot
transfer in the basic DALT tank.

For increased operator safety, the new DALT system has an interlock in its
tank cover, so that the electrical supply is interrupted automatically
when the cover plate is removed. The system also monitors buffer level,
temperature, and buffer flow. As an option, an automatic shutdown at the
end of the run can be achieved either by sensors or by a timer.

Plans are being made for transition from the 7" x 7" format to an 8" x 10"
gel format for compatibility with commercial photographic equipment and
materials. This format change will also enable somewhat higher resolution
and an increase in the capacity of the DALT tank from four to six transfer
gels per tank.

In summary, the present paper has summarized the evolution of a scientific
idea from a small research laboratory, to a major project at a National
Laboratory, to commercialization in industry. This commercial ISO-DALT
system, as modified from the national laboratory prototype equipment, has
been used successfully in many laboratories and countries (including such
diverse places as Norway and Kenya).

Acknowledgement

The work at Argonne was supported by the U.S. Department of Energy under
contract No. W-31-109-ENG-38.

References

1. Anderson, N. G., Anderson, N. L.: Behring Inst. Mitt. 63, 169-210
   (1979).
2. Stegemann, H.: Angew. Chem. 82, 640 (1970).
3. O'Farrell, P. H.: J. Biol. Chem. 250, 4007-4021 (1975).
4. Anderson, N. G., Anderson, N. L.: Anal. Biochem. 85, 331-340 (1978).
5. Anderson, N. L., Anderson, N. G.: Anal. Biochem. 85, 341-354 (1978).
6. Anderson, N. L., Nance, S. L., Pearson, T. W., and Anderson, N. G.:
   Electrophoresis 3, 135-142 (1982).

Cell Electrophoresis

PREPARATIVE FREE-FLOW ELECTROPHORESIS OF PROTEINS, PEPTIDES AND RELATED COMPOUNDS

Horst Wagner, Volker Mang, Reinhard Kessler, Ariane Heydt and
Roger Manzoni
Fachrichtung Anorganische Analytik und Radiochemie der Universität des
Saarlandes
D-6600 Saarbrücken

For preparative electrophoretic separation methods both the required
resolution and the throughput play a decisive role. Depending on the
separation method used, the throughput is in the range of some nanograms
per hour up to some grams per hour. The range of application for a given
method is determined by the required throughput and resolution[1].

The scientific applications are often the separation of a multicomponent
mixture into its single components. The quantities of these components
range between microgram and milligram with a total sample volume of a
few milliliters[2,3].

The industrial application is quite a different thing. Here the problem
is often to isolate a definite component from a mixture. Moreover, as is
the case with interferon for example, the substance to be isolated is in
a very small concentration so that a very high throughput of the sample
solution must be provided for.

For such preparative problems only continuous methods can be applied,
since they guarantee the required high throughput. One of these methods
is the free-flow electrophoresis. In this case a steady fluid film is
pumped through between two cooling plates. The electrical field is
applied perpendicular to the flow direction. The sample feeding and
collecting are done continuously. The electrical field forces ionic

particles to migrate perpendicular to the flow direction, and the angle
of deviation depends on the respective mobilities. With regard to the
separation principle this is a zone electrophoresis. By the previous
apparative developments it is now possible to adapt the system of the
free-flow electrophoresis to other electrophoresis methods as well. Such
an apparatus is produced by Bender & Hobein, Munich[4].
The special feeding system, the exchangeability of the cover plates and
the possibility of applying a counter flow allow the use of the different
electrophoretic methods, as mentioned before (fig. 1).

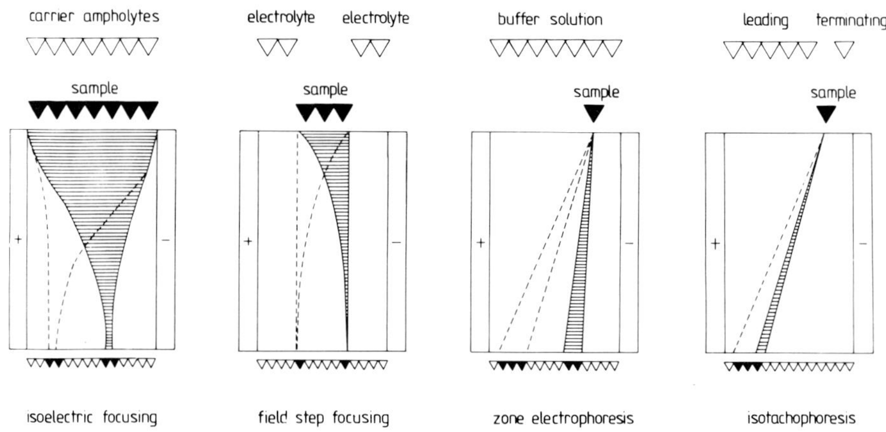

Fig. 1: Preparative free-flow electrophoretic separation methods

So far the main field of application has been the separation of cell
organelles or whole cells by zone electrophoresis, done mainly by
Hannig[5,6], but there are examples of isotachophoresis[7,8] and isoelectric
focusing[9-11] as well.

Here we want to introduce field step focusing, a separation method
developed for the free-flow system[12,13]. The very simple separation
principle can be easily explained by looking at the electrolyte system
used (fig. 2).

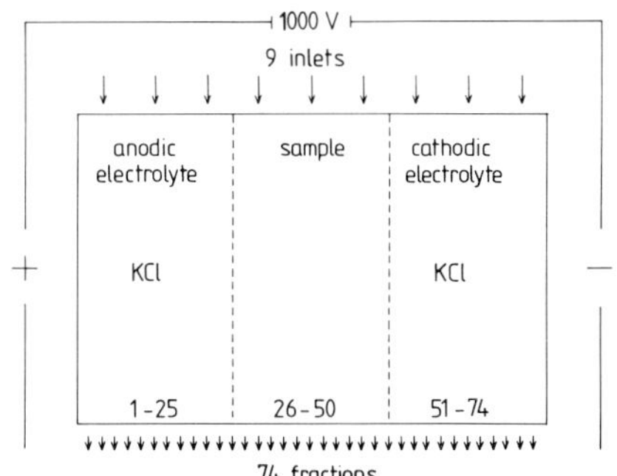

Fig. 2: Electrolyte system used in field step focusing

Two electrolyte solutions and the sample solution are fed into the separation slit by the feeding system. The conductivity of the two electrolyte solutions must be more than 20 times higher than that of the sample solution.

This results in a step-like profile for the conductivity and the field strength (fig. 3). On account of the different conductivities we find different migration velocities for the ions in the respective solutions. For the given example the migration velocity of an ion in the sample solution is about 60 times higher than in each of the electrolyte solutions. Ionic particles of the sample solution will migrate to the respective electrode according to their electrical charge. At the boundary to the solution of high conductivity the migration velocity will be reduced to one sixtieth of the previous value. Thus the focusing at the boundary between the solutions of different conductivities is caused by the change in the migration velocity, and not by the loss of the electrical charge as is the case with isoelectric focusing.

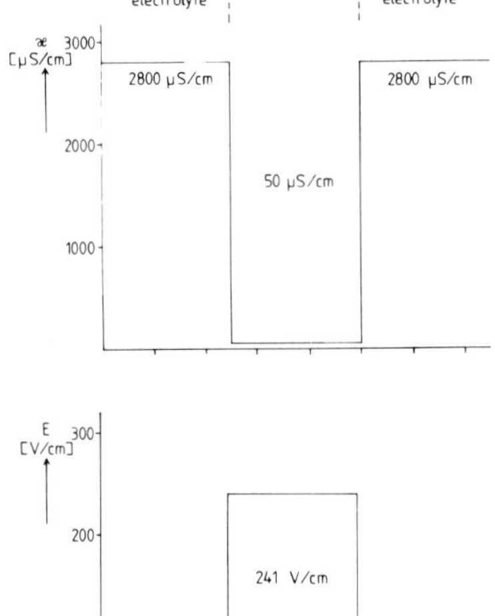

Fig. 3: Step-like profile for the conductivity and the field strength

For the separation of amphoteric substances, for example proteins or peptides, the direction of the migration can be regulated by adjusting a suited pH. In order to stabilize pH of the sample solution it is necessary to provide a continuous supply of H^+ or OH^-. For the same reason, the pH of the electrolyte solutions plays an important role. The different migration velocities of H^+ and OH^- in the different solutions must be considered. In many cases a further stabilization is achieved by adding buffer substances. The optimum pH for a separation problem can be easily determined by plotting the mobility against pH[14]. Another possibility is to run series of experiments with different pH of the sample solution[13].

The enrichment factors achieved by the focusing are about 10 to 15, if the sample solution is fed into the separation slit over a third of the slit width. Higher factors can be realized by enlarging the inlet width of the sample solution. At the same time the voltage applied has to be raised in order to keep the field strength in the sample range constant.

By this way enrichment factors of up to 60 are possible.

The separation possibilities[12,13] can be extended, by having several conductivity steps instead of a single step on each side. In this case substances with same sign of charge can be separated.
A further extension of the separation possibilities is achieved by subdividing the sample range into several ranges of different pH. Then we have a combination of focusing at conductivity steps and isoelectric focusing. These pHs are determined by the amino acids or peptides in order to keep the conductivity of the sample solution low.

Fig. 4 shows the curves for pH and conductivity of an sample electrolyte system consisting of one zone with aspartic acid and one of arginine in the sample range. Such profiles of pH are stable during the whole duration of the experiment, and can easily be varied. An example for separation shows fig. 5.

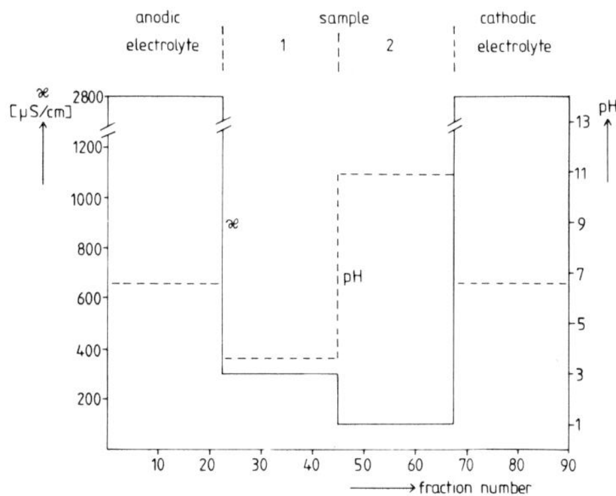

Fig. 4:

anodic electrolyte	KCl $2 \cdot 10^{-2}$ mol/l	pH = 6,6
sample 1	Asp $2 \cdot 10^{-2}$ mol/l	pH = 3,6
sample 2	Arg $2 \cdot 10^{-2}$ mol/l	pH =10,9
cathodic electrolyte	KCl $2 \cdot 10^{-2}$ mol/l	pH = 6,6

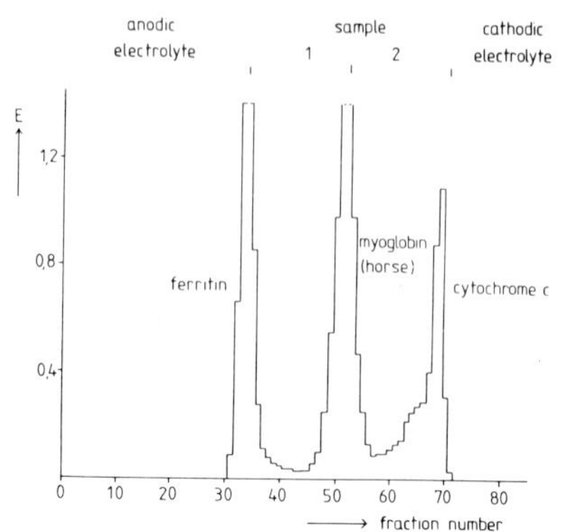

Fig. 5:

anodic electrolyte	KCl 2 · 10^{-2} mol/l	pH = 6,5
sample 1	Glygly 2 · 10^{-2} mol/l	pH = 6,0
sample 2	His 2 · 10^{-2} mol/l	pH = 7,5
cathodic electrolyte	KCl 2 · 10^{-2} mol/l	pH = 6,5
separation parameters	voltage 800 V	
	separation time 200 s	

In this case the step in pH measures only 1.6 units. Ferritin and cyto-chrom c are focused at the conductivity steps. For myoglobin we have isoelectric focusing. Several pH steps can be built up without difficulties.

With field step focusing also peptide separations are possible. Peptides are often prepared by the Merrifield method. There is the risk, especially in the last step of the synthesis, that the peptide bonds just formed will break again and that the peptide will contain impurities. It is therefore of interest to purify the peptides by separating them from the starting materials.

The next two pictures show the purification of lysyl-glutamic acid by separating it from starting amino acids lysine and glutamic acid.

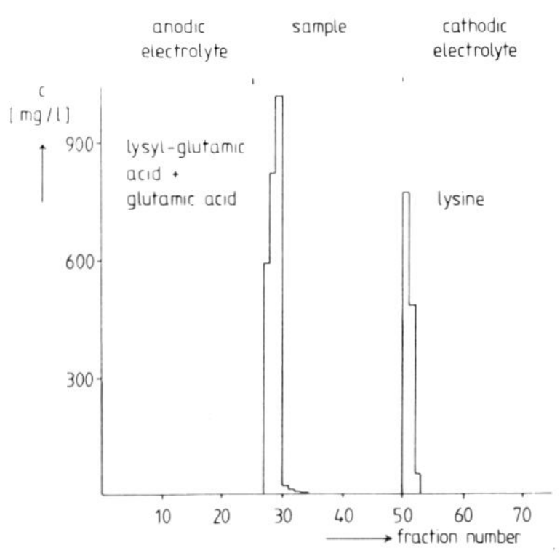

Fig. 6:

anodic electrolyte 0,1 mol/l phosphate buffer pH = 5,9

sample 52 mg/l lysine, 50 mg/l lysyl-glutamic acid and 49 mg/l glutamic acid pH = 7,5

cathodic electrolyte 0,1 mol/l phosphate buffer pH = 7,3

separation parameters:voltage 1000 V

separation time 3 min

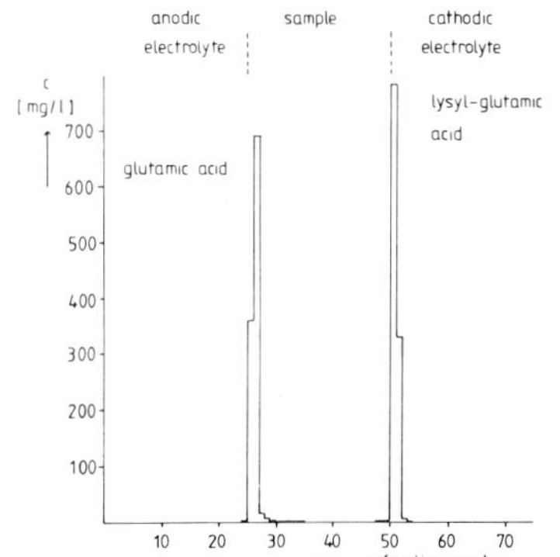

Fig. 7:

anodic electrolyte 0,1 mol/l KCl pH = 2,8

sample 43 mg/l glutamic acid and 45 mg/l lysyl-glutamic acid pH = 4,0

cathodic electrolyte 0,1 mol/l phosphate buffer pH = 7,3

separation parameters:voltage 900 V

separation time 3 min

Another example is the separation of penicillin and cysteine (fig. 8).

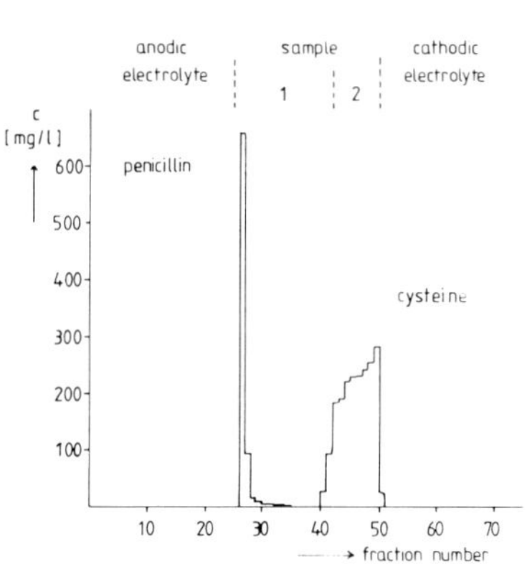

Fig. 8:
anodic electrolyte 0,1 mol/l
phosphate buffer pH = 5,9

sample 1 H_2O dest pH = 2,8

sample 2 100 mg/l penicillin
from isocillin and 250 mg/l
cysteine pH = 2,8

cathodic electrolyte 0,1 mol/
phosphate buffer pH = 5,9

separation parameters:voltage
800 V
separation time 3 min

Cysteine is a biogenetic group in penicillin. This separation is achieved by a modified electrolyte system. The sample is put only into the sample range 2. Penicillin migrates as an anion to the boundary of the anode range and is focused there, whereas cysteine virtually does not migrate at all.

Summary:

A particularly important point is that an equipment for free-flow electrophoresis is now commercially available. This is a prerequisite for a more common application of the preparative methods. The main advantage of this apparatus is its versatility because there is no universal method neither for analytical nor for preparative separation problems, the suited method must first selected for each problem.

The advantage of focusing at conductivity boundaries possibly combined with isoelectric focusing, consists in the possibility of easily adjusting

the experimental parameters including the electrolyte system to the
separation problem. Moreover the high throughput and the concentration of
the sample must be mentioned.

References

1. Righetti, P.G., Van Oss, C.J., Vanderhoff, J.W.: Electrophoretic
   Separation Methods, Elsevier/North-Holland (1979)

2. Everaerts, F.M., Beckers, Th.P., Verheggen, J.W.: Isotachophoresis-
   Theorie, Instrumentation and Applications, Elsevier, Amsterdam-Oxford-
   New York (1976)

3. Radola, B.J.: Biochim. Biophys. Acta 386, 181 (1974)

4. Fa. Bender & Hobein, GIT-Fachz. Lab., 12, 1145 (1982)

5. Hannig, K., Z. Anal. Chem. 181, 244 (1961)

6. Hannig, K., Kolloid.-Z.Z. Polym., 227, 37 (1968)

7. Prusik, Z., J. Chromatogr. 91, 867 (1974)

8. Wagner, H., Mang, V., in: Analytical Isotachophoresis, 41, Ed.:
   F.M. Everaerts, Elsevier, North-Holland (1981)

9. Seiler, N., Thobe, J., Werner, G., Z. Anal. Chem. 252, 179 (1970)

10. Seiler, N., Thobe, J., Werner, G., Z. Physiol. Chem. 351, 865 (1970)

11. Wagner, H., Speer, W., J. Chromatogr. 157, 259 (1978)

12. Wagner, H., Kessler, R., in: Electrophoresis 1982, 303, Ed.:
    D. Stathakos, Walter de Gruyter, Berlin-New York (1983)

13. Kessler, R., Wagner, H., in: Isotachophoresis 1982, Ed.: C.J. Holloway,
    Elsevier, Amsterdam 1983

14. Rosengren, A., Bjellquist, B., Gasparic, V., in: Electrofocusing and
    Isotachophoresis,165, Ed.: B.J. Radola, D. Graesslin, Walter de Gruyter,
    Berlin, New York (1977)

SCALE-UP OF THE FREE FLOW ELECTROPHORESIS DEVICE

Cornelius F. Ivory, William Gobie, Richard Turk
Department of Chemical Engineering, University of Notre Dame
Notre Dame, Indiana, USA

Introduction

The Continuous Flow Electrophoresis Device (CFE) consists of a narrow slit
bounded on the front and back transverse axes by rigid, diathermal walls.
A buffered fluid is introduced at the top of the device and quickly assumes a parabolic flow profile. A continuous, multicomponent stream of
solute is injected into the fluid and, as the solute passes through the
chamber, its components are displaced by a lateral electric field [1].
Individual solute components are displaced in proportion to their
electrophoretic mobilities and, when sufficiently separated, they are
collected at the device exit (see Figure 1).

The CFE has several inherent
advantages [2] over other large
scale biological separations
techniques: The electrophoretic
method is gentle, the device runs
continuously and the carrier
fluid contains neither stabilizing gels nor other modifiers.
As a result one might expect the
CFE to purify solute at high
activity and with low capital and
operating expenses. However,
scale-up of the device to meet
industrial demands is hindered by
two important obstacles: The ther
mal destabilization of the flow

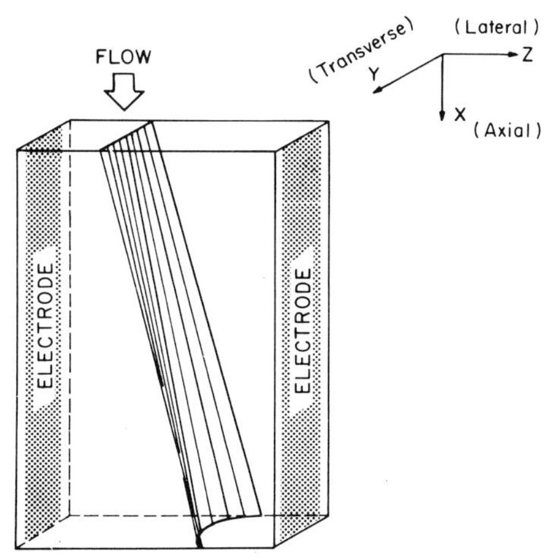

Fig 1. Schematic of the unmodified CFE showing electrode
placement, flow direction and
solute dispersion due to
crescent formation.

field and the low resolving power of the applied electric field.

To a large extent these problems can be eliminated from the device by implementing monor changes in the CFE design. In this paper two alterations are proposed and their effects on the performance of the device are discussed.

Thermal Destabilization

Wherever an unfavorable density gradient occurs in the CFE, the device is susceptible to buoyant destabilization [3]. In an aqueous buffer this occurs above 4°C in the presence of a thermal inversion. The thermal inversion is the result of two conflicting effects: The work done by the electric field in producing an ionic current is dissipated as Joule heat and, to control the temperature of the buffer fluid, heat is removed from the fluid at the front and back transverse walls. This produces closed isotherms (see figure 2) which have their corresponding thermal gradients at right angles and pointing into the center of the device. So long as Joule heat is removed in this fashion the CFE must contain a thermal

inversion irregardless of device orientation and it is therefore vunerable to a buoyancy induced destabilization. The first part of this paper proposes a modification of the thermal field in the CFE which reduces the tendency of the device to suffer flow destabilization.

The thermal field is modified in the following manner. First, the CFE is rotated ninety degrees about its lateral axis so that the gravitational field is aligned with the transverse axis of the device. Then the upper

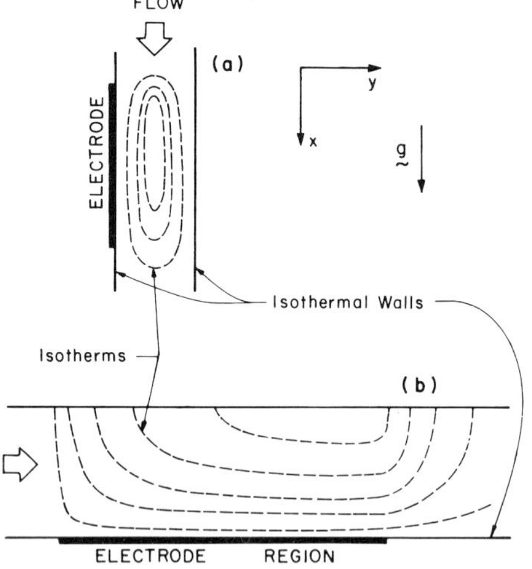

Fig 2. Isotherms in the CFE (a) and the modified CFE (b). Since thermal gradients point at right angles to the isotherms the CFE always maintains a thermal inversion.

Table 1: Parameters used in Model Calculations [3].

Electric field strength, E	$-$ 70 volts/cm
Gravitational acceleration, g	$-$ 980 cm/sec^2
Thermal conductivity, k	$-$ 5.71x10^{-3} watts/cm$^\circ$C
Thermal expansion coefficient, b	$-$ 8.62x10^{-5} ($^\circ$C)$^{-1}$
Thermal diffusivity, κ	$-$ 1.365x10^{-3} cm^2/sec
Fluid density, ρ	$-$ 1.0 g/cm^3
Electrical conductivity, σ	$-$ 6.9x10^{-4} (ohm-cm)$^{-1}$
Kinematic viscosity, ν	$-$ 1.208x10^{-2}/sec
Heat generation/unit volume, σE^2	$-$ 3.38 watts/cm^3
Electrophoretic mobilities, μ	$-$ 2.15, 2.58 m-cm/volt-sec
Chamber length, L	$-$ 16 cm
Chamber thickness, 2b	$-$ .15 cm

transverse wall is insulated to make it adiabatic and a conductive
material is used to make the lower transverse surface isothermal. A model
of the modified thermal field has been solved analytically [4] and the
results are used to calculate the lower bound on the transverse thickness
required for stable operation. In contrast to the unmodified device for
Pe < 14.8 there is no thermal inversion in this device. In this case it
can be shown from the analysis of Nakayama et al. [5] that

$$Ra_{cr} = \frac{g\,\beta\sigma E^2 (2b)^6}{k\kappa\nu} > 10^7. \tag{1}$$

Using the values for the parameters listed in Table 1, we predict that the
modified device is stable for a transverse thickness up to 1.25cm, an
order of magnitude thicker than the terrestrial CFE chambers presently in
use.

296

For Pe > 14.8 a thermal inversion occurs in the entrance region of the
device and a rigorous stability analysis is required to determine the
exact conditions for stable operation. However, the restriction imposed
by an upper limit of Pe = 14.8 is mitigated by two factors:
a) Distortion of the electric field at the entrance and exit of the CFE
    decreases the axial thermal gradients and increases the range of
    Peclet number allowed in chamber design, and,
b) superposition of an appropriate thermal field at the entrance and exit
    of the device will further reduce axial gradients.
In addition, we note that the magnitute of the inversion which is found in
the modified device for Pe > 14.8 is small compared with the inversion
found in the unmodified CFE and should not strongly affect operation of
the device. Future work in this area is directed at determining the exact
criteria for stability in the entrance region of both the modified and the
unmodified CFE taking into account the distortion of the electric field
outside of the electrode region.

Solute Resolution

The electric field in the CFE is used more efficiently by continuously
recycling solute through the device. This modification acts to increase
the resolution of the device and to lower the power requirements by
allowing the chamber to operate as a series of staged transfer units.
Using this modification the column is capable of separating two components
to any specified purity. In the CFE this modification is made in the
following manner. First, the chamber inlet and outlet are divided into
discrete, rectangular ports.

As the column operates, effluent is taken from each of the contiguous
exit ports and is reinjected into each of the corresponding ports at the
entrance of the device. This correspondence between ports is determined
by a specific lateral shift, Δ, between the upper and lower port positions. The magnitude of this shift is given by the formula

$$\Delta = \tfrac{1}{2}(El_1 + El_2)L/Pe. \tag{2}$$

When the effluent is recycled through the CFE with this shift. The solutes overlap in the central portion of the modified device as a result of the recycle, but after a sufficient number of passes through the device they separate and may be removed at the far left and right outlets. Solute behavior in the modified CFE is modeled using the usual convection diffusion equation [6]

$$D \frac{\partial^2 C}{\partial x^2} + D \frac{\partial^2 C}{\partial y^2} + D \frac{\partial^2 C}{\partial z^2} = u(y) \frac{\partial C}{\partial x} + (v(y)+w) \frac{\partial C}{\partial z} \qquad (3)$$

with boundary conditions:

$$J_x(0,y,z) - \int_{-b}^{+b} J_x(L,y,z+\Delta) \, dy/2b = u(y)C_0 \, S_0(y,z) \qquad (3a)$$

$$x = 0 \quad \frac{\partial C}{\partial x} = 0 \qquad (3b)$$

$$z \to \pm \infty \quad C \text{ is finite} \qquad (3c)$$

$$y = \pm b \quad \frac{\partial C}{\partial y} = 0 \qquad (3d)$$

where $C(x,y,z)$ is the solute concentration in the device, D is the solute diffusion coefficient, $u(y)$ is the axial velocity profile, $v(y)$ is the electroosmotic velocity profile and w is the electrophoretic velocity. S_0 represents inlet source distribution at the column entrance and J_x is the solute flux in the axial direction. Condition (3a) indicates that solute is recycled

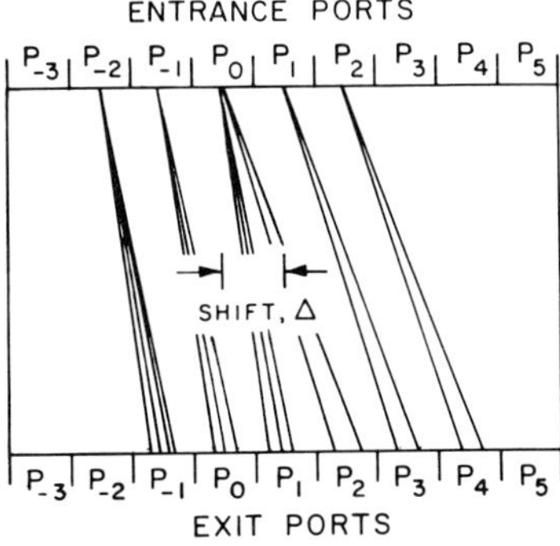

Fig 3. Schematic of the modified CFE showing discrete ports, recycle shift, Δ, and solute displacement to left and right of the feed port, P_0.

298

from the exit ports to the entrance ports after being displaced along the lateral axis by the shift, Δ. In this model the ports are <u>infinitely resolved</u> along the lateral axis, that is, the ports are assumed to be continuous along the entrance and exit of the device. This is a mathematical idealization which simplifies solution of equation (3) without changing its physical interpretation. Equation (3) has been solved in the limit, $Pe = u_{max}b/D \ll 1$ and $El = w\,b/d \ll 1$, which is valid for high solute diffusivities or thin CFE chambers. In figures 4 and 5 the results from this model are compared with results from the unmodified CFE operating under the same conditions.

Figure 4 shows the effluent flux profiles for two solutes after a single pass through the <u>unmodified</u> CFE. The chamber parameters used in these calculations are given in Table 1. For this case, the solute mobilities differ by 20% and therefore the effluent profiles overlap considerably. Figure 5 shows the effluent flux profiles for the same two-component separation performed in the modified CFE. These results clearly demonstrate the effect of internal recycle with shift on the performance of the modified CFE.

Equation (3) has also been solved in the limit $Pe \gg 1$ and $El \gg 1$ [7] using discrete ports each with a lateral breadth of 0.375cm and again using the parameters given in Table 1. The results of this model are given in Table 2 for 10, 20 and 40 passes through the chamber. These results indicate that as the solutes are repeatedly recycled through the device, the separation proceeds exponentially.

The calculations performed for high and low diffusion coef-

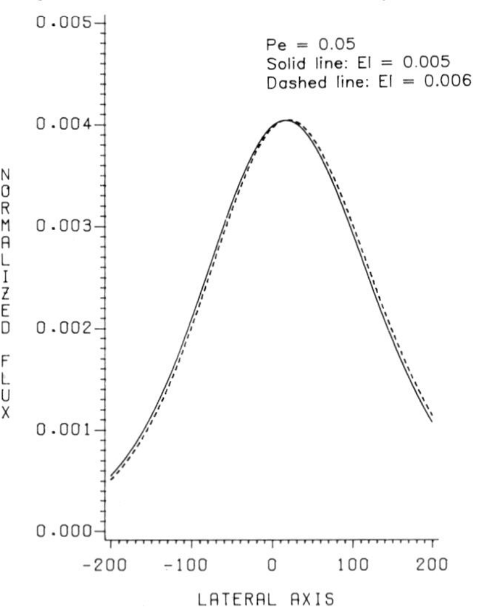

Fig 4. Effluent flux profiles in the unmodified CFE at low Pe and El. The solutes overlap considerably and cannot be separated in the chamber.

Fig 5. Effluent flux profiles in the modified CFE at low Pe and El. When the shift is correctly adjusted, the solutes separate completely in this chamber.

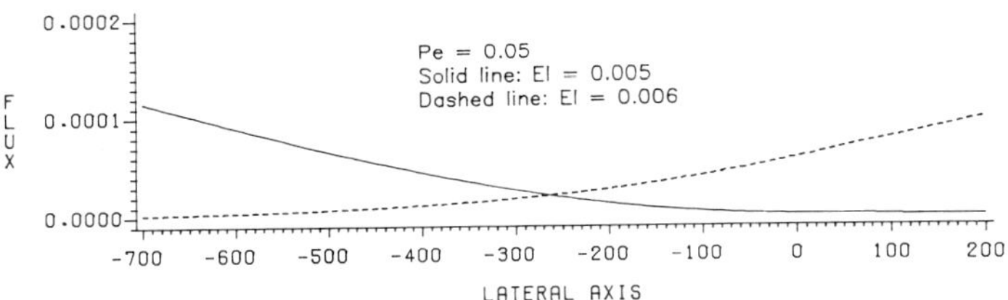

ficients and with continuous and discrete ports strongly suggest that the modified CFE will have the following advantages and disadvantages over the unmodified CFE:

Advantages of the Modified CFE
a) Unfavorable thermal gradients are largely eliminated from the modified CFE and their influence on device stability is diminished.
b) Dispersion effects due to diffusion and crescent formation in the device are reduced by recycle with shift.
c) The solute is not diluted from its feed concentration.
d) Purification of solutes is exponential. That is, if 10 passes through the modified CFE yields a 90% pure product, then 20 passes yields about 99% purity, 30 passes yields 99.9% purity and so on.
e) The modified CFE can be scaled along the lateral axis as well as the transverse axis. Note that throughput increases with port size.
f) The electric field strength can be reduced in proportion to the number of passes through the chamber, N. Thus the power dissipated as Joule heat can be reduced by the factor, N^2.

Disadvantages of the Modified CFE
a) Chamber holdup times are increased by the factor, N.
b) The apparatus required for recyle is expensive and complex.

The modification of the thermal field and the technique of internal recycle with shift are not mutually exclusive and can therefore be combined in a single device. When this is done the problems associated with thermal destabilization and low chamber resolution are markedly reduced.

300

Table 2: Fractional Purification in Modified CFE

Total Number of Recycle Ports	Component #1 μ_1 = 2.15	Component #2 μ_2 = 2.58
11	.8658	.9214
21	.9765	.9928
41	.9994	.9999

By itself the modification of the thermal field allows the chamber to be scaled O(10x) along the transverse axis. The second modification, recycle with shift, allows scale-up along the transverse axis because power consumption is reduced and it also allows scale-up in the lateral direction because dispersion effects are reduced. Scale factors of O(100x) in device throughput are predicted in this case. Together, these two alterations in the CFE design indicate that chamber throughput can be increased by a factor of 1000 or more, making the device a strong candidate for the industrial purification of biological materials.

References

1. Strickler, A.: Sep. Sci. 2;3, 335 (1967).

2. Bier, M., ed: Electrophoresis: Theory, Methods and Applications, vols. I & II, Academic Press, New York 1959.

3. Saville, D., Ostrach S.: Final Report, Contract #NAS-8-31349 Code 361 (1978).

4. Ivory, C., Turk, R.: submitted to Chem. Engg. Sci.

5. Nakayama, W., Hwang, G., Cheng, K.: J. Heat Transfer, Trans. ASME, Series C, 92, 61 (1970).

6. Bird, R., Stewart, W., Lightfoot, E.: Transport Phenomena, John Wiley and Sons, New York 1960.

7. Ivory, C.: J. Chrom. 195, 165 (1980).

AN EVALUATION OF A VIDEO IMAGE CORRELATION TECHNIQUE FOR THE ESTIMATION OF ELECTROPHORETIC MOBILITIES OF HUMAN BLOOD CELLS

A.J.Bater, J.O.T.Deeley, J.A.V.Pritchard
Immunology Department, Velindre Hospital, Cardiff, CF4 7XL,
UK.

Introduction

The adoption of particle electrophoresis as a general
technique has been hampered by technical difficulties. The
standard method of electrophoretic mobility measurement uses
a microscopic time-of-flight determination of the particles
moving in a known electric field. The method is unfortunately
slow, tedious, subjective, and only suitable for the analysis
of a small sample from a given population. These problems can
be alleviated by the automation of the measuring process.

Several approaches to the automation of electrophoretic
mobility measurement have been proposed. These include laser
doppler velocimetry (1), image transduction techniques (2),
and image analysis procedures (3). There are many advantages
of using direct imaging. True image analysis techniques
require the use of large and expensive computers. In
contrast, the analysis of video signals by a correlation
procedure is simple, quick, and relatively inexpensive. The
software implementation of the correlation technique has been
described previously (4). A dedicated hardware cell
detection and correlator unit has been developed by Malvern
Instruments, UK. (patent applied for). The application of the
prototype apparatus to the electrophoretic analysis of blood
cells is described in this paper.

Materials and Methods

Human erythrocytes were obtained by finger puncture and washed
in phosphate-buffered saline before use. Electrophoretic
measurements were made on freshly prepared cells. A model
bimodal population was prepared by treating a portion of the
cells with neuraminidase and mixing control and treated cells
in the required proportions. Lymphocytes were prepared from
heparinised peripheral blood using a standard 1.077 g/ml
Ficoll/Isopaque cushion (5). Manual and automated
electrophoretic measurements were made using a Rank Mk.III
microelectrophoresis cell (6). Mobilities were expressed in
Tiselius Units or TU (7). The measuring process for the
automated system is described below.

Technical Principles

The microscope image is viewed with a standard 625 line
monochrome television camera. The camera is mounted
horizontally so that the electrophoretic movement of the cells
is perpendicular to the camera scan lines. The composite
video signal from the camera is processed by the
electrophoresis correlator, which comprises a cell detection
unit and a correlation processor. Scan lines crossing cells
in the selected optical plane have a characteristic waveform
and are "tagged" by the cell detector. The tag identifies
acceptable cells on the monitor image. The tag signals and
the associated field and line synchronising pulses are
separated from the composite video signal by the correlation
processor which acquires tag information from a single field.
A preset number of frames later further tag information is
acquired from the following field. The horizontal image shift
is computed using a simple cross-correlation procedure on the
information from the two fields. The peak correlation
coefficient represents the position of the most probable
picture shift averaged over all the tagged cells in the

processed picture area. A number of cross-correlation functions computed from pairs of frames are integrated and a continuously updated function is displayed on the monitor. At the end of the measuring process the integrated correlation function is transferred into the microcomputer for further analysis. The approximate peak position is selected by the operator; accurate peak mobility is then determined using a nine-point, second order polynomial curve fitting routine.

Results

The accuracy and reproducibilty of the system was demonstrated with repeat measurements of erythrocytes from the same donor. Automated determinations had a mean value of -10.91 ± 0.16 TU compared with -10.74 ± 0.23 TU obtained manually. These values are in agreement with published data. The smaller standard deviation of the automated values is probably due to the larger number of cells (at least 200 cells per sample) analyzed by the correlation technique.

The correlation function of a single cell is a triangle having a base width equal to twice the "tagging width" of the cell. The correlation function of a population of particles with normally distributed electrophoretic mobilities is the product of many such triangles. Consequently, the width is much greater than that of the frequency distribution. Figure 1a shows a smoothed cross-correlation function and a smoothed manually-determined mobility distribution for samples from the same batch of human erythrocytes. The peak of both curves is in the same position. It is difficult, if not impossible, to deconvolute a correlation function into the underlying frequency distribution. Therefore, it may be impossible to estimate the standard deviation of the population from the correlation function.

304

The curve broadening effect of the correlation procedure need
not preclude the use of the technique for estimating the
relative proportions of two or more sup-populations in a
multi-modal mixture. Figure 1b shows a smoothed cross-
correlation function and a smoothed manually-determined
mobility distribution for the same batch of human lymphocytes.
The overall shape of the correlation function resembles that
of the frequency distribution, with the peaks and shoulders
occurring in the same places in both curves. This suggests
that the peak broadening effect occurs equally across the
whole curve.

Figure 2 shows the correlation functions obtained for control
cells, treated cells, and various mixtures of the two. The
measured mobilities of the two cell types are unaltered in the
mixtures. Visually dividing the correlation functions of the
mixtures into two peaks gives relative areas which agree well
with the theoretical proportions of the two cell types.

This process works when the two sub-populations have widely
separated mean mobilities. When the average mobilities are
similar, or when one or both populations have a very wide

(a)   (b)

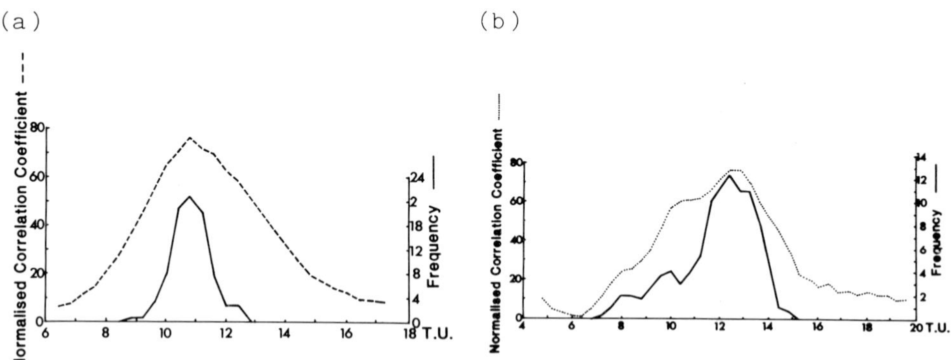

Figure 1.   Smoothed mobility distributions (solid lines) and
            smoothed correlation functions (broken lines) for
            (a) erythrocytes, and (b) lymphocytes.

mobility distribution, the composite function has the form of
a peak and shoulder making it difficult to identify the mid-
point between the two component peaks. Two mathematical
models have been applied to the correlation function: normal
distributions, and triangles.

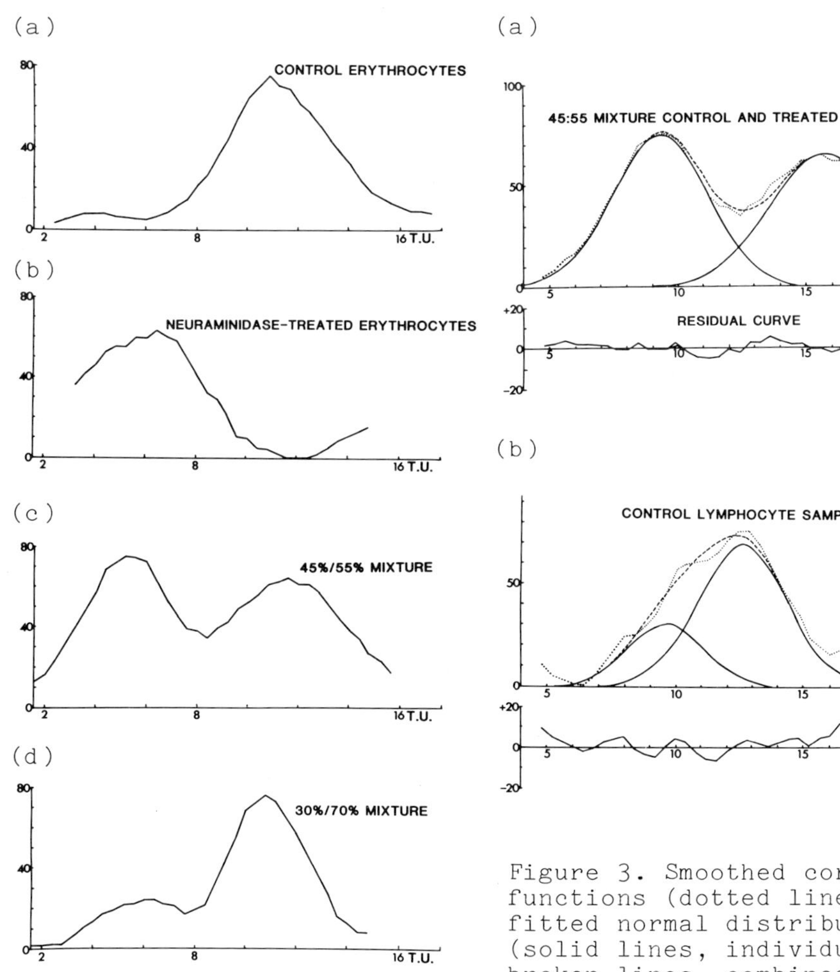

Figure 2. Smoothed correlation
functions for the following
mixtures of treated and control
erythrocytes; (a) 0%/100%,
(b) 100%/0%, (c) 45%/55%,
and (d) 30%/70%.

Figure 3. Smoothed correlation
functions (dotted lines) and
fitted normal distributions
(solid lines, individualcurves;
broken lines, combined curves)
for (a) 45%/55% erythrocyte
mixture, and (b) lymphocytes.
Residual curves are the
difference between correlation
functions and combined normal
distributions.

306

A normal distribution is described by the three parameters,
the height, the mean, and the standard deviation. If one or
two baseline parameters are to be included, the description of
the curve consisting of two normal distributions plus a
sloping baseline requires eight parameters. A non-linear
least squares algorithm to derive the values of these
parameters is being investigated. Figure 3a shows the result
of fitting two normal distributions to smoothed correlation
functions from an erythrocyte model mixture; the background
has been ignored for the analysis. The relative proportions
of the sub-populations derived from the integration of the
normal curves are 49% slow cells and 51% fast cells, compared
with the theoretical 45:55 ratio. Similarly, integration of
lymphocyte curves (Figure 3b) yields values of 26% slow cells
and 74% fast cells; the generally accepted proportions of B-
cells (slow) and T-cells (fast) are approximately 30:70. The
residual curves included in Figures 3 and 4 indicate the
goodness-of-fit of the mathematical models.

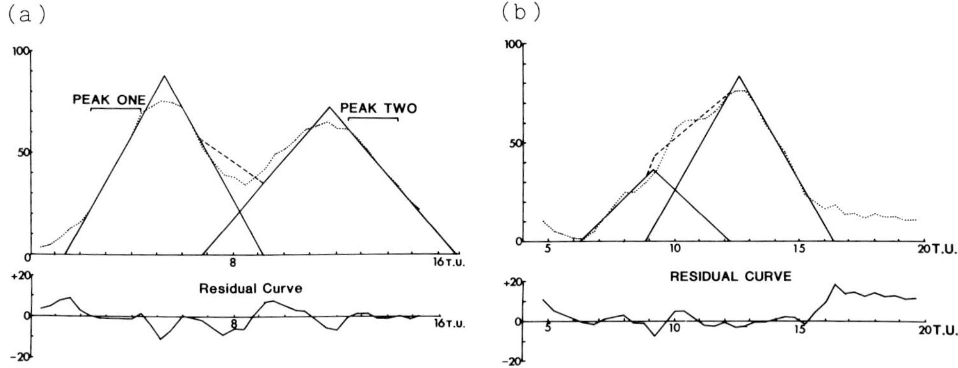

Figure 4. Smoothed correlation functions (dotted lines) and
           fitted triangles (solid lines) for (a) 45%/55%
           erythrocyte mixture, and (b) lymphocytes. Broken
           lines are the sum of the individual triangles. The
           residual curves are the difference between the
           correlation function and the combined triangle
           curve. The regions of the correlation function
           labelled Peak One and Peak Two were those used for
           fitting the slopes of the triangles in (a).

The result of fitting triangles to the correlation function
obtained with the erythrocyte model is shown in Figure 4a.
The peak mobility of the fast population was first obtained
using the polynomial function. Six points on the fast side of
the peak were used for a least squares linear regression
analysis to derive the slope of the fast triangle. The
regression line was extrapolated to give half of the triangle,
which was then reflected on the slow side of the peak. The
process was repeated for the slow peak using the slow side of
the peak for linear regression analysis. The proportions
calculated from the areas of the two triangles, 49% slow and
51% fast, are in close agreement with the theoretical values.
Figure 4b shows the same procedure applied the control
lymphocyte data; the calculated proportions are 26% slow cells
and 74% fast cells.

Concluding Remarks

Using the image correlation technique the mean electrophoretic
mobility of a sample can be obtained accurately in two
minutes. The short measuring time allows a large number of
samples to be studied. Manual determinations based on the
timing of a similar number of cells take between one and two
hours. Mathematical modelling of correlation functions from
mixed populations has been used to estimate the mean mobility
and relative proportions of each component. This analysis
provides little information about the distribution of
individual components since the width of the fitted curve is
not an estimate of the population standard deviation. It is
possible that the width of the correlation function is related
to the standard deviation of the population and the tagging
width of the cells.

Acknowledgements

We are grateful to South Glamorgan Health Authority, South
Wales Cancer Research Council, and Tenovus for their financial
support.

References

1. Uzgiris,E.E.: Optics Commun. $\underline{6}$, 55-57 (1972)

2. Goetz,P.J.: Cell Electrophoresis, (Preece and Sabolovic,
   Eds.), Elsevier/North Holland, pp. 445-461 (1979)

3. Brooks,D.E., Seaman,G.V.F., Olson,G.B., Bartels,P.H.:
   Cell Electrophoresis, (Preece and Sabolovic, Eds.),
   Elsevier/North Holland, pp. 409-419 (1979)

4. Callan,R., Johnson,M., Jones,R., Bater,A.J.,
   Pritchard,J.A.V., Smith,J. and Sutherland,W.H.:
   Cell Electrophoresis, (Preece and Sabolovic, Eds.),
   Elsevier/North Holland, pp. 431-434 (1979)

5. Bøyum,A.: Scand. J. Clin. Lab. Invest. $\underline{21}$,
   (Suppl.97), 31-38 (1968)

6. Sutherland,W.H., Pritchard,J.A.V.: Cell Electrophoresis,
   (Preece and Sabolovic, Eds.), Elsevier/North Holland,
   pp. 421-430 (1979)

7. Catsimpoolas,N., Hjerken,S., Kolin,A., Porath,J.: Nature
   $\underline{259}$, 264 (1976)

APPLICATIONS OF AN AUTOMATED CELL ELECTROPHORESIS EQUIPMENT WITH HIGH RESOLUTION - AN OVERVIEW

Wolfgang Schütt,Uwe Thomaneck,Eberhard Knippel,Joachim Rychly
Clinic of Internal Medicine of the Wilhelm-Pieck-University
Rostock, GDR-25 Rostock

Introduction

At the beginning of the nineteen-sixties increasing use was
made of electrophoresis to investigate biomedical problems.Its
introduction to clinical applications since 1970 (see 1,2,3,
for reviews) showed that the application of cell electrophoresis as an experimental examination technique would be
possible only if a more powerful instrument became available.
Since 1978 we have been using the automatic measuring microscope PARMOQUANT (4,5), which was developed jointly with VEB
Carl Zeiss JENA. It permits many cells to be studied at the
same time with great accuracy. The advantages of being able
to obtain a representative frequency distribution within a
short time and the high resolution of the instrument become
particularly apparent when characterising non-uniform cell
populations. Systematic studies have been performed with the
automatic measuring system in a variety of biomedical
research fields with the aim of establishing the power of
this classical biophysical measuring technique, but also to
identify its limits in order to avoid misapplications.

Methods

Automatic single cell electrophoresis as performed in the
PARMOQUANT is based on the opto-electronic tracing of cells

310

concentrated in a stationary layer in a rectangular cuvette.
The suspension medium used as a rule was phosphate-buffered
saline with an ionic strength of 0.15 and a pH of 7.2. The
optimum cell concentration is 2 - 6 x 10^6/ml. The volume of the
sample required for the measurements is 0.5 - 2 ml.

Results

1) Test results. Repeated measurements on single cells can be
   used to asses an electrophoretic system. Two important
   prerequisites are given in order to determine the actual
   biological scatter by finding the frequency distribution of
   a representative number of individual mobility observations:
   1) the mobilities of the different cells are measured with
   an accuracy of $\pm$ 1 % during a migration time of 2-3 seconds,
   and, 2) due to the extremly small gradient of the velocity
   curve in the measuring plane and the additional electronic
   restriction of the depth of focus, only cells moving at
   approximately the same speed are measured. The lowest relative standard deviations repeatedly measured for erythrocytes from certain donors was below $\pm$ 2 %. The experimental
   reproducibility is $\pm$ 0.6 %. The measuring time necessary for
   obtaining a frequency distribution for 500 mobility values
   is some 10-12 minutes. The achievable resolution for a
   histogram is shown on defined mixtures of erythrocytes from
   various species (Fig. 1).

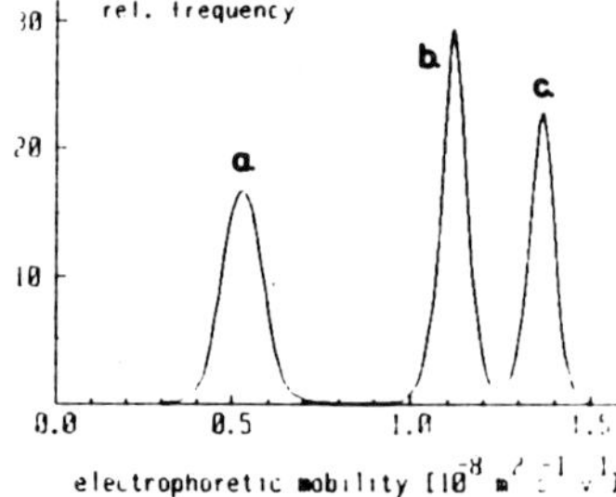

Fig. 1
Mixture of erythrocytes of
different species,
a) rabbit, b) human,
c) dog

2) Frequency distribution of lymphoid cells. In a histogram,
   lymphoid cells yield a bimodal distribution. This analytic
   separation becomes particularly apparent as a result of the
   high resolution of the automatic cell electrophoresis device
   (Fig. 2c). Histograms produced after rosette and adherance
   separatiom show clearly the accumulation of B and T cells,
   as was subsequently proved by measuring the surface immunoglobulins and E-rosettes.

3) Estimation of B/T cell relation. Simple histograms were used
   to estimate the ratio between the numbers of circulating B
   and T cells in
   - 60 patients during recovery after kidney transplantations,
   - 30 tumour patients before, during and after radiological
     treatment,
   - 132 female subjects at different stages of pregnancy.
   The B/T cell ratios of all these subjects were different to
   those of the control group (n = 110). The results obtained
   from the kidney transplant and tumour patients correlated
   with the results yielded by immunological methods of determining the B/T cell ratio. The proportion of T cells, for
   instance, decreased in the kidney transplantation patients
   due to treatment with immunosuppressive drugs and in the
   tumour patients due to radiological treatment. In the second
   and third trimesters of pregnancy the reduction in the number
   of T cells was slight but significant. This observation
   implies qualitative rather than quantitative alterations in
   lymphocyte function to explain the lower cell-mediated
   immune response observed during pregnancy.

4) Characterisation of cell - antiserum interaction. Electrophoretic mobilities of erythrocytes are specifically reduced
   by blood group sera. Specific changes in the electrophoretic
   mobilities of lymphocytes caused by interactions with anti-

lymphocyte sera (ALS) correspond to antibody effects evaluated by immunological in vitro tests (cytotoxicity, E- and
EAC-rosette inhibition). The specific effects of anti-B and
anti-T sera on B and T lymphocytes in the mouse and man are
detectable as decreased electrophoretic mobilities.

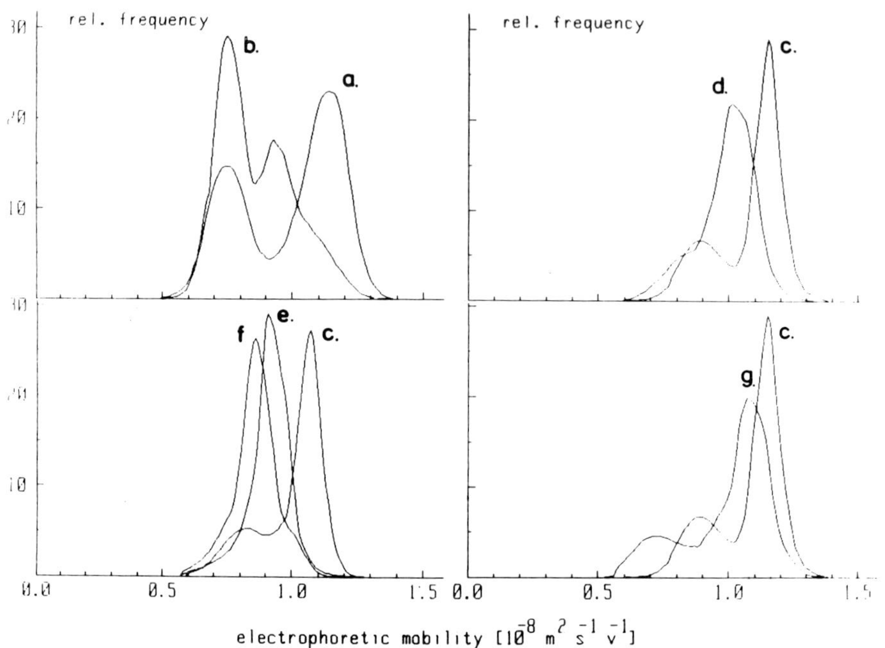

Fig. 2
Effects of anti-lymphocyte sera on electrophoretic mobilities of lymphocytes. a) spleen cells of mouse, control; b)
+ anti-T-serum; c) human peripheral blood lymphocytes,
control; d) + anti-T-serum; e) + monoclonal anti-T-antibody
+ second antibody; f) + monoclonal anti-T-antibody + second
and third antibody; g) + anti-B-serum

The correlation between electrophoretic mobility slowing and
monocyte spreading inhibition caused by the same antiserum
suggest that electrophoretic results seem to be associated
with the expected immunosuppressive action of ALS. The effect
of ALS on the EPM of platelets in vitro corresponds to the
thrombocytopenic effect in man. The analytical cell
electrophoresis is able to detect interactions of cells with
monoclonal antibodies. As a rule the sandwich and double sand-

wich methods have to be used to compensate for low antigen
density. Because cells are selectively open to influences by
specific antibodies detectable as changes in the mean EPM and
cytopherograms, the rapid and easy cell electrophoretic method
can be used to characterise cells and anti-sera.

5) Interactions with lectins. The good analytical separation of
   lymphocyte populations in the histograms yielded by the PAR-
   MOQUANT permits the influence of lectins to be ascertained
   on different populations. The changes in the electrophoretic
   mobilities of mouse thymocytes and spleen cells induced by
   incubation with peanut lectins were investigated.

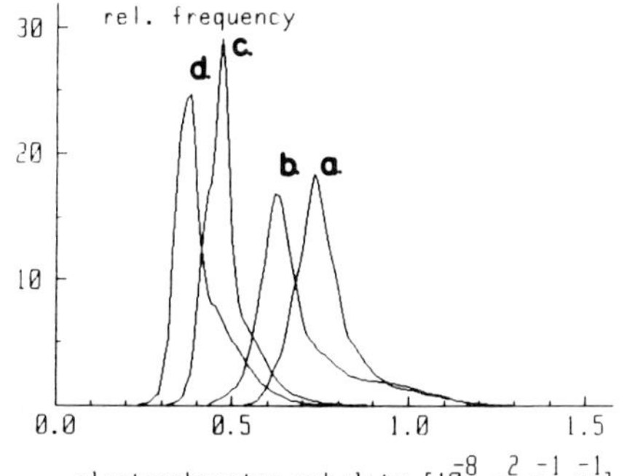

Fig. 3
Effect of PNA and neuraminidase on mouse thymocytes, a)
control; b) + PNA; c) + neuraminidase; d) + both

Treatment with peanut agglutinin (PNA) reduces the mobility
of the slow thymocyte population (immature cortical thymocytes) under capping conditions. It does not, in contrast,
affect the mobility of the mature thymocytes (fast fraction).
No reduction in the mobility of the highly mobile thymocyte
and spleen cell populations occurs under the influence of PNA
until they have been treated with neuraminidase.

6) Electrophoretic characterisation of leukemic cells. The neoplastic cells from the peripheral blood have been characte-
   rised electrophoretically for various forms of acute and
   chronic leukemia. The investigations performed so far have

confirmed that malignant blood cells exhibit unimodal frequency distributions with a standard deviation of about $\pm$
3 %. The electrophoretic mobility of these cells is different
to that of the corresponding normal cells. Our results show
that even leukemias of different origins can be distinguished.
For instance, mobilities are higher in the case of acute
myeloid leukemia (AML) than in the case of ALL. The electrophoretic mobilities of the myoblasts in cases of ALM are
similar to those of normal lymphocytes. The EPM of the mye-
loblasts could be measured more accurately after separation
of the T-lymphocytes by rosette formation. Forms of leukemia
that can be attributed to a cell series can also be differentiated by means of their EPM. These differences seem to
depend on the degrees of maturity and differentiation of the
malignant cells. In the case of leukemias of the B-cell
series, hair cell leukemia has a smaller EPM than CLL. Cells
associated with macroglobulinemia, which are differentiated
plasma cells, have a far higher EPM than resting B-cells.

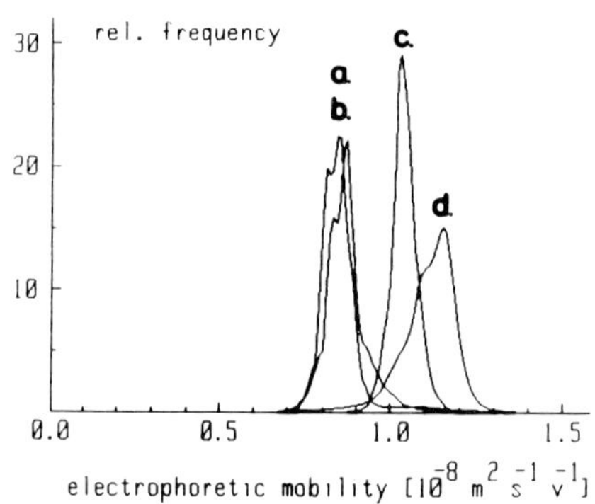

Fig. 4
Examples for histograms of
different leukemic cells
a) Hairy cells (blood),
b) Hairy cells (spleen),
c) CLL, d) macroglobulinemia

The electrophoretic mobility can, together with other morphological, biochemical or immunological methods, be considered a parameter that is suitable for use for differential
diagnosis.

7) In vivo effects on lymphoid cells. Investigation of the relative proportions of mouse thymocyte subpopulations is a good way of studying how mature and immature thymocytes are affected by
   - immunosuppressive drugs: dose-time effects, coergistic effects and
   - transplantations.

Fig. 5
Coergistic effect of cyclophosphamide and prednisolone on percentage of thymocyte subpopulations

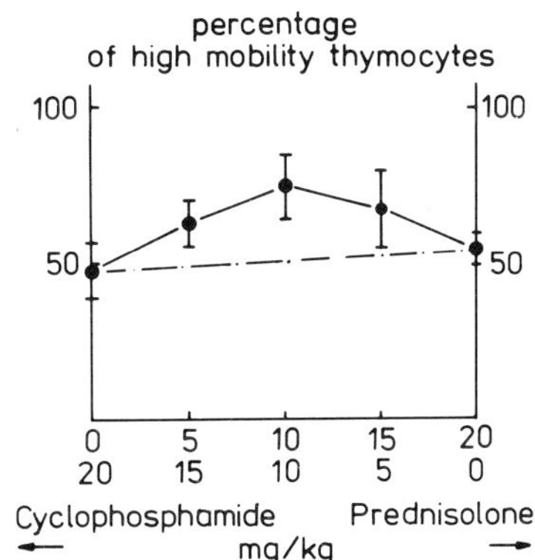

8) Thrombocytes. The PARMOQUANT permits the exact measurement of the EPM of thrombocytes by both manual subjective measurement and, in the case of thrombocyte suspensions containing no aggregations, automatically. Thrombocytes from patients with various disorders showed a different electrophoretic behaviour.

Fig. 6
Histograms of thrombocytes
a) 8 healthy;
b) 4 diabetics;
c) 7 ALL-patients

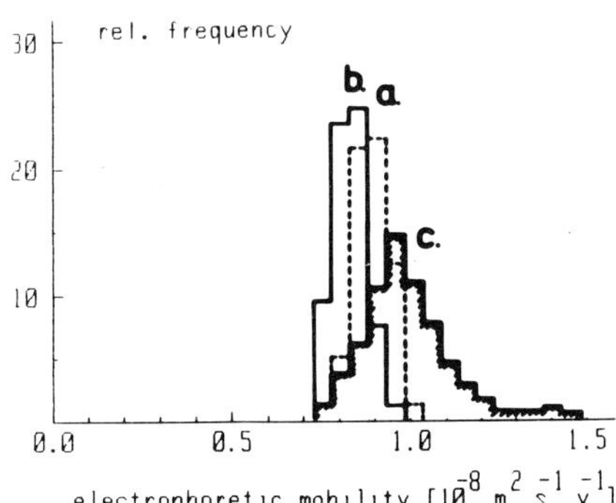

316

9)Liposome electrophoresis. Cell electrophoresis is a suitable
  method for differentiation of liposomes. As shown by electrophoretic measurements of 172 samples from 132 pregnant women,
  amnioticfluid contains at least two kinds of liposomes with
  different electrophoretic mobilities.

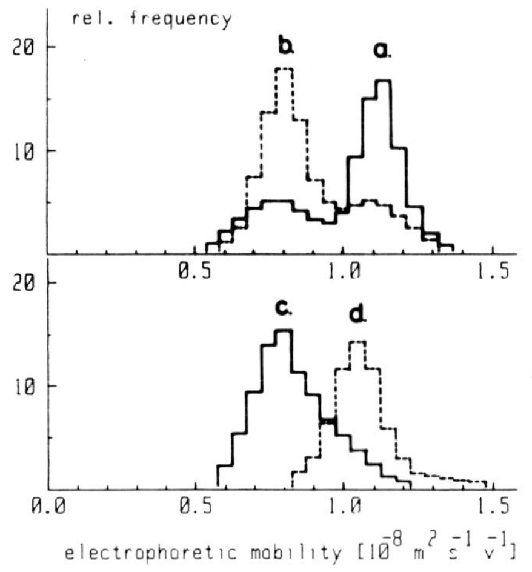

Fig. 7
Liposomes from amniotic fluid:
a) mature fetal lung;
b) immature fetal lung and
fluid obtained by bronchoalveolar lavage; c) lung cancer;
d) pneumonia

The frequencies of the low mobility and high mobility liposomes
were found to be related to apparent fetal lung maturity. In
other words, the electrophoretic investigation of amniotic
fluid seems to be a simple and rapid method for confirming
fetal pulmonary maturity. Liposomes with different electrophoretic mobilities were also found in fluids obtained by
bronchoalveolar lavage. Preliminary results suggest that the
frequencies of several liposomes are likely to correspond to
lung diseases such as lung cancer, sarcoidosis and pneumonia.

We would like to thank our co-operating groups.

ANALYSIS OF LYMPHOCYTE MOBILITY IN TUMOR BEARER BY A FULLY AUTOMATED ANALYTICAL INSTRUMENT

Takao Iwaguchi, Motomu Shimizu
Division of Cancer Therapeutics, Tokyo Metropolitan
Institute of Medical Science

Takeo Mori
Department of Surgery, Tokyo Metropolitan Komagome Hospital

Toshifusa Nakajima
Department of Surgery, Cancer Institute Hospital, Tokyo.

Introduction

Since it was reported by Sabolovic et al. (1,2) that lymphocytes were separable by cell electrophoresis into two main
peaks showing high mobility for T cells and the low mobility
for B cells, there have been several reports in this field
(3,4). Recently Kaplan et al. (5,6) observed in T cell
subsets of normal subjects that the T lymphocyte with greater
affinity for sheep erythrocyte possessed the higher mobility,
while that with weaker affinity possessed the lower mobility.
In studies of electrophoresis of lymphocytes in cancer
patients (7,8,9,10,11) the ratio of the high mobility
lymphocyte to the low has been reported to decrease.
Meanwhile, cell electrophoretic instrumention has been fully
automated from which the mobility histogram as well as the
mean mobility can be obtained by counting objectively through
the optoelectronic image analytical system and rendered
capable of quantitative analysis. The purpose of this study
is to analyze the mobility histogram pattern of lymphocytes
in the host with malignancy from an immunological point of

318

view through basic research by means of the fully automated
cell electrophoretic instrument (Kureha Chemical Ind. Co.,
Ltd.), and then apply the technique to cancer diagnosis and
management of cancer patients.

Results

1) Electrophoretic patterns of mouse lymphocyte in tumor-
    bearing mice.

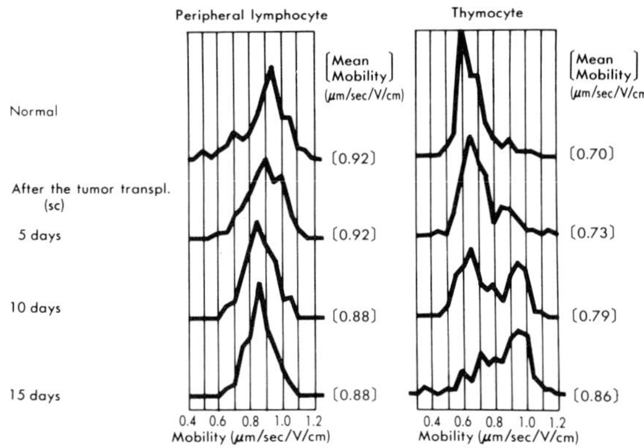

Fig. 1 Changes in the histogram pattern of electrophoretic
mobility of peripheral lymphocytes and thymocytes in
X-5563-bearing C3H/He mice

Figure 1 shows changes in the mobility histograms and mean
mobilities of peripheral lymphocytes and thymocytes from
C3H/He mice given plasmacytoma, X5563, subcutaneously. As
the tumor grew, a shift to the lower zone of the mobility
pattern of peripheral lymphocyte ocurred, along with a
decrease in mean mobility from 0.92 to 0.88 μm/sec/V/cm. In
contrast, the mobility histogram of thymocytes showed a
reverse shift. Monophasic mobility histogram of the thymocyte was displayed with a peak at 0.7 μm/sec/V/cm prior to
the tumor implantation. Five to 10 days after the tumor
implantation, another distinct peak appeared in the region

of 0.9 to 1.0 μm/sec/V/cm. This latter peak became higher
and the pattern broader. The mean mobility also obviously
increased from preimplantation level of 0.70 to 0.86
μm/sec/V/cm. To elucidate the mechanism of electrophoretic
behavior of thymocyte in the tumor bearer, the whole
thymocyte was separated by peanut-lectin (PNA) into
PNA-agglutinated (PNA$^+$) and PNA-nonagglutinated (PNA$^-$)
cells (12). Immature PNA$^+$ cells corresponding to cortical
thymocytes were Lyt-1$^+\cdot$2$^+$ positive cell with low mobility
of 0.7 μm/sec/V/cm. With the progress of tumor growth, there
were few changes in the mobility of PNA$^+$ cells. On the
other hand, mature PNA$^-$ cells in the medullary zone, consisted of cell populations with a high mobility of
1.0 μm/sec/V/cm and a low mobility of 0.7 μm/sec/V/cm, the
ratio of which was different between normal and the tumor-
bearing mice as shown in Figure 2. In any case, the increased ratio of the high to low mobility thymocyte with the
progress of tumor growth is considered to elicit the
appearance of the high mobility peak in the mobility
histogram of the whole thymocyte. From further study of
treatment with αLyt-1$^+\cdot$2$^-$ or αLyt-1$^-\cdot$2$^+$ monoclonal antibody
and complement, the cell with the high mobility of 1.0
μm/sec/V/cm was demonstrated to consist mainly of Lyt-1$^+\cdot$2$^-$
positive, but not Lyt-1$^-\cdot$2$^+$ positive cells.
In pattern analysis of electrophoretic mobility of splenocytes, a new monophasic peak with 0.85 μm/sec/V/cm, was
displayed, while T cells with mobility higher than 1.0
μm/sec/V/cm and B cell peak at 0.65 μm/sec/V/cm were
reduced with the progress of tumor growth, although there
were few changes in the mean mobility (Fig.3). With tumor
growth, the ratio of Lyt-1$^+\cdot$2$^-$ positive cell in thymus
was elevated by a decrease in immature cells, while the low
mobility cell appeared among peripheral and splenic
lymphocytes. These tendencies were evident not only in the
X5563-C3H/He mice system, but also in other systems such as
MH134-C3H/He, MM102-C3H/He, sarcoma-180-ICR, Ehrlich ascitic

tumor-ICR and Lewis lung tumor-C57BL/6 mice (data not shown).
These results suggest that these phenomena occur commonly in
the tumor bearer.

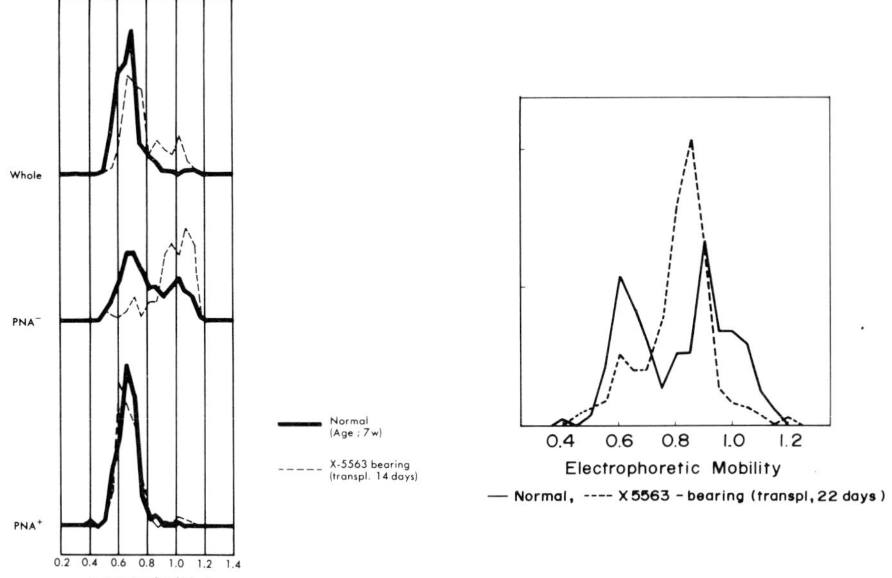

Fig.2 Histogram pattern of
PNA⁻, PNA⁺ and whole thymocytes in C3H/He mice

Fig. 3 Histogram pattern of
splenocytes in C3H/He mice

2) <u>Electrophoretic patterns of peripheral lymphocyte in
    cancer patients.</u> In cancer patients, the whole lympho-
    cytes showed decreased mobility as a result of increased
    numbers of low mobility cells (Fig. 4). Compared with
    that of normal subjects, the mobility pattern of cancer
    patients shifted to the left with three peaks. The ef-
    fluent cell from a nylon wool column decreased markedly in
    mobility, with a single major peak at 0.95 μm/sec/V/cm and
    a slight shoulder at 1.15μm/sec/V/cm. The pattern of the
    adherent cell had a minor single peak at 0.80 μm/sec/V/cm.
    Thus, changes in the lymphocyte mobility histogram of
    cancer patients were mainly attributed to decreased
    mobility of part of the effluent cell. In contrast to

normal subjects, there was a marked increase in the number

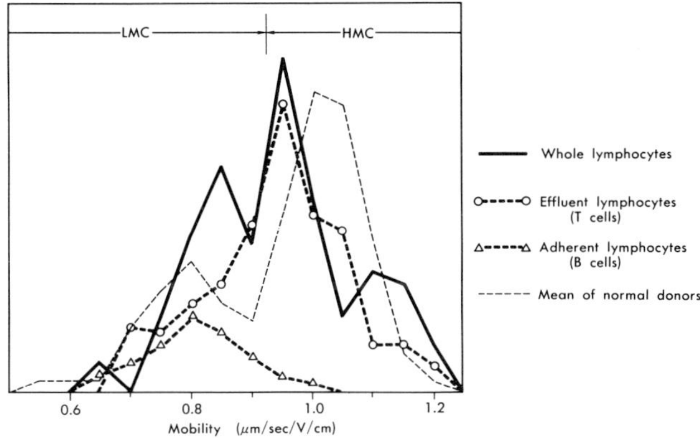

Fig. 4 Histogram pattern of human peripheral lymphocyte
fraction separated by nylon wool column method

of low mobility cells in cancer patients, although there
were few changes in the ratio of T to B cells as determined by rosette-formation assay.

3) Relationship between electrophoretic mobility of lymphocyte
subpopulation and other immune parameters in cancer
patients. A characteristic abnormality of electrophoretic
patterns of lymphocytes in cancer patients seems to be for
the increase in the number of low mobility T cells (LMT).
Thus, the relationship between LMT and other immune
parameters was studied in cancer patients. The percentage
of LMT was calculated by the following formula: LMT (%) =
low mobility cell(%) − B cells(%). High mobility T cells
(HMT) were designated as those at or higher than 0.95
μm/sec/V/cm, and low mobility cells those lower than 0.95
μm/sec/V/cm. The percentage of the low mobility cells was
obtained from electrophoresis of whole lymphocytes, and
that of B cells was determined by EAC rosette-formation
assay. IgG·FcR(+) T cells and immunosuppressive (IS)
substance in the serum were determined as cellular and
humoral parameters of suppressed immunity.

322

Figure 5 shows a significant correlation between the percentage of low mobility T cells and both the frequency of IgG·FcR(+) T cells (p<0.01) and the serum level of IS substance (p<0.005).

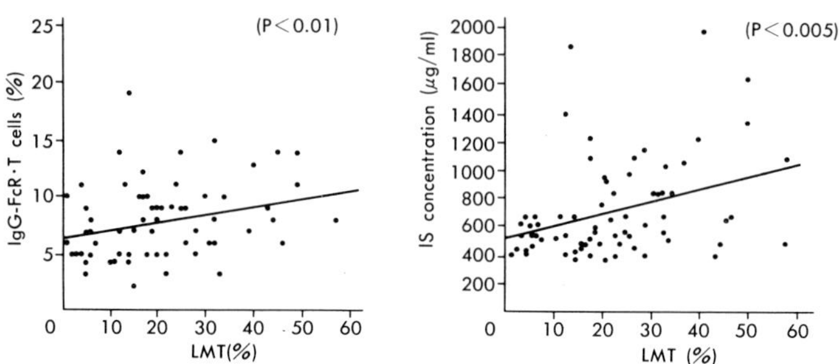

Fig. 5 Correlation between LMT(%) and IgG FcR(+) T cell(%), immunosuppressive substance in the serum

4) <u>LMT/HMT ratio in relation to the types of tumors and clinical stages.</u> Variation in LMT/HMT ratio was examined with respect to the type of tumors and clinical stages. As illustrated in Figure 6, LMT/HMT ratio ranged from 0 to 60 and averaged $21^{\pm}17$ (mean $^{\pm}$S.D.) in the control subjects. When the normal limit was set at 40, 80% of controls fell in normal range. Abnormal LMT/HMT ratios seem to be associated with the tumor burden, because they show a parallel correlation with the clinical stage in three types of cancer. These findings suggest the usefulness of LMG/HMT ratio in the mobility histogram for follow up and early detection of cancer recurrence after surgery.

Conclusions

By pattern analysis of the mobility histogram obtained with fully automated analytical electrophoresis, it was shown that the ratio of Lyt-1^{+}·2^{-} positive cells in the mouse thymus was elevated and low mobility cells appeared among peripheral lymphocytes in animal or human malignancy.

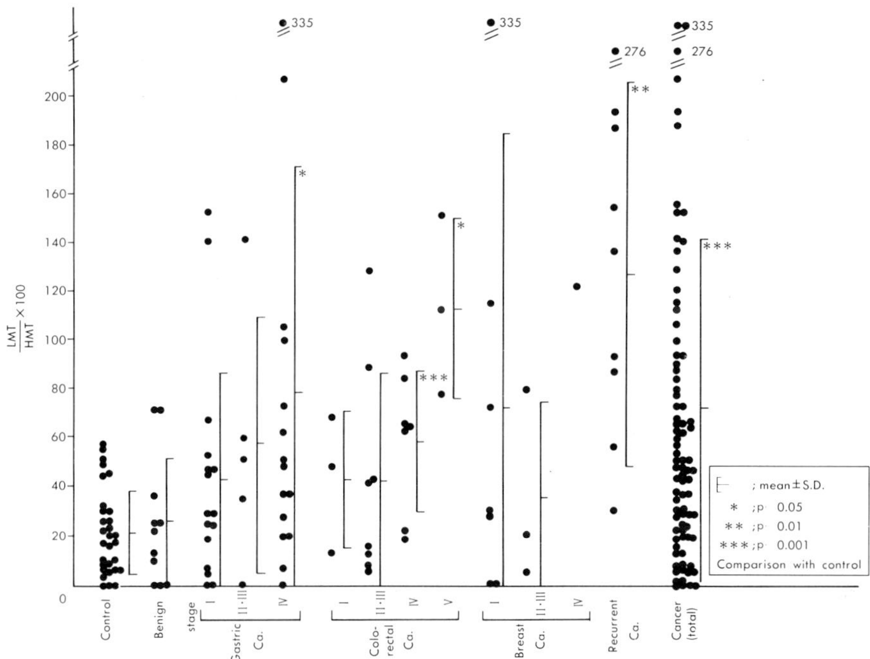

Fig. 6 LMT/HMT ratio in control, benign and cancer patients
in relation to the location and clinical stages

In addition, for humans it was shown that the highest LMT/HMT
ratio (80%) was observed in the group at risk for recurrent
cancer. These facts suggest the usefullness of the LMT/HMT
ratio for early detection of cancer recurrence after surgery.
It is well known that tumor tends to evoke immunosuppression
in the host. Although further investigation is required for
the elucidation of this mechanism, these observations suggest
the possible involvement of an immunosuppressive cell among
the low mobility cells.

Acknowledgements

We would like to thank the Biomedical Res. Labo., Kureha
Chemical Ind. Co., Tokyo, for scientific support.

324

References

1. Sabolovic D., Sabolovic N., Dumont F.: The Lancet <u>28</u>, 927 (1972).
2. Wioland M., Sabolovic D., Burg C.: Nature New Biol. <u>237</u>, 274-276 (1972).
3. Vassar P.S., Levy E.M., Brooks D.E.: Cell. Immunol. <u>21</u>, 254-271 (1976).
4. Chollet Ph., Herve P., Chassagne J., Masse M., Plagne R., Peters A., Biomedicine <u>28</u>, 119-124 (1978).
5. Kaplan J.H., Uzgiris E.E.: J. Immunol. <u>117</u>, 1732-1740 (1976).
6. Kaplan J.M., Uzgiris E.E., Lockwood S.H.: J. Immunol. Methods., 27, 241-255 (1979).
7. Plagne R., Chollet Ph., Chassagne J., Bidet J.M., Sauvezie B.: Biomedicine <u>23</u>, 447-451 (1975).
8. Chassagne J., Chollet P., Vuillaume C., Plagne R.: Biomedicine <u>27</u>, 93-94 (1977).
9. Amiel J.L., Sabolovic D.: Cell Electrophoresis: Clinical application and methodology 119-130, North-Holland Publishing Company, Amsterdam 1979.
10. Kaplan J.H., Uzgiris E.E., Lockwood S.L., Cunningham T.J., Steiner D., Cell Electrophoresis: Clinical application and methodology, 107-117, North-Holland Publishing Company, Amsterdam 1979
11. Nakajima T., Mori T., Iwaguchi T.: Int. J. Cancer <u>31</u>, 6, 681-686 (1983)
12. Iwaguchi T., Shimizu M., Mori T., Nakajima T.: submitted

THE APPLICATION OF CELL ELECTROPHORESIS TO RENAL TRANSPLANTATION

Brian Shenton, Peter Veitch, David Francis, Peter Donnelly,
Ala Alomran, George Proud, R.M.R. Taylor
University Department of Surgery
Royal Victoria Infirmary
Newcastle upon Tyne NE1 4LP, UK

Introduction

Assessment of the functional immune status of renal transplant
patients by _in vitro_ techniques has proved of little value in
routine clinical practice due to the time taken for their
completion. Over the last few years we have developed an
electrophoretic technique for studying lymphocyte function
which takes four hours to perform. We have assessed the
following parameters of the immune status of transplant
patients:
1) the prognostic value of the pretransplant lymphocyte
   reaction to donor cells and renal antigen,
2) the prognostic value of the TEEM test in transplantation,
3) the diagnostic value of the MEM test in monitoring of
   acute rejection.

Materials and Methods

Patients
All renal patients in this study were the recipients of first
renal cadaver allografts performed between 1973 and 1980.

Graft Success
Patients we followed for a maximum of 6 years and the graft
was considered successful if it was able to sustain life

without the need for dialysis. Both immunological and all
failures were included in the analysis. Rejection was
diagnosed independently by the renal physicians.
Electrophoretic Tests of Lymphocyte Function
The pre- and post transplant lymphocyte response of graft
recipients (in donor-recipient mixed lymphocyte reaction (MLR),
to renal antigen,purified protein,derivative of Mycobacterium
tuberculosis (PPD) were measured in the macrophage electrophoretic mobility (MEM) test (1)). The immunosuppressive effect
of plasma was determined in the tanned erythrocyte electrophoretic mobility (TEEM) test (2). In blind studies, both the
MEM test (3, 4) and TEEM test (5) have been found to correlate
with other in vitro tests of lymphocyte function. In
principle both the MEM and TEEM tests depend upon the
liberation of lymphokines from stimulated lymphocytes, which
reduce the electrophoretic mobility of indicator particles -
either guinea pig peritoneal exudate cells (PEC) in the MEM
test or freshly prepared tanned sheep red blood cells (TSRBC)
in the TEEM test (see Fig. 1).

Figure 1 - Scheme for MEM and TEEM testing

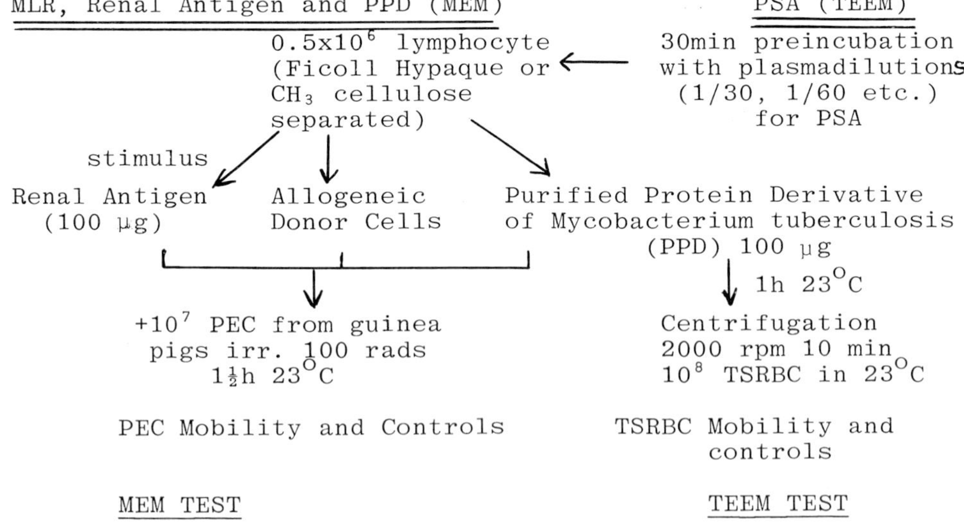

All lymphocytes were washed in TC 199 three times before use
and viability was checked by using trypan blue dye exclusion.
Mobility of the indicator cells was determined visually in a
Zeiss cytopherometer using electronic timing to measure
indicator cell migration over 16µ in the cuvette under a
current of 7.4ma. Automation of the machine produces
increased objectivity (6). Results were expressed as % slowing of the indicator cells compared with controls. In lymphocyte titrations with plasma, results were expressed graphically
and by extrapolation the volume of plasma (in µl) causing 50%
inhibition (Inhibitory Dose 50, ID^{50}) was calculated.

Results

a) The Prognostic Value of MEM-MLR and Response to Renal
   Antigen

Tables 1 and 2 show that the success of renal grafts 6 years
after transplantation was related both to MEM = MLR and renal
antigen responses before operation:

Table 1 – Relationship between 6y.-graft survival and MEM-MLR

Patient Groups	No.	Mean 2 way MLR (%) ± ISD	t test
Successful grafts	30	6.38 ± 2.3	p<.002
Failure grafts	30	9.17 ± 4.1	

Table 2 – Relationship between 6y.-graft survival and
response to renal antigen

	No.	Mean % slowing ± ISD	t test
No rejection	8	12.0 ± 2.4	p<.025
Total rejection	13	14.8 ± 2.7	

328

Both the MEM-MLR and response to renal antigen were significantly higher in those patients who had lost their grafts.
Just as Cochrum (7) and Opelz (8) have divided patients into
groups based on MLR we have divided kidney recipients into a
high MLR (8-16%) and low MLR (0-7.9%) group. Figure 2 shows
that when all cases of graft were considered 59% of low MLR

Fig. 2

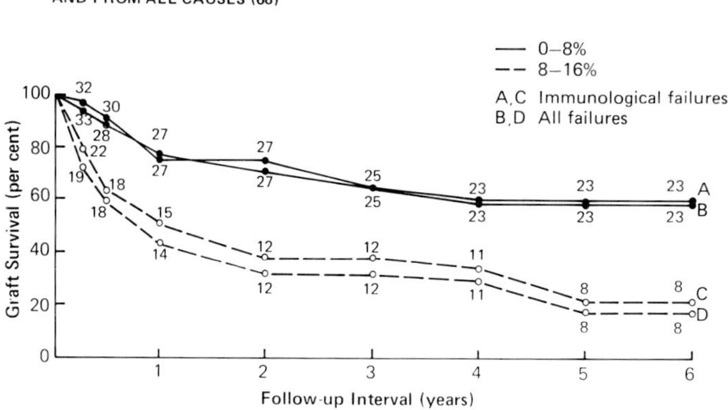

patients (Group B) retained their grafts for 6 years compared with 18% of high MLR patients (Group D). When non-immunological failures were excluded 59% of low MLR patients (Group A) retained their grafts for 6 years compared with 27% of high MLR patients (Group C) (p<.01 Log Rank Sum Test).

b) The Prognostic Value of the TEEM test in transplantation

Over the past few years it has been a universal finding that
blood transfusions given to renal transplant patients are
beneficial to subsequent graft survival. Results from this
centre have shown that an elevated ability of plasma to
suppress lymphocyte reaction to antigen is associated with the
naturally occurring plasma protein α_2Macroglobulin (9). The
relationship between pretransplant PSA and kidney graft
survival has been studied prospectively in 60 renal transplant
patients. Graft survival up to 2 years was compared between

those patients with a high PSA (<5ul) and those with a low PSA
(>5ul). The PSA was expressed as the number of ul of plasma
required to produce 50% inhibition of lymphocytes from a panel
of three normal adult volunteers to PPD antigen in the TEEM
test. As can be seen in Fig. 3, all failures in the low PSA
group occurred within 6 months of transplantation, whereas the
2 failures in the high PSA group occurred after 1 year. The

Fig. 3.

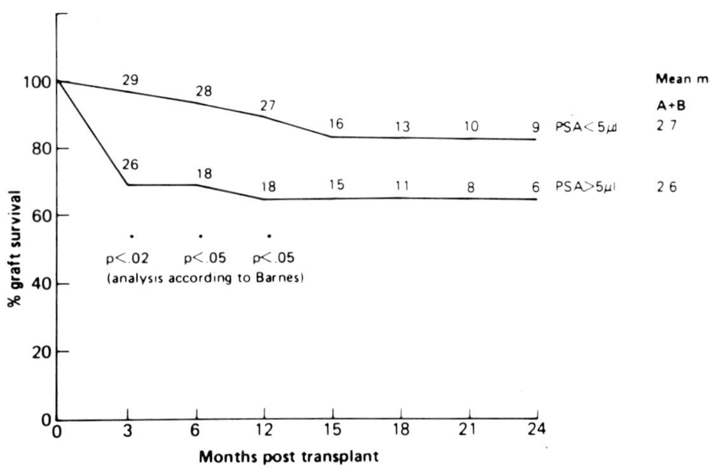

groups were
significantly
different at
3, 6 and 12
months after
transplan -
tation·
Assessment of
the PPD
response of
patients
before trans-
plantation has
shown better
graft survival

in the low responders (10).

c) The Diagnostic Value of the MEM test in Monitoring
   Acute Rejection

In order to determine if cellular reactivity in the post-
transplant period was correlated with clinical rejection,
daily lymphocyte (0.5×10^6) responses of patients were
measured to 0.5×10^6 donor cells (stored in liquid nitrogen),
100Ug of a prepared renal antigen and 100µg of PPD in the MEM
test. The results are shown in Tables 3, 4 and 5.

Table 3 - Peak MEM-MLR Response

Group	State	No.	Mean % MLR ± ISD	t-test
1	Never rejected	5	5.0 ± 0.9	
2	Stable	37	5.6 ± 2.9	p<.001
3	Rejection for Group 2	9	10.8 ± 3.2	
4	Anuric patients	18	7.6 ± 2.1	p<.001
5	Rejection	11	14.3 ± 2.8	

Table 4 - Peak Renal Antigen Response

Group	State	No.	Mean % ± ISD	t-test
1	Never rejected	22	5.9 ± 0.5	
2	Stable	148	4.4 ± 2.7	p<.001
3	Rejection Group 2	24	12.2 ± 2.9	
4	Anuric patients	131	8.2 ± 3.1	p<.001
5	Rejection Group 4	44	13.4 ± 3.3	
6	Infections - no rejections	18	6.9 ± 4.5	

Table 5 - PPD Response

Group	No.	Mean %±ISD	IS daily dose mg
1 No immunosuppression (IS)	51	16.8±0.9	NIL
2 Low IS pre rejection	56	8.5±2.1	>150mg. Imuran > 40mg. Prednisone
3 High dose IS rejection	70	7.1±5.5	>200mg. Imuran >300mg. Prednisone
4 Low IS post rejection	94	7.8±2.8	>150mg. Imuran >100mg. Prednisone
5 No IS post nephrectomy	7	14.5±5.0	NIL

t test 'P' 1 x 2 <.001 1 x 3 <.001 1 x 4 <.001 1 x 5 <.001
2 x 3 <.004 2 x 4 N.S. 2 x 5 <.001 3 x 4 <.005 4 x 5 <.001

In the MEM-MLR, patients (Table 3) either with (Group 2) or
without (Group 4) renal function showed highly significant
elevation of their lymphocyte responses to donor cells during

kidney rejection. With renal antigen tests, rejections were also found to be associated with increased lymphocyte response (Table 4). With PPD, lymphocyte response was depressed during immunosuppressive therapy (Table 5). Thus a comparison between non-specific lymphocyte stimulation (to PPD) and donor specific stimulation (MEM-MLR on Renal Antigen) provided a monitor of how far the sensitised lymphocyte population had escaped from the immunosuppressive regime. The fact that the rise in lymphocyte reactions to donor cells or renal antigen preceded signs of clinical rejections by 4 days allows the test to be of clinical value. No rises in lymphocyte reaction were found after the signs of clinical rejection. We are now evaluating the value of the TEEM test in diagnosing acute rejection.

Discussion

Few reports have shown a positive correlation between pre-transplant MLR results and the clinical success of renal allo-grafts. Whilst some workers (7, 8, 11, 12) have found a high correlation for living donor transplants only Cochrum (7) has found a correlation in unrelated cadaver transplants. The results of this study establishes the MEM-MLR test as a valid measure of compatability between donor/recipient pairs and its use may be particularly important in cadaver transplantation where it is more difficult to obtain a good HLA match. As the test can be completed in 3-4 hours it is particularly suited to prospective studies. The level of sensitisation of recipient lymphocytes before operation to renal antigen was related to ultimate success of the graft and indicated that cellular (13) as well as humoral presensitisation contribute to graft survival. Results have also shown that the measurement of PSA may be of prognostic value, and indeed these studies have also proved a previously hypothetical mode

of action concerning the method of graft protection by blood
transfusion in a large majority of chronic renal failure
patients. The use of the MEM test in the study of lymphocyte
response in the post transplant period has shown that
rejection may be predicted and thus a more effective post-
operative management of clinical rejection episodes may be
achieved.

Conclusion

The use of the MEM and TEEM tests in the field of renal
transplantation not only greatly assists the selection of the
most suitable potential recipients for a given kidney but also
allow an early diagnosis of rejection to be made.

References

1. Shenton, B.K., Field, E.J.: J.Imm.Methods 7, 149 (1975).
2. Shenton, B.K. et al.: J.Imm.Methods 14, 123 (1977).
3. Hughes, D. et al.: Z.Immunitäts. 141,14 (1970).
4. Caspary, E.A.: Immunological techniques for detection of
   cancer (1973).
5. Donnelly, P.K. et al: Europ.Surg.Research 14, 79 (1982).
6. Shenton, B.K. et al. In 'Cell Electrophoresis' Elsevier
   99 (1981).
7. Cochrum, K.C. et al. Transpl.Proceed. 5, 391 (1973).
8. Opelz, G. et al. Transpl. 23, 375 (1977).
9. Shenton, B.K., et al. Transpl.Proceed. 11, 171 (1979).
10. Veitch, P.S., et al. Brit.J.Surg. 68, 813 (1981).
11. Bach, J.F. et al. Clin.Expt.Imm. 6, 821 (1970).
12. Hamburger, J. et al. Transpl.Proceed. 3, 1 (1971).
13. Garovoy, M.R. et al. Lancet 1, 573 (1973).

THE ROLE OF SURFACE NEGATIVE CHARGE ON PLATELET FUNCTION

Kenjiro Tanoue, Stephanie May Jung, Naomasa Yamamoto and
Hiroh Yamazaki
Division of Cardiovascular Research, Tokyo Metropolitan
Institute of Medical Science
3-18 Honkomagome, Bunkyo-ku, Tokyo 113, Japan

Introduction

It has been previously suggested that platelet surface
negative charge was involved in platelet function, especially
in restocetin-induced agglutination (1,2). However, the
precise mechanism is not yet elucidated, at least partially
because of the difficulty in measuring the electrophoretic
mobility of a large number of platelets with a standard
technique. Recently, we have used an automatic Laser Zee
System 3000 (Pen Kem, Bedford Hills, New York) to measure
platelet electrophoretic mobility and obtained good
reproducivility with human platelets (3).
This paper reports the relationship between platelet electrophoretic mobility and sialic acid content and the role of the
negative charge on platelet in ristocetin-induced agglutination.

Materials and Methods

Blood was obtained from young healthy adults and patients
with various prostatic diseases and mixed with 0.1 vol of
3.8% sodium citrate. Platelets were extensively washed and
resuspended as indicated in Results. The sialic acid
contents of the washed whole platelets were determined by
the method of Warren (4). Partially purified von Willebrand

334

factor was prepared by the method of Olson et al (5).
Electrophoretic mobility of platelets was measured with the
Laser Zee System 3000 as described previously (3,6).

Results and Discussion

1) Relationship between sialic acid and electrophoretic
    mobility of platelets.
    Platelets were extensively washed with Phillips' buffer
    finally resuspended into Hanning's solution and measured
    for electrophoretic mobility with the Laser Zee System
    (7). Fig.1 shows a linear relationship between whole
    platelet sialic acid content and the electrophoretic
    mobility of platelets obtained from the patients with
    various prostatic diseases. The relationship was highly
    significant with a correlation coefficient of 0.97.
    We then observed the effect of neurminidase on the
    relationship between sialic acid content and the electro-
    phoretic mobility of washed platelets.
    Neuraminidase-treated platelets showed a dose-dependent
    decreases in both platelet bound sialic acid and platelet
    electrophoretic mobility (Fig.2) which was accompanied by
    a concomitant appearance of sialic acid in the
    supernatant (Fig.3).

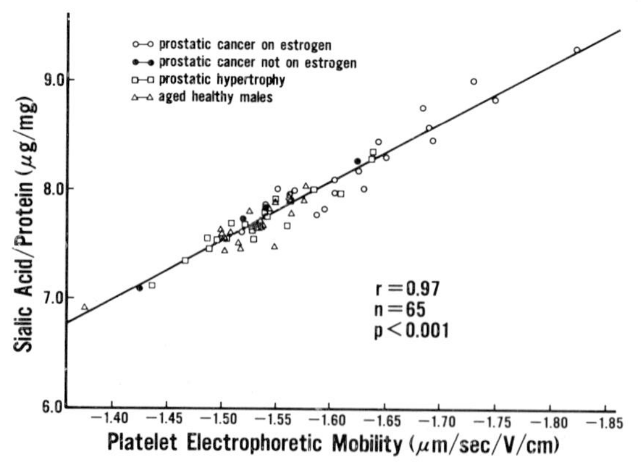

Fig.1.
Relationship be-
tween sialic acid
content and plate-
let electrophoretic
mobility.

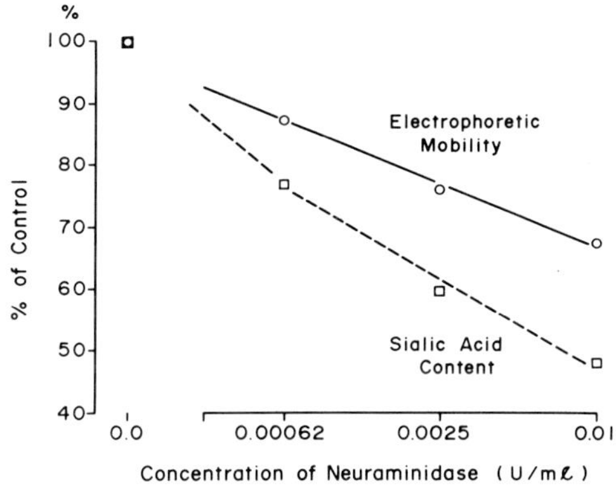

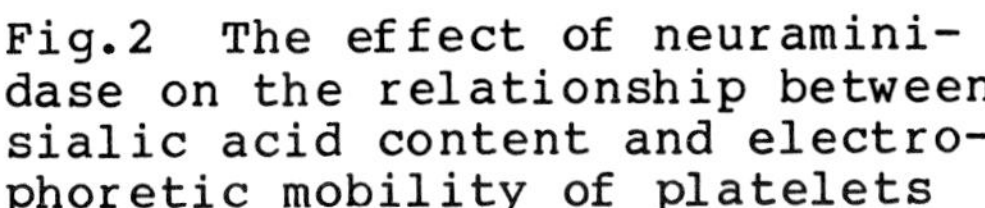

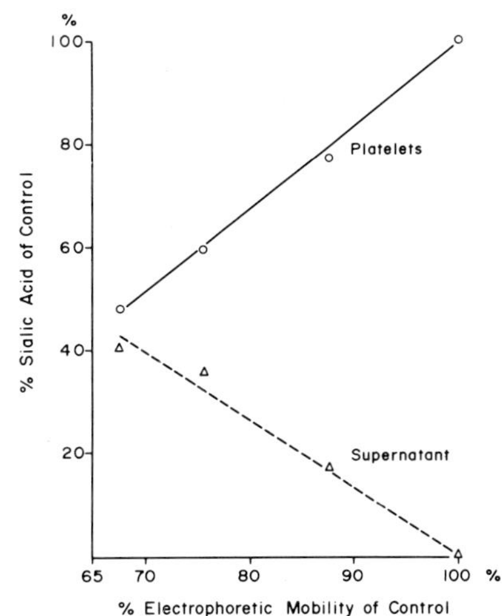

Fig.2 The effect of neuraminidase on the relationship between
sialic acid content and electrophoretic mobility of platelets

Fig.3. The effect of neuraminidase on sialic acid content of
platelets.

These results reconfirmed the previous reports that the
negative charge of platelets was contributed mainly by
the surface exposed sialic acid of the membrane-located
gloycoproteins and gangliosides.

2) Effect of ristocetin on the electrophoretic mobility of
platelets.

Ristocetin, an antibiotic, can induce agglutination of
human platelets in the presence of von Willebrand factor.
This agglutination is analogous to the adhesion of
platelets to subendothelium, which is the first step in
the formation of hemostatic plug and thrombosis (8).
Human platelets were washed and finally resuspended in
Zeiller-Hannig's solution. Normal Saline or irstocetin
dissolved in either normal saline or distilled water was
added to the plitelet suspension.
The electrophoretic mobility and conductance of the medium

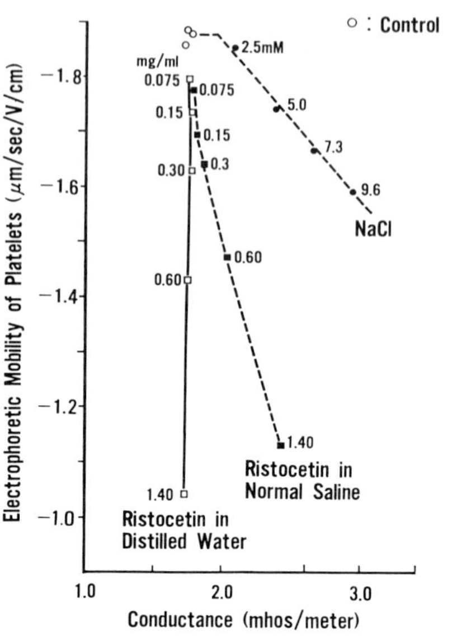

Fig.4. Effect of ristocetin or sodium chloride on electrophoretic mobility of platelets. When ristocetin dissolved in distilled water was added to the platelet suspension to the concentrations as noted on the left of the solid line, platelet electrophoretic miblity was does-dependently decreased without change in the conductance. When normal saline (0.15M NaCl) was added to the concentrations as noted, the conductance was increased in association with apparent decrease in platelet electrophoretic mobility. When ristocetin dissolved in normal saline was added to the platelet suspension, a substantial decrease in electrophoretic mobility was associated with change in the conductance.

were measured with the Laser Zee System. The result (Fig.4) showed that platelet surface negative charge was decreased by ristocetin (9).

3) Role of surface negative charge on ristocetin-induced platelet agglutination.
    Human platelets were washed once by Mastard's method, suspended into Tyrode's buffer and incubated with chymotrypsin or neuraminidase at 37°C for 30 minutes. These treated platelets were then washed 3 times with Philipps Buffer and measured for electrophoretic mobility with the Laser Zee System and for ristocetin-induced agglutination. These platelets were also analyzed for glycoprotein patterns by Laemmli SDS-gels (10). Other portions of the enzyme-treated platelets were fixed with 1% formalin for at least 1 day then washed and the effect

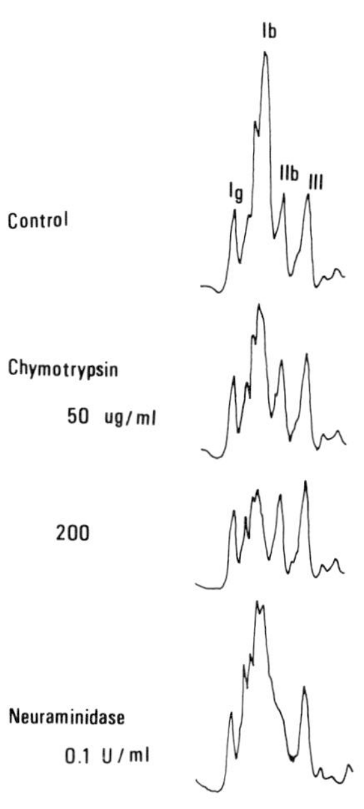

Fig.5 Densitometric patterns of PAS-stained SDS-PAGE of platelets treated with chymotrypsin or neuraminidase. Washed platelets were incubated with chymotrypsin or neuraminidase, then washed 3 times with Philips' buffer and solubilized. Approximately the same numbers of platelets were applied.

of ristocetin with or without von Willebrand factor on electrophoretic mobility was studied.

Fig. 5 shows densitometric patterns of platelet glycoprotein after the treatment with chymotrypsin or neuraminidase.

Treatment with chymotrypsin resulted in almost selective removal of glycoprotein-Ib. Since glycoprotein-III was not changed by this treatment it was used as a standard for quantitative estimation of glycoprotein-Ib.

Fig.6 shows the relationship between electrophoretic mobility and the amounts of glycoprotein Ib of chymotrypsin-treated platelets. This correlation curve suggested that complete removal of glycoprotein-Ib from the platelet membrane caused about a 15% decrease in electrophoretic mobility.

The ristocetin-induced agglutination of the enzyme-treated platelets was does-dependently decreased in chymotrypsin-treated platelets, while neuraminidase-treated platelets

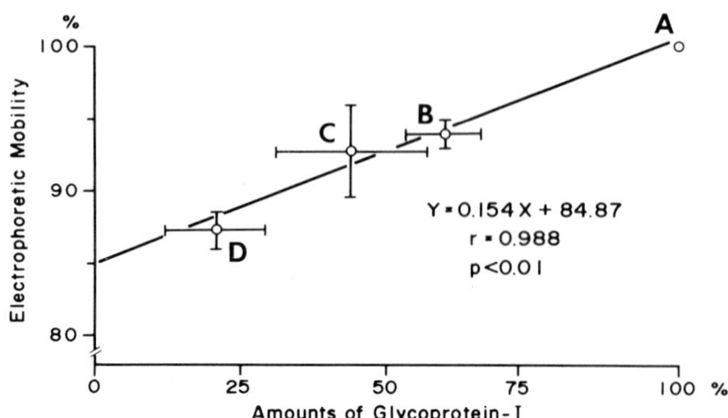

Fig.6   The relationship between electrophoretic mobility
and amounts of glycoprotein Ib of chymotrypsin-treated
platelets.   Concentrations of the enzyme were: A;O,
B;50, C;100, D;200 ug/ml.   M $\pm$ SE from 5 or 6
experiments.

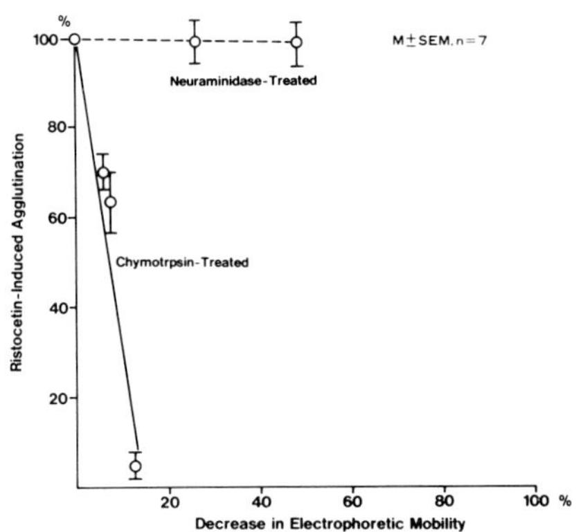

Fig.7.   The relationship between electrophoretic mobility
and ristocetin-induced agglutination of platelets treated
with chymotrypsin or nueraminidase.

showed almost normal agglutination.

Ristocetin-induced agglutination and electrophoretic

mobility of the enzyme-treated platelets were plotted as

shown in Fig.7.

Neuraminidase-treated platelets lost 50% of their electrophoretic mobility, however, their agglutination was not
changed. On the other hand, chymotrypsin-treated
platelets lost only 15% of their electrophoretic
mobility, but they lost almost all their ability to
agglutinate. This finding suggested that the whole
negative charge on the platelet surface is not important
in ristocetin-induced agglutination.
We then observed the effects of ristocetin with or
without von Willebrand factor on the electrophoretic
mobility of the platelets. As shown in Fig.8, both
control and chymotrypsin-treated platelets decreased
their electrophoretic mobility to the same extent in
response to 0.75mg/ml of ristocetin alone. When a small
dose of von Willebrand factor, 2.9 µg/ml, was added to
the platelet suspension in the presence of 0.75 mg/ml
ristocetin, nontreated platelets further decreased their

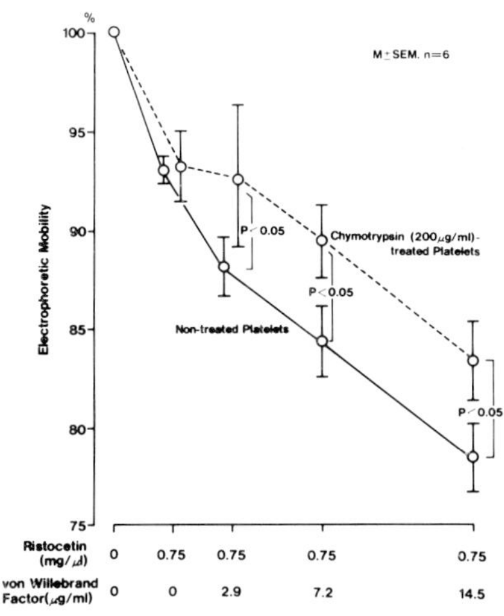

Fig.8. Effects of ristocetin
with or without von Willebrand
factor on electrophoretic
mobility of platelets.

electrophoretic mobility. In contrast,
chymotrypsin-treated platelets showed no decrease in
electrophoretic mobility. With increasing doses of von
Willebrand factor, the enzyme-treated platelets showed
significantly less decreases in electrophoretic mobility
than control platelets. This finding suggested that the
negative charge on glycoprotein-Ib was neutralized by the
binding of von Willebrand factor in the presence of
ristocetin. This change may be critical in the
triggering mechanism for ristocetin-induced agglutination
(to be published).

References

1. Seaman, G.V.F.: Platelets. Their role in haemostasis and
    thrombosis. Brinkhaus, K.M., Wright, I.S., Soulier,
    J.P., Robert, H.R. and Hinnom, S. (Eds). pp.53-68,
    Stuttgart, Schattauer-Verlag, 1967.

2. Coller, B.S.: J. Clin. Invest. $\underline{61}$, 1168-1175, 1978.

3. Shimizu, M., Iwaguchi, T., Kikutani, N., Motomiya, T.,
    Yamazaki, H.: Proc. Japan Acad. $\underline{55}$(B), 418-423, 1979.

4. Warren, L.: J. Biol. Chem. $\underline{234}$, 1971-1975, 1959.

5. Olson, J.D., Brockway, W.J., Fass, D.N., Bowie, E.J.W.,
    Mann, K.G.: J. Lab. Clin. Med. $\underline{89}$, 1278-1294, 1977.

6. Yamazaki, H., Tsukui, R., Motomiya, T., Jung, S.M.,
    Sonoda, M., Watanabe, C., Ogino, M., Miyagawa, N.: Thromb.
    Haemostas. $\underline{44}$, 43-45, 1980.

7. Jung, S.M., Kinoshita, K., Tanoue, K., Ishohisa, I.,
    Yamazaki, H.: Thromb. Haemostas. $\underline{47}$, 203-209, 1982.

8. Phillips, D.R.: Prog. Hemost. Thromb. $\underline{5}$, 81-109, 1980.

9. Tanoue, K., Ishohisa, I., Jung, S.M., Sakakibara, C.,
    Ariga, S., Yamazaki, H.: Blood and Vessel $\underline{13}$, 543-546,
    1982.

10. Tanoue, K., Jung, S.M., Isohisa, I., Yamamoto, N.,
    Yamazaki, H.: Acta Haemat. Jpn. $\underline{45}$, 1391-1400, 1982.

CELL ELECTROPHORETIC ANALYSIS OF POLYMORPHONUCLEAR CELLS IN
COLLAGEN DISEASES

Nobuya Hashimoto, Masakazu Horita, Shunichi Nose, Naomi
Matsumoto, Toshiko Kobayashi, Masaki Ageshio and Masakazu Abe
Department of Internal Medicine, Jikei University School of
Medicine, Tokyo

Introduction

As polymorphonuclear cells have functions such as chemotaxis
and phagocytosis, it is important to investigate change of
surface charge of polymorphonuclear cells. Brown recently
reported electrophoretic analysis of polymorphonuclear cells
from rheumatoid arthritis patients (1). The experiments
reported here were performed to investigate whether the
electrophoretic mobilities of peripheral blood polymorphonuclear cells from collagen diseases patients differ from
those of normal subjects.

Materials and Methods

Patients: Fifteen patients with rheumatoid arthritis (RA),
satisfying the criteria for " classical " of the American
Rheumatism Association (ARA), were examined in this study.
All patients were seropositive according to the results of
a latex fixation test for rheumatoid factor. They were on
cmbinations of several drugs, including non steroid anti-
inflammatory drugs (NSAID). Ten patients with systemic
lupus erythematodes (SLE) who have been diagnosed as
" definite " for criteria of the ARA were studied. On this
experiment some of them were on remission stage. Most of

the patients received prednisolone. Seven patients with
Behçet's syndrome, four patients with progressive systemic
sclerosis (PSS) and three patients with polymyositis (PM)
were also examined. The controls consisted of 10 healthy
adults who were studied comparing with the patients.
Preparation of peripheral blood polymorphonuclear (PMN)
cells: PMN cells were separated from blood by the method of
Chenoweth (2). Briefly, erythrocytes were removed by sedi-
mentation in the presence of Dextran. The leukocyte-rich
supernatant was taken and the cells were centrifuged at
150 xg for 10 min. Residual erythrocytes were removed by
0.83% ammonium chloride solution. Leukocytes were harvested
by centrifugation at 150 xg for 10 min and resuspended in
TC 199. Highly purified neuthrophils were separated from
mononuclear cells by Ficoll-Hypaque centrifugation.
Cell electrophoretic analysis: Analytical cell electrophoresis was performed as described previously (3). All
measurements were made on samples containing 10^6 cells/ml
in a cylindrical electrophoretic chamber (manufactured by
Sugiura Lab. Co., Tokyo). Electrophoretic buffer was
triethanolamine solution devised by Prof. Hannig. All
measurements were carried out at a potential difference of
40 V and current of 200 μA and at 25°C. Each cell's electrophoretic mobility was expressed in terms of velocity (μ/sec)
and field strength (V/cm). The cells in focus in the
stationary layer were measured.
The time for each of 50 cells from each sample to cross one
square of the microscope grid was determined, each cell
being timed in both directions of the field, and recorded by
an electric digital stop-watch with an automatic printer.
Measurements were performed on image projected by TV monitor.

Results and Discussion

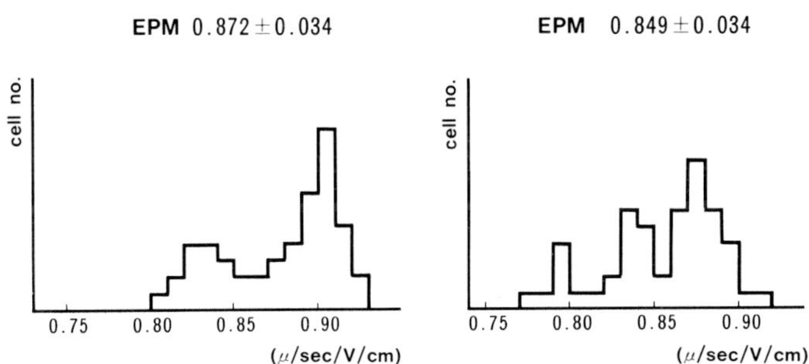

Fig. 1   Typical electrophoretic histograms of peripheral PMN
         cells from normal subjects

Typical histograms of electrophoretic mobility of PMN cells
from two normal subjects are shown in Fig. 1.   Mean electrophoretic mobility value was 0.872 ± 0.034, and the histogram
showed bimodal pattern ((a) of Fig. 1).   The pattern of
histogram shown in (b) of Fig. 1 was three modal.
Electrophoretic mobility values of PMN cells from 10 normal
subjects are presented in Table 1.   Mean electrophoretic
mobility value was 0.867 ± 0.011 μ/sec/V/cm.   Most of electrophoretic histograms showed bimodal pattern, but occasionally
three modal or undifferentiated pattern.
Electrophoretic analysis of PMN cells from RA patients are
presented in Fig. 2 and Table 2.   RA PMN cells had electrophoretic mobilities lower than that of normal PMN cells.
Electrophoretic values of PMN cells from RA patients are
presented as one distribution (Table 2) though occasionally
in RA histogram profile a bimodal pattern was evident.   Mean
electrophoretic mobility of PMN cells from 15 RA patiens was
0.801 ± 0.015 μ/sec/V/cm.   In the histogram bimodal pattern

Table 1 Electrophoretic mobilities of PMN cells from normal
subjects

| subjects | | EPM ± SD | histogram pattern |
age	sex	(μ/sec/V/cm)	
38	M	0.872 ± 0.034	two
40	M	0.849 ± 0.034	three
19	M	0.866 ± 0.051	two
22	M	0.854 ± 0.038	two
31	M	0.880 ± 0.044	*
26	F	0.879 ± 0.060	two
29	M	0.870 ± 0.030	*
41	M	0.862 ± 0.057	three
20	M	0.860 ± 0.041	two
25	M	0.878 ± 0.038	two
mean EPM ± SD		0.867 ± 0.011	

(* not differentiated)

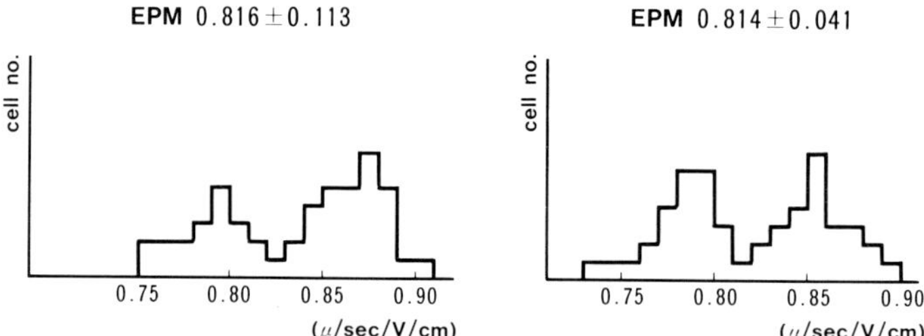

Fig. 2 Typical electrophoretic histograms of peripheral PMN
cells from RA patients

Table 2   Electrophoretic mobilities of PMN cells from RA
          patients

| patients | | EPM ± SD | histogram pattern |
age	sex	(μ/sec/V/cm)	
48	F	0.816±0.113	two
46	F	0.814±0.041	two
58	F	0.790±0.124	two
50	F	0.772±0.108	*
46	M	0.798±0.122	*
19	F	0.808±0.084	two
52	F	0.820±0.062	two
50	F	0.785±0.140	*
56	F	0.794±0.082	two
35	F	0.790±0.065	two
58	F	0.818±0.088	two
28	F	0.782±0.110	two
32	F	0.810±0.044	*
34	M	0.796±0.136	two
36	F	0.825±0.132	two
mean EPM ±SD		0.801±0.015	

which had two peaks of around 0.79 and 0.85 to 0.88 were
shown (Fig. 2).   Brown et al. reported that probit analysis
of electrophoretic distribution of RA PMN cells suggested
the presence of three main sub-populations (1).
Electrophoretic mobilities of PMN cells from 10 SLE patients
who were mostly treated with prednisolone and NSAID were
investigated (Table 3).   Mean electrophoretic value of SLE
PMN cells was 0.832 ± 0.029, not significantly different from
normal PMN cells.   In the electrophoretic histogram a definite
modal distribution was not evident.   In a few case of SLE
patients mean electrophoretic values were lower than normal
control value, however correlation between electrophoretic
values and clinical characteristics was obscure.

Table 3 Electrophoretic mobilities of PMN cells from SLE
        patients

| patients | | clinical | treatment | EPM ± SD |
age	sex	phase		(μ/sec/V/cm)
23	F	active	PSL, NSAID	0.820 ± 0.160
32	F	active	PSL, NSAID	0.774 ± 0.046
19	F	active	PSL, NSAID	0.797 ± 0.033
34	F	active	PSL, NSAID	0.818 ± 0.052
37	F	remitted	——	0.867 ± 0.020
40	F	remitted	PSL, NSAID	0.850 ± 0.018
29	F	remitted	PSL,NSAID	0.862 ± 0.011
38	F	remitted	PSL, NSAID	0.824 ± 0.022
27	F	remitted	NSAID	0.858 ± 0.028
25	F	remitted	——	0.846 ± 0.129
mean EPM ± SD				0.832 ± 0.029

For electrophoretic analysis of PMN cells from PSS and PM
patients no alteration was shown comparing with normal
controls (Table 4). As only 4 PSS and 3 PM patients were
experimented, this investigation remains to be resolved.
In electrophoretic mobilities of PMN cells from Behçet's
syndrome patients diagnosed as clinically complete and
incomplete type, the mean value as a whole was 0.833 ± 0.032
μ/sec/V/cm (Table 5). Comparing with the normal controls,
it was not significant. In a few case, however, electrophoretic values of PMN cells from Behçet's syndrome were
lower than that of the normal controls. As it has been
reported that PMN function in Behçet's syndrome was changed
(4), relationship between function and decrease of electrophoretic mobility of PMN cells from Behçet's syndrome should
be more investigated.

Table 4 Electrophoretic mobilities of PMN cells from PSS and
PM patients

| patients | | EPM $\pm$ SD |
age	sex	(μ/sec/V/cm)
PSS		
44	F	0.860 $\pm$ 0.062
60	F	0.872 $\pm$ 0.025
52	M	0.844 $\pm$ 0.037
38	F	0.855 $\pm$ 0.052
mean EPM $\pm$ SD		0.859 $\pm$ 0.010
PM		
49	F	0.886 $\pm$ 0.027
51	F	0.806 $\pm$ 0.042
40	M	0.825 $\pm$ 0.141
mean EPM $\pm$ SD		0.839 $\pm$ 0.034

Table 5 Electrophoretic mobilities of PMN cells from Behçet's
syndrome patients

| patients | | clinical type | treatment | EPM $\pm$ SD |
age	sex			(μ/sec/V/cm)
40	F	complete	PSL, Colch.	0.810 $\pm$ 0.182
39	M	complete	PSL, NSAID	0.868 $\pm$ 0.084
42	M	complete	PSL, NSAID	0.790 $\pm$ 0.038
48	M	complete	PSL, NSAID	0.862 $\pm$ 0.114
32	F	incomplete	PSL, NSAID	0.872 $\pm$ 0.056
36	M	incomplete	NSAID	0.832 $\pm$ 0.202
54	F	incomplete	NSAID	0.798 $\pm$ 0.048
mean EPM $\pm$ SD				0.833 $\pm$ 0.032

Conclusion

Cell electrophoresis was performed on peripheral blood PMN
cells obtained from collagen disease patients and normal
controls.
Electrophoretic histogram of normal PMN cells showed two or
three modal patterns (occasionally, undifferentiated).
Electrophoretic mobilities of PMN cells from RA patients were
lower than that of normal controls.
Electrophoretic mobilities of PMN cells from SLE, PSS, PM and
Behçet's syndrome patients as a whole showed no alteration.
However, reduction of electrophoretic mobility was found in
a few case of SLE and Behçet's syndrome. Relation between
electrophoretic change of PMN cells and clinical characteristics in these diseases remains to be more studied.

References

(1)   Brown, K. A. et al., : Cell Electrophoresis in Cancer
      and other Clinical Research, (ed. by A. W. Preece and
      P. Ann Light), 209-216, Elesevier / North-Holland,
      Amsterdam (1981)

(2)   Chenoweth, D. E. et al., : J. Immunol. Methods $\underline{25}$:
      337-353. (1979)

(3)   Hashimoto, N. et al., : Electrophoresis '81 (ed. by
      Allen and Arnaud) 983-993, Walter de Gruyter, Berlin
      (1981)

(4)   Fordham, J. N. et al., : Ann. Rheum. Dis., $\underline{41}$, 421-425.
      (1982)

ELECTROPHORETIC MOBILITY TEST FOR GYNECOLOGICAL MALIGNANCY

Masaoki Yamada, Ryoki Ohkawa, Kimiyasu Ohkawa
Department of Obstetrics and Gynecology
Nippon Medical School, Tokyo Japan

Introduction

Field and Caspary (1) reported that lymphocytes from patients
with malignant disease can be stimulated by specific antigen
to release a macrophage slowing factor (MSF) or an electrophoretic slowing factor (ESF) which reduces the electrophoretic mobility of quinea pig peritoneal macrophage. With few
exceptions, the effect was not found in lymphocytes from
nonmalignant controls. This technique, which is known as the
macrophage electrophoretic mobility (MEM) test (2,3), is
possible in-vitro cellular immunological-test for cancer. MEM
test results have been shown to be highly correlated with the
results of macrophage migration inhibition and lymphocyte
transformation assays of cell mediated (4). However, if
constant electrophoretic mobility is important, the use of
tanned sheep red blood cells (TSRBC) has many advantages over
the use of quinea pig peritoneal macrophages as reported by
Shenton et al (5,6). This technique, called the tanned sheep
erythrocyte electophoretic mobility (TEEM) test, has been
shown to correlate with the MEM test. In our present study
the TEEM test was used to study malignancies in gynecological
patient and mechanism of MSF (ESF) production was
investigated. Finally we make another attempt to discern a
relationship between the TEEM test and the macrophage
migration inhibition test (MIT).

Electrophoresis '83
© 1984 Walter de Gruyter & Co., Berlin · New York

Material & Methods

A. Mononuclear cells

Serum free cells were separated from 10-20 ml of heparinized venous blood by differential centrifugation and the isolated cells were washed three times with RPMI-1640 medium. These cells were then suspended in the same medium at a concentration 1×10^6/ml as measured with a coulter counter. Additional studies were performed using the lymphocyte rich fraction from which macrophages were removed adhesion.

B. Antigens

The main antigen used Encepalitogenic Protein (EP) of E.A. Caspary & E.J. Field (7) (Fig. 1), Procedure used to extract EP, which is known to be one of the cancer basic protein (8).

First, adult normal brain tissue was homogenized and lipid was removed.

Secondly, EP was precipitated by saturated ammonium sulphate in the solution of substance incapable of liquifying PH 2.5 HCl, soluble at PH 1.5 HCl.

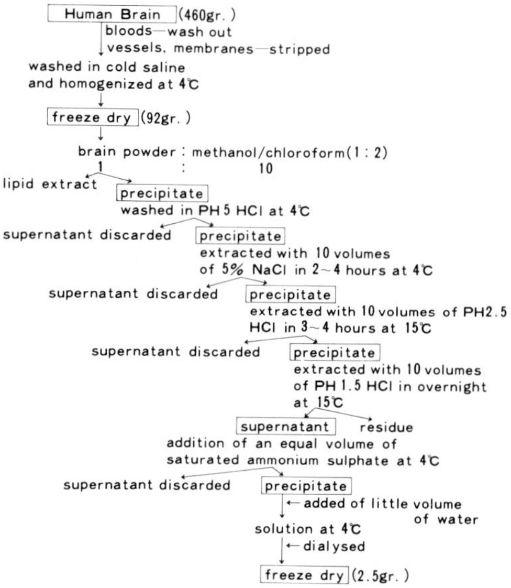

Fig.1 Extracation of encephalitogenic protein

C. The method of Stavifsky[9] was used to prepare sheep red blood cells (TSRBC)

2.5 g% of Freshly drawn sheep red blood cells in Hanning triethanolamine buffer (10) were incubated with an equal volume of dilute tannic acid (usually 1/40000, freshly made in the same buffer) for 10 min. at 37°C. The cells were washed three times and then resuspended in a PH 7.2 buffer solution. A stock solution containing 4 x 10^8 erythrocytes/ml.

D. Cell electrophoresis

The mobility of the tanned sheep red blood cells was measured. A constant current of 2.0 mA was applied at a voltage 300 volts and the observation chamber was maintained at a constant temperature of 23°C by a thermostat.

E. Test system

In the first stage of the TEEM test, 1x10^6 human mononuclear cells from healthy women and women with gynocological malignanciss were mixed with 100 μg EP in a volume of 2 ml at 37°C. Three incubation procedures were used: 1 hour, 24 hours and lymphocyte rich fraction without MØ by adhesion. Human blood cells in this solution were removed by centrifution. The supernatant was added to 4 x 10^8 (in 1 ml) TSRBC. Solution and incubated for 60 min. at 37°C. The order in which samples were measured was random.

F. Measurements

The mobility of TSRBC over 16 μ in the stationary layer of the electrophoresis cells was recorded by an electric timer. Results are reported as the percentage change in mean migration time for antigen treated cells relative to untreated cells (migration index).

That is, $\frac{Tt - Tc}{Tc}$x100 Where Tt is the mean migration time of cells in the presence of the antigen and Tc is the mean for untreated cells from the same subject.

352

G. Macrophage migration inhibition test (MIT)
    Additional studies were performed using the indirect MIT
    for ESF. 20ml for this test, parafin were injected into
    the intraperitoneum of a quinea pig. Exdate cells were
    extracted from the intraperitoneum after 72 hours. Red
    blood cells then washed out, we used exdate cells for
    macrophage (MØ) and so packed microcappillary.
    Electrophoretic slowing factor (in a 2 ml), was added and
    the cells incubated for 24 hours at 37°C. The migration
    index was calculated of area ratio from each test system.

Results

A. TEEM test
    Both of healthy women and malignancy electrophoretic
    mobility was not significantly P=0.01 greater for cells
    incubated for 1 hour than those incubated for 24 hours.
    No significant differences were seen for the cases.
    Results for lymphocytes only, decrease of percentage
    reductive was confirmed in both healthy women and woth
    malignancy (Fig.2). The TEEM test shows MØ significant
    in this regard. And so we examined on our test system,
    1×10^{6} human mononuclear cells/ml with MØ were incubated
    for 60 min. in the first stage of the TEEM test. More
    than 7 percentage reduction was regarded as positive, we
    were tried to diagnosis of gynecological malignancy at
    the clinically. We were investigated the positivity
    ratio in this test for 35 controls, 23 cases with benign
    tumors (uterus 10 cases; ovaries 13 cases) and for 105
    cases with gynecological malignancies that is carcinoma
    of cervix ut., carcinoma of ut. body, carcinoma of ov.,
    others gynecological carcinoma and terminal stage on the
    several stages in each one (Table 1). A significant
    ratio of mobility reduction is seen for cases with
    malignant tumors. The reduction was particularly high

for recurrent carcinoma cases. Post operative malignant
tumors have a tendency to delay in some periods. On the
other hand, this tendency to delay is not observed in
the terminal stage, suggesting a lowered sensitivity in
cellular immunological function. The TEEM test was found
to be an efficient as a parameter to clinically
distinguish in immunity.

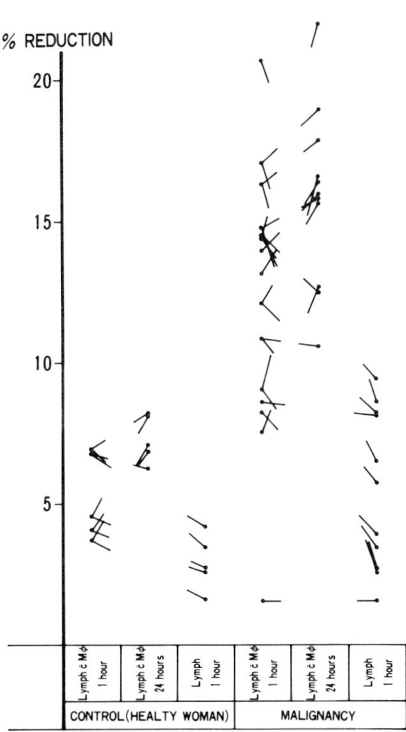

Fig.2 Basic TEEM test

Table 1 TEEM test of gynecological malignancy

	DATA No.	NEGATIVE $<7\%$	POSITIVE $\geq 7\%$	POSITIVE RATE(%)
Normal	35	34	1	2.86
Benign Tumor	23	21	2	8.70
Malignant Tumor	105	12	93	88.57

				CASES	NEGATIVE $<7\%$	POSITIVE $\geq 7\%$	POSITIVE RATE(%)	
CONTROL (HEALTHY WOMAN)				35	34	1	2.9	
BENIGN TUMOR OF UT.				10	10	0	0	*, **
BENIGN TUMOR OF OV.				13	11	2	15.4	
CARINOMA OF UTERI	Ca. of Cervix ut.	preoperative	early *	21	1	20	95.2	
			invasive **	8	0	8	100 *	
		postoperative		17	2	15	88.2	
		recurrence		12	0	12	100	
	Ca. of uterine body preoperative			7	0	7	100	
	Ca. of uterine body postoperative			3	2	1	33.3	**
CARCINOMA OF OV.	preoperative			7	1	6	85.7	
	postoperative			7	2	5	71.4	
	recurrence			12	0	12	100	
OTHERS CARCINOMA				7	0	7	100	
TERMINAL STAGE				4	4	0	0	*, **

* early include   T₁ & T₂
** invasive include   T₃ & T₄
* P<0.05   N.S.
** P<0.01   N.S.

B.   Comparision of TEEM test and MIT

We used the standard index for MIT, which was extracted for electrophoretic slowing factor from healthy women and so standard migration index was calculated for the mean migration ± 2 SD.

Well, we judged that less than standard migration index was positive in MIT. We found the significant migration index for seven out of ten cases (Table 2). So electrophoretic slowing factor not reducted electrophoretic mobility but also migration rates in the of macrophage migration inhibition test. In other words, percent reduction of TEEM test results are correlated with results from the macrophage migration inhibition test.

Case	Condition	slowing of TEEM test		MIT	
		60 min	24 hours	60 min	24 hours
S.K.	Cervix Ca T_1 preoperative	14.75		57.8	46.5
M.K.	Cervix Ca T_1 preoperative	8.21	10.30	34.9	12.7
Y.E.	Cervix Ca T_1 preoperative	13.96	15.96	78.0 *	94.0 *
F.Y.	Cervix Ca T_1 preoperative	12.75	16.62	42.4	22.6
S.A.	Cervix Ca T_2 preoperative	15.65	14.76	77.7 *	66.0
T.S.	Cervix Ca T_2 preoperative	8.75	7.80	55.9	40.1
S.I.	Cervix Ca T_1 preoperative	13.14	15.18	65.3	36.4
S.O.	Cervix Ca T_2 postoperative	12.07	15.65	101.2 *	103.9 *
M.N.	Cervix Ca T_2 postoperative	9.67	12.51	42.6	27.2
H.K.	Ov. Ca Ⅲ preoperative	14.82	16.39	62.5	61.3

* N.S.

MI(%) : Standard   MEAN$\pm$2SD   69.56~88.44

Table 2   TEEM test makes a comparative study of MIT

References

1. Field, E.J. & Caspary, E.A.: Lancet, ii, 1337 (1970)
2. Pritchard, J.A.V. et al: Brit.J.Cancer, 28 (suppl.1), 229 (1973)
3. Bagschawa, K.D.: A tibiotics Chemother, 25, 155 (1978)
4. Chanegue, P.R. et al: Clin. Exp. Immunol. 14, 37-45 (1972)
5. Shenton, B.K. et al: Brit.J.Cancer, 28 (suppl.1), 215 (1973)
6. Shenton, B.K. et al: J.Immunol Methods, 14, 123 (1977)

7. Caspary, E.A. & Field, E.J.: Ann. N.Y. Acad. Sci, 122, 182-193 (1965)

8. Muller, M. et al: Cancer Letters, 2, 139-146 (1977)

9. Stavitsky, A.B.: J. Immunol, 72, 360 (1954)

10. Hannig, K.: Method of Microbiology, vol.5B, N.Y. Acad. Press, 513 (1973)

RELATION BETWEEN AGGLUTINATION AND ELECTROPHORETIC MOBILITY
OF SHEEP ERYTHROCYTES

Nobuya Hashimoto, Masakazu Horita, Shunichi Nose, Naomi
Matsumoto, Toshiko Kobayashi, Masaki Ageshio and Masakazu Abe
Department of Internal Medicine, Jikei University School of
Medicine, Tokyo

Introduction

In electrophoretic mobility test tanned sheep erythrocytes
are used as indicator cells. It is important to investigate
electrophoretic change of indicator cells which were treated
with various materials. Otherwise, it has been shown that
agglutination of lymphocytes was induced by particulate
materials such as histone (1) and synthetic peptide (2) in
the malignancies. The purpose of this study is to measure
charge of the sheep erythrocytes incubated with histone and
poly-l-lysine, and to investigate ability to agglutinate with
these materials.

Materials and Methods

Sheep erythrocytes, tanned and stabilized with sulphosalicylic
acid (EIC) were obtained as a lyophilized preparation from
Behringwerke, West Germany, and adjusted to 10^7 cells in
0.1 ml. Also freshly drawn sheep erythrocytes (f-SRBC) and
polystylene latex particles (Nakarai) which have diameter
of 1.0 μ were used.
Electrophoretic mobility of erythrocytes was determined in a
analytical cell microelectrophoretic apparatus, manufactured
by Sugiura Laboratory, Co., Tokyo. Electrophoretic time was

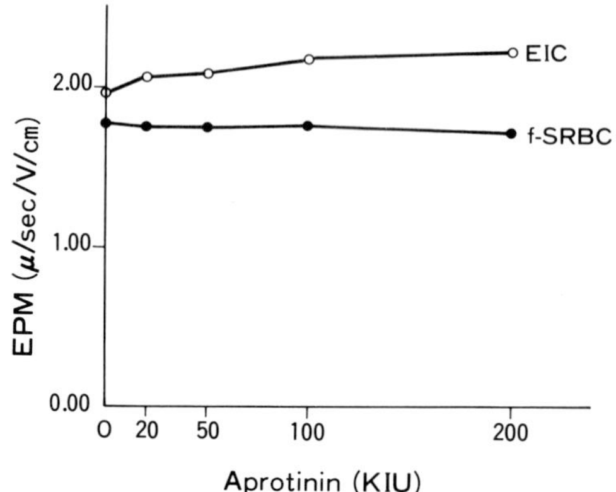

Fig. 1   Effect of aprotinin on cell electrophoretic mobility

recorded by an electric digital stop-watch with an automatic
printer.
Electrophoretic buffer was triethanolamine solution, devised
by Prof. Hannig. Measurements were carried out at temperature
of 25°C, at a potential difference of 40 V, 200 µA and 15 V,
60 µA for latex particles.
As reagents which have an effect on electrophoretic mobility,
Aprotinin, proteinase inhibitor (Trasylol), Poly-l-lysine
4000 (Sigma), Histone from calf thymus, F 2a (Sigma) and
Neuraminidase (Nakarai) were used.

Results and Discussion

Effect of aprotinin on cell electrophoretic mobility was
investigated. In f-SRBC, addition of aprotinin of concentration from 20 to 200 KIU did not influence on electrophoretic
mobility, but in EIC, treatment of aprotinin slightly
increased electrophoretic mobility (Fig. 1). This means

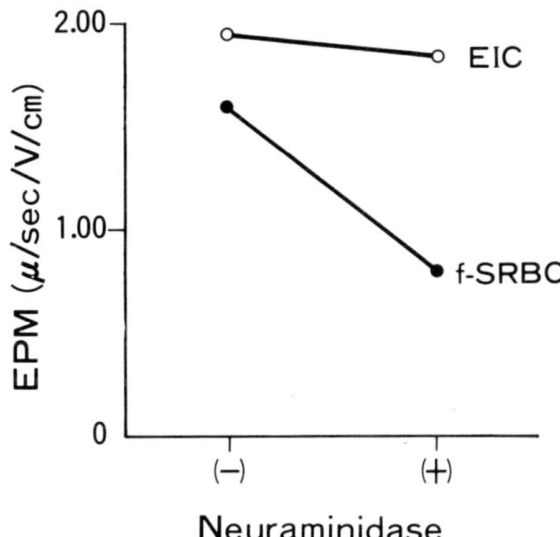

Fig. 2   Effect of neuraminidase on cell electrophoretic
        mobility

that aprotinin was not adsorbed to f-SRBC but to EIC. Latner
(3) reported that increase of negative charge on surface of
tumor cells treated with aprotinin was to prevent loss of
surface protein by inhibiting proteiase activity of cells.
Treatment of sheep erythrocytes with neuraminidase made
electrophoretic mobility reduced (Fig. 2). The reduction
was striking in f-SRBC. As negative charge of surface on
the cells depends on sialic acid, it is reasonable that
reduction of cell electrophoretic mobility was induced by
treatment of neuraminidase.
Effect of poly-l-lysine on electrophoretic mobility was
investigated (Fig. 3). In f-SRBC, treatment of 10^{-5}mg
of poly-l-lysine did not induce change of electrophoretic
mobility, but reduction of electrophoretic mobility was
observed in EIC. It is interesting that agglutination induced
by poly-l-lysine was different in f-SRBC and EIC. In addition

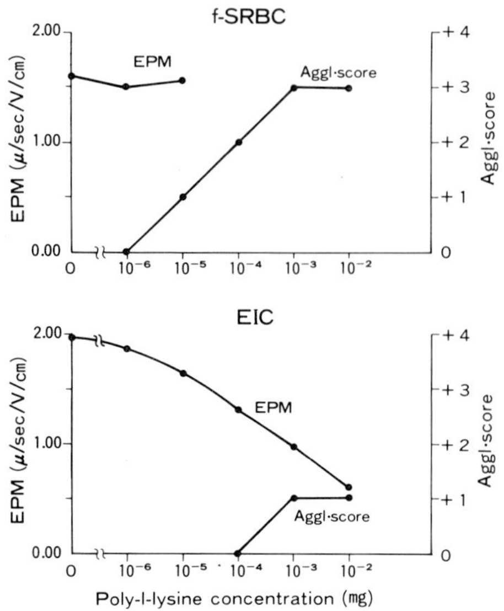

Fig. 3   Effect of poly-1-lysine on cell electrophoretic
mobility and agglutination

of f-SRBC with 10^{-4}mg of poly-1-lysine, agglutination score
was +2 and it was unable to electrophoreze the cells due to
strong agglutination. In EIC, however, addition of 10^{-2}mg
in higher concentration induced only agglutination score +1,
and it was possible to measure electrophoretic time.
Treatment with histone from calf thymus to f-SRBC did not have
an effect on electrophoretic mobility in any concentration
(10^{-3} μg to 1 μg) of histone (Fig. 4). In EIC, however,
1 μg of histone reduced electrophoretic mobility. Addition of
5 μg of histone could not be electrophorezed due to strong
agglutination in both f-SRBC and EIC. Otherwise, agglutination of f-SRBC induced by histone was stronger than that in
EIC. By addition of 1 μg of histone EIC did not agglutinate,

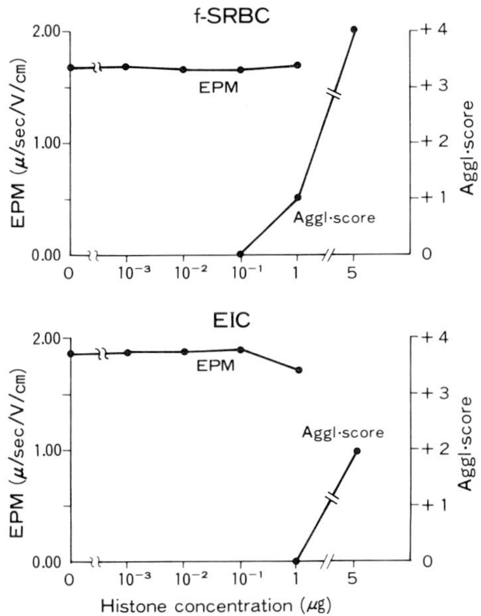

Fig. 4　Effect of histone on cell electrophoretic mobility
and agglutination

while f-SRBC showed agglutination score +1. Addition of 5 μg
of histone showed agglutination score +4 in f-SRBC, and +2 in
EIC. This result was similar to that of poly-l-lysine and
means that reduction of zeta potential on the cells did not
correlated to agglutinability.
Polystylene latex particles were used in place of f-SRBC and
EIC (Fig. 5). By addition with 5×10^{-4}mg of poly-l-lysine,
electrophoretic mobility of polystylene latex particles was
decreased, and extremely reduced in 5×10^{-3}mg of poly-l-
lysine. Although decrease of the electrophoretic mobility
was observed, agglutination of latex did not take place in
both 5×10^{-4}mg and 5×10^{-3}mg of poly-l-lysine.
To 10^{-1} μg of concentration of histone, the electrophoretic

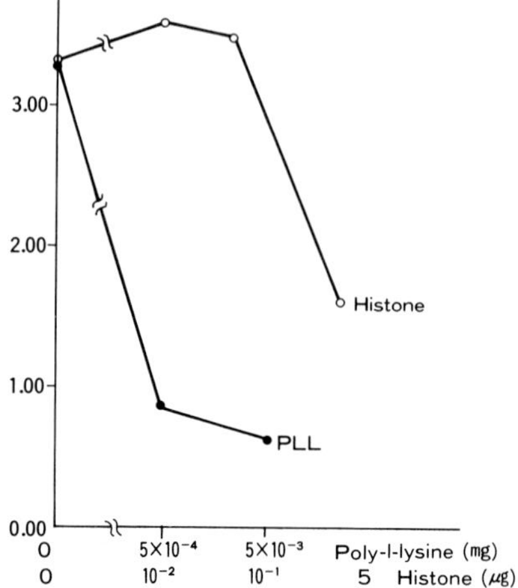

Fig. 5　Effect of poly-l-lysine and histone on electrophoretic mobility of latex particles

mobility of latex was not changed, but it was remarkably reduced in treatment with 5 μg of histone. In the higher concentration of histone, electrophoretic time was extremely delayed and not reproducible.

Although reduction of the electrophoretic mobility of latex was observed, latexes were not always agglutinated by addition of these materials.

Conclusion

Relation between reduction of electrophoretic mobility and ability of agglutination of sheep erythrocytes and latexes was investigated by using poly-l-lysine and histone. These materials induced change of surface charge on the cells and particles, and reduction of electrophoretic mobility and strength of agglutination were not always correlated.

References

(1)   Sabolovič, D., et al., :   Brit. J. Cancer <u>32</u>, 28-33 .
       (1975)

(2)   Bauer, H. W. and Ax, W. :   Z. Krebsforsch. <u>92</u>, 35-40
       (1978)

(3)   Latner, A. L. and Sherbet, G. V. :   Exp. cell Biol. <u>47</u>,
       392-400 (1979)

A FUNCTION TEST OF CHORIONIC VILLI

Yoshida, Y., Yamada, M., Hatano, H.,
Shimizu, Y., Ohkawa, R., Ohkawa, K.
Department of Obsterics and Gynecology, Nippon Medical
School, Tokyo, Japan

Introduction

Viewing pregnancy from the viewpoint of immunology, the
embryo can be regarded as a sort of transplanted tissue
homogeneous with those of the mother's body. The fact that
the embryo can continue growing within the mother for forty
weeks without rejection is a peculiar immunological
phenomenon. If the embryo dies in the uterus, it is rejected
by means of an immunological method which dissolves the
villi. Because it is primarily accomplished by cellular
immunological mechanisms, this protective process can be
diagnosed during the early stage of pregnancy using MSF
(Macrophage Slowing Factor), but this is a complex testing
procedure. However, Shenton and others have proposed the
TEEM (Tannic acid Erythrocytes Electrophoretic Mobility)
Test, which tests the electrophoretic mobility of cells using
tanned sheep red blood cells to absorb the MSF. We report
here our investigations using the TEEM test to analyze
changes in the level of mobility by studying villous function
as a means of prognosis for threatened abortion.

Methodology

(1) Method for extracting villous antigen; (E.A. Caspary
method) Homogenized villous is freeze-dried and the lipid is
removed. Protein sediments of 5 pH is removed and then

highly alkali protein deposits of less than pH 2.5 are
extracted and dissolved in water. Saturated ammonium
sulphate is added to this solution, and the sediment obtained
is dialyzed and freeze-dried.
(2) Experiments for the culture of lymphocytes: Pregnant
woman's lymphocytes are dissolved and cultured in RPMI with 1
x 10^6 lymphocytes /ml. to which 100 μg of villous antigen
are added. The supernatant fluid is removed from the
solution. Tannic acid treated sheep red blood cells are
added to the supernatant fluid, and this solution is then
incubated at 370 C for minutes.

Results

Table I shows results using the TEEMTest in normal pregnancies, and pregnancies with both good and poor prognoses for
threatened abortion. We consider a retarded mobility rate of
about 7% as positive. For normal pregnancies, the positivity
rate was 10%; the rate for pregnancies with good prognoses
for threatened abortion was 28.6%, while cases with poor
prognoses showed a 70% positivity rate, indicating a significant difference in the retardation rate of mobility.
Table II showing test results for patients with poor
prognoses for threatened abortion indicates a significant difference in the mobility retardation rate for this group.

Case 2 is a patient with a history of habitual abortion who
entered the hospital at seven weeks for medical treatment
after being diagnosed for threatened abortion. Because the
clinical diagnosis showed HCG 4,800 iu/l, ultrasound B mode
scan and a positive HCG pattern, the clinical diagnosis was
good, but Teem Test clearly showed a mobility retardation
rate at 16.2%. When the patient was reexamined at 10 weeks,
the TEEM Test at 21.5% showed a worsening condition, even
though the HCG level had risen to 6,400 iu/l. This case

TEEM test of normal pregnancy and threatened abortion

Condition			Data. No.	Negative >7%	Positive ≦7%	% Positive
Normal Pregnancy (first trimester)			10	9	1	10.0%
Abnormal Pregnancy	Threatened abortion	good prognosis	14	10	4	28.6%
		poor prognosis	10	3	7	70.0%

Table I.

Poor prognosis of threatened abortion

	Name	Years	Diagnosis	G.P.A.	TEEM	H.C.G (iu/ℓ)	Ultrasonic diag	
							G.S.	H.B.
1	T.U	31	13W	4-1-3	8.2%	4,800	(+)	
2	K.O	38	7W Habitual abortion	4-0-4	16.2%	4,800	(+)	(−)
			10W		21.5%	6,400	(−)	
3	M.T	22	7W	1-0-1	−3.5%	1,600	(−)	
4	H.T	27	6W	0-0-0	6.1%	800	(−)	
5	F.I	39	15W	1-1-0	7.9%	9,600	(−)	
6	Y.T	26	14W	0-0-0	12 %	2,400	(+)	
7	M.S	33	14W	4-1-3	21.4%	1,600	(−)	
8	T.K	27	16W	0-0-0	12.9%			
9	S.O	30	9W	1-1-0	9.8%	19,200	(+)	(−)
10	K.Y	34	11W	3-2-1	0 %	51,200	(+)	(−)

Table II.

later became an inevitable abortion.

Conclusion

(1) The TEEM Test is a clinically useful method to test villous function.
(2) The TEEM Test is useful for the prognoses of threatened abortion.
(3) It is recognized that cellular immunology plays a role in the mechanism for removing villi.

References

1) Caspary, E.A. & Field, E.J.: Ann. N.Y. Acad. Sci, 122:182, 1965.
2) Shenton, B.K. et al: J. Immunol. Methods, 14:123, 1977.

DIFFERENT DISTRIBUTION OF THE ELECTROPHORETIC MOBILITY OF PERIPHERAL BLOOD
LYMPHOCYTES IN 3 BABIES WITH DI GEORGE SYNDROME.

Michel Wioland

Laboratoire de Biophysique, Faculté de Médecine Saint-Antoine,
27 rue Chaligny, 75012 Paris, France, in collaboration with

Hôpital Saint-Antoine, Department Professor G. Milhaud
Hôpital des Enfants Malades, Department Professor C. Griscelli
Hôpital Trousseau, Department Professor G. Lasfargues
Hôpital Antoine Beclère, Department Professor J.C. Gabilan.

Introduction

The Di George syndrome is a congenital aplasia or hypoplasia of thymus and
parathyroid glands generally associated with unusual facies, cardiac defects
and other congenital abnormalities, probably related to some intrauterine
insult occuring before the 8th week of gestation. Many patients die in the
neonatal period from their cardiac malformation. However, those who survive,
among other difficulties, suffer from an immune deficiency affecting almost
exclusively the cell mediated immunity (CMI). In some cases, a spontaneous
development of CMI may be observed over months and years (partial Di George
syndrome) (1). Therefore, recognizing the magnitude of the immune deficiency is essential to prescribe the appropriate treatment. To help in the evaluation of the immunological status of these infants, the determination with
a moderately expensive analytical microelectrophoresis apparatus (Cam Apparatus, Impington, Cambs, U.K.) (2) of the distribution of the electrophoretic mobility (EPM) of lymphocytes isolated from the peripheral blood of 3
babies with either complete or partial Di George syndrome is reported.
Normally, age-dependent changes in the distribution of the lymphocyte EPM
is observed : the distribution is bimodal for lymphocytes isolated from
cord blood and from blood of infants up to 2 years of age with a small percentage of cells with a mobility of 0.95 μm.sec^{-1}.V^{-1}.cm, this value having
been chosen to discriminate between low mobility cells (LMCs) and high mobility cells (HMCs). Thereafter, the distribution is unimodal and asymetric
in children and adults, and nearly Gaussian in aged people (3) (Fig. 1).

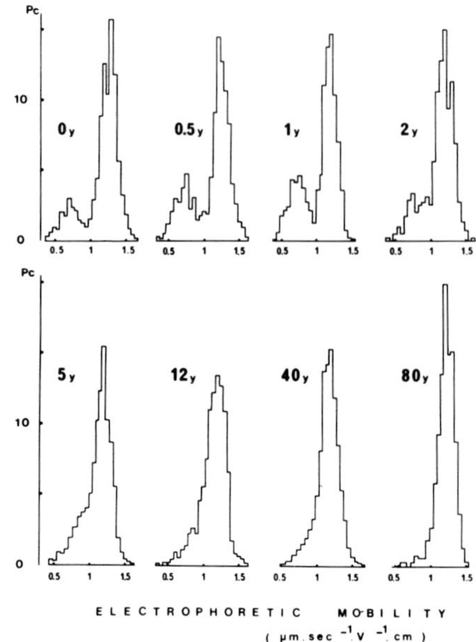

Fig. 1 : Distribution of the electrophoretic mobility of human peripheral
blood lymphocytes (cord blood lymphocytes and lymphocytes isolated from
blood of 6 months and 1, 2, 5, 12, 40 and 80 years old individuals). Pc
represents the percentage of the total cells in each class of mobility.
(reprinted from Scand. J. Immunol. 10, 453-463, 1979).

Results

Case 1 concerns a 3 weeks old baby with a renal polykystic disease associated with the classical features of Di George syndrome. Bovine thymic extracts prepared in this laboratory (4) were used to treat the baby until
he was 18 months of age. A favourable evolution of the illness was noticed.
No change in the proportions of low and HMCs was observed up to now, when
the baby is 4 years of age (manuscript submitted and fig. 2). Case 2 concerns a baby admitted to the hospital at the age of 16 days for severe cardiac difficulties. The diagnosis of Di George syndrome was established in
view of an associated hypocalcemia and an impaired CMI. The distribution
of the lymphocyte EPM, determined at the age of 1 month was bimodal with a
small (2) percentage of cells of mobility 0.95 µm.sec-1.V-1.cm, and a greater percentage of LMCs (49) compared to controls (16 ± 6 ; p < 0.05). After
treatment with thymosin (3 mg/day for 3 months) the distribution was unimodal and nearly symmetrical with the highest (12) percentage of cells of
mobility 0.95 µm.sec-1.V-1.cm (Fig. 3).

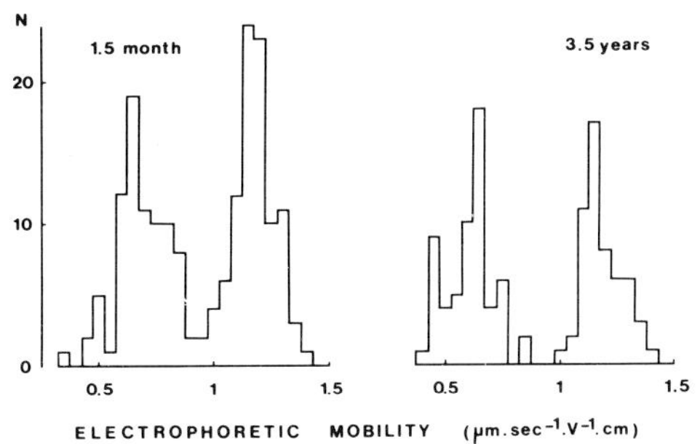

Fig. 2 : Distribution of the electrophoretic mobility of peripheral blood lymphocytes from a baby with Di George syndrome, at the age of 1.5 month and 3.5 years. N represents the number of cells in each class of mobility.

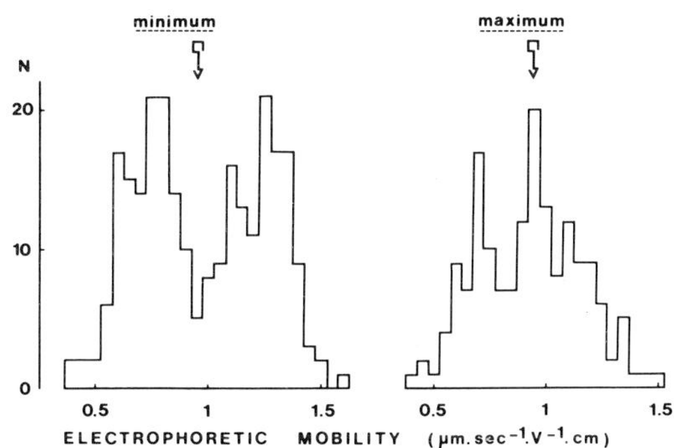

Fig. 3 : Distribution of the electrophoretic mobility of peripheral blood lymphocytes from a baby with Di George syndrome, before and after treatment with thymosin. N represents the number of cells in each class of mobility.

Case 3 related to a baby admitted to the hospital for a suspicion of partial Di George syndrome, kept in a sterile atmosphere for 3 months and grafted with a foetal thymus at the age of 4.5 months. The distribution of the lymphocyte EPM, determined at the age of 1.5 month, was unimodal with a mean value of the lymphocyte EPM of 0.84 µm.sec-1.V-1.cm, i.e., very different from that of the LMCs (0.70 µm.sec-1.V-1.cm) and HMCs (1.26 µm.sec-1.V-1.cm) (3) (Fig. 4).

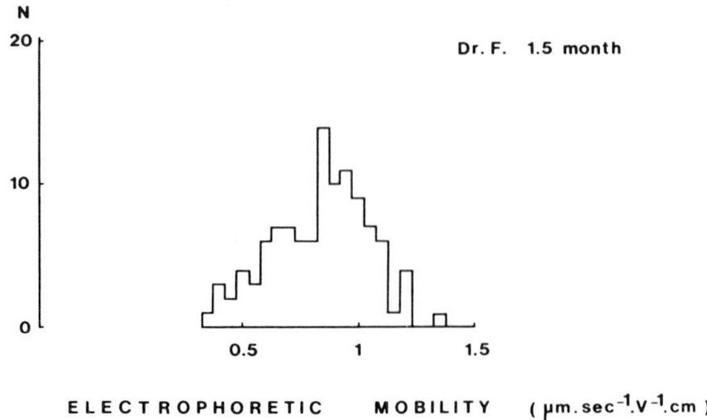

Fig. 4 : Distribution of the electrophoretic mobility of peripheral blood lymphocytes isolated from a baby 1.5 month of age with partial Di George syndrome. N represents the number of cells in each class of mobility.

Summary and conclusion

An abnormal distribution with an abnormal mean value of the lymphocyte EPM seems to be associated with a severe disease. No definitive correlation between the severity of the immune deficiency involved in Di George syndrome and the distribution of the EPM of blood lymphocytes has been established as yet. The bimodal nature of the distribution of the lymphocyte EPM in infants is especially sensitive to alteration in pathological conditions. The potential of cell electrophoresis for the diagnosis of immune deficiencies is neonatalogy is emphasized.

References

1. Cooper, M.D., Lawton, A.R., Miescher, P.A., Mueller-Eberhardt, H.J. : Immune deficiency, Springer-Verlag, Berlin (1979).

2. Mehrishi, J.N. : Molecular aspects of the mammalian cell surface, in Progress in Biophysics and Molecular Biology, vol. 25, edited by Butler, J.A.V. and Noble, D., Pergamon Press, Oxford 1-70 (1972).

3. Wioland, M., Mehrishi, J.N. : Scand. J. Immunol. 10, 453-463 (1979).

4. Kelemen, N., Lasmoles, F., Milhaud, G. : Nature 272, 65-66 (1978).

MEASUREMENT OF MIXED CELL POPULATIONS BY AN AUTOMATED CELL ELECTROPHORETIC INSTRUMENT

Haruhisa Hayashi, Naomi Toyama, Yoshiharu Oguchi, Kenichi Matsunaga, Masahiko Fujii, Humio Hirose, Chikao Yoshikumi, Tetsuya Hotta and Masaaki Yanagisawa
Kureha Chemical Ind. Co., Ltd., Tokyo, Japan

Introduction

Although several kinds of automated cell electrophoretic instruments have been developed, it apparently is difficult to measure accurately cell populations, especially mixtures of cells of different size, with instrument using the principle of light scattering. We attempted to measure cell populations using an automated cell electrophoretic instrument (Parmoquant) that measures individual cell mobility using the same principle as the manual instrument.

Instrument

The fully automated cell electrophoretic instrument (Parmoquant; produced by Kureha Chemical Ind. Co., Ltd., Tokyo, Japan) consists of opto-electronic image analysis and computer (Fig.1). Ten to 20 migrating cells in the microscopic field are momentarily recorded by image processing and the data are stored in a computer and printed out. Characteristics are the following, (1) high accuracy and reproducibility, since individual particles are measured automatically and objectively, (2) high speed [100 particles/5 min], (3) histogram pattern analysis, (4) measurement in physiological medium, (5) small amount of cells [$\simeq 10^6$ cells].

374

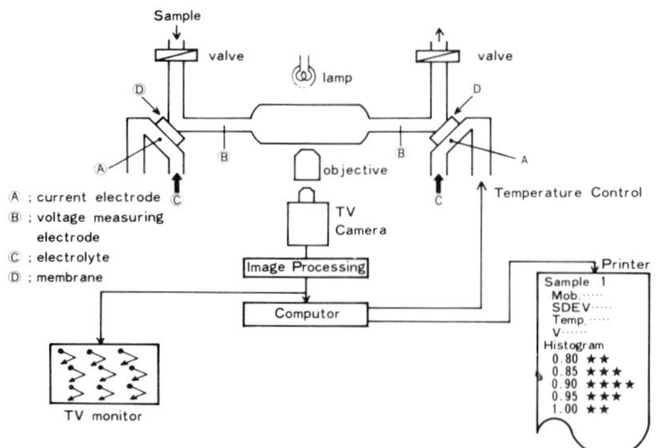

Fig. 1 Block diagram of fully automated cell electrophoretic instrument (Parmoquant)

Results

1) Mixture of sheep red blood cell (SRBC) and rabbit red
    blood cell (RRBC)
    SRBC and RRBC were washed twice
    and suspended in PBS. Mixed
    ratios of SRBC and RRBC between
    0 to 100% were used. The mean
    mobilities of SRBC and RRBC
    were 0.42 and 1.05 μm/sec/V/cm,
    respectively, and the histograms
    of high (HMC) and low (LMC)
    mobility cell populations were
    separated completely (Fig. 2).
    The ratio of SRBC to RRBC and the
    ratio of HMC to LMC were identical
    (R=0.998). And the mean mobility
    of whole cells changed linearly
    according to the ratio of SRBC to
    RRBC (Fig. 3(a)).

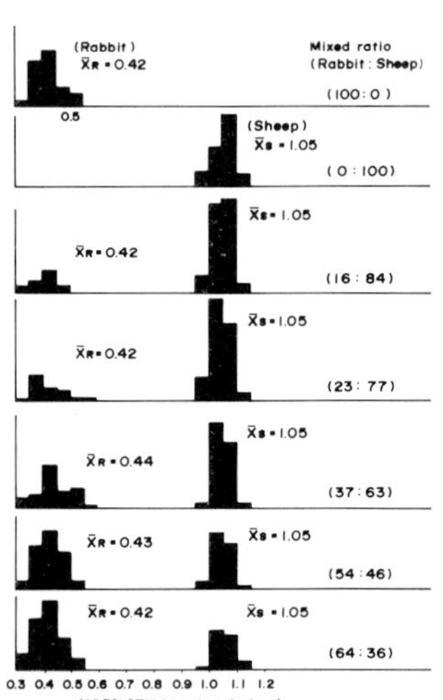

Fig. 2 Electrophoretic
mobility histogram of a
mixture of sheep RBC and
Rabbit RBC

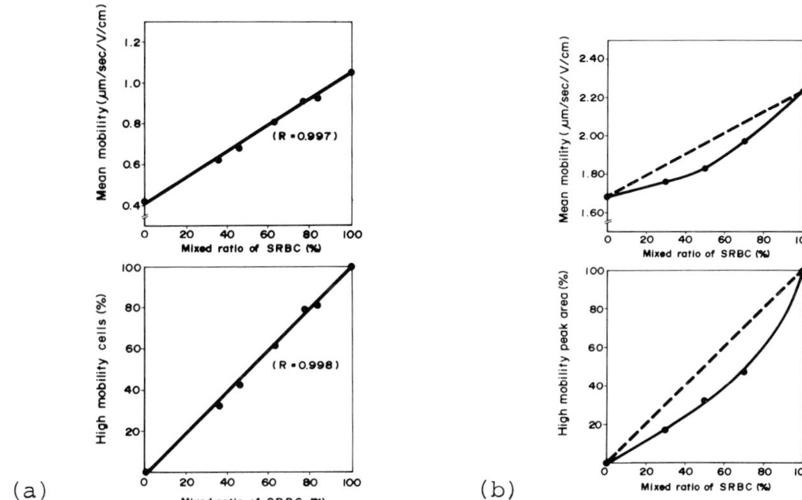

Fig. 3 Correlation between mixed ratio of RBCs and
mean mobility or high mobility peak measured by (a)
Parmoquant, (b) laser doppler instrument

The same samples suspended in a medium of low ionic
strength (I=0.05) were measured by laser doppler
instrument. But, the results did not show satisfactory
linear correlations (Fig. 3(b)).

2) Peripheral lymphocytes of normal donors
In the case of normal peripheral lymphocytes suspended in
Eagle's MEM, two prominent peaks were observed in the
electrophoretic mobility histrogram and cells were
divided into two types; low mobility cell (LMC), less
than 0.95 and high mobility cell (HMC), more than 0.95
μm/sec/V/cm (Fig. 4). T cells, passed through a nylon
wool column, were shown to be the HMC, and B cells,
adherent to it, to be the LMC. And correlation
coefficients between HMC and T cell (assayed by E rosette
formation) as well as LMC and B cell (by EAC) were both

0.99 (Fig. 5). The percentage of the T cell population ranged from approximately 60 to 80% in the 30 healthy individuals, which coincided reasonably well with that of the high mobility cells (Fig. 6).

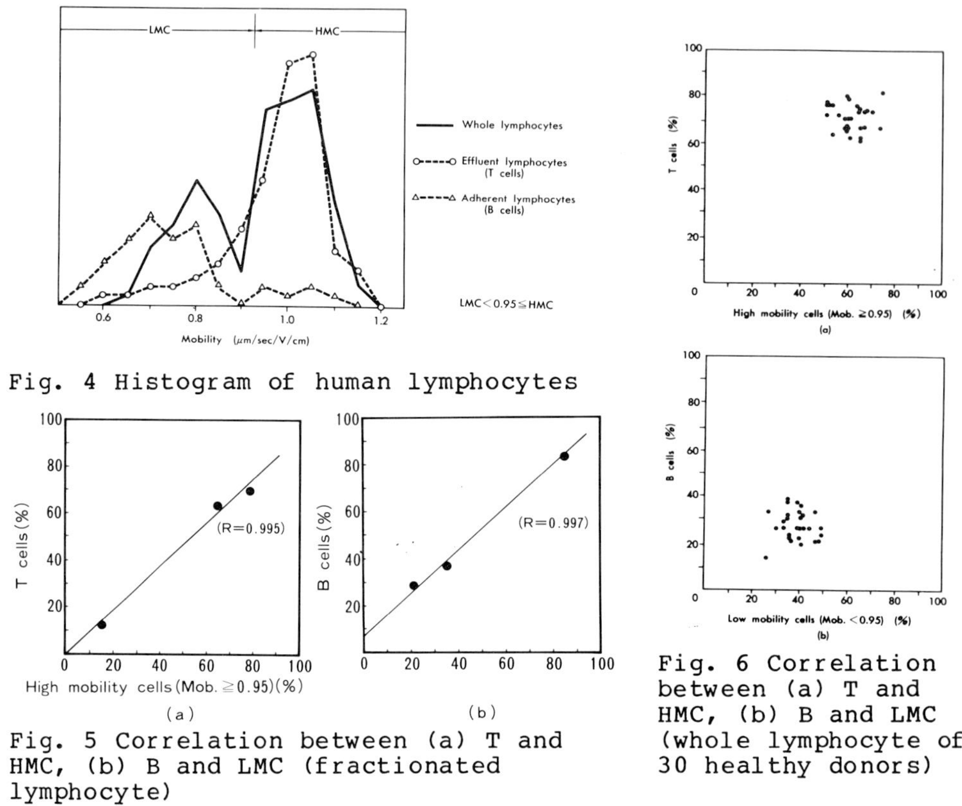

Fig. 4 Histogram of human lymphocytes

Fig. 5 Correlation between (a) T and HMC, (b) B and LMC (fractionated lymphocyte)

Fig. 6 Correlation between (a) T and HMC, (b) B and LMC (whole lymphocyte of 30 healthy donors)

Conclusion

1. When a mixture of SRBC and RRBC was analyzed by the fully automated cell electrophoretic instrument (Parmoquant), a clear bimodal electrophoretic mobility pattern was obtained and the ratio of SRBC to RRBC and the ratio of HMC to LMC were quite similar.

2. Analysis of peripheral lymphocytes from normal donors showed two peaks and each peak corresponded to T or B lymphocyte, which coincided well with the result of rosette formation assay.
3. Consequently, Parmoquant was considered to be able to analyze quantitatively mixed cell populations and this method may prove useful for experimental research and clinical application.

SEPARATION OF RAT LIVER CELL ORGANELLES RESPONSIBLE FOR HEMOGLOBIN-HAPTOGLOBIN METABOLISM BY MEANS OF CARRIER-FREE ELECTROPHORESIS

Satoru Oshiro, Masao Takami, Hiroshi Nakajima
Department of Biochemical Genetics, Medical Research
Institute, Tokyo Medical and Dental University, Tokyo 113,
Japan

Introduction

Hemoglobin-haptoglobin complex (Hb-Hp) formed by intravascular
hemolysis is removed by liver parenchymal cells through the
receptor specific for the molecule (1). The Hb-Hp taken up
by the liver cells is first incorporated in organelles of low
density (density range, 1.05 - 1.07g/ml), and particularly
concentrated is Golgi fractions in a substantially intact
form. The distribution of radioactive labeled Hb-Hp shifted
progressively from organelles of low density to those of high
density (density range, 1.07 - 1.15g/ml). The Hb-Hp was
dissociated symmetrically into two 82,000-dalton subunits (half
Hb-Hp) at an early stage of degradation and then degraded
into small peptides in organelles with high density (2,3).
In the present paper, the intracellular site of the incorporation and degradation of heme and globin moiety of Hb-Hp in
rat liver cells has been investigated by means of carrier-free
electrophoresis and Percoll density-gradient centrifugation
after intravenous administration of (^{3}H-heme, ^{14}C-globin)
Hb-Hp to rats.

Results

The density of total microsomal and light lysosomal fractions
is low, and that of mitochondrial-lysosomal fraction is high in
Percoll density-gradient centrifugation(4). The heavy
lysosomal fraction has a high anodic mobility, and mitochon-
drial-microsomal fraction a lower anodic mobility in carrier-
free electrophoresis at pH 7.4(5). Therefore, we were able
to separate these three fractions by a combination of the two
methods. At 10 min after administration of the doubly labeled
Hb-Hp into rats, total microsome and light lysosomal fraction
contained approximately 60 and 40% of the radioactivities,
respectively. And then, both radioactivities shifted gradually from microsomal fraction to heavy lysosomal fraction with
time. At 40 min postinjection, the ratio of ^{3}H/^{14}C in the
heavy lysosomal fraction was twice that of intact (^{3}H-heme,
^{14}C-globin)Hb-Hp(Fig.1).
Furthermore, 60 min or later, the heavy lysosomal fraction had
only the ^{3}H-radioactivity. The foregoing evidence suggests
that (^{3}H-heme,^{14}C-globin)Hb-Hp was first incorporated in
receptosomes (5) with lower anodic mobility and low density
(density range, 1.03 - 1.05g/ml), and that these organelles
then progressively acquired a higher anodic mobility as well
as high density (density range, 1.06 - 1.12g/ml) during transport process across the liver cells

References

1. Kino, K., Tsunoo, H., Higa, Y., Takami, M., Hamaguchi, H.,
   Nakajima, H.: J. Biol. Chem. 255, 9616-9620 (1980).
2. Higa, Y., Oshiro, S., Kino, K., Tsunoo, H., Nakajima, H.:
   J. Biol. Chem. 256, 12322-12328 (1981).
3. Oshiro, S., Nakajima, H.: Proc. Jpn. Acad. 57B, 222-227
   (1981).

4. Pertoft, H., Wärmegard, B.: Biochem. J. 174, 309-317
    (1978).

5. Stahn, R., Maier, K.P., Hanning, K.: J. Cell Biol. 46,
    576-591 (1970).

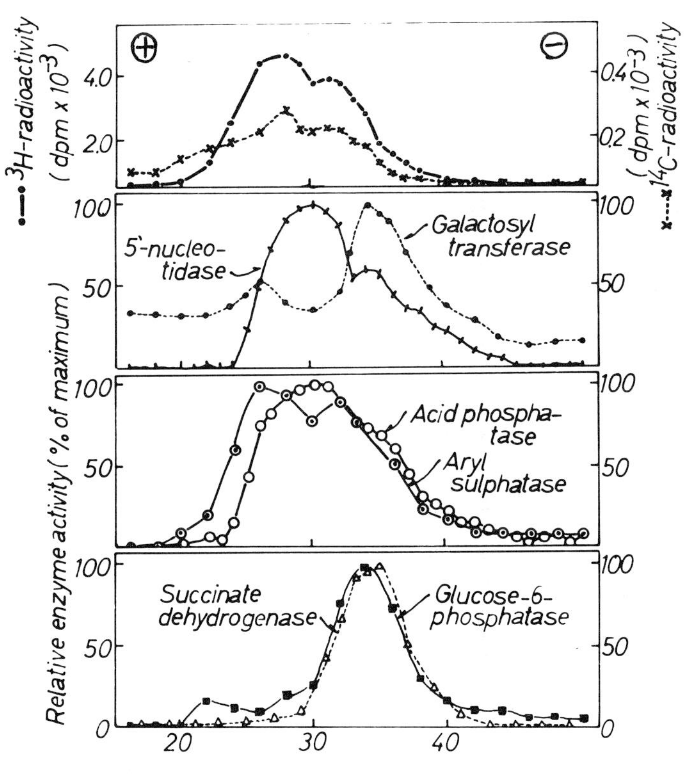

Fig.1. Distribution of ^{3}H-and ^{14}C-radioactivities and
various marker enzyme activities in fractions obtained by
carrier-free electrophoresis of mitochondria-lysosome fraction
40 min after administration of (^{3}H-heme, ^{14}C-globin)Hb-Hp
to rats. The anodic and cathodic side are indicated by ⊕ and
⊖, respectively.

A RAPID IN VITRO ASSAY OF LYMPHOCYTE DOSE RESPONSE TO
IMMUNODEPRESSANT AGENTS

Brian Shenton, Peter Donnelly, Craig Parker, Peter Friedman,
Ala Alomran, David Francis, Tom Lennard, George Proud,
R.M.R. Taylor.
University Department of Surgery
Royal Victoria Infirmary
Newcastle upon Tyne NE1 41P UK

Introduction

Over the past few years increasing attention has been given to
the use of _in vitro_ assays to determine lymphocyte response to
immunosuppressive drugs (1). This may be particularly
relevant to renal patients awaiting transplant. Despite
improving success with renal transplantation up to 30% of all
grafts are lost through rejection and in a study by Dumble (1)
50% of irreversible rejection crises were found in renal
patients who showed lymphocyte resistance (in A.D.C.C.) to
steroid drugs _in vitro_. The aims of the study were: i) to
describe the Tanned Erythrocyte Electrophoretic Mobility
(T.E.E.M.) test for lymphocyte sensitisation, ii) to compare
the effect of hydrocortisone on normal lymphocyte response to
P.P.D. in the T.E.E.M. and transformation tests.

Methods and Results

The principle of the test we have used depends upon the fact
that when lymphocytes came into contact with antigen to which
they are sensitive, they produce soluble effector lymphokines.
Our interest over the past 10 years has been upon those
lymphokines which reduce the surface charge of indicator test

particles. Due to several disadvantages of using macrophage
indicator cells we now use tanned sheep red blood cells (T.S.
R.B.C.) (Fig. 1). Accordingly we have called the test system
the Tanned Erythrocyte Electrophoretic Mobility (T.E.E.M.)
test (3) (Fig. 1). In this study we used a Zeiss Cytophero-
meter with visual measurement of mobility and electronic timing. In the calculation of results, lymphocyte responses were expressed as %ge slowing of the T.S.R.B.C. When the effect of immunosuppressive

Fig. 1

drugs is measured the % reduction (or inhibition) of the
maximal response to antigen (P.P.D.) is calculated.

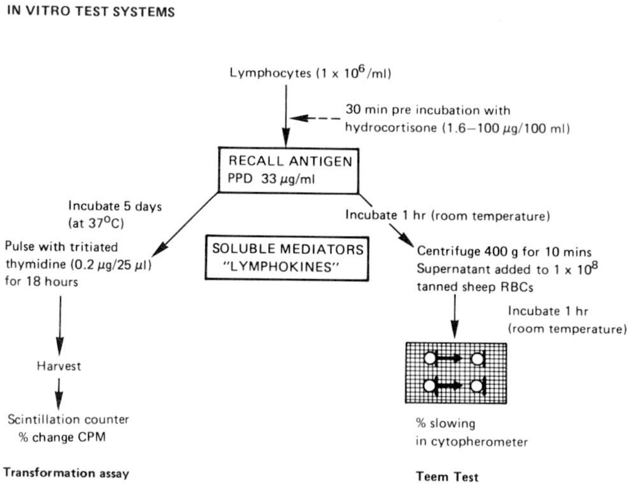

A) <u>Reproducibility</u>

To test the reproducibility of the test system, blood was
collected from 10 subjects with a variety of pathological
conditions. Lymphocytes were isolated, the samples were
split into 2 aliquots and coded so that the observer did not
know which cells were which. The lymphocytes were then
reacted with P.P.D. (Experiment A). One week later the
experiment was repeated with fresh blood from the same
original 10 subjects (Experiment B). Only when both parts of
the experiment were completed was the code broken. Good
correlation was found between the pairs in both Experiment A
(r= 0.969 p<.001) and Experiment B (r= 0.941 p<.001). Whilst

results did vary slightly when Experiment A was compared to B, correlation was again high (r= 0.938 p<.001)

B. <u>Comparison with Lymphocyte Transformation Studies</u>
To compare the T.E.E.M. test result for normal lymphocyte response to immunosuppressives, a parallel experiment was carried out. Whilst our laboratory studied the immuno-suppressive effect of hydrocortisone on the lymphocytes of 6 normal subjects in the T.E.E.M. test, an independent group examined the effect on lymphocyte transformation. The same blood samples were used in each study. Methods are summarised in Figure 1. Figure 2 illustrates the results. Sigmoid lymphocyte dose response curves were found in both tests and the amount of hydrocortisone to inhibit maximal response by 50% was identical in the two test systems.

Fig. 2

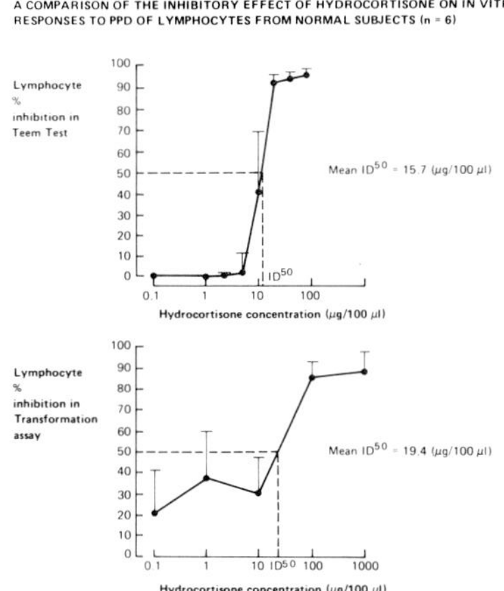

Further analysis showed the kinetics of the test systems were different, it being more difficult to maximally or minimally inhibit lymphocyte transformation than the T.E.E.M. test. This may be due to different lymphokines being measured in the two test systems.

C) The Effect of Immunosuppressive Drugs on Lymphocyte
    Response to P.P.D.

To examine the effect of conventional immunosuppressive
agents on normal lymphocyte (from 5 subjects) response to
P.P.D., dose response curves were constructed. As can be seen

Fig.3

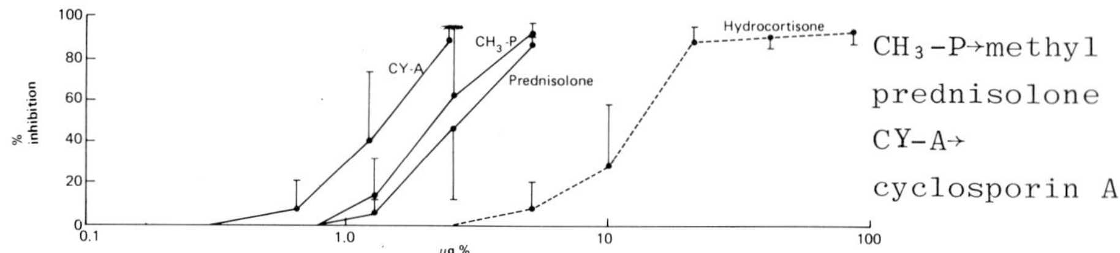

CH3-P→methyl
prednisolone
CY-A→
cyclosporin A

in the Figure a gradation of effect was found. With
reference to hydrocortisone, prednisolone was 6.5 times as
effective, methyl prednisolone 8.4 times, and cyclosporin A
12.6 times as effective.

Summary

1) The T.E.E.M. test is a reproducible and rapid test of
    lymphocyte response.
2) The dose-dependent immunosuppressive effect of
    different drugs can be measured.
3) The test may provide a convenient and safe means of
    testing patient immune response to a variety of natural
    and synthetic immunomodulatory agents.

References

1. Dumble, L.T., et al. Tpl.Proceed. 8, 1569-1571 (1981).
2. Shenton, B.K., et al. In 'Proceed. 2nd Workshop in
   Cell Electrophoresis'. Ed. Preece, A.W., Ann. Light 99-115
   (1981).

ISOLATION AND CHARACTERISATION OF SUPPRESSIVE PEPTIDE FROM α_2MACROGLOBULIN–PROTEASE INTERACTION

Ala Alomran, Brian Shenton, Peter Donnelly, David Francis,
George Proud, R.M.R. Taylor
University Department of Surgery
Royal Victoria Infirmary
Newcastle upon Tyne NE1 4LP UK

Introduction

Over the past decade many workers have shown that the αglob-
ulin fraction of plasma may inhibit many immunological
reactions both <u>in vivo</u> and <u>in vitro</u> (1, 2). Recently Hubbard
(3) has suggested that α_2Macroglobulin (α_2M) interaction with
proteases (α_2M-P) may represent an important immunological
mechanism. The aim of this project was to investigate the
association between the properties of α_2M as a protease
inhibitor and also as an immunoregulatory protein.

Results

Pure α_2M was purified from the plasma of normal healthy adult
volunteers as described by Virca (4) and reacted with
increasing amounts of trypsin and many other proteases
(elastase, collagenase, cathepsin D) in the presence of Tris-
HCl buffer pH 8 containing 20mM $CaCl_2$. α_2M-P complexes were
examined for their suppressive effect on a panel of normal
lymphocytes stimulated with PPD of myobacterium tuberculosis
in the Tanned Erythrocyte Electrophoresis Mobility (TEEM)
test (5). Aliquots of α_2M-P were fractionated on Sephadex
G75 (Pharmacia) (2.6 x 38cm) columns equilibrated with NH_4-
acetate buffer pH 8. The fractions collected were tested for

their suppressive activity in the TEEM test. Our results showed a significant increase in the suppressive activity of $\alpha_2 M$ after protease binding. Fractions from the G75 column showed two regions of protein - a high molecular weight region of $\alpha_2 M$ and a low molecular weight region (2-3000d). Both of these regions were highly suppressive to lymphocytes in the TEEM test. The low molecular weight region was called suppressive peptide from $\alpha_2 M$ (SP-$\alpha_2 M$). The liberation of the peptide from $\alpha_2 M$ showed a dose dependent response to the amounts of protease added to the $\alpha_2 M$ (Table 1).

EFFECT OF TRYPSIN ADDITION TO 500 μg $\alpha_2 M$ ON THE SUPPRESSIVE ACTIVITY OF THE $\alpha_2 M$ AND PEPTIDE PEAKS AFTER GEL FILTRATION ON G75

Weight of Trypsin added (μg)	Weight of $\alpha_2 M$ required to inhibit lymphocyte-PPD response (μg)	Volume of peptide peak required to produce 50% inhibition of lymphocyte-PPD response (μl)
0	38.0	—
10	18.0	520
20	14.2	300
30	9.4	160
50	4.8	63

The peptide was found to be a single component when fractionated on a high pressure liquid chromatography column (μ Bondpack C18), and had an isoelectrofocusing point (PI) of 5.6 in the chromatofocusing technique (Pharmacia). Biologically the peptide was a potent inhibitor of lymphocyte transformation to phytohaemoglutinin (PHA) and Mixed Lymphocyte Reaction (MLR), and to the polymorph stimulation by latex particles in the chemiluminescence assay (LKB). These findings allowed us to study and compare peptides isolated from the plasma of patients with a variety of clinical pathologies (uremia, thermal burns and malignant diseases). The peptides of interest were isolated from the group of patients by a two step fractionation method of 1) sephacryl S300

(Pharmacia). The protein region of molecular weight >10000
was collected and concentrated. 2) After fractionation on a
Sephadex G75 column, the region containing the peptide
eluated in a 2-3000d molecular weight region with absorbance
at 206nm, was suppressive to lymphocyte reactivity in the
TEEM test. An I^{125}-labelled aliquot of these peptides was
eluted at PI of 5.6 and was biologically active (Fig. 1).

CHROMATOFOCUSING OF I^{125} SUPPRESSIVE PEPTIDES

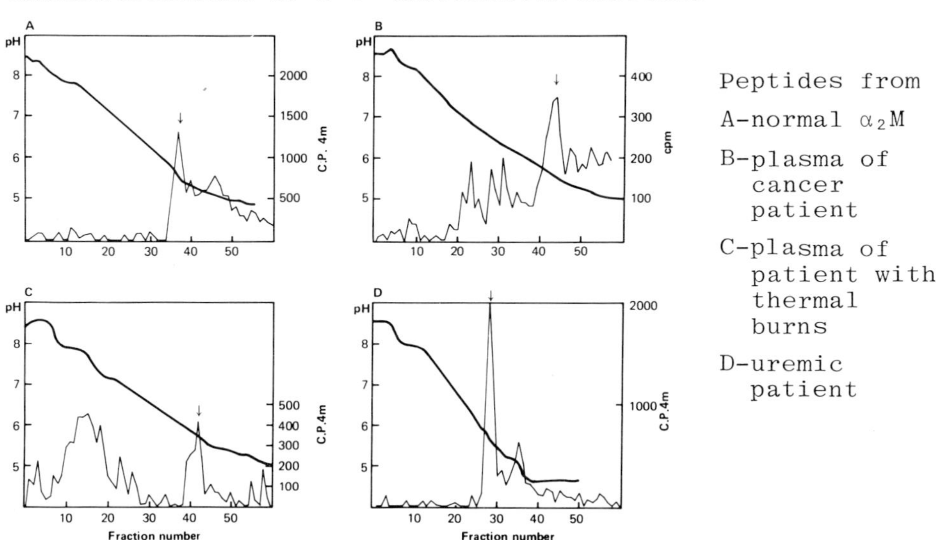

Peptides from

A-normal $\alpha_2 M$

B-plasma of
   cancer
   patient

C-plasma of
   patient with
   thermal
   burns

D-uremic
   patient

Discussion

In this work we have successfully isolated suppressive peptides from previously purified $\alpha_2 M$ after binding with physiological levels of proteases. The peptide was shown to be a
part of the $\alpha_2 M$ molecule by isolating a radioactive peptide
from the I^{125}-labelled $\alpha_2 M$. We compared this peptide with
other suppressive peptides which have been reported in
patients with uremia, burns and cancer (6, 7, 2). All these
conditions are associated with a degree of anergy expressed
by the susceptibility of these patients to sepsis, viral

infections and most of them showed negative immunological skin test. We have found that a suppressive peptide with mol. wt. 2-3000d, and a PI of 5.6 was found in their plasma. The groups of patients mentioned above are known to show considerable release of lysosomal enzymes and other proteases as a result of massive cellular proliferation and destruction. These naturally produced proteases could bind to $\alpha_2 M$ and release this peptide (SP-$\alpha_2 M$). A preliminary study of groups of patients whose diseases are associated with increased protease activity (acute pancreatitis, rheumatoid arthritis, and patients after major surgical trauma) has shown a high level of this peptide in their calculations. The association of the presence of suppressive substances in the circulation of many pathological conditions and the immunological surveillance of the body have been investigated (2, 7). Our results on the biochemical mechanism to explain increased immunosuppressive properties of $\alpha_2 M$ and the liberation of SP-$\alpha_2 M$ provide additional evidence for the hypothesis proposed by Hubbard (3). This process may as a naturally occurring "switch off" mechanism to protect the body from unwanted hyperimmune reactions which might harm the body.

References

1 Mowbray, J.F.: Transplant 1, 16 (1963).

2 Cooperband, S.R., Nimberg, R., Schmid, K., Mannick, A.:
  Transplant Proc. 82, 225 (1976).

3 Hubbard, W.J.: Cell Immunol. 39, 388 (1978).

4 Virca, G.D., Travis, J., Hall, P.K., Roberts, R.C.:
  Ann.Biochem. 89, 274 (1978).

5 Shenton, B.K., Jenssen, H.L., Werner, H., Field, E.J.:
  J.Imm.Methods 14, 123 (1977).

6 Proud, G., Shenton, B.K., Smith, M.B.:
  Br.J.Surg. 66, 678 (1979).

7 Constantin, M.B.: Ann.Surg. 188, 209 (1978).

ISOENZYME

CHARACTERIZATION OF HIGH Km ALCOHOL DEHYDROGENASE EROM MOUSE
LIVER

Takeshi Haseba,   Keiko Hirakawa,   Yukari Tomita,
Tokinori Watanabe
Department of Legal Medicine,   Nippon Medical School
1-1-5,   Sendagi,   Bunkyo-ku,   Tokyo,   113   (Japan)

Introduction

Mammalian alcohol dehydrogenase (ADH; E.C.1.1.1.1.) generally
exhibits a low Km, which is below 3 mM for ethanol.   It is
inhibited by pyrazole or pyrazole derivatives and has a basic
pI above 8 (1,2,3).   However, recently high Km ADHs have been
found in human liver (4,5,6) and in rat stomach (7). The high
Km ADHs in human liver are called π-ADH and χ-ADH.   The
sensitivity of these ADHs to the pyrazole derivatives is low.
There are fewer studies of mouse ADH isozymes than of the other
mammalian ADHs.   In recent years, Holmes and others found
three ADH isozymes; $ADH-A_2$, $ADH-B_2$ and $ADH-C_2$, in several
mouse tissues, using cellulose acetate membrane electrophoresis
at pH 8.5   (8, 9, 10).   According to their reports, the
most cathodally migrating $ADH-A_2$, found predominantly in liver,
exhibits a low Km (0.29 mM) for ethanol and is sensitive to a
pyrazole derivative (aminopropyl pyrazole).
On the other hand, $ADH-C_2$, which migrates less toward the cathode, exhibits a high Km (133 mM) for ethanol and is insensitive
to the pyrazole derivative.   This $ADH-C_2$ is found mainly in
the stomach, but not in the liver.   The most widely distributed $ADH-B_2$, which migrates toward the anode, shows a high
activity with hexenol, but shows no activities with ethanol as
substrate.   Further characterization of $ADH-B_2$ has not been
performed.
In this paper, we report that in mouse liver there are three
ADH fractions, separable by CM-cellulose column chromatography

and four bands detected on disc electrophoresis(pH 9.7) by a
new ADH staining method.   Also, we present data on the enzymic
characteristics of these ADH isozymes. Finally, these isozymes
are compared with the known ADH isozymes from mice and other
mammalian species.

Materials and Methods

The livers and stomachs from ddY strain male mice(25-30g B.W.)
were homogenized in 3 vol. of 5 mM Tris-HCl(pH 8.5), 0.5 mM NAD,
0.7 mM DTT, 0.25 M sucrose buffer using a teflon homogenizer,
and the extracts were collected after centrifugation at 105,000g
for 1 hr.   The liver extract was dialyzed against 5 mM phos-
phate(pH 7.0), 0.5 mM NAD, 0.7 mM DTT buffer and applied at a
flow rate of 126 ml/hr onto a CM-cellulose(Whatman CM-52)column
(2.7 x 43 cm) equilibrated with the same phosphate buffer.
Enzymes were eluted with a linear gradient of the buffer con-
centration(5-100 mM).   The unadsorbed first ADH peak was further
purified by DE-chromatography under the same conditions as CM-
chromatography.   The second and the third ADH peaks were purified by gel-filtration.
Disc electrophoresis was performed at pH 9.7 by Takeo's method
(11).   The buffer was supplemented with NAD(0.5 mM).
Enzymes were stained by a new method(NT staining method) as
follows: neotetrazolium-Cl(NT) was used as a color reagent
because it was found to be more stable than other color reagent
at the high alkaline pH (12), optimal for mouse ADH. Gels were
incubated at 37°C in a 0.1 M glycine buffer(pH 10.7) containing
2.2 mM NAD, 0.8 mM NT, 1.4 mM PMS and various concentrations of
ethanol or hexenol.   The standard ADH staining was performed
by the method of Smith et al.(13).
Isoelectrofocusing was performed according to the method of
Righetti et al.(14).
For the enzyme assay, the rate of NADH production was measured
at 37°C by spectrophotometer using ethanol or hexenol as substrate.

Results

Three ADH fractions were obtained by CM-chromatography of mouse liver extract (Fig.1). ADH-Ⅱ showed the most prominent activity at 15 mM ethanol concentration, whereas with 15 mM hexenol, ADH-I showed the greatest activity among the three ADH fractions. The activity of ADH-Ⅲ with ethanol was almost the same as that with hexenol.

The optimum pHs of these ADHs were between 10.5 and 11.0 (Fig. 2). For the various kinds of saturated alcohols, the Kms of ADH-I were greater by three orders of magnitude than those of ADH-Ⅱ and ADH-Ⅲ. Saturation of this enzyme was not seen even when 3 M of ethanol as substrate was tested. The Km of ADH-I with hexenol, an unsaturated alcohol, was only 30 times higher than the other two ADHs. The Km of ADH-Ⅱ was low as similar to ADH-Ⅲ, but the Km of ADH-Ⅲ was somewhat lower than that of ADH-Ⅱ (Table 1).

The rates of inhibition of ADH-Ⅱ and ADH-Ⅲ by 0.04 mM of 4-methylpyrazole(4-MP) with 0.4 mM hexenol as substrate were similar to each other. ADH-Ⅲ, however, slightly more sensitive to 4-MP. ADH-I was not inhibited at all by 4-MP (Table 1). When the activity of the three ADHs was measured using various concentrations of hexenol as substrate, ADH-I was saturated at hexenol concentrations over 15 mM. On the other hand, ADH-Ⅱ was strongly inhibited at higher concentrations of the substrate; its activity at 15 mM hexenol concentration was only about 18% of the maximum activity. A similar trend was seen for ADH-Ⅲ up to 3 mM substrate concentrations, but its activity increased above 3 mM (Fig.3).

On disc electrophoresis of liver extract, one band was detected at 120 mM ethanol as substrate and two bands were detected at 7 mM hexenol as substrate by standard ADH staining. When the NT staining method was employed with either 120 mM ethanol or 7 mM hexenol as substrate, two more bands were detected between the double bands seen by the standard staining technique. These additional two bands are indicated as α and β in Figure 4. The anodal band was stained more intensely at

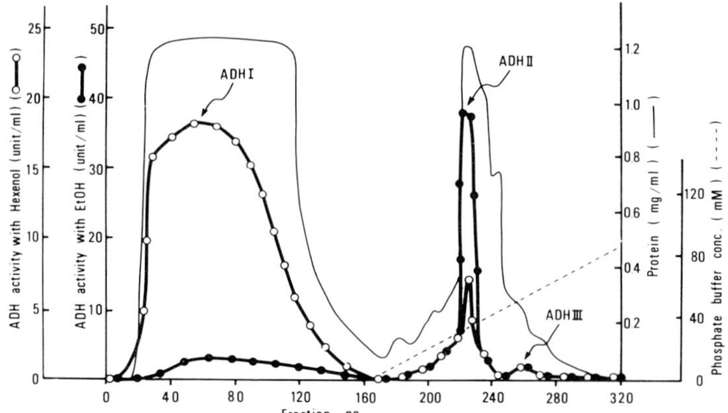

Fig. 1. CM-cellulose chromatography of mouse ADH of liver extract.

Table 1. Michaelis constant of three ADH fractions for various kinds of alcohol and the rate of inhibition by 4-methylpyrazole.

	MeOH*	EtOH*	Km (mM) n-PrOH*	n-BuOH*	Hexenol	Rate of inhibition** by 4-MP (%)
ADH I	non	6700	2000	400	1.4	0
ADH II	31	1.1	0.7	0.2	0.05	78
ADH III	22	1.0	0.2	0.1	0.04	98

The enzyme was assayed at 37°C in 100 mM glycine/NaOH buffer (pH 10.7) with 1.7 mM NAD.

* MeOH: Methanol, EtOH: Ethanol, n-PrOH: n-Propanol, n-BuOH: n-Butanol.

** The activity was measured with or without 0.04 mM 4-Methylpyrazole (4-MP), using 0.4 mM Hexenol as a substrate.

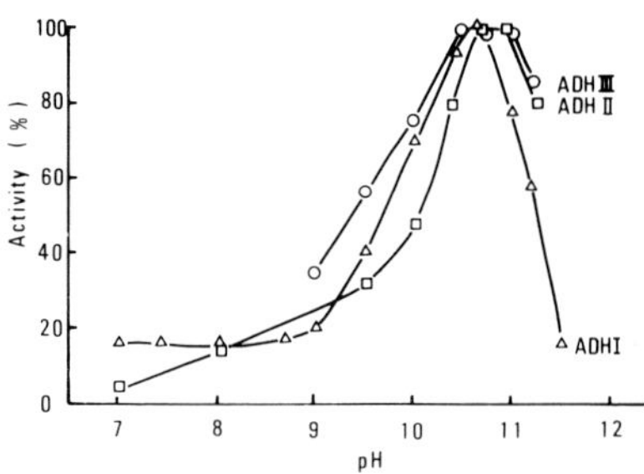

Fig. 2. pH-activity profile of three ADH fractions. (15mM ethanol was used as a substrate.)

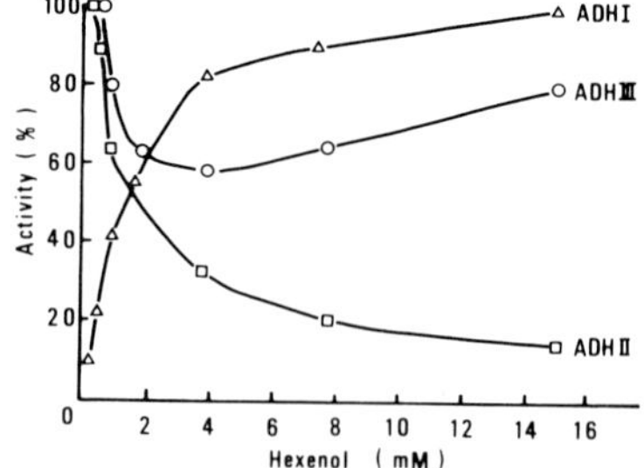

Fig. 3. Activity of three ADH fractions at the various concentrations of hexenol.

2.5 M of ethanol as substrate.   However, at this ethanol
concentration, the α and β bands were difficult to detect (Fig.4).
As shown in Figure 5, ADH-I corresponds to the anodal band and
ADH-II to the cathodal band of liver extract.   ADH-III was
separated into two bands and the mobility of these bands was
similar to that of the α and β bands from the liver extract.
After disc electrophoresis of liver extract or ADH-III, when the
gels were stained with either low or high concentration of hexenol as substrate, the α band of the extract and the broad band
of ADH-III were stained more faintly at 15 mM than 0.4 mM hexenol.
On the contrary, the β band of the extract and the sharp band
of ADH-III were stained more intensely at 15 mM than at 0.4 mM
hexenol (Fig.6).   By adding 4-MP (0.02 mM) to the staining
buffer with 0.4 mM hexenol as substrate, the broad band completely disappeared, while the sharp band remained (Fig.6).
This indicates that the sharp band is not as sensitive to 4-MP.
Isoelectrofocusing of the mouse ADHs is shown in Figure 7.
ADH-I has an acidic pI of 5.3-6.4.   ADH-II has the most basic
pI, which is above 9.0.   ADH-III, which contains the broad and
the sharp band, has a pI of about 9.0.   The pI of the stomach
ADH was found to be 5.7-7.5.

Discussion

In this study we have presented data showing that there are
three ADH fractions separated by CM-chromatography and four
ADH bands detected on disc electrophoresis using a new NT staining technique.   When the substrate saturation curves (Fig.3),
banding pattern on disc electrophoresis (Fig.5) and the mode of
staining of the gels with different concentrations of substrate
(Fig.6) are taken into consideration, it is reasonable to conclude that the ADH-III fraction contains low Km ADH-α and high
Km ADH-β.   The low Km value and the high sensitivity of ADH-III
(as shown in Table 1) can be considered as a reflection of ADH-α
fraction in ADH-III.   Thus, this is the first time that four
ADH isozymes have been detected in mouse liver.   Comparing

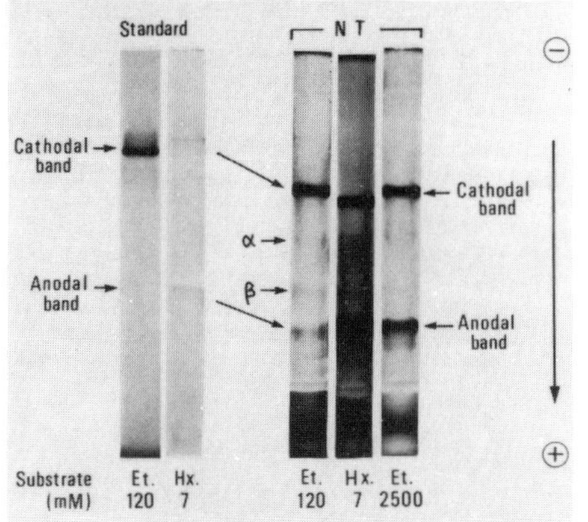

Fig. 4. Comparison of the standard and NT methods of staining after disc electrophoresis(pH9.7) of liver extract. (Et.:Ethanol, Hx.: Hexenol.)

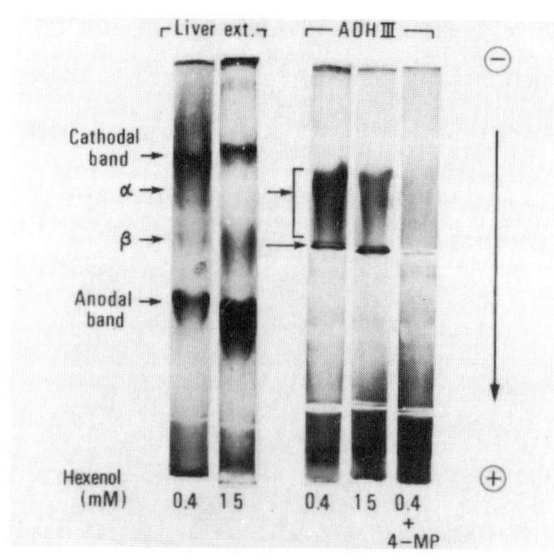

Fig. 6. ADH activity after disc electrophoresis(pH9.7) of liver extract and ADHIII fraction stained by NT method with or without 0.02mM of 4-methylpyrazole(4-MP).

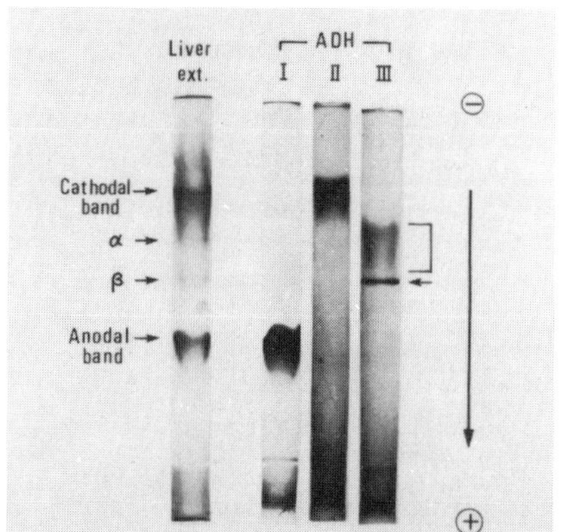

Fig. 5. Disc electrophoresis(pH9.7) of liver extract and three ADH fractions stained by NT method. (7mM hexenol was used as substrate.)

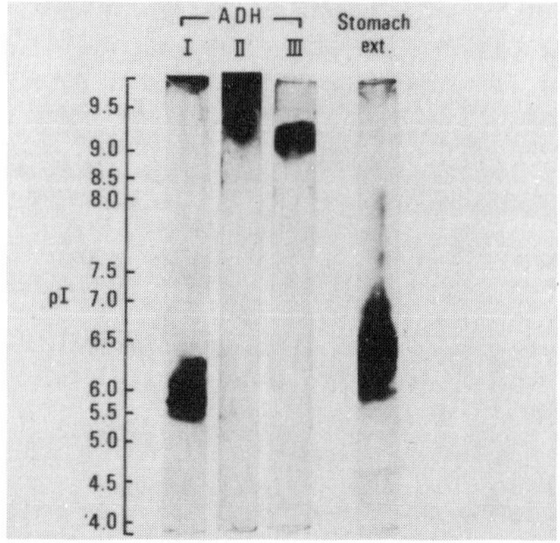

Fig. 7. Isoelectrofocusing of three liver ADH fractions and stomach extract stained by NT method. (7mM hexenol was used as substrate.)

these four ADH isozymes with mouse ADH isozymes reported by other authors, ADH-I seems to correspond to ADH-B_2 and ADH-II to ADH-A_2. Also, ADH-α seems to resemble a slow ADH found in the deer mouse as a genetic variant of liver ADH (15). ADH-β appears to be a new mouse ADH isozyme in liver found only by our NT staining method. Moreover, when we compared these ADH isozymes in this study with ADH isozymes of the other mammalian

species, ADH-I was found to be similar to χ-ADH of human liver. In contrast to all other ADHs, χ-ADH moves toward the anode on starch gel electrophoresis (around pH 8), and is not saturated even at 2 M of ethanol and also is not inhibited by high concentrations of 4-MP with octanol as substrate (6). ADH-Ⅱ is similar to the classical ADHs of the various mammalian species. The Km of ADH-β appears to be high (Fig.3 and 6) but lower than that of ADH-I. Since ADH-β was inhibited but ADH-I was activated at 2.5 M of ethanol (Fig.4).
In addition, only ADH-β has a basic pI (9.0) among the mouse high Km ADHs, which also include ADH-I and stomach ADH-C_2 (Fig. 7). Therefore, the newly found ADH-β seems to be equivalent to π-ADH, the high Km ADH in human liver, which has a basic pI about 9.0 (4).

Summary

Disc electrophoresis (pH 9.7) of mouse liver extract, using a novel staining method, enabled us to separate and detect four ADH isozymes; ADH-I, ADH-Ⅱ , ADH-α and ADH-β, with a broad range of pI values. This staining method was useful at a high alkaline pH optimal for the detection of mouse ADHs with very low activities. A new ADH isozyme, ADH-β, has been found in mouse liver by this new staining method. ADH-β exhibits a high Km for alcohol, has a basic pI and is resistant to 4-methylpyrazole. The similarity of the four mouse liver ADH isozymes in this study to the other known ADH isozymes is discussed.

Acknowledgements

We thank the other members in our laboratory for their technical assistance and Drs. T.Sudo, M.Miki and N.Nakazawa for their useful advice.

References

1. Brändén,C.I., Jörnvall,H., Eklund,H., Furugren,B. : The
   Enzymes, third edition, *11*, 103-190, Academic press, New
   York · London 1975

2. Pietruszko,R. : Biochemistry and Pharmacology of Ethanol,
   1, 87-106, Plenum Press, New York · London 1979

3. Deis,F.H., Lester,D.: Biochemistry and Pharmacology of
   Ethanol, *2*, 303-323, Plenum Press, New York · London 1979

4. Li,T.K., Mangnes,L.J. : Biochem. Biophys. Res. Commun. *63*,
   202-208 (1975)

5. Borson,W.F., Li,T.K., Dafeldecker,W.P., Vallee,B.L. : Biochemistry *18*(6), 1101-1105 (1979)

6. Parés,X., Vallee,B.L. : Biochem. Biophys. Res. Commun. *98*,
   122-130 (1981)

7. Cederbaum,A.I., Pietruszko,R., Hempel,J., Becker,F.F.,
   Rubin,E. : Arch. Biochem. Biophys. *171*, 348-360 (1975)

8. Holmes,R.S. : Genetics *87*, 709-716 (1977)

9. Holmes,R.S. : Comp. Biochem. Physiol. *61B*, 339-346 (1978)

10. Holmes,R.S., Albanese,R., Whitehead,F.D., Duley,J.A. : J.
    Exp. Zool. *217*, 151-157 (1981)

11. Takeo,K. : Annu. Rep. Soc. Protein Chem. Yamaguchi Univ.
    Sch. Med. *4*, 1-36 (1970)

12. Harris,H., Hopkinson,D.A. : Handbook of Enzyme Electropho-
    resis in Human Genetics, Chapter 2, 1-10, North-Holland
    Publishing Company, Amsterdam 1976

13. Smith,M., Hopkinson,D.A., Harris,H. : Ann. Hum. Genet. *34*,
    251-270 (1971)

14. Righetti,R.G., Drysdale,J.W. : Ann. N.Y. Acad. Sci. *209*,
    163-186 (1973)

15. Burnett,K.G., Felder,M.R. : Biochem. Genet. *16*(5/6),
    443-454 (1978)

ISOELECTROPHORETIC ACIDIC ISOENZYME OF HUMAN RIBONUCLEASE

Shuji Hishiki, Takashi Kanno*, Kayoko Sudo*, and Shukichi
Sakaguchi
Second Department of Surgery, *Department of Laboratory
Medicine, Hamamatsu University, School of Medicine,
Hamamatsu, Japan.

Introduction

As we previously reported(1), abnormal elevations of serum
ribonuclease(RNase) activity were observed in the sera of the
patients with various cancers and benign hepatobiliary diseases. The specificity of serum RNase for the diagnosis of
pancreatic cancer was relatively low and clinical assessment
of the elevation of serum RNase is still obscure. Recently,
isoelectrophoretic heterogeneity was observed in serum and
tissue RNase, and the presence of three major pI isoenzymes
(basic, neutral and acidic) were demonstrated using isoelectric focusing.
Moreover, the acidic pI isoenzyme was observed to predominate
in fetal and cancerous tissues, and was frequently detected
in the sera of the patients with pancreatic and liver
cancers. In this paper, the characteristics and clinical
significances of this unique acidic RNase are studied and
discussed.

Materials and methods

Samples of human tissues were obtained from a number of surgical specimens and four fetuses of 13- to 19-week gestation,
and 100,000xg supernatants of crude homogenates were used for
experiments. Pure pancreatic juice was obtained from an

external drainage tube left in the pancreatic duct as part of
surgical procedure.
The RNase activity was measured by the modified method of
Reddi (2). The standard assay system included 0.1 ml of 1.0
mg/ml polycytidylic acid, 0.3 ml of 0.1 mole/l phosphate-
borate buffer pH 6.5 and 0.1 ml of samples, containing one
mmole/l of spermidine phosphate and 50 mmole/l of sodium
chloride as activators.
Separation of RNase isoenzyme was performed by isoelectric
focusing using an LKB 8100-1 Ampholine electrofocusing column
and the focusing was carried out for 24 hours in the cold
with carrier Ampholine pH range 3.5 to 10.0. Enzyme activity
and pH of each 1.5ml fraction collected were measured.

Results

1)   RNase levels in human adult and fetal tissues. The
      levels of RNase in various tissues (pancreas, liver,
      spleen and lymph node) were analyzed. Table 1 shows the
      results. In normal pancreas and pure pancreatic juice,
      the levels of RNase activities were extremely high, as

Table 1. Tissue RNase

Adult

Organ		n	mean (units/g)	S.D	range
Pancreas	normal	5	742.6	432.1	355.7 − 1340.8
	chronic inflammation	2	175.6	143.0	74.5 − 276.7
	carcinoma	5	49.9	36.9	10.1 − 101.1
Liver	normal	3	6.6	8.3	1.1 − 16.1
	cirrhosis	2	5.8	3.2	3.5 − 8.0
	hepatoma	5	1.4	0.9	0.3 − 2.4
Spleen		3	6.3	1.2	5.0 − 7.4
Lymph node	metastatic	3	28.1	22.7	2.4 − 45.1
	non-metastatic	2	22.4	20.1	8.2 − 36.6
Pure pancreatic juice		1	1140.3		

Table 2. Tissue RNase   Fetal

Organ	n	mean (units/g)	S.D	range
Pancreas	4	17.0	24.1	3.3 − 53.0
Liver	3	0.7	0.4	0.3 − 1.1
Spleen	1	3.6		
Stomach	1	5.8		
Small bowel	2	12.7	6.1	8.4 − 17.1
Brain	1	0.9		
Heart	1	26.6		
Skeltal muscle	2	7.9	1.3	7.0 − 8.8
Lung	1	1.0		
Thymus	2	1.6	0.5	1.3 − 1.9
Kidney	3	202.6	176.2	25.2 − 377.5
Adrenal gland	3	1.6	2.0	0.5 − 3.9
Placenta	2	9.4	7.6	4.0 − 14.8

much 20- to 80-fold those of serum. In contrast, the
levels of the enzyme markedly decreased in the tissues of
chronic pancreatitis and pancreatic cancer as compared
with normal. Table 2 shows the tissue RNase levels of
various fetal organs. The enzyme activities of examined
tissues were very low except kidneys. In particular, the
concentration in the fetal pancreas was 50-fold less than
in the adult pancreas.

2) Separation of RNase isoenzymes by isoelectric focusing.
In this study, three major groups of pI isoenzymes were
obtained and classified as basic, neutral and acidic
RNase. The pI values of these three major groups were
9.6 (basic range 8.70 to 10.16), 6.3 (neutral range 5.77
to 6.66) and 4.6 (acidic range 4.32 to 4.95), respectively.
Figure 1 shows the isoelectrophoretic patterns of normal
serum, tissues of normal liver and pancreas and pure
pancreatic juice. In normal serum, two major peaks of
basic and neutral RNase were observed. On the other
hand, two major peaks of basic (pI 9.64) and acidic
isoenzyme (pI 4.95) were detected in normal pancreatic
tissue. In contrast, isoelectric focusing of RNase in
pure pancreatic juice produced only the single major peak

404

Figure 1

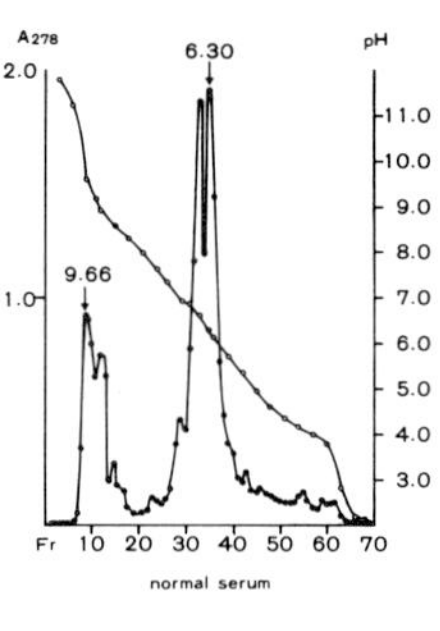

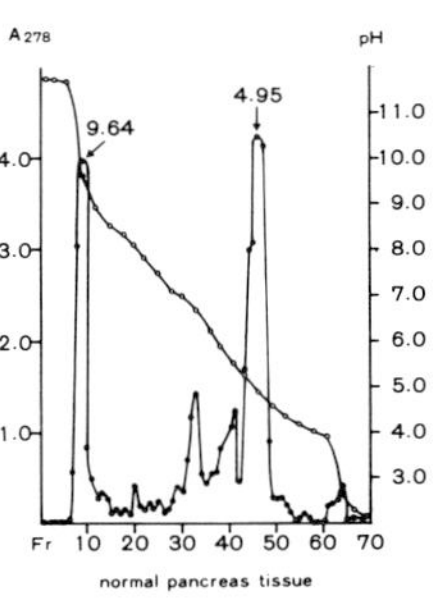

Figure 2

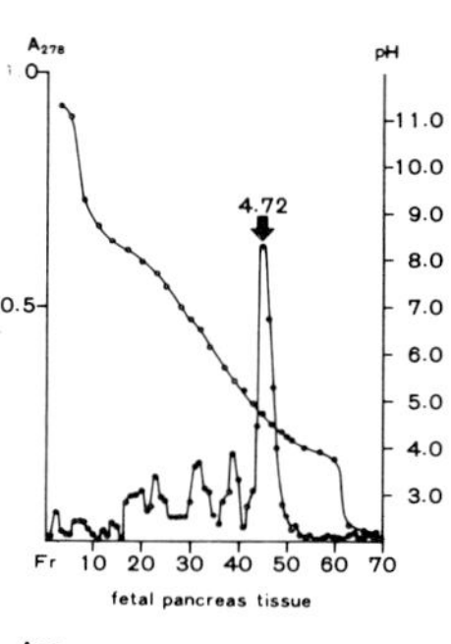

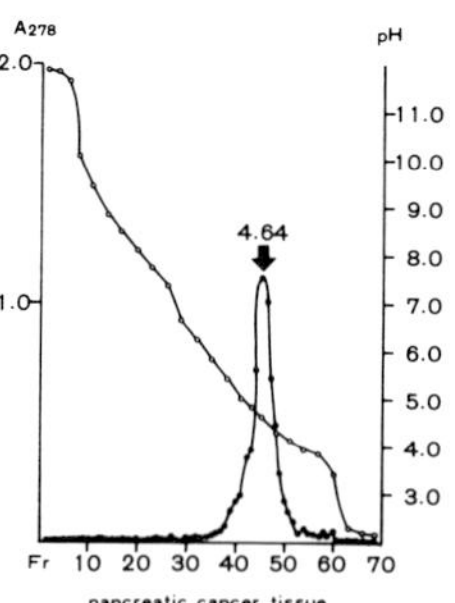

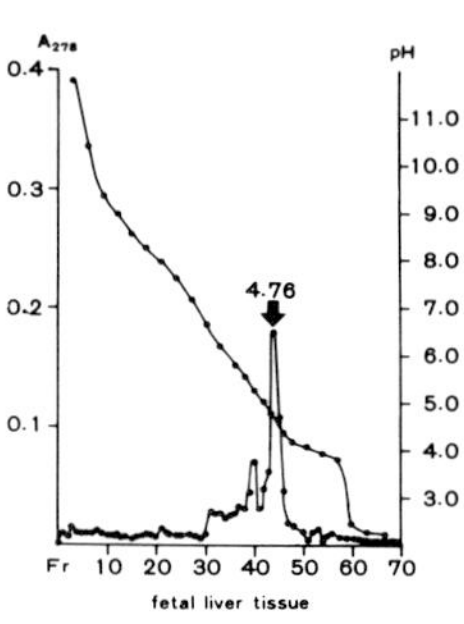

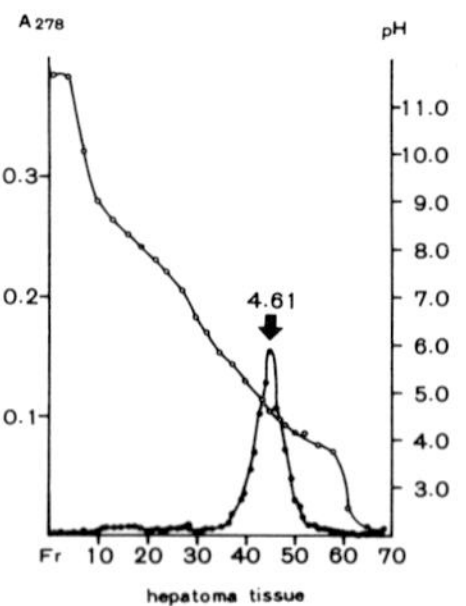

of basic isoenzyme (pI 10.14), we suggest that only a
basic isoenzyme is secreted into a pancreatic duct and
the neutral isoenzyme observed in serum is different from
that of pancreatic tissue.
Figure 2 shows the isoelectrophoretic patterns of fetal
tissues and cancer tissues of pancreas and liver. In
both fetal pancreatic and liver tissues, a single
predominant peak of acidic pI isoenzyme (pI 4.7) was
observed. Moreover, this acidic isoenzyme was detected
exclusively in the tissues of pancreatic cancer and
hepatoma. It is noteworthy that these patterns were
strikingly similar to those obtained from fetal pancreas
and liver.
The sera from four patients with pancreatic cancer and
four patients with liver cancer were examined by
isoelectric focusing. All patients with pancreatic cancer
and three of four patients with liver cancer (hepatoma)

Figure 3

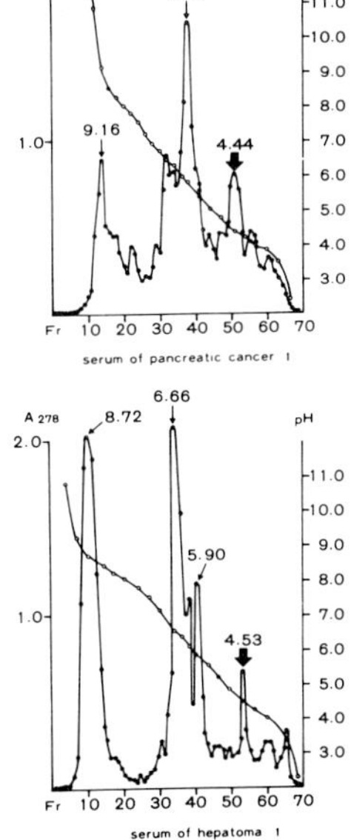

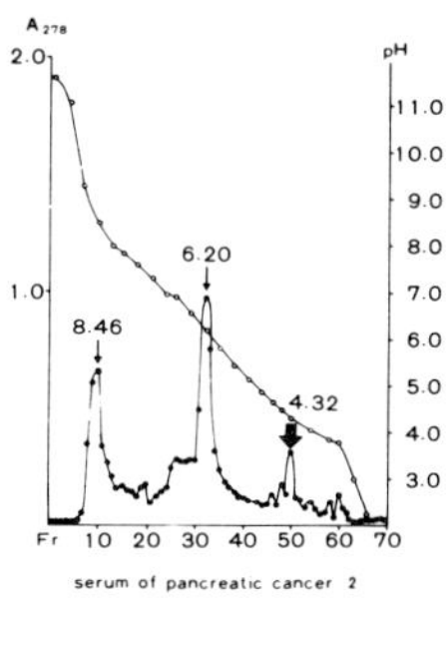

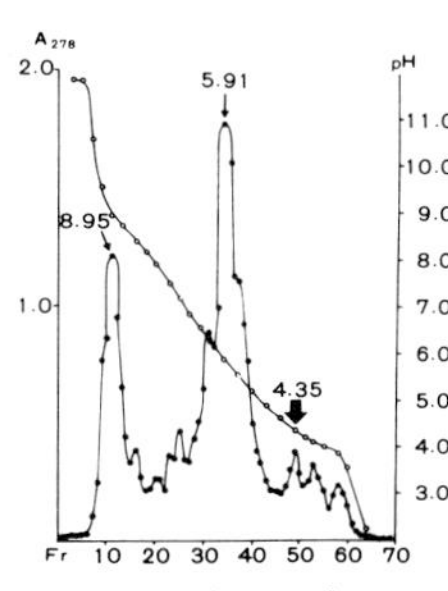

demonstrated the characteristic acidic pI isoenzyme as a
minor peak.

Figure 3 shows the isoelectrophoretic patterns of two
patients with pancreatic cancer and two with hepatoma.

Discussion

Attempts to separate the isoenzymes of RNase by isoelectric
focusing technique have been made by several investigators
(3,4,5,6,7), using polyacrilamide gel rods or thin layer
polyacrilamide gel, and staining the RNase activity with
toluidine blue or measuring the eluted enzyme activity of
divided segments of gel. However, no detailed or clear
isoenzyme separation using isoelectric focusing has yet been
achieved. In the present study, three major pI isoenzymes
were demonstrated in human serum and various tissues using
isoelectric focusing with an Ampholine column. These three
pI isoenzymes were classified as basic, neutral and acidic
isoenzyme.

Employing the method, the clinical significance of acidic pI
isoenzyme of RNase was determined, 1) Acidic pI isoenzyme was
found in the normal pancreas, in the pancreatic and hepatic
cancerous cells, and in the fetal pancreas and liver. 2) In
the sera of non-cancerous patients this acidic pI isoenzyme
could not be detected, even in chronic pancreatitis. 3) In
pancreatic cancer tissue, pI isoenzymes other than acidic
RNase were markedly suppressed and the pI isoenzyme pattern
closely resembled that obtained from fetal pancreatic tissue.
Moreover, in fetal tissues, only an acidic pI isoenzyme was
present. Pathological changes such as fibrosis or carcinous
changes in the tissues of pancreas and liver might suppress
the production of basic RNase. 4) This acidic pI isoenzyme
was detected at a high frequency in the sera of the patients
with pancreatic and liver cancer. The present data suggest
that this acidic pI isoenzyme might be characterized as one

of the carcinofetal proteins. Although detailed mechanisms
of the elevation of serum RNase in various pathological
conditions are still obscure, the detection of this acidic pI
isoenzyme in the patients serum appears to be useful as a
biochemical tumor marker for the diagnosis of pancreatic and
liver cancer.

Conclusion

The characteristics and clinical significance of a unique
isoenzyme of ribonuclease are presented and discussed. This
acidic pI RNase was found in the tissues of pancreatic
carcinoma, hepatoma and fetal organs, and appears to be a
carcinofetal protein. The detection of this isoenzyme may be
useful for the disgnosis of pancreatic cancer and hepatoma.

References

1. Hishiki, S., Kanno, T., Sudo, K., Sakaguchi, S.:
    Abstract book of 7th World Congress of Gastroenterology
    (OMGE), 332 (1982)
2. Reddi, K.K., Holland J.F.: Proc. Natl. Acad. Sci. USA
    73, 2308-2310 (1976)
3. Kottel, R.H., Hoch, S.O., Parsons, R.G., Hoch, J.A.: Br.
    J. Cancer 38, 280-286 (1978)
4. Warshow, A.L., Lee, K.H., Wood, W.C. Cohen, A.M.: Am. J.
    139, 27-32 (1980)
5. Tournut, R., Allan, B.J., White, T.T.: Clin. Chim. Acta
    88, 345-353
6. Thomas, J.M., Eigen, H., Hodes, M.E.: Clin. Chim. Acta
    111, 199-209 (1981)
7. Renner, I.G., Mock, A., Reitherman, R., Douglas, A.P.:
    Gastroenterology 74, 1142 (1978)

INCIDENCE OF ENZYME-LINKED IMMUNOGLOBULIN IN HUMAN SERUM

Hiroshi Shibata, Tatsuo Tozawa, Kaoru Taishi, Keiko Hayashi,
Sachiyo Morita, Hitomi Satoh and Rumi Okasaka
Section of Clinical Laboratory, Hyogo College of Medicine,
Hyogo, Japan

Introduction

The incidence of enzyme-linked immunoglobulins is estimated by the total
frequency in isoenzyme analysis or based on the activity of each enzyme as
a result of mass screening.[1] The incidence of lactate dehydrogenase (LDH)-,
alkaline phosphatase (ALP)- and amylase (AMY)-linked immunoglobulin has
been reported to be about 0.1%, in Japan.[2,3] Since the incidence of these
enzyme-linked immunoglobulins have been based on varying populations, it is
impossible to compare an incidence of an enzyme-linked immunoglobulin with
another. It is known that patients with each enzyme-linked immunogloblins
are among the middle- and advanced-age groups.[2]

We detected LDH-, ALP- and AMY-linked immunoglobulin in the same
population that has been frequently encountered on a daily bases due to
widespread use of isoenzyme analysis, and we confirmed the incidence of
these enzyme-linked immunoglobulins. We also disclosed the relationship
with aging. Furthermore, evaluation was made of the distribution of isotypes of each enzyme-linked immunoglobulin and the correlation with diseases, especially with autoimmune disease. The incidence by-age of benign
monoclonal gamma-pathy (BMG) was also estimated for comparative purpose.

Materials and Method

Specimens used were sera drawn from 13,098 out- and in-patients (AMY,9,611;
BMG, 1,086). The ratio of male and female patients was almost identical,
though female patients were nearly twice more than male patients in the
25 - 34 year age group. Variations in number of males and females by age

are shown in Fig. 1.

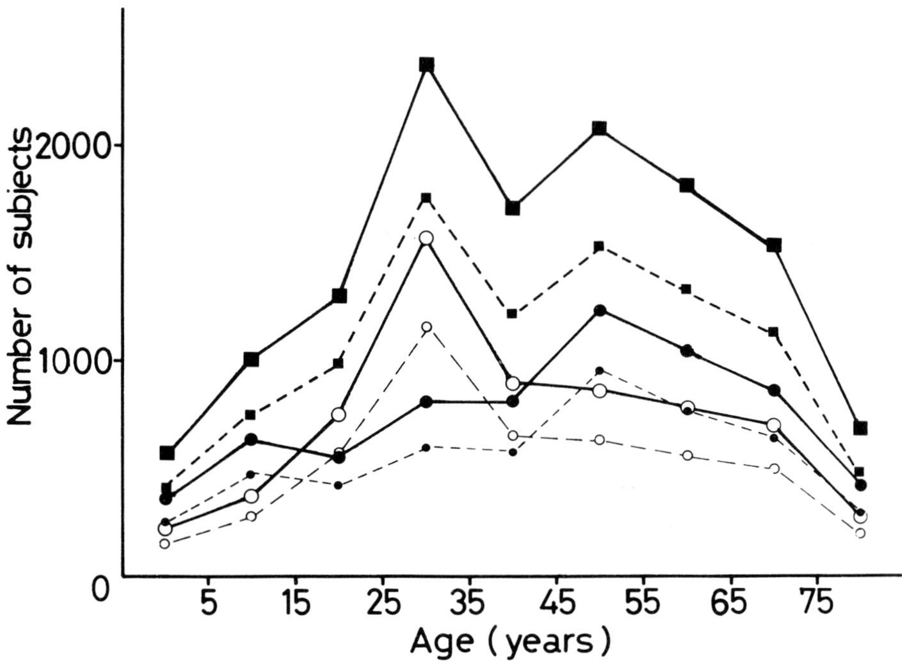

Fig. 1   Population by age and sex.

Number of subjects with LDH & ALP (——■——;total, ——●——;male, ——○——;female)   and with AMY (--■--;total,--●--;male,--○--; female).

Reagents and method:

Antisera used were anti-γ, anti-α, anti-μ, anti-κ and anti-λ serum manufac-tured by DAKO.

Enzyme staining reagents:

LDH was stained using LDH NEO D (EIKEN CO., LTD); ALP, by ALP ISOENZYME STAIN (WAKO PURE CHEMICALS CO., LTD.); and AMY, by BLUE STARCH POLYMER (DAIICHI PURE CHEMICALS CO., LTD.).

Screening of enzyme-linked immunoglobulin:
LDH- and AMY-linked immunoglobulins were detected by isoenzyme analysis on
cellulose acetate membrane, while ALP-linked immunoglobulin was determined
by electrosyneresis with agar gel as carrier.

Identification of enzyme-linked immunoglobulin:
LDH- and AMY-linked immunoglobulins were identified by immunoprecipitin reaction in free liquid media.[4] ALP-linked immunoglobulin was subjected to
electrosyneresis using agarose gel as carrier.

Results

1) Number of Patients and Incidence of Each Enzyme-Linked Immunoglobulin
   and BMG (Fig. 2)

LDH-linked immunoglobulin was found in 45 patients (0.34%). The number
of male patients was 1.65 times more than that of female patients. With
the advancement of age, patients increased in number. Most patients were
in the 65 - 74 year age group.
      The number of patients with ALP-linked immunoglobulin was 37 (0.28%)
and males were 1.47 times More numerous than females. Patients increased
with age and most patients were 75 years of age or older.
      AMY-linked immunoglobulin was detected in 25 patients (0.26%). Males
were 2.13 times more numerous than females. The number of patients were
in the 55 - 64 year age group.
      However, BMG occurred in 64 cases (0.64%), and males were 2.56 times
more numerous than females. Most patients with BMG were in the 45 - 64
year age group.

2) Incidence by Age of Patients with Each Enzyme-Linked Immunoglobulin and
   BMG (Fig. 3).

The incidence of each enzyme-linked immunoglobulin increased with advancing age. Patients with BMG also increased in number with advancing
age. The pattern of increases was the most similar to that of patients

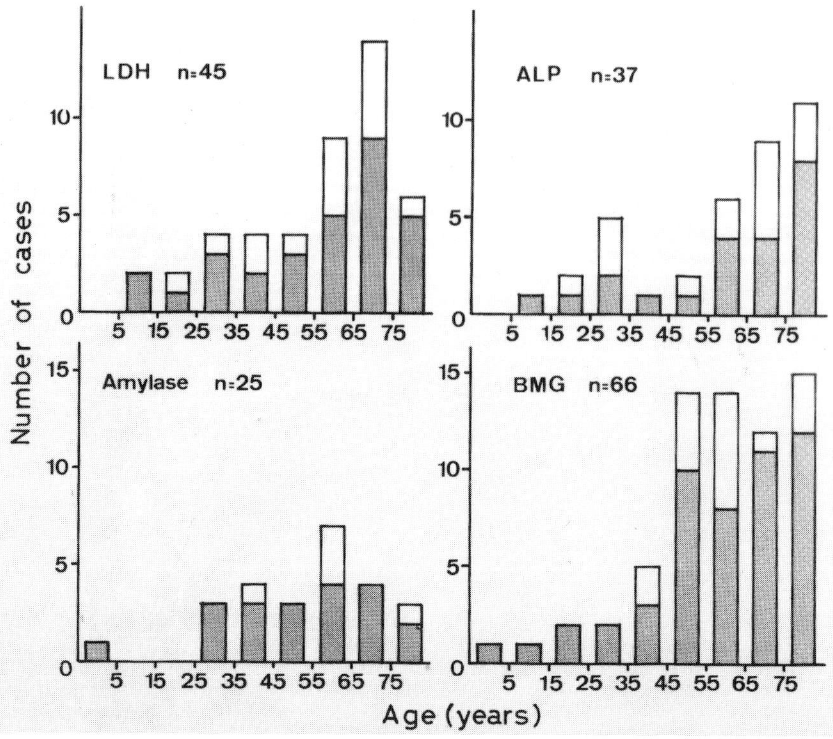

Fig. 2   Cases of each enzyme-linked immunoglobulins
and BMG by age and sex.

with ALP-linked immunogloblin.

3) Distribution of Isotypes of Each Enzyme-Linked Immunoglobulin and BMG
   (Table 1)

The most frequently encountered isotype of LDH-linked immunoglobulin was
IgG (κ + λ), found in 15 patients (33.3%); that of alp-linked immunoglobulin was IgG (λ), found in 20 patients (54.1%). IgA (κ) was the most
frequently encountered isotype of AMY-linded immunoglobulin, and it was
detected in 10 patients (40%). IgG was the most frequency encountered
in 56 BMG patients (87.5%) and IgG (κ) was found in 34 patients (53.1%).

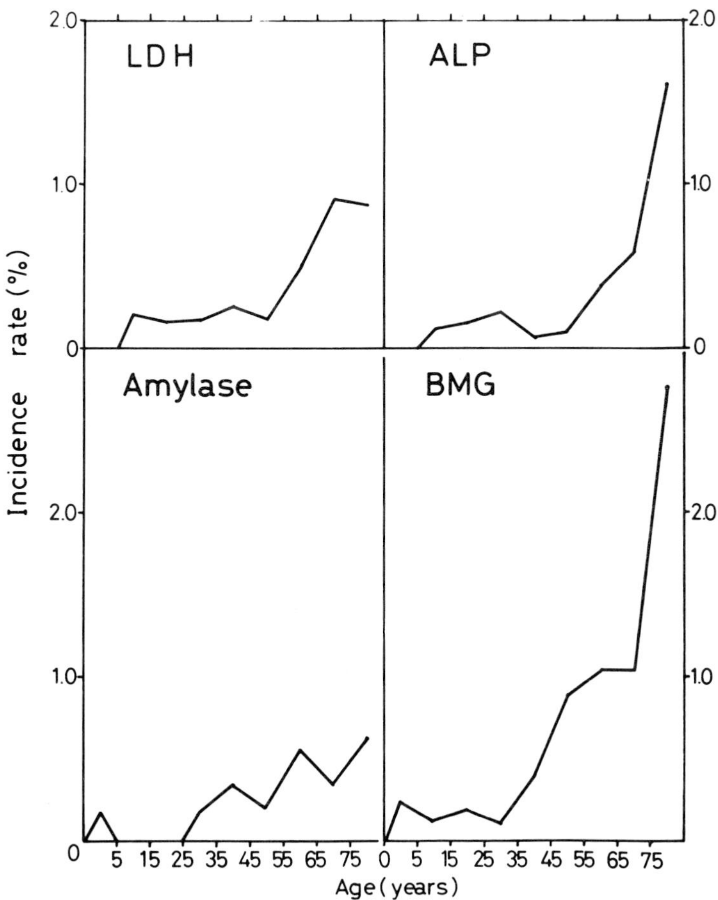

Fig. 3   Frequency of each enzyme-linked immunoglobulin
           and BMG.

4) Disease Classification of Each Enzyme-Linked Immunogloblin

Diseases in more than 20% of the patients with either of enzyme-linked
immunoglobulins were related to neoplasms and the digestive system.

Correlation with autoimmune disease was observed in nine patients with ALP-linked immunoglobulin (23.7%) (ulcerative colitis, 3; chronic rheumatoid arthritis, 3; SLE, 1; AIHA, 1; and polymyositis, 1), and in 6 patients with LDH-linked immunoglobulin (13.3%) (ulcerative colitis, 3; collagen disease, 1; and chronic rheumatoid arthritis, 2). None of the patient with AMY-linked immunoglobulin had autoimmune disease.

Table 1   Isotype of enzyme-linked immunoglobulin and BMG.(%)

Heavy chain	Light chain κ	$\kappa+\lambda$	λ	Total
LDH IgG	6(13.3)	15(33.3)	9(20.0)	30(66.7)
IgA	9(2.2)	1(2.2)	1(2.2)	11(24.4)
IgM	1(2.2)			1(2.2)
IgG+A		2(4.4)	1(2.2)	3(6.6)
Total	16(35.6)	18(40.0)	11(24.4)	45(100.0)
ALP IgG	3(8.1)	9(24.3)	21(56.8)	33(62.2)
IgA	1(2.7)	1(2.7)	1(2.7)	3(8.1)
IgG+A		1(2.7)		1(2.7)
Total	4(10.8)	11(29.7)	22(59.5)	37(100.0)
Amylase IgG	3(12.0)	1(4.0)		4(16.0)
IgA	10(40.0)	1(4.0)	5(20.0)	16(64.0)
IgG+A	2(8.0)	2(8.0)	1(4.0)	5(20.0)
Total	15(60.0)	4(16.0)	6(24.0)	25(100.0)
BMG IgG	34(53.1)		22(34.4)	56(87.5)
IgA	3(4.7)		3(4.7)	6(9.4)
IgM	1(1.6)		1(1.6)	2(3.1)
Total	38(59.4)		26(40.6)	64(100.0)

Discussion

LDH-, ALP- and Amy-linked immunoglobulins were detected in the same population to compare their incidence. The incidence of LDH-linked immunoglobulin was highest, i.e., 0.34%, followed by that for ALP-linked immunoglobulin, 0.28%. The incidence of AMY-linked immunoglobulin was lowest, 0.26%.

It was demonstrated that each of enzyme-linked immunoglobulins increased as the age of patients advanced. It was confirmed that the occurrence of enzyme-linked immunoglobulins was part of the aging process.

Correlation of enzyme-linked immunoglobulins with disease was such that more than 20% of patients with each enzyme-linked immunoglobulin had neoplasms or disorders of the digestive system.

Many current reports indicated that enzyme-linked immunoglobulins may be immunocomplexes.[5,6] It is of interest, therefore, to relate them to autoimmune disease: of the cases with enzyme-linked immunoglobulins, autoimmune disease patients were mostly found in those with ALP-linked immunoglobulin; whereas none of the patient with this disease was noted in cases with AMY-linked immunoglobulin.

It is known that isotypes of enzyme-linked immunoglobulins are characterized by predominance of IgG[3] and IgA[7,8] in ALP- and LDH-linked immunoglobulins, respectively. Current results were to the contrary in that there were many LDH-linked IgG cases. Many of these isotypes were monoclonal. Moreover, even if the number of L-chains were the enzyme activity due to enzyme-linked immunoglobulins is limited on the immunoprecipitation line of immunoglobulin by enzyme immunoelectrophoresis. It is presumed, therefore, that these enzyme-linked immunoglobulins were monoclonal or of extremely adjacent clone. The aging phenomenon associated with immunogloblins was evaluated by comparing results for enzume-linked immunoglobulins and BMG. The incidence of BMG was 0.64%, which was higher than the incidence for the enzyme-linked immunoglobulins. IgG was the most frequently occurring isotype (89.1%), and the pattern of increases of BMG with advancing of age was similar to that ALP-linked immunoglobulin, and less like that of AMY-linked immunoglobulin.

Since the occurrence of BMG is considered an immune phenomenon associated with aging, of the enzyme-linked immunoglobulins, ALP-linked immunoglobulin has the highest possibility of creating the immune phenomenon which accompanies aging, and AMY-linked immunoglobulin has the lowest possibility.

AMY-linked immunoglobulin predominate the IgA isotype, differs from the other two enzyme-linked immunoglobulin. Moreover, LDH-linked IgA has been observed in many healthy persons and broad range from 20 to 79 years of age. Therefore, it has been stressed that LDH-linked IgA and LDH-linked

IgG should be evaluated separagely.[7] In addition, a close relationship
has been reported between enzyme-linked immunoglobulins and ulcerative
colitis which is regarded as one of the autoimmune disorders. The enzyme-
linked immunoglobulins that frequently occur in ulcerative colitis are
ALP-linked IgG[9,10] and LDH-linked IgG.

   As a consequently, it is suggested that enzyme-linked IgG has different characteristics from those of enzyme-linked IgA. This is a question to
be addressed in the future.

References

 5) Crofton P.M.: Clin. Chim. Acta, 112, 33-42,1981.

10) Crofton P.M.: and Smith A.F.: Clin. Chim. Acta, 83, 235 - 247, 1978.

11) Tatsuo Tozawa and Yaeko Fujiwara: Physico-Chem. Biol.,26, 79 - 84,
    1982.

(In Japanese)
 1) Yasuhide Tsuzumi and Mitsutaka Nagamine: Phy sico Chem. Biol.,26, 49,
    1982.

 2) Takashi Kanno: Physico-Chem. Biol.,26, 411 - 413, 1982.

 3) Shojiro Kano: Physico-Chem. Biol.,26, 423 - 428, 1982.

 4) Tatsuo Tozawa, Kaoru Taishi and Junko Kuwahara: Physico-Chem. Biol.,
    26, 243 - 248,1982.

 6) Osamu Sugita and Minoru Yakata: Physico-Chem. Biol., 22, 151-156,1978.

 7) Mitsutaka Nagamine: Physico-Chem. Biol.,26,415 - 422,1982.

 8) Minoru Yakata and Osamu Sugita: Physico-Chem. Biol.,26, 435-439,1982.

 9) Kazumasa Miki, Hiroshi Suzuki, Shirou Iino, Hirofumi Niwa ·and Toshiji
    Oda: Jap. J. Gastroenterol, 73, 162-168, 1976.

MITOCHONDRIAL CREATINE KINASE IN HUMAN TISSUE

Tatsuo Yasui, Ryuichi Uzawa, Shuji Ishizawa, Yasushi Takagi,
Tadayoshi Hayama, Kunihide Gomi, Toru Ishii
Department of Clinical Pathology, Showa University School of
Medicine, Tokyo, Japan

Introduction

Creatine Kinase (CK; ATP: creatine N-phosphotransferase, EC
2.7.3.2) is an extremely important enzyme in muscle enzyme
metabolism. This enzyme is a dimer consisting of two subunits, M(muscle) and B(brain). Therefore, three types of
isoenzymes, BB(CK_1), MB(CK_2), MM(CK_3), are widely found in
sera, and the other type of isoenzyme called CKmi has been
found in cellular mitochondria which migrates at anodal site
with electrophoresis.
Recently, an atypical CK with phisico-chemical properties to
be distinguished from these three isoenzymes has been found in
serum from patients with advanced diseases. The relationship
of this atypical CK with mitochondria is currently under investigation.

Materials and Methods

1.Isolation of mitochondrial CK The tissue specimens were
obtained from autopsy materials within 12 hours of death and
kept frozen at -70°C until use. In the preliminary experiment,
we tested mitochondrial fractions obtained by usual procedure.
These mitochondrial fractions had how poor purity for our CK
experiment as shown Figure 1. Then, samples were treated by
following procedures.

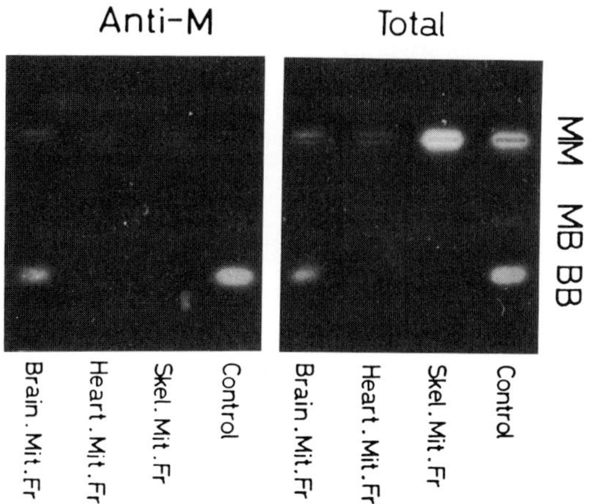

Fig. 1   The electrophoretogram of CK in the mitochondrial fraction of the skeletal muscle, heart and brain.

    upper : ordinary CK stain

    lower : CK stain with anti-CK-M antibody

mitochondrial fraction of skeletal muscle and heart were not inhibited completely by anti-CK-M antibody, and that of brain was not inhibited and had the similar fluorescent density compared with ordinary stain.

Figure 2 shows the electrophoretogram of CK of the purified mitochondrial CK described latter. The CK in mitochondria migrated more cathodic than CK-MM position, and the mobility of skeletal muscle and heart were different from that of brain and stomach. Brain and stomach had only one band and it migrated between the two bands of skeletal muscle and heart. The physico-chemical properties of mitochondrial CK were summarized in Table 1. Estimating of molecular weight with Sephadex G-200 column chromatography, the skeletal muscle, heart and stomach revealed to have two kinds of CK, one was 80 000 and the other was 370 000, but brain had 370 000 only.

Mitochondrial CK was isolated by the method of Webers et al.
(11). The tissues were homogenized in 10 mmol/l Tris-HCl (pH
7.4), and the mitochondrial pellets obtained as described
above were washed with 10 mmol/l Tris-HCl buffer (pH 7.4) until the CK activity in the 13 000G supernatant disappeared
The pellet was suspended in 50 mmol/l phosphate buffer (pH 7.4
) containing 0.2% BSA and 5 mmol/l glutathione. The supernatant thus obtained by ultracentrifugating the suspension at
105 000G was concentrated with Amicon® (Amicon Corp. Lexington
, Mass. 02173) and CK in the concentrated solution was purified by DEAE Sepharose CL-6B and Sephadex G-200 (Pharmacia
Fine Chemicals, Uppsala, Sweden) column chromatography.
2.Assay methods CK activity was estimated at 37°C with CK
reagent (Mercko-test, E.Merck, Darmstadt, F.R.G.). CK isoenzymes were separated electrophoretically and detected fluo-
rometrically. Electrophoresis was carried out with Corning
Special Purpose Film Agarose (Corning, Palo Alto, CA 94306)
with 4-morphorinopropanesulfonic acid buffer (50 mmol/l, pH
7.8, containing 3 mmol/l of sodium azide as preservative) at
90 V for 25 min. The gel were overlaid with one vial of CK
reagent (Merck-1-Test, E.Merck) or CK-MB reagent (Merck-1-
Test CK-MB, E.Merck) which were dissolved in one vial in 1 ml
of accompanying buffer with 20% sucrose, incubated at 37°C for
30 min, and dried at 50°C for 20 min. Fluorescence elicited
by ultraviolet light was quantitated with a Corning 720
densitometer.

Results

Figure 1 shows the electrophoretogram of CK in mitochondrial
fraction. As control, CK-MM, MB and BB were extracted from
the soluble fraction of heart and brain. In control, CK at
ordinary CK-MM position was completely inhibited by anti-CK-M
antibody, and that at MB position was decreased its fluorescent density by it. On the other hand, CK at MM position in

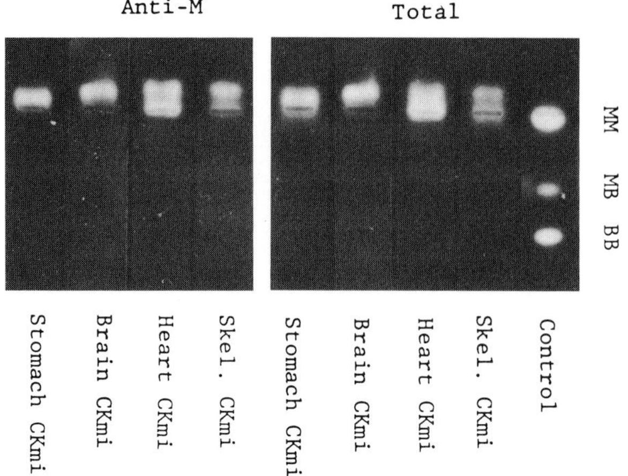

Fig. 2　The electrophoretogram of the skeletal muscle, heart, brain and stomach mitochondrial CK (M.W. 370 000).
upper : ordinary CK stain
lower : CK stain with anti-CK-M antibody

Table　1　Chemical Properties of Creatine Kinase

	M.W.	Km(mmol/l)	Ea(KJ/mol)
CK in Cytoplasm			
MM	80.000	2.66	54.9
MB	80.000	1.67	66.5
BB	80.000	0.98	59.4
CK in Mitochondria			
Skeletal Muscle	370.000	1.28	101.6
Heart	370.000	1.03	110.8
Braim	370.000	0.51	103.9
Stomach	370.000	0.58	110.3

The CK of 80 000 were completely inhibited by anti-CK-M antibody. Therefore, we considered only CK of 370 000 was mitochondrial CK. The activation energy of mitochondrial CK in five organs calculated by Arrhenius plot (7) was all above 100 KJ/mol, which was significantly higher than that of ordinary CK in soluble fraction.
The apparent Michaelis constant (Km) for creatine phosphate was 1.28 and 1.03 mmol/l in skeletal muscle and heart respectively, whereas that in the brain and stomach was 0.51 and 0.58 mmol/l.

Discussion

In recent years, macromolecular types of CK compared with ordinary one were reported in patients with advanced diseases (6-10, 12-17). Some of them were binding with immunoglobulin, and the others were considered to be CK from the mitochondria. The mitochondrial CK migrated at cathodic to CK-MM on agarose electrophoresis. But the mobility of mitochondrial CK is easily changed by the modificating factors in serum, or urea and 2-mer-captoethanol(11.18.19). Therefore mitochondrial CK is readily confused with CK-MM. Since the mitochondrial CK is immunologically different from CK-MM(11.18.19), it can be identified by anti-CK-M antibody in this study.
Other approaches to differentiate these two CK have been reported, such as activation energy (7) or Km for creatine phosphate (19.20). In our study, the activation energy of mitochondrial CK is significantly higher than that of ordinary CK, the former one was over 100 KJ/mol and the latter one was 60 KJ/mol. Apparent Km value of mitochondrial CK was approximately 1.2 mmol/l for skeletal muscle and heart, and approximately 0.5 mmol/l for brain and stomach. These data suggests the presence of microheterogeneity of mitochondrial

422

References

1. James, G.P., Harrison, R.L.: Clin. Chem. 25, 943-947 (1979).
2. Liu, T.Z., Shen, J.T., Lee, Y.-T.N.: Clin. Chem. 26, 1765 (1980) Letter.
3. Bark, C.J.: J. Am. Med. Assoc. 23, 2058-2060 (1980).
4. Stein, W., Bohner, J., Eggstein, M.: Clin. Chem. 28, 1641 (1982).
5. Nakagawa, H., Kida, N., Maeda, M., Wakura, Y., Ohtaki, S.: Clin. Chem. 28, 723-725 (1982) Letter.
6. Heintz, J.W., O'Donnel, N.J., Lott, J.A.: Clin. Chem. 26, 1908-1911 (1980).
7. Stein, W., Bohner, J., Steinhart, R., Eggstein, M.: Clin. Chem. 28, 19-24 (1982).
8. Bohner, J., Stein, W., Steinhart, R., Würzburg, U., Eggstein, M.: Clin. Chem. 28, 618-623 (1982).
9. Wu, A.H.B., Herson, V.C., Bowers, Jr.G.N.: Clin. Chem. 29, 201-204 (1983).
10. Kanemitsu, F., Kawanishi, I., Mizushima, J.: Clin. Chim. Acta 122, 377-383 (1982).
11. Webers, R.A., Mul-Steinbusch, M.W.F.J., Soons, J.B.J.: Clin. Chim. Acta 101, 103-111 (1980).
12. Yuu, H., Takagi, Y., Senju, O., Hosoya, J., Gomi, K., Ishii, T.: Clin. Chem. 24, 2054-2057 (1978).
13. Cheminiz, G.: J. Clin. Chem. Clin. Biochem. 17, 725-729 (1979).
14. Bohner, J., Stein, W., Kuhlmann, E., Eggstein, M.: Clin. Chim. Acta 97, 83-88 (1979).
15. Jochers-Wretow, E., Plessing, E.: J. Clin. Chem. Clin. Biochem. 17, 731-739 (1979).
16. Urdal, P., Landaas, S.: Clin. Chem. 25, 461-465 (1979).
17. Yuu, H., Ishizawa, S., Takagi, Y., Gomi, K., Senju, O., Ishii, T.: Clin. Chem. 26, 1816-1820 (1980).
18. Webers, R.A., Reutelingsperger, C.P.M., Dam, B., Soons, J.B.J.: Clin. Chim. Acta 119, 209-223 (1981).
19. Kanemitsu, F., Kawanishi, I., Mizushima, J.: Clin. Chim. Acta 119, 307-317 (1982).
20. Desjardins, P.R.: Clin. Chim. Acta 121, 67-78 (1982).

SEPARATION OF MOLECULAR FORMS OF RAT BRAIN SOLUBLE ACETYL-
CHOLINESTERASE (AChE) BY POLYACRYLAMIDE GEL ELECTROPHORESIS
FOR THE STUDY OF THE MODIFICATIONS DURING INTOXICATION BY DFP

Guillermo Mario Bisso, Gianni Marcacci, Hanna Michalek
Lab.of Pharmacology, Istituto Superiore di Sanità, 00161-Roma

Introduction

The multiplicity of rat brain AChE molecular forms has been
shown by chromatography, ultracentrifugation in density
gradient and gel electrophoresis. The latter method appears
the simplest and the most convenient when many samples have to
be analyzed. In order to exclude possible artifacts, especially dissociation of high- into low-molecular weight forms, mentioned by some authors (1,2), checks concerning the recovery
of the same molecular forms by different separation methods
are presented here. This enabled the use of gel electrophoresis for evaluation of the role of individual molecular forms of
soluble brain AChE during intoxication by DFP at various ages
and under various physiological conditions (3, 4, 5).

Methods and results

Preparation of enzyme extracts. The brain of Wistar rats was
homogenized for 3 min in a Potter with a teflon pestle using
Tris-HCl 0.038M buffer at pH 8.5 (1:6) and centrifuged at 100,
000 g for 1 hr in a Beckman L2 65B ultracentrifuge.
Polyacrylamide gel electrophoresis was carried out using cylindrical double layer-gels (7.5% separating- and 3% spacer gel).
Samples containing 1-5 mU of enzymatic activity, nmoles of

acetylthiocholine -AcThCh- hydrolyzed/min,(6) were run for 150 min at 1 mA per gel. The enzymatic reaction was carried out at 30°C for 60 min in a solution of 0.2M maleic acid, 0.002M $CuSO_4$, 0.1M glycine and 5.2 mM AcThCh at pH 6.5. The gels were fixed in a 30% solution of Na_2SO_4 and scanned at 600 nm in a Gilford 2500 spectrophotometer. The peak areas were integrated using a Hewlett-Packard 9804 digitizer.

<u>Gel filtration</u> was carried out on Sephacryl S 300 in a LKB 2137 column buffered with 0.1M NaCl-0.038M Tris-HCl at pH 8.1. Samples containing 250 mU were eluted with the buffer, fractions collected, AChE determined (6) and those corresponding to each peak pooled and run on gel electrophoresis.

<u>Ultracentrifugation in density gradient</u> was performed using sucrose gradient from 5% to 20% and samples containing 15 mU centrifuged in a Beckman L2 65B at 85,000 g for 18 hr. Fractions were collected and processed as described above.

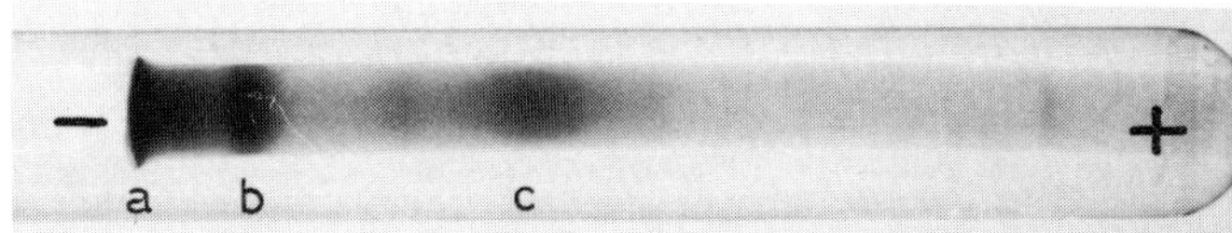

Fig.1. Molecular forms of soluble brain AChE on polyacrylamide gel

Fig.1 shows three main molecular forms of brain soluble AChE separated by gel electrophoresis and their typical densitometric recording (Fig. 2C). There are slowly migrating forms (a) near to the top (MW over 740,000), medium-migrating (b) in a sharp and intense band, (MW 340,000), and fast migrating ones (c) in a rather broad band (MW 115,000). Up to 6 mU the reaction showed satisfactory linearity and reproducibility. The 14-day old rats were used in these experiments since a considerably lower contribution of heavy forms was previously shown (7). Fig.2A and 2B show that by means of the two independent methods three main peaks of enzymatic activity were separated. In fact each of the peaks, when subsequently processed by gel electrophoresis, behaved as a single band corresponding to heavy, medium and light forms detected under standard experi-

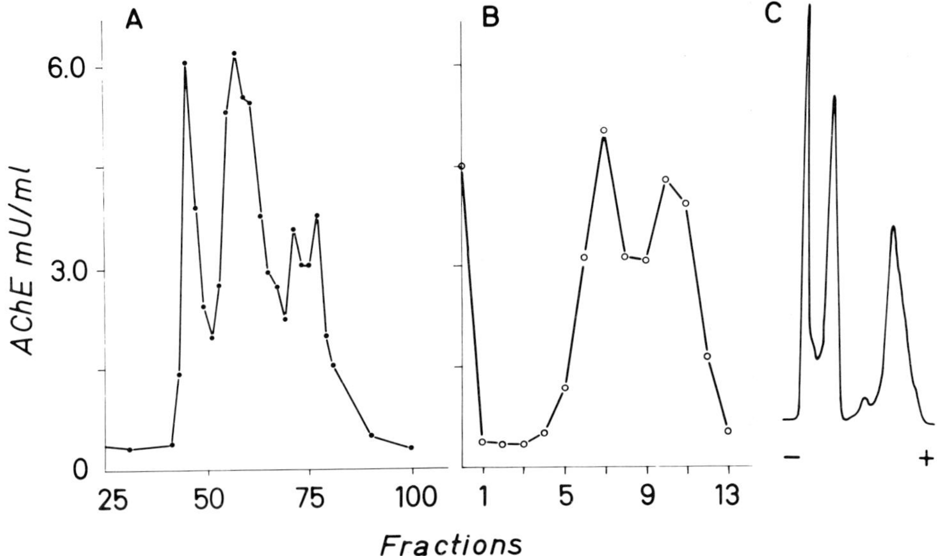

Fig.2. Molecular forms of soluble brain AChE separated by:
A) gel filtration B) ultracentrifugation in density gra
dient C) gel electrophoresis (densitometric recording)

mental conditions thus excluding dissociation phenomena and other artifacts (3).

As concerns the role of the individual molecular forms during acute intoxication by DFP (in arachis oil, 1.1 mg/kg s.c.) Fig.3 shows a transient increase of medium forms during the initial phase of recovery after maximum inhibition of AChE in adults rats (in which the % contribution of heavy, -medium- and light-forms is different from that of 14-day ones).Similar transient increases were found also in various brain regions of adult rat, especially in striatum (4) as well as in whole brain of weanling rats (5). Long-lasting increase of medium-forms was observed during chronic intoxication by DFP of adult rats (8). These increases may be due to their accumulation as precursors of heavy forms in the biosynthesis of enzyme molec-cules or to their higher turnover number with different metab-olic activity and different physiological significance.

426

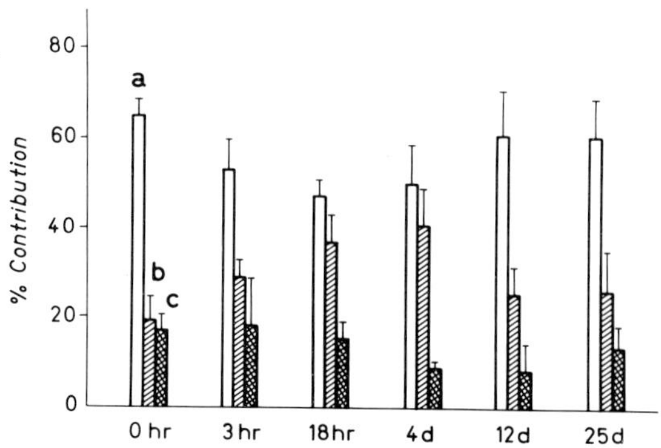

Fig.3. Modifications of the individual molecular forms during
acute intoxication by DFP in adult rats

References

1. Massoulié, J., Bon, S., Rieger F., Vigny, M.: Croat. Chem.
   Acta 47, 163-180 (1975).

2. Gisiger, V., Venkov, L., Gautron, J.: J. Neurochem. 25, 737-
   748 (1975).

3. Michalek, H., Meneguz, A., Bisso, G.M.: Neurobehav. Toxicol
   Teratol.: 3,303-312 (1981).

4. Meneguz, A., Bisso, G.M., Michalek, H.: Clin. Toxicol. 18,
   1443-1451 (1981).

5. Bisso, G.M., Meneguz, A., Michalek, H.: Develop. Neuroscien
   ce 5, 508-519 (1982).

6. Ellman, G.L., Courtney, K.D., Andres, V. Jr., Featherstone,
   R.M.: Biochem. Pharmacol. 7, 88-95 (1961).

7. Bisso, G.M., Nemesio, R., Michalek, H.: Multidisciplinary
   Approach to Brain Development, Elsevier North Holland,
   Amsterdam 1980, pp. 235-236.

8. Michalek, H., Meneguz, A., Bisso, G.M.: Archs. Toxicol.
   Suppl. 5, 116-119 (1982).

The authors thank Dr. Annarita Meneguz, Dr. Annita Pintor and
Mr. Stefano Fortuna for their collaboration. Partially suppor-
ted by the Commission of the European Communities - Contract
N. ENV-548-I (S).

DETECTION OF ALKALINE PHOSPHATASE-LINKED IMMUNOGLOBULIN A IN HUMAN SERUM

Yaeko Fujiwara, Tatsuo Tozawa and Junko Kuwahara
Section of Clinical Laboratory, Hyogo College of Medicine,
Hyogo, Japan

Introduction

Alkaline phosphatase(ALP)-linked immunoglobulin(Ig) was found
in sera of 39 out of 14,648 patients by immunoelectrosyneresis
(ES) on agar-agarose gel. In the sera of five, ALP-linked IgA
was detected. ALP-linked IgG is often found by polyacrylamide
disc gel electrophoresis as ALP VI. However, ALP-linked IgA
has not been observed by isoenzyme analysis. In the present
study, the electrophoretic mobility and column chromatographic
pattern of ALP-linked IgA were investigated.

Materials and Methods

The sera of seven patients, in which the presence of ALP-linked
IgA was suspected, were examined (Table 1).

| Patient | Sex | Age | Diagnosis | Clinical findings | | | | | | | | ES & IEP | IPR |
				GOT (K.U)	GPT (K.U)	LDH (W.U)	ALP (BLU)	T-Bil (mg/dl)	IgG	IgA (mg/dl)	IgM		
Kk	m	9	Urethral stricture	31	19	440	10.6	10.3	1490	273	292	A-K	n.d.
Tt	m	83	Sigma cancer	11	6	304	1.3	0.5	775	388	49	A-L	A-K.L
Is	f	22	Healthy	19	14	365	0.7	—	—	—	—	A-L	A-L
Fk	m	78	Diabetes melltus Arteriosclerosis obliterants	9	7	263	3.0	0.4	1075	269	275	G.A-K.L	G.A-K.L
Ts	m	64	Pancreas cancer	30	40	481	16.1	6.7	1744	390	152	A-K	A-K.L
Sk	m	63	Cholestasis	54	59	290	16.5	17.8	626	160	74	A-K.L	Neg.
It	m	35	Hepatoma	400	74	595	32.7	14.1	1700	740	96	A-K	Neg.

Table 1. Clinical findings and results of
detecting ALP-linked Ig.

Electrophoresis '83
© 1984 Walter de Gruyter & Co., Berlin · New York

1) ES was performed on 1.2% agar-agarose gel with veronal buffer using antisera (DAKO, Denmark).
2) ALP isoenzyme was analysed on 7% polyacrylamide disc gel with Tris-boric acid buffer (pH 8.9). ALP isoenzymes were stained with ALP isoenzyme reagent (WAKO, Japan).
3) Gelfiltration was performed on Sephadex G-200 column with Tris-boric acid buffer.
4) Immunoelectrophoresis(IEP) and immunofixation(IEF) were performed on 1% agarose film (CORNING, U.S.A.) with veronal buffer.
5) Immunoprecipitin reaction in free liquid media(IPR) was according to the method of Taishi and Tozawa.[1]

Results

By ES, ALP activity was found on the immunoprecipitin line of IgA in seven patients (Fig. 1). In patients Kk, Tt, Is and Fk, main ALP activity was observed in the G fraction and slight activity in the G-M fraction by gelfiltration. ALP activity in the G-M fraction was found on the immunoprecipitin line of IgA after ES. In patients Ts, Sk and It, two peaks were observed in the G and M fractions. ALP activity in the M fraction was found on the immunoprecipitin line of IgA after ES. Patient Ts had some ALP activity in the G-M fraction in addition to those in the G and the M fractions described above.
The electrophoretic patterns of these seven patients are shown in Fig. 2. The main ALP activity of patient Kk was in the bone isoenzyme position. Patients Tt, Is and Fk had liver isoenzymes with some diffusion toward the cathode side. Patient Fk also had ALP VI corresponding to ALP-linked IgG. Three other patients had high-molecule ALP(ALP I) which remained in the sample gel and high ALP activity in the liver isoenzyme position. To investigate the electrophoretic mobility of ALP-linked IgA, IEF was performed ; ALP activity remained immediately adjacent to the cathode side of the liver isoenzyme posi-

tion. By IPR, ALP-linked IgA was not identified in 2 patients SK and It.

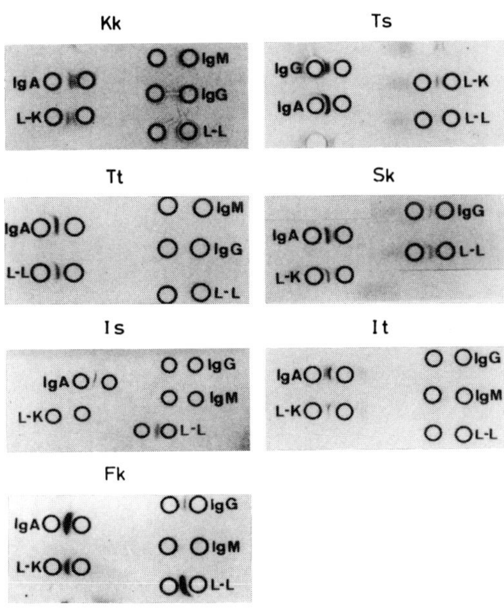

Fig. 1. Results of ES for seven patients.

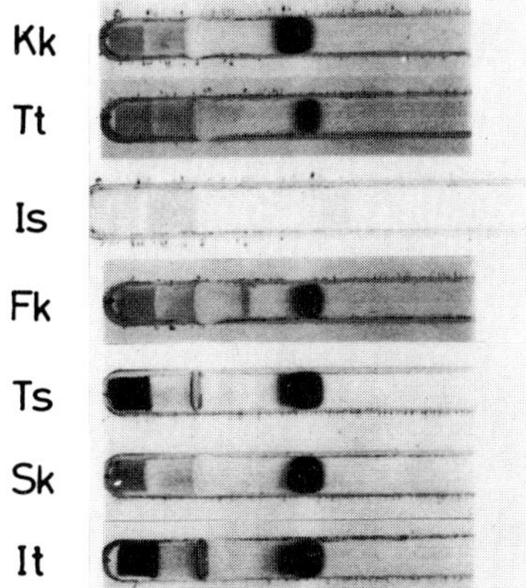

Fig. 2. Electric patterns of ALP on polyacrylamide disc gel.

Discussion

Saga and Kano reported that high-molecule ALP(ALP I) was a complex of ALP and plasma proteins such as IgA.[2] In our laboratory, ALP activity was found on the immunoprecipitin line of IgA by ES in seven patients. In five of them, ALP-linked IgA was demonstrated from the results of gelfiltration and IPR. ALP-linked IgA was eluted in the G-M fraction as well as ALP-linked IgG. The electrophoretic pattern of patients with ALP-linked IgA showed the diffusion pattern of liver isoenzyme, but it was not a clear band contrary to that of ALP-linked IgG as ALP VI. From the results of IEF, the mobility of ALP-linked IgA was found in the bone isoenzyme position as a diffuse band. therefore, it was impossible to detect ALP-linked IgA by isoenzyme analysis. Thus, immunological methods are necessary and more efficient in the detection of ALP-linked IgA. In patients Sk and It having high-molecule ALP, ALP activity on the immunoprecipitin line of IgA was found as in the other five patients. However, it was only in the M fraction and ALP-linked Ig could not be identified by IPR. Therefore, we presumed that a complex of ALP and IgA in the M fraction might not be an immunocomplex. These two patients were suffering from obstructive jaundice. So, high molecule ALP remaining on the immunoprecipitin line of IgA could not be washed out thoroughly. Patient Ts seemed to have both ALP I and ALP-linked IgA judging from the results of the gelfiltration pattern.

References

1). Taishi, K., Tozawa, T.:Jap. J. Clin. Chem.,12,30-38,(1983).
2). Saga, M., Kano, S.:Clin. Chim. Acta,95,512-524,(1979).

ELECTROPHORETIC ANALYSIS OF α-GLYCEROPHOSPHATE DEHYDROGENASE
AND ITS ROLE IN METABOLIC REGULATION

Kayoko Sudo, Shuji Hishiki* and Takashi Kanno
Department of Laboratory Medicine, *Second Department of
Surgery, Hamamatsu University School of Medicine, Hamamatsu,
Japan

Introduction

Glycolysis in the muscle of a patient with lactate
dehydrogenase(LDH) M subunit deficiency is markedly retarded
at the position of glyceraldehyde 3-phosphate
dehydrogenase(GAPD), and half of the glycogen of glucose-
derived triose phosphates are converted to glycerol and
α-glycerophosphate due to coupling reoxidation of NADH by
cytosol α-glycerophosphate dehydrogenase(GPD) (1,2). In this
paper the heterogeneity of cytosol GPD in various tissues is
analyzed electrophoretically. The abnormal participation of
GPD to NADH reoxidization is studied.

Results and Discussion

1) Multiple forms of GPD in human tissues: GPD is a dimeric
    protein composed of two different subunits, H(heart type)
    and L(liver type) (3). Figure 1(A) illustrates zymograms
    of GPD in various tissues. Three forms of isoenzymes,
    HH, HL and LL are observed on Cellogel and polyacrylamide
    disc electrophoresis. HH is the main isoenzyme in heart
    muscle and erythrocytes. LL and HL are detected in
    almost all of the tissues analysed. HL in liver tissue
    is very low. The affinity for NADH of these three

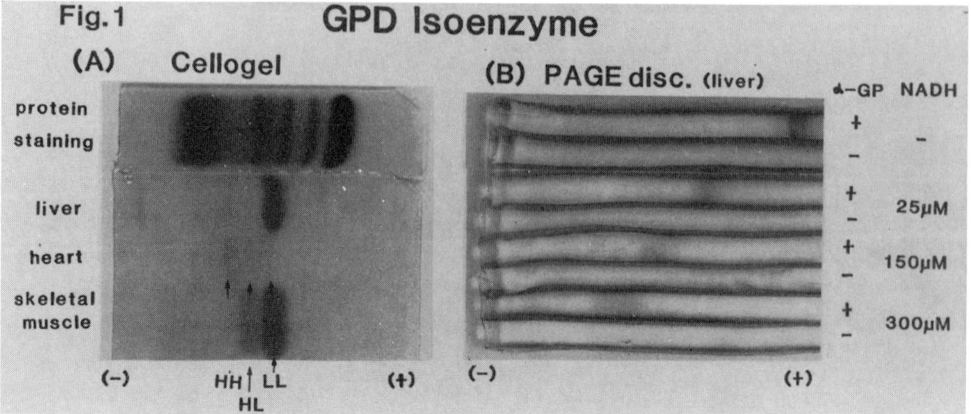

isoenzymes was examined using the method described by
Takeo and Nakamura(4).

Figure 1(B) shows the zymograms of liver tissue GPD using
polyacrylamide gel containing various concentrations of
NADH, and retardations of the mobilities were observed
with increasing concentration of NADH. Employing this
technique, the differences of affinity were studied for
three isoenzymes of GPD, and the apparent affinity of HH
enzyme for NADH was five times higher than that of LL
enzyme.

2) GPD activities in human tissues: The specific activities
    observed in 100,000 x g supernatants of tissue homogenates
    are summarized in Figure 2(●). The activities in adult
    skeletal muscle, heart and liver are higher than those in
    other tissues, and the activities of GPD in fetal tissues
    are relatively lower.

3) LDH/GPD ratio in human tissues: LDH/GPD ratio indicates
    an abnormal participation of GPD in the reoxidizing of
    NADH in cases of LDH subunit deficiencies. In Figure 2(o),
    the ratios between LDH and GPD in various tissues are
    illustrated. Higher ratios are observed in erythrocyte,
    fetal muscle and in fetal lung tissue. In contrast,
    relatively lower ratios are obtained from adult skeletal
    muscle and the liver.

Fig.2   **Specific activities of GPD and LDH/GPD ratios in human tissues   (100000xg sup.)**

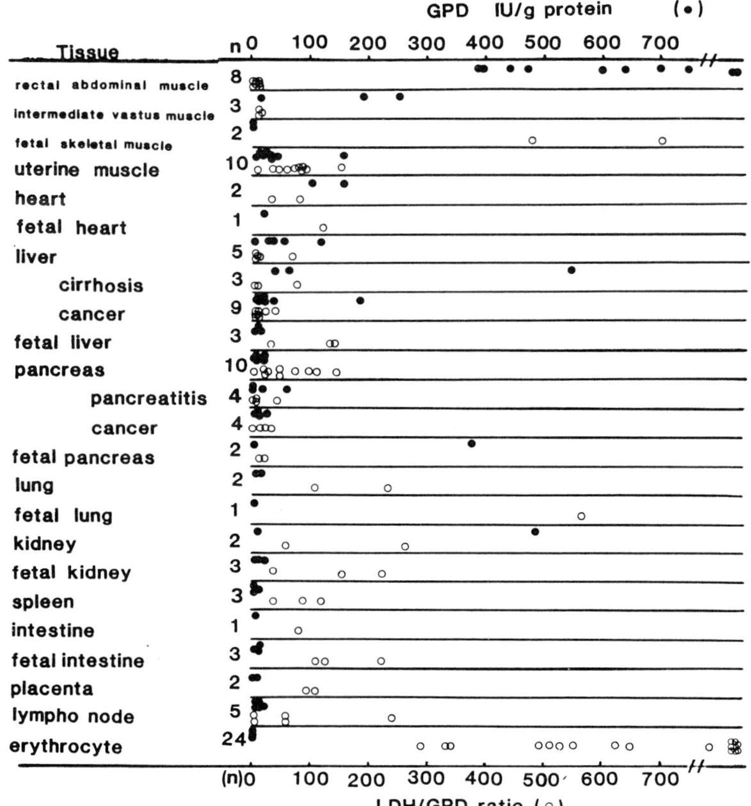

Fig.3 LDH/GPD ratios in the cases of LDH subunit deficiencies.

434

4) Participation of GPD in the conversion of triose
   phosphate: Assumed LDH/GPD ratios in the cases of LDH
   subunit deficiencies are summarized in Figure 3. In the

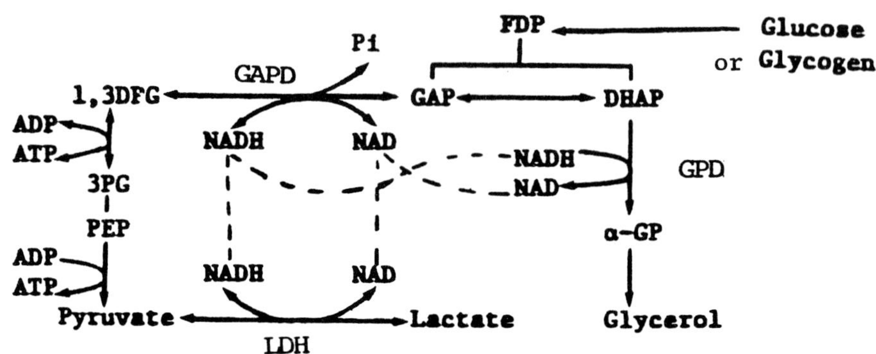

case of LDH M subunit deficiency, LDH/GPD ratios are
markedly decreased in skeletal muscle and liver tissue.
The LL isoenzyme of GPD has a higher affinity for
NADH(5). The affinity of HH isoenzyme for NADH is lower
than that of LL. The abnormal participation of GPD to the
coupling NADH reoxidation in human tissues is studied in
this paper. In anaerobic conditions, the abnormal
accumulation of triose phosphates might be expected to
occur in skeletal muscle and in liver in the case of M
subunit deficiency. On the other hand, abnormal accumulation would occur in heart muscle and pancreas in the case
of H subunit deficiency.

References

1. Kanno, T., Sudo, K., Takeuchi, I., Kanda, S., Honda, N.,
   Nishimura, Y., Oyama, K.: Clin. Chim. Acta 108, 267-276
   (1980).
2. Sudo, K.: J. Keio Med. Soc. 59, 833-848 (1982).
3. Mcginnis, J.F., Vellis, J.D.: J. Molec. Cel. Cardio. 11,
   795-802 (1979).
4. Takeo, K., Nakamura, S.: Arch. Biol. Biophy. 153, 1-7
   (1972).
5. Ostro, M.J., Fondy, T.P.: J. Biol. Chem. 252, 5575-5583
   (1977).

ELECTROPHORETIC SEPARATION OF TRIPEPTIDE AMINOPEPTIDASE AND ITS DISTRIBUTION IN HUMAN TISSUES

Shinji Kanda, Kayoko Sudo, Takashi Kanno and Shuji Hishiki*
Department of Laboratory Medicine, *2nd Department of Surgery,
Hamamatsu University School of Medicine, Hamamatsu, Japan

Introduction

Recently we reported the presence of tripeptide aminopeptidase (EC 3.411.4, TAP) in human serum[1]. In this paper, the electrophoretic separation of TAP from other aminopeptidases and the subsequent activity staining were established. By using these techniques, we studied the distribution of TAP in various human adult, fetal and cancerous tissues.

Materials and Methods

Adult human tissues obtained during surgical procedures and fetal tissues at 13- to 19 weeks gestation were homogenized and fractionated by the method of Hogeboom[2]. L-Leucylglycylglycine(LGG) hydrolytic activity was assayed by the method we have previously described[1]. Electrophoresis was performed on a Cellogel(Chemetron) using the continuous barbital buffer (μ =0.06, pH 8.6) and the discontinuous buffer system described by Kohn[3]. The quantitative activity staining of aminopeptidases was based on the principle previously described[4].

Results and Discussion

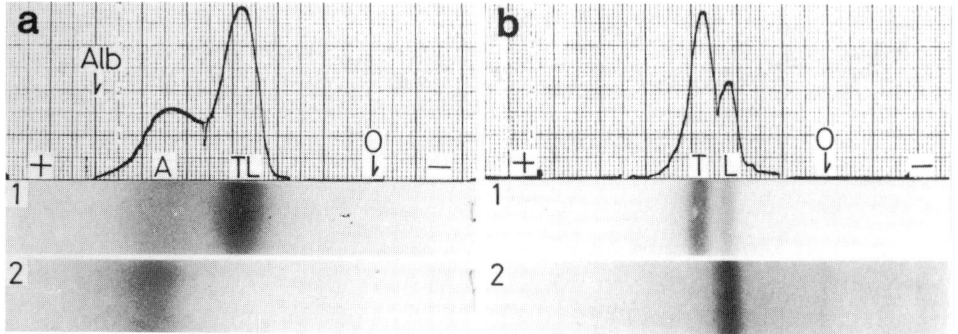

Fig.1 Zymograms and densitograms of aminopeptidases in serum
of a patient with liver disorders. Electrophoresis was performed with continuous(a) and discontinuous(b) buffer system.
Enzyme activities were stained with the following substrates:
a-1; L-leucylglycylglycine(4mM), a-2; L-leucine-p-nitroanilide
(4mM), b-1; L-leucylglycylglycine(4mM), b-2; L-leucinamide
(23.4mM)

1) Electrophoretic separation and subsequent activity stain-
    ing of TAP; Fig.1 shows zymograms and densitograms of the
    aminopeptidases in the serum of a patient with liver dis-
    orders. LGG hydrolytic aminopeptidases were roughly sepa-
    rated into two bands, A and TL, by the continuous buffer
    system(a-1). Band A shows arylamidase activity, since
    L-leucine-p-nitroanilide was hydrolyzed on the same
    band(a-2). The other band, TL, was further separated
    into two bands of T and L by the discontinuous buffer
    system(b-1). Band L also hydrolyzed L-leucinamide(b-2);
    we identified it as cytosol leucine aminopeptidase. On
    the other hand, band T, which specifically hydrolyzed
    LGG, results in TAP. By using these procedures, TAP
    activity was clearly distinguished from other amino-
    peptidase activities.

2) Subcellular localization of TAP; Table 1 shows the acti-
    vities of TAP and marker enzymes in subcellular fractions
    prepared from human liver tissue. Sixty-five percent of
    the TAP activity in whole homogenates was recovered in
    the supernatant fraction while no significant activity
    was observed in other fractions.

Table 1 Distribution of TAP and marker enzyme activities in subcellular fractions prepared from human liver

Table 1 Distribution of TAP and marker enzyme activities in subcellular fractions prepared from human liver

	Specific activity; I.U./g protein (% recovery)					Overall recovery (%)
	Homogenate	Nucleus	Mitochondrion	Microsome	Supernatant	
TAP	228.2 (100.0)	37.8 (0.1)	8.3 (0.1)	6.9 (0.2)	183.1 (65.0)	(65.4)
LDH	956.6 (100.0)	291.8 (0.2)	35.6 (0.2)	88.7 (0.6)	677.3 (57.4)	(58.7)
G6Pase	73.1 (100.0)	28.3 (0.3)	384.1 (21.7)	363.1 (30.7)	0.6 (0.7)	(53.4)
GLDH	9.4 (100.0)	53.3 (4.3)	157.2 (69.7)	neg. –	neg. –	(73.3)
Total protein; mg	77.6 (100.0)	0.6 (0.8)	3.2 (4.1)	4.8 (6.2)	62.9 (81.1)	71.5 (92.2)

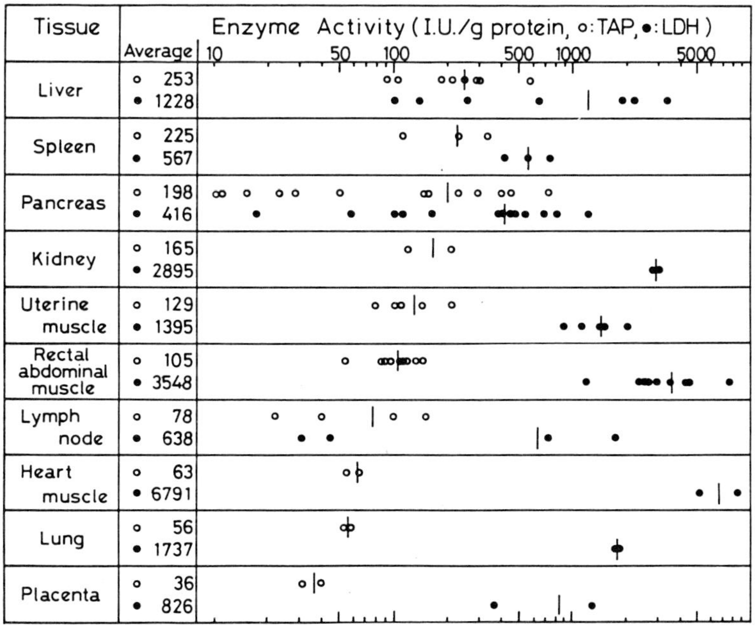

Tissue	Average	Enzyme Activity (I.U./g protein, o:TAP, •:LDH)
Liver	o 253 • 1228	
Spleen	o 225 • 567	
Pancreas	o 198 • 416	
Kidney	o 165 • 2895	
Uterine muscle	o 129 • 1395	
Rectal abdominal muscle	o 105 • 3548	
Lymph node	o 78 • 638	
Heart muscle	o 63 • 6791	
Lung	o 56 • 1737	
Placenta	o 36 • 826	

Fig.2 Distribution of TAP and LDH activities among adult human tissues.

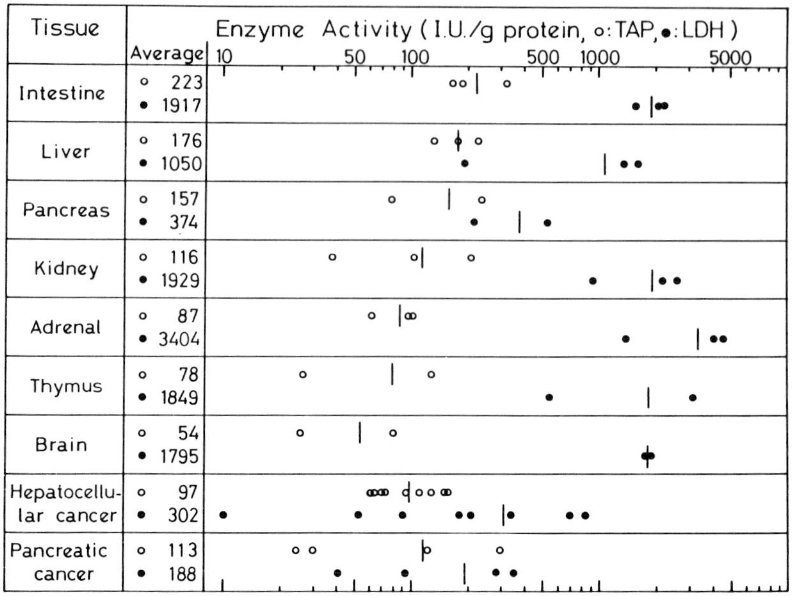

440

Tissue	Average	Enzyme Activity (I.U./g protein, o:TAP, •:LDH)
Intestine	o 223 • 1917	
Liver	o 176 • 1050	
Pancreas	o 157 • 374	
Kidney	o 116 • 1929	
Adrenal	o 87 • 3404	
Thymus	o 78 • 1849	
Brain	o 54 • 1795	
Hepatocellu- lar cancer	o 97 • 302	
Pancreatic cancer	o 113 • 188	

Fig.3 Distribution of TAP and LDH activities among fetal and
cancerous human tissues.

3) Distribution of TAP among human tissues; Fig.2 and 3 show
    the activities of TAP and LDH obtained from the 105,000 x
    g supernatant of various tissue homogenates. TAP was
    widely distributed in adult, fetal and cancerous tissues.
    No electrophoretic heterogeneity of TAP could be found in
    any of these tissues.

References

1. Kanda, S., Sudo, K. and Kanno, T.: Jap. J. Clin. Chem.
    11, 314-325 (1982).
2. Hogeboom, G.H.: Method in Enzymology, Academic Press. New
    York, Volume 1, 16-19 (1955).
3. Koh, J.: J. Clin. Pathol. 22, 109-111 (1969).
4. Kanda, S., Manabe, M., Sudo, K., Kanno, T. and Shibata,
    S.: Physico-Chem. Biol. 25, 167-171 (1981).

LIVER MITOCHONDRIAL SPECIES OF CREATINE KINASE

Fusae Kanemitsu, Isami Kawanishi, Jun Mizushima
Clinical Laboratories, Kurashiki Central Hospital, Kurashiki, Japan

Tohru Okigaki
Division of Cell Biology, Shigei Medical Research Institute, Okayama, Japan

Introduction

Following the detection of creatine kinase (ATP:creatine N-phosphotransferase, EC 2.7.3.2, CK) isoenzyme in mitochondria by Jacobs et al (1), various physico- and immunochemical properties of the enzyme have been reported (2-4). These studies have been performed with mitochondrial CK mainly prepared from brain, heart or skeletal muscle. Although liver cells are rich in mitochondria, CK activity is very low per unit weight of mitochondria (5), thus there is little information on liver CK. We recently studied the enzyme from normal and metastatic tumor liver tissues, and found there were four forms for the isoenzyme (6, 7). The present study describes the properties of the mitochondrial CK from liver tissue.

Results

1) Electrophoretic mobility. Creatine kinase from liver mitochondria was separated into four forms by agar gel electrophoresis. Two bands were at the cathodal side of CK-MM (CKmL1 and CKmL2 from the cathodal side), the third is at the MM position (CKmL3) followed by one between CK-

442

MM and CK-MB (CKmLT). CKmLl and CKmL2 were obtained directly from mitochondria of normal liver, and CKmLT was from
metastatic tumor liver. CKmL3 was obtained after 2 mole/L
urea treatment of the above three forms at 26°C for 30 min.
None of the forms are adenylate kinase because they were
not seen when creatine phosphate was omitted from the
reaction mixture.

2) Antigenicity. Electroimmunofixation against anti-human
    mitochondrial CK antibodies was performed on agar plates.
    CKmLl, CKmL2 and CKmLT reacted with the antibodies, and
    were positively identified on the precipitation lines
    after CK staining. None of the four forms reacted with
    anti-CK-M and -CK-B subunit antibodies. Therefore the
    forms were mitochondrial CK and not cytoplasmic.

3) Relative molecular masses. The relative molecular masses
    of the CKs were estimated by Sephadex G-200 superfine
    columns. The chromatogram of mitochondrial extracts from
    normal liver tissue produced two CK peaks; one between
    7s and 19s, and the other around 4s. They were estimated
    to be approximately 350,000 and 80,000 respectively.
    Electrophoresis showed that the larger fraction contained
    predominantly CKmLl and the smaller fraction contained
    CKmL2. CKmL3 and CKmLT were similarly estimated to be
    320,000 to 350,000 and 80,000 respectively. From these
    results it appears that CKmLl and CKmLT are oligomeric
    forms of mitochondrial CK, and CKmL2 and CKmL3 are mono-
    meric.

4) Affinity for concanavalin A. The concanavalin A affinity
    chromatograms showed that CKmLl, CKmL2 and CKmLT were
    eluted by the starting buffer, and have no affinity for
    concanavalin A. Therefore the above CKs have no carbohyd-
    rate moieties which could react with concanavalin A.

5) Heat stability. The residual enzyme activities of CKmLl
    and CKmLT after heating at 56°C for 4 min. were 79% and
    75% respectively, so that they are heat stable. CKmL2
    was almost completely inactivated by heating at 56°C for

TABLE I. FOUR FORMS OF MITOCHONDRIAL CREATINE KINASE FROM
          LIVER TISSUE

	Origin	Mobility	Anti-CKm Antibodies	Form
CKmL1	Normal liver mitochondria	Cathodal side of CKmL2	Reacted	Oligomer
CKmL2	Normal liver mitochondria	Cathodal side of CK-MM	Reacted	Monomer
CKmL3		CK-MM position	ND	Monomer
CKmLT	Tumor liver mitochondria	Between CK-MM and CK-MB	Reacted	Oligomer

	2 M/L Urea Treatment	Heat stability (56°C, 4 min)	Affinity to Con A	Neuraminidase Treatment
CKmL1	Altered to CKmL3	Stable	None	Not reacted
CKmL2	Altered to CKmL3	Weak	None	Not reacted
CKmL3	Altered from CKmL1,2,T	ND	ND	Not reacted
CKmLT	Altered to CKmL3	Stable	None	Not reacted

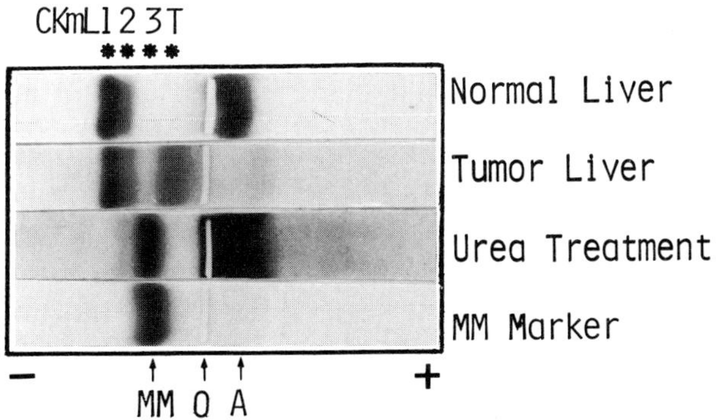

Fig. 1. Four forms of mitochondrial CK from normal and meta-
         static tumor liver by agar gel electrophoresis.
         0, sample apprication point; A, adenylate kinase.

1 min.

6) Additional characteristics. Electrophoretic mobilities of
the CKs were unstable. CKmL1, CKmL2 and CKmLT migrated at
CKmL3 position after urea treatment as described in 1).
From this we conclude that CKmL3 is thought to be the
final form of the other three CKs. Further the electrophoretic pattern of CKmLT changed after storage at -20°C
for 3 years. Before storage the enzymogram showed a
single band of CKmLT, but after storage CKmL1 and CKmL3
appeared with the disappearance of CKmLT. It seems that
CKmLT was altered to CKmL3 through CKmL1 and CKmL2.

7) From the data 1) to 6), we conclude that CKmLT from tumor
liver is a modified form of CKmL1 from normal liver with
a negative charge inasmuch as the characteristics of CKmLT
are similar to those of CKmL1 with exception of the electrophoretic mobility. The negative charge of CKmLT,
however, is not due to N-acetylneuraminic acid or phosphoric acid because CKmLT had no sensitivity to neuraminidase or phosphorylase treatment. CKmL1 is an oligomeric
form of CKmL2, and CKmL3 is the final form of the liver
mitochondrial CK.

References

1. Jacobs, H., Heldt, H.W., Klingenberg, M.: Biochem. Biophys. Res. Commun. 16, 516-521 (1964).
2. Sobel, B.E., Shell, W.E., Klein, M.S.: J. Mol. Cell. Car-
   diol. 4, 367-380 (1972).
3. Scholte, H.R., Weijers, P.J., Wit-Peeters, E.M.: Biochem.
   Biophys. Acta 291, 764-773 (1973).
4. Booth, R.F.G., Clark, J.B.: Biochem. J. 170, 145-151 (1970).
5. Jacobus,W.E., Lihninger, A.L.: J. Biol. Chem. 13, 4803-
   4810 (1973).
6. Kanemitsu, F., Kawanishi, I., Mizushima, J.: Clin. Chim.
   Acta 119, 307-317 (1982).
7. Kanemitsu, F., Kawanishi, I., Mizushima, J.: Clin. Chim.
   Acta 128, 233-240 (1983).

LDH ISOENZYME OF RABBIT GRANULOCYTES

Tadayoshi Imaizumi and Masaharu Horiguchi
Internal Medicine, The Jikei University School of Medicine
Komae-shi, Tokyo

Introduction

In our previous studies, some differences in LDH isoenzyme
composition between circulating blood granulocytes(1) and pus
corpscles(2) in human were observed. In the present study,
the LDH isoenzyme composition of rabbit granulocytes was
examined.

Materials and Methods

Male rabbits weighing 1.5-2.0 kg were used. Blood
granulocytes: The Buffy coat was take from heparinized
centnfuged blood. The cells were obtained by
gravity-layer(1077) centrifugation and 6% dextran treatment
of the buffy coat suspension. Peritoneal exduate
granulocyte: Rabbits were injected intraperitoeally with
physiological saline. After 24 hrs., the peritoneal cavity
was washed by saline. The saline washing was centrifuged and
peritoneal exduate cells (granulocytes) were collected.
After freezing and thawing the cells were used for electrophoresis. Electrophoresis: The method of thin layer
polyacrylamide-gel electrophoresis was used(3), at 200 volt,
20 mA, for 3 hrs., with veronal buffer (PH 8.6, ion strength:
0.06). After electrophoresis, LDH staining was performed(4).

446

Results

Experiment 1. LDH isoenzyme composition of blood
granulocytes in sample No. 1 and peritoneal exduate
granulocytes in sample No. 2 are shown in Table 1, Exp. 1.
the LDH_2 content was 24.2% in No. 1, while LDH_4 was 24.5%
in No. 2.
Experiment 2. Blood granulocytes following bacterial
injection. E. coli ($5x10^8$ p.f.u.) was injected
intravenously into rabbits. After 72 hrs blood granulocytes
were collected. The white blood cell count was $17000/mm^3$.
Sample No. 3 consisted mainly of stab cells and sample No. 4
of segment cells. The LDH_1 content was 23.2% and LDH_2
23.2% in No. 3. In No. 4. LDH_1 was 46.5% and LDH_2 32.4%,
shown in Table 1, Exp. 2.
Experiment 3. Peritoneal exduate granulocytes after adjuvant
or bacterial treatment: Freund's incomplete adjuvant (FIA)
(10ml) or E. coli ($1x10^8$ p.f.u.) was injected into rabbits
intraperitoneoally. After 24 or 72 hrs. peritoneal exduate
granulocytes were collected. As shown in Table 1, Exp. 3,
sample No. 5 (24 hrs. after FIA injection) the LDH_4 content
was 23.7%. In sample No.6 (24 hrs. after E. coli injection),
and LDH_5 was 14.8%, while in sample No. 7 (72 hrs. after E.
coli injection), LDH_5 was 23.3%.
Experiment 4. For the study of pus and pus corpscles, human
muscle homogenate (3 ml) was injected into rabbits
subcutaneously. After pus formation, pus and pus corpscles
were collected. Table 1, Exp. 4 shows that the LDH_5
content in pus was 12.0% while in pus corpuseles it was 21.3%.

Table 1. LDH isoenzyme composition of the samples

Exp.	Sample	Material	Snear (main cell)	LDH$_1$	LDH$_2$	LDH$_3$	LDH$_4$	LDH$_5$
Exp. 1	No. 1	Blood	Granulocyte	16.4	24.2	27.3	20.0	12.0
	No. 2	PE*	Granulocyte	11.3	18.9	30.2	24.5	15.1
Exp. 2	No. 3	Blood	Stab cell	23.2	23.2	23.2	18.5	12.0
	No. 4	Blood	Segment cell	46.5	32.4	14.5	5.6	2.8
Exp. 3	No. 5	PE	Granulocyte	10.1	18.6	23.7	23.7	23.7
	No. 6	PE	Granulocyte	9.9	21.0	27.2	27.2	14.8
	No. 7	PE	Granulocyte	8.1	19.7	24.5	24.5	23.7
Exp. 4	No. 8	Pus		7.9	19.0	31.7	28.5	12.0
	No. 9	Pus	Pus corpscle	10.4	19.8	28.7	19.8	21.3

* PE: peritoneal exduate

Exp. 1: Granulocytes from blood (No. 1) and peritoneal exduate (No. 2) of normal rabbit.

Exp. 2: Blood granulocytes after bacterial injection (No. 3, (No. 4).

Exp. 3: Peritoneal exduate granulocytes after injection of Freund's incomplete adjuvant (no. 5) or bacterial suspension (No. 6, No. 7)

Exp. 4: Subcutaneous pus (No. 8) and pus corpscles (No. 9).

Discussion

From the result of Exp. 1, a slight difference may be seen in LDH isoenzyme composition between blood granulocytes and peritoneal exduate granulocytes. After E. coli-injection, LDH$_1$ and LDH$_2$ were markedly elevated in blood granulocytes in Exp. 2. On the other hand, intraperitoneal stimulation by FIA or E. Coli caused some LDH$_5$-elevation in peritoneal exduate granulocytes in Exp. 3. In pus corpscles by comparison with granulocytes of Exp. 1, LDH$_4$ and LDH$_5$ increased slightly, while in the pus corpscle of human there was a

448

marked increase in LDH_4 and LDH_5 (2).
There may be some problems on the materials. That is,
a) cell destruction during sample collection, b) contamination by other components, and c) cell population of the
sample, and so on.
It is thought that LDH isoenzyme composition does not change
by physiological condition in organ and tissue cells, but it
is indicated that granulocytes change LDH isoenzyme
composition according to its conditions. One of main
condition will be a degree of oxygen consumption of the
granulocytes.

Summary

It was observed that rabbit granulocyte shows varieties in
LDH isoenzyme composition according to its conditions.

References

1. Imaizumi, T., Horiguchi, M.: The physicochemical Biology
    (Japanese) 26, 44 (1982).
2. Imaizumi, T., Horiguchi, M.: The physicochemical biology
    (Japanese) 26, 99 (1982).
3. Ogita, Z., et al.: SABCO J. 2, 58 (1966).
4. Ogihara, M., et al.: The Physicochemical biology 21,
    15-19 (1977).

PROPERTIES OF NEWLY DISCOVERED LDH-X 5 SUBFRACTIONS OF HUMAN
SEMEN USING ISOELECTRIC FOCUSING

Mitsutaka Yoshida, Toshio Imai
Dept. of Biology, Faculty of Science, Toho University, Chiba,
Japan

Makoto Hara, Kohichiro Isurugi
Dept. of Urology, School of Medicine, Tokyo University, Tokyo,
Japan

Tadao Kinoshita
National Medicine Central Hospital, Tokyo, Japan

Yasuko Higashi
Special Reference Laboratories Inc., Tokyo, Japan

Introduction

Sperm specific LDH-X in human semen has been isolated as one
band by conventional electrophoretic methods. We undertook to
separate LDH-X into 5 fractions by Isoelectric Focusing and
studied of the properties of each fraction which might lead to
findings of genetic disorder of male infertility, and along
with chemical properties of total LDH-X and relationship
between the number of the spermatozoa and the motility. We
were able to succeed to separate 5 fractions of LDH-X (Xa,
Xb, Xc, Xd, Xe). Methodology is as described in a previous
report, see reference.[1]

Electrophoresis '83
© 1984 Walter de Gruyter & Co., Berlin · New York

450

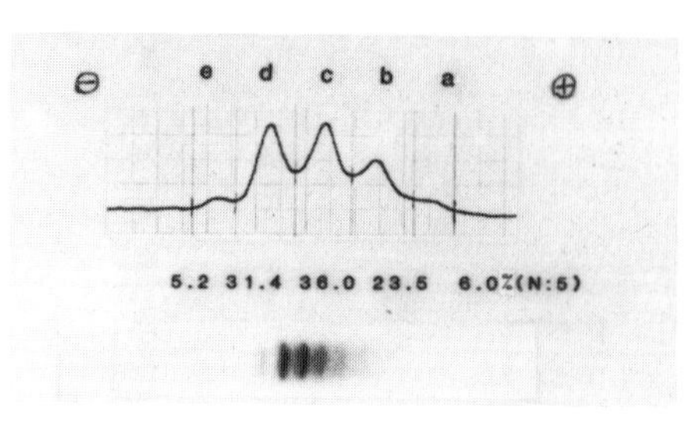

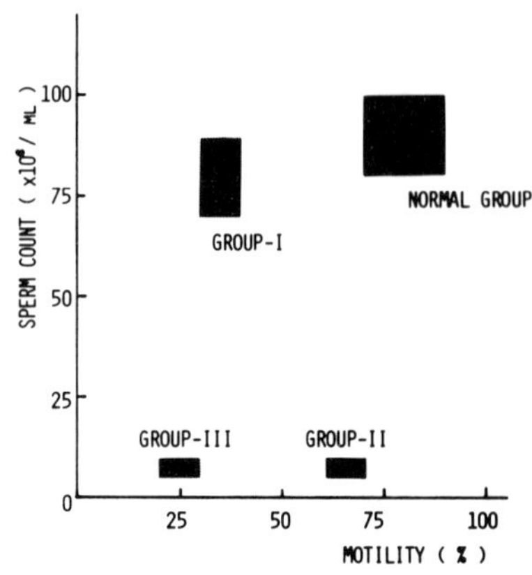

Fig 1   5 Fractions of sperm specific LDH-X (From anode LDH-Xa, -Xb, -Xc, -Xd, -Xe)

Fig 2   Relationship between sperm count and motility of human semen in each group

Results

1) We were able to separate sperm specific LDH-X into 5 fractions by Isoelectric Focusing. (Fig 1)
   (From anode LDH-Xa, -Xb, -Xc, -Xd, -Xe)

2) Decrease in total LDH-X and Xc fraction in lower sperm count groups I, II, III were obtained. (Fig 2)

3) The normal group and group II with low sperm counts and close to normal motility had Xb < Xd group I and III with low motility had Xb > Xd   (Fig 3, 4)

4) In inhibition studies, the order of inhibitory action of inhibitors to total LDH follows:
   $CoCl_2$ > Oxalic acid > PCMB > CH_2ICOOH

5) In the lower sperm count groups, such as azospermia, groups II and III, $CoCl_2$ was the strongest inhibitor.

6) The thermal stability of LDH-X is stronger in group I with low motility and sperm counts close to normal, than in the other groups.

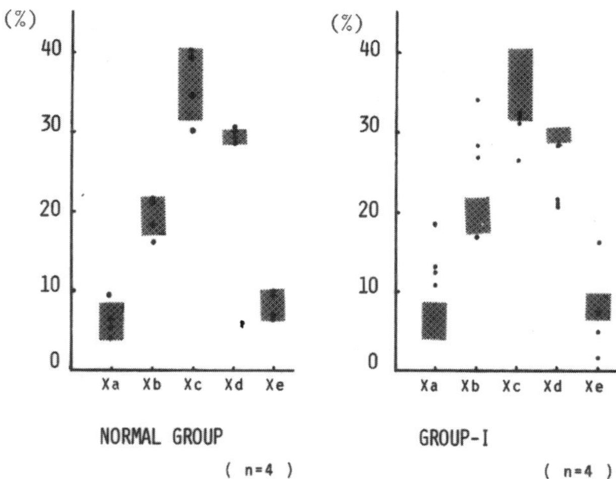

Fig 3 Percentage of LDH-X subfractions activity to total
LDH-X activity. (Normal and Group I) Shaded areas show
percentage in normal group.

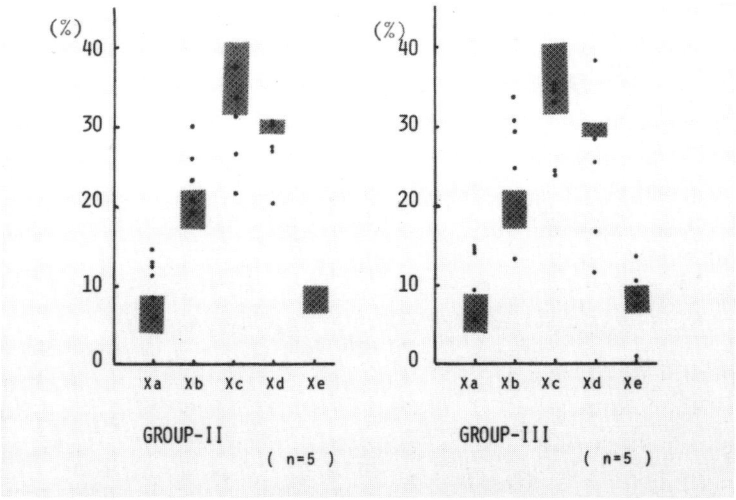

Fig 4 Percentage of LDH-X subfractions activity to total
LDH-X activity. (Group-II and Group-III)

452

Discussion

Those findings that lower sperm counts related to decrease of
total LDH-X and Xc subfraction in groups I, II, III, and
higher motility related to Xb $<$ Xd in normal and group II, are
new properties of subfractions of LDH-X.
To achieve final conclusion related to genetic disorder of
male infertile, the further investigations of wide variety of
properties in each subclass will be needed using the advanced
techniques, and with close observation of each male infertile
patient's clinical history.

Reference

*1.   Yoshida, M., Imai, T., et al: J. Clin. Chem. Clin.
                           Biochem.
                           19. 883, 1981

STUDIES OF A SHIET OF LDH-ISOZYME DURING HYPOXIA

Taizo Hayashi*, Takao Tanaka**

Department of Clinical Pathology*, Department of the 3rd
Internal Medicine**, Osaka Medical College, 569 Takatsuki,
Osaka, Japan.

Introduction

In general, in tissues exhibiting aerobic metabolism H-
subunits predominate, while in tissues exhibiting anaerobic
metabolism there are more M-units. However, an increase in
the proportion of M- to H-subunits of LDH-isozyme has been
reported in the heart muscles of patients with ischemic heart
disease[1]. It has been suggested that the regulation of
the LDH-isozyme pattern is in some way dependent on oxygen
tension[2]. We induced a relative hypoxic state by the
administration of thyroxine and an absolute hypoxemia by the
low oxygen loading in rats, and using electrophoresis studied
the alteration of LDH-isozymes in heart muscle, liver and
blood.

Material and Methods

Male Wistar strain rats and ICR mice were used. Tissue
homogenates of the heart and liver were prepared by the
method of Solaro et al[3]. Hyperthyroidism as a relative
hypoxic state were induced by the intraperitoneal injection
of 50ug thyroxine per 100g body weight daily for 14 days. An
absolute hypoxemia was induced by low, about 8% O_2, ambient
oxygen.

Electrophoresis '83

454

Results

1) Changes of LDH-isozyme in the heart of rats during
    hypoxia.
    A decrease of LDH-1 and an increase of LDH-2 in the
    heart muscles were observed during the relative hypoxic
    state of hyperthyroidism, comparing with the control. In
    addition to the same changes of LDH-1 and LDH-2, further
    marked increases of LDH-3 were recognized in myocardial
    LDH-isozyme in absolute hypoxemia induced by low oxygen
    loading.
2) Changes of LDH-isozyme in the liver of rats during
    hypoxia.
    Usually LDH-isozyme in the liver consists almost
    completely of LDH-5. There were no changes in hepatic
    LDH-isozyme pattern in the absolute hypoxic state.
3) Changes of LDH-isozyme in the heart of mice during
    hypoxia.
    The decreases of LDH-1 and LDH-2, and the increases of
    LDH-3 and LDH-4 were observed in the hearts of mice during
    hypoxia, compared with controls.
4) LDH-isozyme of plasma compared with that of the heart and
    the liver in the hypoxic state.
    Generally very high values of LDH-5 and low values of
    LDH-1, LDH-2, LDH-3 and LDH-4 were observed in the plasma
    of rats. During low oxygen loading, plasma LDH-isozyme
    did not show such marked changes. The patterns were
    approximately like those in the liver, unlike the heart.
    Therefore, LDH-isozyme is not a useful indicator of
    myocardial damage in the hypoxic state. After recovery
    from the hypoxia, however, LDH-1 and LDH-2 in plasma
    markedly increased after about 5 minutes, suggesting
    myocardial damage.

Summary

A decrease of H-monomer and an increase of hybrid isozymes of
LDH-isozyme of the heart muscles were found in hypoxic
states. A shift toward an anaerobic isoenzyme distribution
was observed in the myocardial LDH-isozyme during acute
hypoxia. The related increases of H-subunits were also seen
in plasma LDH-isozyme coincident with elevation of the oxygen
tension during recovery from hypoxic states.

References

1.   Hammond, G.L. et al.: Circulation, 53:637, 1976.
2.   Johansson, G.: Exp. Cell Research, 43:95, 1966.
3.   Solaro, R.J. et al.: Biochim. Biophys. Acta, 245:259,
     1971.

CHANGES OF PROTEASE INHIBITORS IN PATIENTS WITH DISSEMINATED
INTRAVASCULAR COAGULATION

Takashi Kageyama, Hiroshi Oyabu, Seiji Tsumoto
The 2nd Division, Dept. of Internal Medicine, Osaka Med.
College, Takatsuki, Osaka, Japan

Introduction

Malignant tumor is the commonest disease that underlies disseminated intravascular coagulation(DIC). Recently, the
interrelation between development and growth of mailgnant
tumors and protease inhibitors has been attracting increasing
attention [1) 2) 3)]. Seven protease inhibitors are currently
recognized to occur in the circulating plasma: i.e. α_1-
antitrypsin (α_1AT), α_2-macroglobulin(α_2M), inter-α-trypsin
inhibitor(IαI), α_1-antichymotrypsin(α_1X), antithrombin III(AT
III), C_1-inactivator(C_1INA) and α_2- plasmin inhibitor(α_2PI).
As yet, there is no report of study encompassing all these
seven protease inhibitors. This report presents the results
of simultaneous measurements of the protease inhibitors,
plasminoger(PLg) and fibrinogen(Fg) in the plasma of patients
with malignant tumors having developed a complication of DIC.

Subjects and Methods

Fifty-seven inpatients with DIC complicating various disorders
under treatment at this hospital were studied. The underlying
diseases included acute myelogenous leukemia(AML) in 5 cases,
chronic myelogenous leukemia(CML) in 3, acute promyelocytic
leukemia(APL) in 7, malignant lymphoma in 7, gastric cancer in
15, carcinoma of the pancreas in 9 and other diseases in 11.

458

Twenty-five healthy subjects, male and female, served as
normal controls. Fasting blood specimens were drawn early in
the morning and, after centrifugation, citrated plasma samples
were stored frozen at -80°C. They were assayed by single
radial immunodiffusion for protease inhibitors; α_1AT, α_2M, IαI,
α_1X, AT III, C$_1$INA and α_2PI, as well as for PLg. Plasma Fg
levels was measured with a TG micronephelometer.

Results

1) Changes in plasma protease inhibitor, plasminogen and
fibrinogen levels in patients with DIC.
All patients displayed higher values for α_1AT than normal
controls, with a wide range of distribution. The elevation
of plasma α_1AT was prominent particularly at terminal stage
of DIC associated with APL. Plasma α_2M values of the patients
with DIC were also widely scattered in their distribution, with
a tendency to depression especially in cases of carcinoma of
the pancrease, carcinoma of the large intestine and lung cancer.
Generally low values for IαI were observed while this plasma
protease inhibitor tended to be increased in DIC accompanying
APL. As was the case of α_1AT, all patients with DIC showed
high plasma levels of α_1X and the increase was prone to be
marked in APL. The present series exhibited generally low
plasma levels of AT III, the trend of depression being conspicuous in patients at the terminal stage of DIC and in those with
solid tumors. Increased levels of C$_1$INA in plasma were seen
as in the case of α_1AT and α_1X. The patients with DIC had
remarkably low concentrations of α_2PI in plasma, compared with
the normal controls. PLg levels were generally low with a wide
range of variation in the series of patients with DIC. The
decrease was of greater degree in cases at the terminal stage
of DIC and in those with liver dysfunction. Most cases of DIC
showed low plasma levels of Fg while there was a marked trend
to elevation of this moiety in cases of DIC with solid tumors.

Thus changes in plasma levels of protease inhibitor observed
in the present series of DIC were characterized by elevation
of α_1AT, α_1X and C_1INA and depression of IαI, AT III and α_2PI,
along with lowered PLg concentrations.

2) Comparison of plasma protease inhibitor levels between
DIC and early gastric cancer.

The patients with DIC showed markedly increased plasma
α_1AT and α_1X levels compared to those of patients with
early gastric carcinoma, whereas no significant difference
was noted for plasma α_2M concentration. The former patients
showed a tendency for the plasma IαI levels to be lower than
the cases of early gastric cancer. They also displayed lower
AT III levels, higher C_1INA levels and lower α_2PI values than
those in the latter group. Plasma PLg levels were lower in
DIC. Plasma Fg in DIC was also lower than in early gastric
cancer and prone to approach the normal range, whereas in
early gastric cancer, Fg was markedly increased over the
normal control level. In summary, the patients with DIC had
an elevation of such plasma protease inhibitors as α_1AT, α_1X
and C_1INA and depression of plasma IαI, AT III, α_2PI, PLg and
Fg levels as compared to the patients with early gastric
carcinoma.

3) Plasma protease inhibitor levels in DIC patients classified
by underlying diseases.

Plasma protease inhibitor levels of patients with DIC are
classified into two groups: patients with APL in whom, in
particular, conspicuous changes were observed, and the rest of
the patients with other underlying diseases. There was ele-
vation of α_1AT, α_2M, IαI and AT III, decrease of α_2PI and a
declining trend of Fg in the cases of DIC associated with APL.
Thus to data analysis disclosed noticealbe differences with
respect to changes of plasma protease inhibitor levels by types
of underlying diseases, especially in cases with solid tumors
and in those with APL. These findings suggest possible differences in the mechanism whereby the underlying diseases lead
to the development of DIC.

460

4)   Comparison of plasma protease inhibitor levels between
groups of patients with mild or severe symptoms of DIC.
Patients were divided according to degree of symptoms, i.e.
patients with mild symptoms of DIC and those with severe symptoms, and compared with respect to plasma levels of protease
inhibitors. In the group with severe clinical manifestations,
plasma levels of α_1AT, α_2M, IαI, AT III, α_2PI, PLg and Fg
were prone to be lowered and those of α_1X and C$_1$INA showed a
tendency to elevation.

Conclusion

The results of the present study indicate that changes in
plasma protease inhibitor levels are in close accord with the
clinical progress of DIC, and, therefore, their measurement
would provide information useful for prognosis and for estimation of the course of disease from the prethrombotic state
of DIC. The disparities observed with respect to changes in
plasma levels of protease inhibitors among the various underlying disorders in patients with DIC suggest the possibility
that different pathogenetic mechanisms for DIC may be
operating, particularly in those cases with solid malignant
tumors and in those with APL.

References

1.   Check, W.A.: Fibrinolysis suggested in tumor metastasis.
      Arch. Intern. Med. 140, 308(1980).
2.   Turner, G.A., Weiss, L.: Analysis of aprotinin-induced
      enhancement of metastasis of Lewis lung tumors in mice.
      Cancer Res. 41, 2576-2580(1980).
3.   Troll, W., Klassen, A., Janoff, A.: Tumorigenesis in
      mouse skin: Inhibition by synthetic inhibitors of
      proteases. Science 169, 1211-1213(1970).

ELECTROPHORETIC ANALYSIS OF ESTERASE ISOZYMES IN ORGANOPHOSPHATE
RESISTANT MOSQUITOES, CULEX PIPIENS

Yukiko Maruyama and Kiyoshi Kamimura

Department of Pathology, Faculty of Medicine
Toyama Medical and Pharmaceutical University, Toyama, Japan

Introduction

To obtain biochemical information on Culex pipiens and to search for biochemical differences between organophosphorus insecticide-resistant and
susceptable mosquitoes, electrophoretical studies of esterase isozymes were
carried out using thin layer agar gels. A relationship between high esterase
activity and insecticide-resistance has been established in certain insects
including mosquitoes. In pursuit of chemical control of insects in the
field, it is important to discover the specific genetic mechanisms of
insecticide-resistance. This report deals with esterase isozymes of Culex
pipiens complex (molestus, pallens and quinquefasciatus), particularly with
the changes in their isozyme pattern that occur at critical stages of development. The relationship between these esterases and insecticide-resistance,
such as to resistance to malathion and temephos is discussed.

Materials and Methods

Electrophoresis. Thin layer agar gel electrophoresis was carried out, using
an improvement on the method reported by Ogita(1,2) in order to separate
esterase isozymes in individual mosquitoes. The buffer system: Electrode
buffer: 0.025M potassium phosphate buffer(pH 6.8, ionic strength 50). Gel
buffer: 7.5mM potassium phosphate buffer(pH6.8, ionic strength 15). Gel
medium: 0.7g of polyvinylpyrolidone(PVP) and 0.7g of agar or agarose were
mixed with 100ml of gel buffer. Sample: A mosquito was minced with a
spatula in one drop of 10% Triton X-100 solution on a glass plate. Supernatant was obtained by hematocrit centrifugation (12,000 rpm, 5min at 4°C),
and absorbed on an oil-free cotton thread(5-7mm). The thread was embedded

462

in the gel layer on the plate. Conditions for electrophoresis: A constant
current of 1mA/cm was applied for 45min to 1 hour at 4°C. Staining for
esterase activity: 1) non-specific esterase activity: 1% β-naphthyl acetate
in acetone was sprayed on the gel surface as a substrate and incubated for
15min at 37°C. The staining dye was naphthanil diazo blue B salt. 2) specific esterase activity: The pH-indicator method(3,4) was used. Components of
the reaction mixture: substrate 0.2-1.0g, $NaHCO_3$ 21mg, 2ml of BTB 1% ethanol solution, and 100ml of H_2O. Esterase activities were revealed as yellow
bands on a greenish blue background as a consequence of the acid produced
by the esterase reaction.

<u>Moaquito strains</u>. Used in these experiments were laboratory strains of
<u>Culex pipiens</u>.1) organophosphate-susceptible strains: 4 strains of <u>molestus</u>,
1 strain of <u>pallens</u>, and 2 strains of <u>quinquefasciatus</u>. 2) organophosphate-
resistnat strains: T strain(<u>molestus</u>, temephos-resistnat), N strain(<u>pallens</u>.
temephos-resistnat , K strain(<u>quinquefasciatus</u>, malathion-resistnat), and B
strain(<u>quinquefasciatus</u>, malathion-resistant).

<u>Population mating</u>. Reciprocal population mating was done between K and T
strains. Individual pupae of each strain were isolated in a test tube.

Results

1) Changes in developmental stages. In eggs, only one weak isozymic band
   was seen. The mobility of this band was different for each strain and
   the band persisted through all developmental stages in each strain. In
   larvae, many additional bands were gradually appeared. They were located
   on both sides of the single band that was present in the eggs. The most
   anodal bands were observed only in larvae, and these bands were lost
   with pupation. Some larvae had such two anodal bands(slow and fast), but
   others had only one band, which could be either fast or slow. No larvae
   without these bands were observed. The mobility of these bands of <u>quin-
   quefasciatus</u> were different from those of <u>pallens</u> and <u>molestus</u>. In the
   F_1 hybrid population between K strain and T strain, those bands were
   segregated. Thus, these esterases were larva-specific esterase(L-Est)
   and were so named L-Est 1 , L-Est 2 and L-Est 3, due to the mobility in
   order from the anode. Based on differences of mobility of L-Est, which

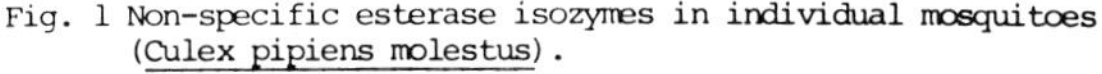

Fig. 1 Non-specific esterase isozymes in individual mosquitoes
(Culex pipiens molestus).

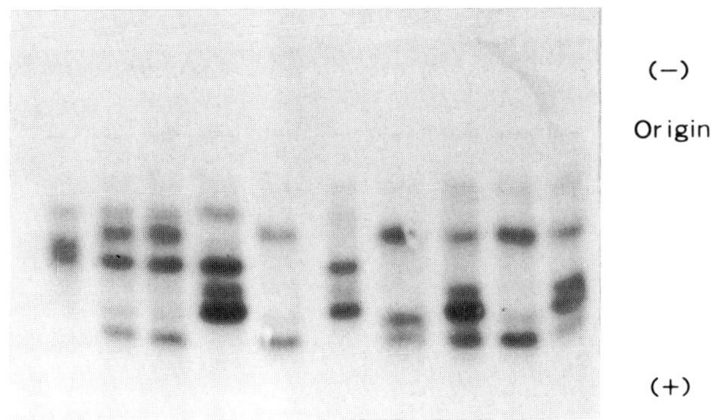

were observed within Culex pipiens complex, it is suggested that L-Est
are controlled by at least three co-dominant alleles. In adults, many
esterase active bands were seen, shown in Fig.1. Each isozyme was assigned to one of seven groups on the basis of mobility, and was named Est-1,
Est-2, Est-3, Est-4, Est-5, Est-6 and Est-7, in order from the anode.
The population was divided into four groups each exibiting a different
pattern of Est-2. In this group, two co-dominant alleles and a recessive
silent gene control Est-2 isozymes.

2) Esterases in resistant-strains. All of organophosphate-resistant strains
we examined had very strong active bands, shown in Fig.2. Even in adults,
the number of the bands was smaller than in susceptible strains, and
these bands had slow mobility. However, malathion-resistant strains had
a different pattern of strong active bands from that of temephos-
resistnat strains. T strain(molestus, temephos-resistant) and N strain
(pallens, temephos-resistnat) had one main strong band and several relatively weak bands. K strain and B strain(quinquefasciatus, malathion-
resistant) had two almost equally strong active bands that were located
on both sides of the main band of the temephos-resistant strains. In the
F_1 hybrid population between K strain and T strain, all offspring had
two strong bands corresponding with those of K strain. Some samples
showed three strong bands, two of which corresponded with those of K
strain and the third corresponded with that of T strain. None of them
exibiting the same pattern of strong active bands seen in T strain. Thus,

these strong active bands were named
Est-R1, Est-R2 and Est-R3, due to
the mobility from the anode, and may
correspond with Est-4, Est-5, and
Est-7 in adult, respectively. Thus,
the malathion-resistant strain had
Est-R1 and Est-R3 and the temephos-
resistnat strain had Est-R2 and
relatively weak Est-R1 and Est-R3.
Analysis of F_1 hybrid population
shows that Est-R2 is independent
of either Est-R1 or Est-E3.
However, it is unknown whether
Est-R1 and Est-R3 are linked. In
any case, Est-R are dominant traits.
However, one question still remains,
whether the difference in pattern
of Est-R depend on the kinds of
strain used or on the kinds of
insecticides used. In the case of
temephos-resistance, since both <u>pallens</u> and <u>molestus</u> had the same pattern,
it is more likely that the insecticides are the key factor. The differences
in isozyme pattern of Est-R between malathion-resistance and temephos-
resistance might be explained by the differences in their detoxification
mechanisms by hydrolytic emzymes.

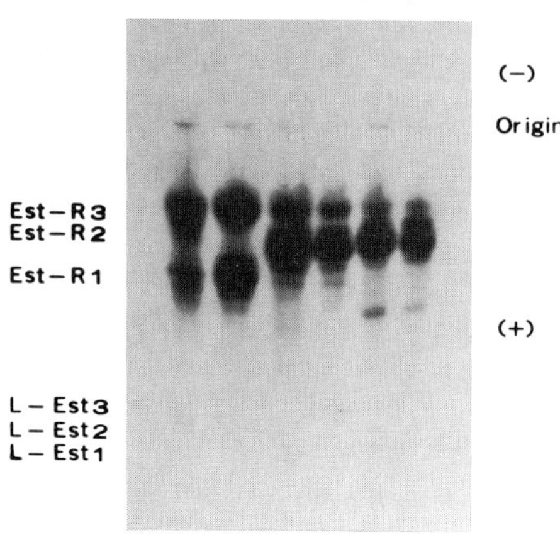

Fig. 2 **Non-specific esterase isozymes** in organophosphate-resistant mosquitoes (individual larvae).

References

1. Ogita, Z.-I., Kasai, T:: Med. J. Osaka University <u>15</u>, 141-153(1964).
2. Ogita, Z.-I., Kasai, T.: SABCO J. <u>1</u>, 1-4(1965).
3. Ogita, Z.-I., Kasai, T.: Japan J. Genet. <u>40</u>, 173-184(1965).
4. Ogita, Z.-I.: Ann. New York Acad. Sci. <u>151</u>, 243-345(1968).

ISOELECTRIC FOCUSING
AND ISOTACHOPHORESIS

TWENTY YEARS OF SCIENTIFIC WORK AND DEVELOPMENT OF ISOELECTRIC
FOCUSING

Olof Vesterberg

Chemistry Division, Research Department, National Board of
Occupational Safety and Health
S- 171 84 Solna, Sweden

Introduction

For all of us most days pass away and give only transient
memories. However, some experiences leave images that could
be of interest to other scientists. Due to space limitation
a complete review of my activities and other publications on
isoelectric focusing (IEF) can't be given here.

After studies in analytical, organic and bio-chemistry at the
University of Stockholm, I studied medicine at the Karolinska
Institute (KI). Early in 1963 I heard that Harry Svensson had
started to work at KI. I decided to contact him, which
resulted in the opportunity for me to join his research group.
He had just published the theoretical background of IEF (1-3).
We strived at improving the buffer substances (carrier
ampholytes, C.A.) and a lot of substances e.g. hydrolysates of
proteins, were tried. Although not perfect the results were
encouraging and demonstrated some unique capabilities of the
new method e.g. in experiments with hemoproteins, nice focused
coloured zones were obtained in density gradient columns.
Unfortunately, as early as September 1963, Harry Svensson
moved to Gothenburg where he was appointed Professor in
physical chemistry. I got his support to continue working on
IEF at the Nobel Medical Institute (at KI) in the laboratories
of Professor Hugo Theorell. In spite of scrutinising most

chemical catalogues and consulting many specialists in organic chemistry, the C.A. desired could not be obtained. This was a great challenge for me and I thought much about it. One day during a cross country mountain ski tour I got a good idea. In my studies I had learnt about the mutual influence of dissociation constants of groups in a single molecule and effects of their distances. It occurred to me that if starting with   aliphatic polyamine, e.g. with ethylene groups between the amino groups, and coupling to this a variable number of carboxylic groups very interesting interactions of the dis-sociation constants could be achieved. I eagerly consulted the scientific literature to learn more about suitable coupling reactions and dissociation constants of compounds related to my ideas. I got some very useful information but even after consulting Chemical Abstract and other sources a lot remained unresearched. To avoid many problems such as removal of salt ions I started to couple different unsaturated carboxylic acids to various amines. The products were sub-mitted to IEF in a multicompartment electrolyzer and the yields in different fractions with different isoelectric points (pI), their conductivity, buffering and other signi-ficant properties (4) were determined. The new approach gave in a few synthetic steps substances fulfilling the properties
   ideal C.A. (5). A lot of work was devoted to trying different synthetic procedures for optimising the combination of substances. Only a minor fraction of this has been published. This was partly due to the fact that I wanted in addition to get more theoretical and practical experience of IEF, e.g. in studies of different proteins. Comparison of the results obtained with independent methods seemed very important. This was also desirable because IEF often separated more forms of a protein than was known earlier. This also brought me to detailed studies of protein structures and reasons for heterogeneity (5-9). Much of that has not yet been published due to lack of time. I have never worked so

hard as during these and following years, partly because I was motivated by the ever increasing success and partly by the overwhelming feeling that IEF with such buffer substances could become very useful and important with numerous applications in laboratories worldwide. Very early I could repeatedly prove the outstanding properties of IEF; a high degree of reproducibility in allowing in a uniquely simple way determination of pI of many proteins in one experiment (which is a physicochemical characterization of high importance), and an ability to separate even very similar proteins (6). IEF was found to have important analytical and preparative capabilities and a high potential for development. My thesis "Isoelectric Focusing of Proteins" was presented in 1968 (5). Meanwhile I also had to devote time to medical studies and practice, examinations, teaching, my family and some other activities. In spite of everything else I was preoccupied with IEF. It was also of importance to improve the facilities for the practical applications of IEF, e.g. columns, gradient mixers (5). Initially IEF was not given general acceptance and recognition. Isoelectric points determined earlier with other methods did not agree very well with those assessed by IEF. This was due to some of the factors that pI is dependent upon e.g. ionic strength and temperature (5). Correction for such factors is important. The multiple molecular forms and heterogeneities of many "pure" protein preparations revealed by IEF caused some problems and conflicts. Thus a "pure" preparation of alcohol dehydrogenase which had been the subject of several studies and enzymological characterization for years by outstanding research groups (even conducted by Nobel Laureates), was found to be heterogeneous. This implied that much of the earlier enzymological work on such enzymes could be questioned. In this context I remember a professor in Biochemistry saying "It seems as if IEF brings confusion. Why should there be more than one form of a crystalized protein?" Now it is known that it is often so.

It can also be mentioned that a well known Nobel Laureate was first against IEF, but later gladly borrowed slides from me on IEF experiments that were shown in different countries. I also remember when professor Alexander Kolin visited me and said "Your column technique seems impractical because in science it is often important to perform many experiments in a day". He was partly right because at that time we used two or more days to achieve good focusing. A few years later we developed IEF in columns requiring only overnight focusing. Harry Rilbe and collaborators developed "rapid" IEF requiring only one hour of focusing (10).

From 1966 - 1970 I worked in the Department of Microbiology at the KI with several interesting proteins including enzymes and toxins produced by micro-organisms (11). This resulted in several publications and IEF constituted important parts of the theses of many of my friends in subsequent years. Another review was also published (12).

The presentations of myself and others on IEF at the Protides of Biological Fluids in Bruges in 1969 attracted the interest of many biochemists. It was fantastic. Thereafter I was invited to lecture by the scientific academies of several countries. On two such occasions I was invited to Moscow. I had several unusual impressions and experiences. It was astonishing to meet people knowing a lot about a swedish king who tried to conquer Russia and died 165 years ago. To experience outdoor swimming in Moscow at -20°C is unforgetable. Nor will I forget a visit to Bucharest in Rumania and the hospitality, the kindness and the tears at departure. The world-wide success and spreading of IEF must be attributed to many including people at LKB. This company has produced and distributed facilities of quality for IEF since 1967 (12). Later other companies also entered that scene. In 1972 I had a visit to my laboratory of a chemist. Some time thereafter the company he worked for started selling C.A. They used the same principle for amine synthesis I developed in 1965. "Rediscovery of Vesterberg's synthetic procedures" also re-occurred later.

Towards the end of the sixties IEF in polyacrylamide gels was investigated (review 14). Some of us chose to develop IEF in horizontal flat beds of polyacrylamide with cooling. More than fifteen thousand copies of apparatus based upon my work have now been distributed. I also devoted a lot of effort to improve focusing and the staining of proteins in gels often with the purpose of lowering the detection limit (15-17). Some advantages of the flat bed technique were stressed at international conferences (16-20). The information obtained with this technique is often found useful and important. This has been reflected in the rapidly increasing number of publications. We started IEF of cerebrospinal fluid proteins in 1972 (21). The high number of protein bands as obtainable with IEF of serum and urine samples, gives a lot of data. This fact and advanced densitometry problems brought me to pattern recognition and computerization for evaluation (20). I tried to contruct a laser scanner in 1973. This was later accomplished by others.
In 1970 I was appointed Associate Professor in a laboratory for occupational medicine. Extensive studies of urinary proteins revealed kidney damage as caused by occupational exposure to cadmium and some other diseases (22). Samples from Japanese patients suffering from the serious Itai-Itai disease were analyzed. We showed that the protein patterns from patients clearly differed from controls and were similar to those seen after extensive cadmium exposure. These studies led us to describe analysis and preparation of β_2-microglobulin (23). This protein is important in immunology and transplantation. Most preparative work on IEF was in density gradient columns before Radola described the use of Sephadex (24). This technique has been improved and used by many (23).

472

I also devised a new principle involving adsorption of proteins
to a solid matrix followed by desorption by IEF for analytical
or preparative purposes (25). This principle is nowadays used
with advantage after attachment of proteins to matrixes used
for hydrophobic interaction, ion exchange chromatography and
immunosorbents. An interesting utilization of C.A. is their
use in displacement chromatography (26) which is used increasingly.

A friend and myself, having experience with proteins and in
microbiology, took the initiative to do part of our military
service as research projects aimed at the detection of toxins.
These projects, although important, were to be carried out
without costs. This was achieved by performing mainly literature
studies which gave us a good knowledge of antibodies, purification procedures and assays. This was a good basis when,
in 1977, I started to work on a new principle for specific and
exact protein quantification, which I later named zone immunoelectrophoresis assay (ZIA). For some years obstacles were
encountered but in overcoming these the technique was improved (27). Even antibodies (28) and proteins separated by
IEF can be determined with ZIA (29). I have written twelve
articles which describe various aspects of ZIA. A review is in
press. For years IEF in agarose gels was tried with limited
success. However, after the development of special procedures
IEF can now be performed in agarose which offers important
advantages (29, 30), especially when using the new agarose
Type Z.

I have also been interested in the metabolism of organic
chemicals and the effects of exposure to chemicals (31). We
also use isotachophoresis for metabolite quantification (32).
I have also devoted a lot of time to development of methods
for trace metal analysis in biological samples with advanced
atomic absorption (33-35). Now we work to detect effects of

chemical exposure. We have developed a specific, effective
system for quantification of certain proteins (and also their
multiple molecular forms) following separation by electrophoretic procedures (29). The importance of and quantification
of multiple molecular forms of proteins has been stressed in
a condensed review on IEF (36). Such studies will give valuable
information in investigations on diseases, cancer and aging.
I have discussed such matters with among others Professor
Linus Pauling. Without doubt IEF and methods combined
therewith can supply important information.

Focusing, in a literal sense, means convegence to a spot, but
for me when I think of conferences it has had also the
opposite meaning in a geographical sense. I have, thanks to the
success with IEF, travelled the globe to take part in scientific
meetings in several countries e.g. the international conferences on IEF and electrophoresis that have taken place almost
every year since 1972. These activities have been very edu-
cational and stimulating. I have also been invited by the
Chinese Academy of Sciences. Furthermore, it has given me the
pleasure of a second visit to Japan. All the travelling has meant
contacts with many people, some of them outstanding personali-
ties. Thus I will never forget discussions with the Nobel
Laureate A.J.P. Martin and many, many others. After having
presented a report at such a conference I was approached by a
tall man who introduced himself: "I am from the FBI". FBI
(Federal Bureau of Investigation, US Police Department) is
for many people associated with the excitement of TV thrillers.
Because of the special circumstances I was quite perplexed.
He said "I want evidence.... and documentation on your method
ZIA". After some hesitation he got information. Whilst
travelling and flying I have experienced a widening of pers-
pectives and had good ideas. I also get excited when obtaining
letters or information from NASA and the European Space Agency.
I get filled with thankfullness for such occasions and

endeavour with a humble spirit to do my best. The main problems on this planet are wars, malnutrition, insufficient food supply and diseases all of which constitute challenges to workers in science. It would be good if IEF and other forms of electrophoresis could be useful in the context of such problems. This requires contributions from many people. IEF can be useful to studies or clinical diagnosis of genetic abnormalities, cancer, arteriosclerosis and diabetes (36). There are some examples where evidence supplied by IEF has revealed unsuitable impure insulin preparations causing their withdrawal from the market or substantial modification. Since several millions of people have diabetes this is important. The total number of published articles where IEF has been used is estimated to be about five thousand to date. The literature computer search supplies me continuously with a flood of references of newly published articles dealing with IEF. There are many branches of the biological sciences and medicine where IEF has been found useful. Even five years ago the method had found so many important applications in widely different areas that it was not easy to review all of them. Some reviews mention more than twelve important fields with numerous references to each subject (36, 37). Some applications seemingly odd may be mentioned here. It is known that IEF is used for optimization of food and beverage quality e.g. in the production of cheese, spaghetti and beer. IEF has also been shown to the general public on TV in some industrialized countries. IEF is used in many state and control laboratories to check meat variety. Thus scandals have been revealed, e.g. expensive meat was in fact from kangaroos. Admixture of plant and fish protein (e.g. in sausage) can easily be revealed by IEF giving characteristic patterns for each variety (38). Even fish species can be differentiated (36, 38). The outstanding ability of IEF to differentiate even genetic variants is routinely utilized in genetics, paternity determinations, and forensic medicine (38, 39). ZIA could prove valuable to many laboratories (particularly those unable to

use RIA and other sophisticated procedures as a means of
quantifying proteins e.g. for diagnosis or surveillance (40).

Increasingly I have become concerned about serious problems
on earth   including those in developing countries. It is
important that we work to transfer knowledge to such countries,
especially when some important new branches of science arise
e.g. protein production, genetechnology and hybridoma techniques.
In this respect IEF, ZIA and other forms of electrophoresis
could be of value for diagnostic purposes or vaccine production. /The
impact upon food supply of developing new genetic varieties
of cultivated plants or animals would be enormous. To help
developing countries I arranged a specialists international
four days conference in 1981 at the KI under the auspices of the
United Nations. The meeting was also taking place because some
of us had the idea that Sweden could spend a fraction of the
financial aid for developing countries on establishing a center
for the training of scientists from such countries to give
them opportunities to learn new and useful methods. How
disappointed I was when we learnt that although the Swedish
Government yearly grants much more money than its agency can
distribute, no money could be allocated for such a center in
Sweden.Ironically, I also had to cover the costs for the
secretararial assistance for the meeting report myself.

In my personality is a striving to make meaningful use of my
time. Improving procedures and facilities gives me satis-
faction. This may be one reason behind some of my inventions,
e.g. a device for supplying water and nutrients to plants.
I also invented a typewriter more optimal than present models
with respect to ergonomy and speed. Curiously some years later
a similar typewriter was "reinvented" in the Soviet Union.
For twenty years I have been fascinated by the phenomenon of
gradients, e.g. the continous increase in concentration in one
dimension, as pH gradients in IEF, density and concentration
gradients. Having worked in microbiology I saw that the proce-

-dures for testing the effects of antibiotics on micro-organisms were not optimal. I developed a new system depending on linear diffusion. I have also developed a simple but effective system (selective binding assay, SEBA) for concentration determination of substances. Most of my inventions are documented by publications or patent applications.

An overview of IEF is nowadays incomplete if not dealing with 2-D electrophoresis, i.e. IEF in a first dimension followed by electrophoresis in a second dimension. The proteins are effectively separated first according to charge differences and then in respect to molecular size. It is the most optimal and high resolving technique available (41). Results of 2-D electrophoresis will have important impacts and open up new epoques in biological sciences. I attended a special international conference on 2-D electrophoresis in 1981 at the famous Mayo Clinic in the USA. This period and the subsequent visit to Argonne National Laboratories outside Chicago was exciting.There 2-D electrophoresis is used to make a catalogue of all human proteins, their changes in diseases and cancer (41). The human protein index is one of the biggest projects in biological sciences.

In our laboratory we use a method of 2-D electrophoreis (42) for detecting adverse effects following exposure of animals and humans to noxious substances. The potential of 2-D electrophoresis largely benefits from the development of silver staining. About 0.1 ng of protein in a spot in a gel can now be detected (42). It is indeed fascinating and exciting to follow the development of biological sciences. Methods for measurements and other achievements become increasingly important for mankind and I should like to end as follows:

"Measured values form the basis of all knowledge" (Lord Kelvin".
Reliable measured values form the basis of all meaningful
knowledge (O.V.).

Dedication

No single person can contribute more than a fraction to the
development of science but some people have had a decisive
influence in this context. I would like to mention my friend
Harry Rilbe and dedicate this article in honour of his
approaching seventieth birthday. I should also like to mention
all who have contributed to the development and progress of
isoelectric focusing and related techniques and/or helped me.

References

1. Svensson, H.: Acta Chem. Scand. 15, 325-341 (1961).
2. Svensson, H.: Acta Chem. Scand. 16, 456-466 (1962).
3. Svensson, H.: Arch. Biochem. Biophys. Suppl. 1, 132-140 (1962)
4. Vesterberg, O.: Acta Chem. Scand. 23, 2653-2666 (1969).
5. Vesterberg, O.: Svensk Kem. Tidskr. 80, 213-225 (1968).
6. Vesterberg, O., Svensson, H.: Acta Chem. Scand. 20, 820-834
   (1966).
7. Vesterberg, O.: Acta Chem. Scand. 21, 206-216 (1967).
8. Flatmark, T., Vesterberg, O.: Acta Chem. Scand. 20, 1497-
   1503 (1966).
9. Vesterberg, O.: VIIth Symposium Chromatography & Electrophoresis (Ed. M. Renoz). Publ. Press. Acad. Europeennes,
   Brussels (1973) pp 81-98.
10. Jonsson, M., Pettersson, S., Rilbe, H.: anal. Biochem. 51,
    557-576 (1973).
11. Vesterberg, O.: In:Methods in Microbiology (Ed. Norris, J.R.
    and Ribbons, D.W.) Vol. 53, Academic Press, New York (1970)
    pp 595-614.
12. Vesterberg, O.: In: Methods in Enzymology (Ed. Jakoby, W.B.)
    Vol. 22, Academic Press, New York (1971) pp 389-412.

13. Rilbe, H.: Science Tools 23, 18-22 (1976).

14. Wellner, D.: Anal. Chem. 43, 59-65 (1971).

15. Vesterberg, O.: Biochim. Biophys. Acta 243, 345-348 (1971).

16. Vesterberg, O., Hansén, L.: In: Electrofocusing and Iso-
    tachophoresis (Ed. Radola, B.J. and Graesslin, D.)
    de Gruyter, Berlin (1977) pp 123-133.

17. Vesterberg, O.: In: Isoelectric Focusing (Ed. Arbuthnott,
    J.P. and Beeley, J.A.) Butterworths, London (1975) pp 78-98.

18. Vesterberg, O.: Ann, N.Y. Acad. Sci. 209, 23-33 (1973).

19. Vesterberg, O.: In: Electrophoresis and Isoelectric
    Focusing in Polyacrylamide Gel (Ed. Allen, R.C. and Maurer,
    H.R.) W. de Gruyter, Berlin (1974) pp 145-158.

20. Vesterberg, O.: In: Progress in Isoelectric Focusing and
    Isotachophoresis (Ed. Righetti, P.G.) North Holland, Amster-
    dam (1975) pp 121-135.

21. Kjellin, K.G., Vesterberg, O.: J. Neurol. Sci. 23, 199-213
    (1974).

22. Hansén, L., Kjellström, T., Vesterberg, O.: Int. Arch.
    Occup. Environ. Hlth. 40, 273-282 (1977).

23. Vesterberg, O., Hansén, L.: Biochem. Biophys. Res. Commun.
    80, 519-524 (1978).

24. Radola, B.J.: Biochim. Biophys. Acta 194, 335-338 (1969).

25. Vesterberg, O., Hansén, L.: Biochim. Biophys. Acta 534,
    369-373 (1978).

26. Leaback, D.H.: In: Progress in Isoelectric Focusing and
    Isotachophoresis (Ed. Righetti, P.G.) North Holland,
    Amsterdam (1975) pp 137-145.

27. Vesterberg, O.: Hoppe-Seyler´s Z. Physiol. Chem. 361, 617-
    624 (1980).

28. Vesterberg, O.: J. Immunol. Methods 37, 311-314 (1980).

29. Vesterberg, O., Breig, U.: In: Electrophoreis ´81 (Ed.
    Allen, R.) W. de Gruyter, Berlin (1981) pp 103-116.

30. Rosén, A., Ek, K., Åman, P., Vesterberg, O.: In: Protides
    of Biological Fluids (Ed. H. Peeters) Pergamon Press,
    Oxford (1979) pp 707-710.

31. Vesterberg, O.: Int. Arch. Occup. Environ. Hlth. 35, 89-115
    (1975).

32. Vesterberg, O., Sollenberg, J.: In: Proc. Internat. Conf.
    in vivo Aspects of Biotransformation and Toxicity of
    Industrial and Environmental Xenobiotics, Prague.
    (Ed. I. Gut) Excerpta Medica Int. Congr. Ser (1977)
    Ser. No. 440, pp 186-189.

33. Nise, G., Vesterberg, O.: Clin. Chim. Acta 84, 129-136
    (1978).

34. Vesterberg, O., Wrangskogh, K: Clin. Chem. 24, 681-685
    (1978).

35. Nise, G., Vesterberg, O.: Scand. J. Work. Environ. &
    Hlth. 5, 404-410 (1979).

36. Vesterberg, O.: International Laboratory May/June 1978,
    pp 61-68 and Am. Laboratory June 1978, pp 13-24 and also
    in Protein Nucleic Acid and Enzyme (Ed. N. Ui) Kyoritsu
    Shuppan Co. Tokyo 1978.

37. Righetti, P.G.: Isoelectric Focusing Theory, Methodology
    and Applications. Elsevier, Amsterdam, 1983 pp 314-385.

38. Bishop, R.: Science Tools 26, 2-8 (1979).

39. Kühnl, P.: In: Electrophoresis '81 (Ed. R.C. Allen and
    P. Arnaud) W. de Gruyter, Berlin 1981, pp 563-571.

40. Vesterberg, O.: Clin. Chim. Acta 113, 305-310 (1981).

41. Anderson, N.G., Anderson, N.L.: Medical Laboratory 11,
    75-94 (1982).

42. Marshall, T., Latner, A.L.: Electrophoresis 2, 228-235
    (1981).

ISOELECTRIC FOCUSING (IEF) ON SUPPORTED CELLULOSIC MEMBRANES

Borek Janik

Gelman Sciences Inc., Ann Arbor, MI 48106, U.S.A.

Jeffrey Ambler

Clinical Chemistry Dept., University Hospital, Nottingham NG72UH, U.K.

Introduction

Cellulose acetate membranes (CAM) are widely used for electrophoresis, but in their non-modified form they are not suitable for IEF. Negatively charged groups present in the CAM after manufacture induce electroendosmosis which results in the cathodal drift of gradient and proteins before focusing is complete. Using non-supported CAM, the problem was solved by methylation of these groups using BF_3 in methanol (1) which produced good results, both in an electrophoresis arrangement using wicks from the electrode solutions (1), and a conventional IEF arrangement with surface electrodes (2). Such treatment, however, impaired the mechanical properties of unsupported CAM so that they had to be stored in methanol until use. Although perfectly satisfactory for self-production by laboratories, these are obvious difficulties for a manufacturer. A newly prepared Mylar-supported CAM can be treated with BF_3 without adverse side effects. We have evaluated the membranes which are being developed at Gelman Sciences for IEF.

Electrophoresis '83

482

Procedures and Results

1. Equilibration with Ampholytes

The membrane (7.5 x 12 cm) was wetted by slowly immersing it in 10
mL of ampholyte mixture solution prepared in 7.5% or 10% glycerol,
and allowing it to equilibrate for at least one hour. About 1 mL
was taken up by the membrane. The 7.5% content of glycerol was
found to be the minimum concentration able to reduce the rate of
evaporation during focusing and improve the mechanical properties
of the membrane to a satisfactory degree. Increase to 10% glycerol
did not have any further improvement effects.

2. Focusing Method

A modified Semi-Micro Chamber (Gelman) with a cooling plate had
adjustable surface electrodes which could be applied firmly (at
about 600 g overall pressure) to the membrane. For wide
gradients, for example, pH 3-10, the Power Supply only needs to be
500 V. The wicks were soaked with electrode solutions and blotted
well. The anolyte was either Anode Fluid 3 (Serva) or 0.75% w/v
glutamic acid in 7.5% glycerol; the catholyte was either Cathode
Fluid 10 (Serva) or 0.75% w/v lysine in 7.5% glycerol. The wet
membrane was placed on the cooling plate, carefully excluding air
bubbles from the underside, and blotted evenly. Focusing was at
500 V (2-3 mA starting current for 1-2 hours). The Sepratek
applicator (Gelman) was used to deliver 4 or 8 samples onto the
membrane simultaneously up to 15 minutes after the voltage had
been applied.

3. Staining

Staining of proteins was achieved by directly immersing the membrane to Coomasie Blue R 0.25 w/v in methanol:water:acidic acid
(5:4:1 v/v) for 15 min. Excess stain was removed by the same solvent. Alternatively, staining for isoenzymes was carried out.

4. Stability of the Gradient

All media, whether cellulose acetate, agarose, or polyacrylamide gel suffer from gradient drift to the cathode. Many studies have been carried out to find a reason why this occurs. Although charged groups in the membrane will cause electroendosmosis manifesting as cathode drift, this is not the sole cause of the problem. The stability of the gradient may be improved by having the electrode solution pH as close as possible to the pH of the end of the gradient. For this reason, glutamic acid and lysine or Serva electrode solutions were chosen for this work.

Distortion of protein bands is another problem often met during focusing. Suggestions to alleviate this problem include adjusting gradient strength, decreasing initial voltage and removing salts from the sample. A uniform strength of the gradient is one of the most important parameters to be considered. Commercial ampholytes have different sources and methods of synthesis and many valuable comparisons have been carried out. However, we found that for the purpose of the work described here, Servalyte (Serva) produced the best results not only in terms of focusing but also in the ease of removal at the protein staining stage. The consequence of this is a rapid wash out of excess stain with minimal background interference.

Gradients may be strengthened in parts by the use of 'narrow' gradient ampholytes. A much more exciting development is the use of polybuffers for flat bed IEF. It is claimed that these will accomplish the task with a lessening of cost.

As part of this work, Poly/Sep 47 (Polysciences) was tried as the sole source and in combination with ampholytes. It became apparent that polybuffers did not have the strength despite high initial current (8 mA at 500 V). By lessening the concentration of Poly/Sep, we found that a combination of 3% Servalyte pH 3-10 and

10% w/v of Poly/Sep produced the best results. An example of Poly/Sep at (90% w/v is shown in Figure 1a. This demonstrates electroendosmosis, distortion and extreme compression of the cathode area of the strip. By decreasing the Poly/Sep to 25% v/v and adding 3% Servalyte, much better focusing is obtained (Figure 1b). However, this was still not satisfactory because the gradient was compressed at the cathode end and much of the myoglobin and hemoglobin had been converted to the met derivative (this did not happen in absence of Poly/Sep).

Poly/Sep was decreased to 10% v/v and ampholytes were maintained at 3% (Figure 1c). Very clear focusing was obtained with minimal distortion. The gradient was compressed at the cathode, but Figure 1c already shows the resolving power of his technique. The electrophoregrams were obtained using Serva Electrode Fluids #3 and 10 (Figure 1a–c) or glutamic acid/lysine solutions (Figure 1d; see para. 2). The effects of these two kinds of electrode solutions on the gradient linearity and its relative positioning are also illustrated in Figure 2.

Poly/Sep alone (i.e., without any additional Servalyte) did not perform satisfactorily at any concentration from 90% to 10% (compression of the gradient at the cathode end). Interestingly, Servalytes alone suffered from similar problems.

The method is in an early stage of development, but preliminary work suggests that it can also be used for isoenzymes. Beta-lactamase, for example, has been detected on the new membrane.

Work with narrow gradients has not yet been carried out, as the early experiments have been concerned with choice of electrode solutions, choice of additives to gradients, etc.; however there is no reason to suppose that, like unsupported membrane, a pH gradient of 3 units cannot be applied to this system.

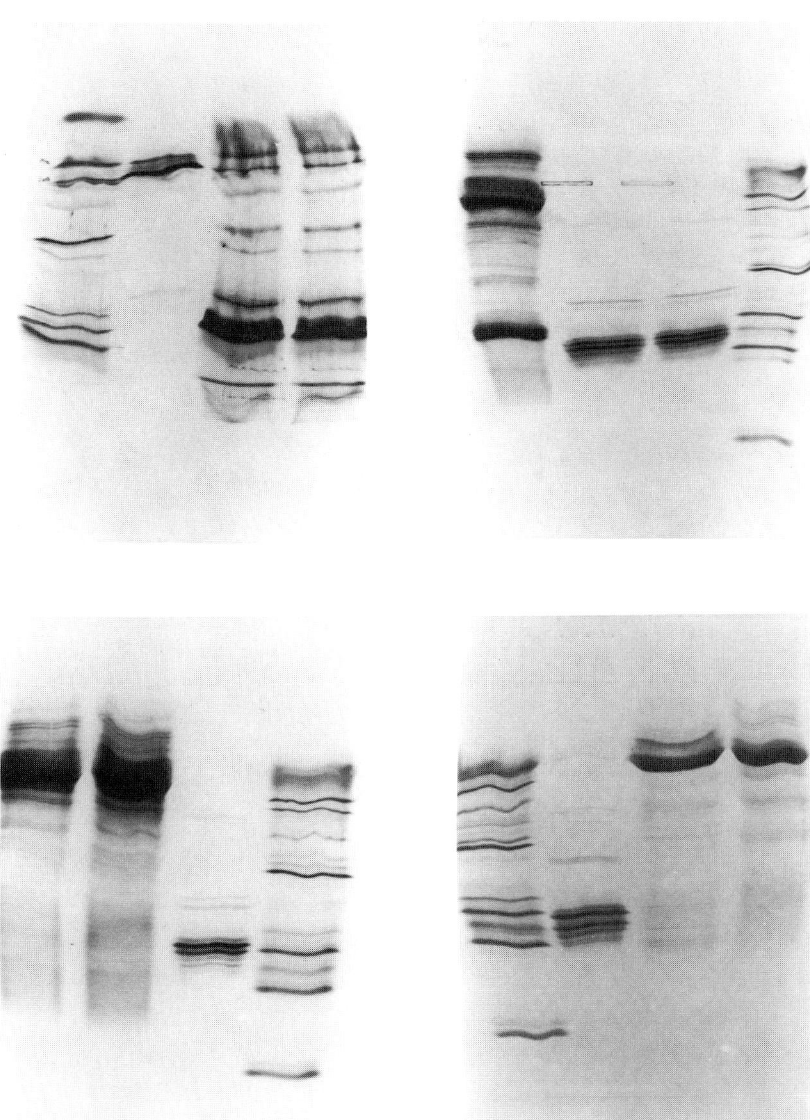

Figure 1. Effect of Poly/Sep-Servolyte composition on the gradient formation: (a) 90% Poly/Sep; (b) 25% Poly/Sep, 3% Servolyte; (c and d) 10% Poly/Sep and 3% Servolyte. All mixtures were in 7.5% or 10% glycerol. Electrode solutions: (a-c) Serva Cathode Fluid 3 and 10; (d) glutamic acid/lysine. Samples (right to left): protein marker mixture (Track No. 1 in 1a-c and No. 4 in d) contained 10 mg/mL each of bovine serum albumin, beta-lactoglobin A & B, bovine carbonic anhydrase, horse myoglobin, whale sperm myoglobin cytochrome C and 1 mg/mL bromophenol blue; red blood cell hemolysate (Tract No. 2 in a-c and No. 3 in b and d); serum samples in the remaining tracks.

486

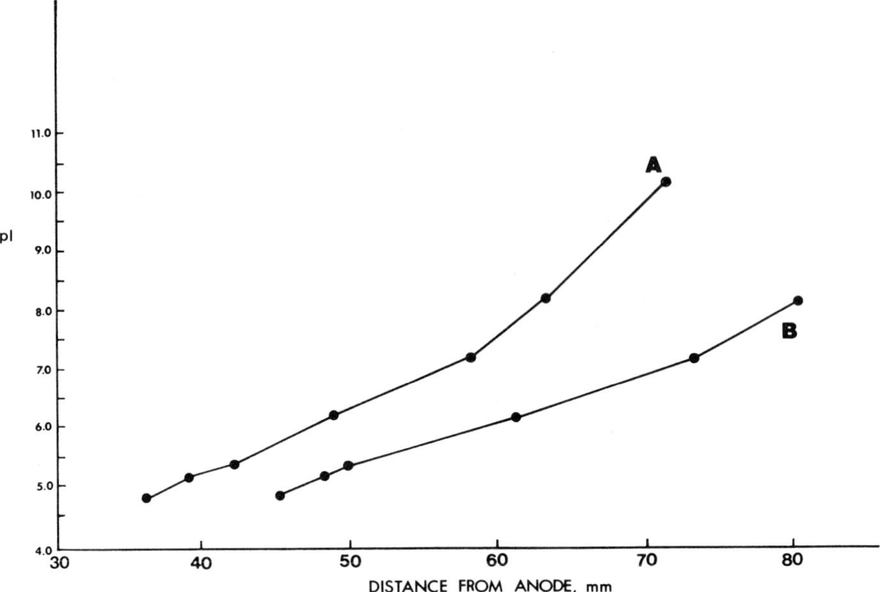

Figure 2. Effect of electrode solutions on gradient linearity. (A) Serva electrode solutions; (B) glutamic acid/lysine. The gradient was formed in a mixutre (v/v) of 3% Servolyte, 10% Poly/Sep and 10% glycerol in both D and B.

Conclusions

Advantages of the new membrane are:

o Technically simple steps involve direct equilibration with ampholytes, conventional IEF equipment and a short direct protein staining procedure.

o No neurotoxicity or other biological hazards in use.

o Pore sizes do not exclude large molecules such as IgM.

o Low voltage only required for wide gradients.

o Small requirement for ampholytes--a major cost item of most focusing systems.

o No interference with proteins by polymerizing agents.

References

1. Ambler, J., Walker, G.: Clin. Chem., 25, 1320-1322 (1979).

2. Janik, B., Dave, R.G.: Electrophoresis '81, Allen and Arnaud, eds., Walter de Gruyter, Berlin, 1981, pp. 77-82.

A TEST FOR BINDING DURING ISOELECTRIC FOCUSING:
BUFFERS VERSUS SYNTHETIC CARRIER AMPHOLYTES

Charles B. Cuono, M.D., Ph.D.
Giselle A. Chapo

Yale University School of Medicine
Section of Plastic and Reconstructive Surgery
Department of Surgery
New Haven, CT   USA

Andreas Chrambach, Ph.D.
Larry Hjelmeland, Ph.D.

National Institutes of Health
Bethesda, MD

Introduction

The binding of proteins and other macromolecules to synthetic
carrier ampholytes during the conduct of electrofocusing has been
widely reported. Such binding appears to be more frequently
encountered in polyanionic macromolecules and it has been
observed[1] that low pI carrier ampholytes are more strongly
complexing than high pI carrier ampholytes. Such observations have
been used to explain the more frequently observed artifactual
binding in the acid pH range.

On purely thermodynamic grounds, one would expect that the
components of synthetic carrier ampholyte mixtures (SCAMs) with
average molecular weights of 700 or more to interact with
polyelectrolytes more strongly than simple buffers with an average
molecular weight of 150. The advent of electrofocusing in simple
buffers[2,3,4] raised the possibility of comparatively
experimentally probing interactions between macromolecules and the
gel matrix during electrofocusing. Such a study was facilitated by
the introduction of a synthetic macromolecular amphoteric dye with
molecular weight of 160,000 (Poly R-480).

Electrophoresis '83

488

Materials and Methods

<u>Polymeric dye</u>: Poly R-480 was obtained from Aldrich Chemical Co. Aqueous solutions of a concentration of 0.5-1.0 mg/ml were prepared freshly prior to electrofocusing.

<u>Synthetic carrier ampholyte mixtures</u>: Four commercial preparations were used: Ampholine (LKB, pH 3.5-10), Bio-Lyte (Bio-Rad, pH 3-10), Pharmalyte (Pharmacia, pH 3-10) and Servalyte (pH 2-11). Solutions as obtained from the manufacturer were stored at 4° C, and used prior to expiration dates specified by the manufacturers. Final carrier ampholyte concentration in these studies was 2% w/v.

<u>Buffer carrier constituents</u>: A mixture consisting of 41 simple amphoteric buffers and six non-amphoteric buffers was used. The composition of this mixture is described elsewhere.[4]

<u>Isoelectric focusing</u>: A horizontal, flat-bed apparatus (Pharmacia, FBE 3000) was used in conjunction with polyacrylamide gels cast on Serva Gel-Fix membranes as previously described.[4,5] Temperature of EF was 10° C. Anolyte and catholyte were 1 M oxalic acid and 1 M KOH, respectively. Gels were pre-focused for 30 min. at 10 W. Samples of 50 µl Poly R-480 solution were applied on pledgets of Whatman GF/B glass fiber paper. EF was then conducted at 15 W until the field strength reached 1000 V, after which time constant voltage conditions were maintained. Preliminary studies demonstrated that the steady state condition was achieved by 6 h of electrofocusing.[6]

The zwitterionic detergent CHAPS was obtained from Pierce Chemical Co. and used at a final concentration of 0.008M. Ultrapure urea was used as freshly prepared solutions for periods not in excess of 8 h.

Results

Electrofocusing of Poly R-480 in the buffer mixture results in resolution of the dye into three discrete bands in the acidic region of the gel (Fig. 1). By comparison, electrofocusing of Poly R-480 in all of the synthetic carrier ampholyte mixtures tested produces only a broad smear of the dye. In Ampholine, Bio-Lyte and Servalyte, the zonal dye distribution was located in the region of pH 6.3-7.2. In the case of Pharmalyte, the dye was distributed in a narrower zonal pattern in the region of pH 7.4-7.6 (Fig. 1).

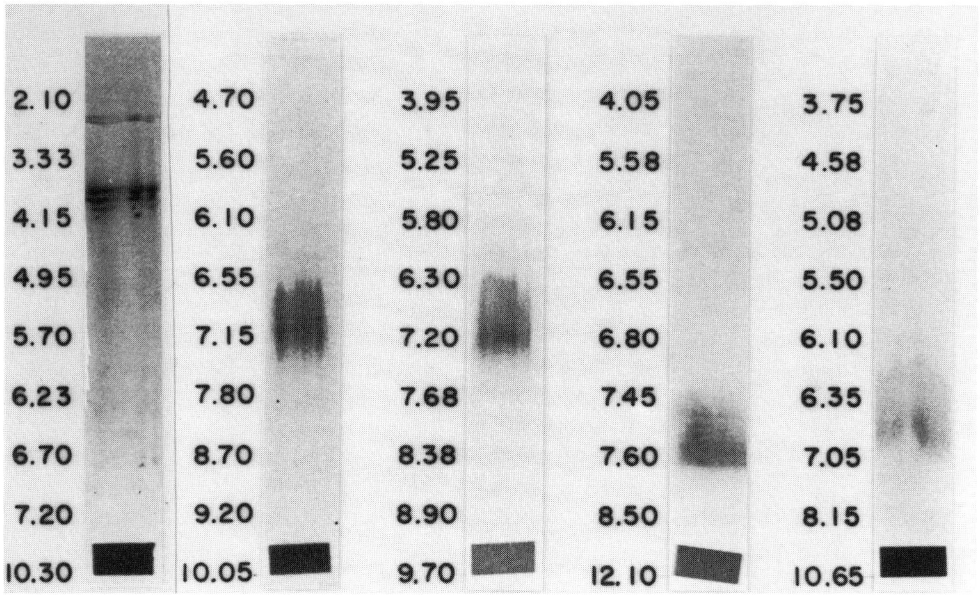

Figure 1: EF pattern of Poly R-480. Left to Right: buffer carrier constituents, Ampholine, Bio-Lyte, Pharmalyte, Servalyte.

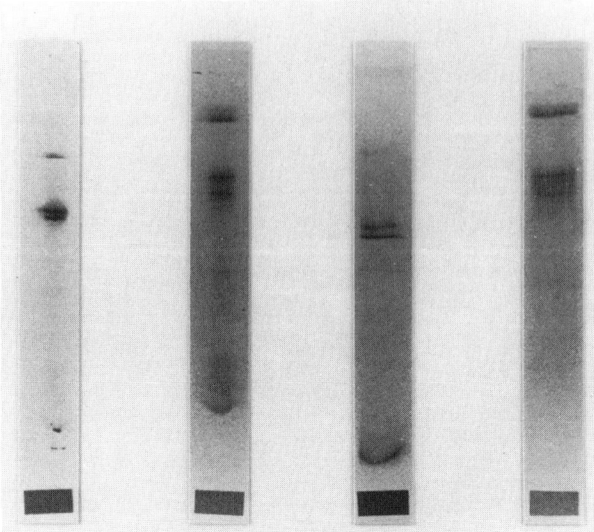

Figure 2: EF pattern of Poly R-480 in buffer gels as influenced by additives. Left to Right: neat gel (no additives), 8M urea, 0.008M CHAPS, 8M urea plus 0.008M CHAPS.

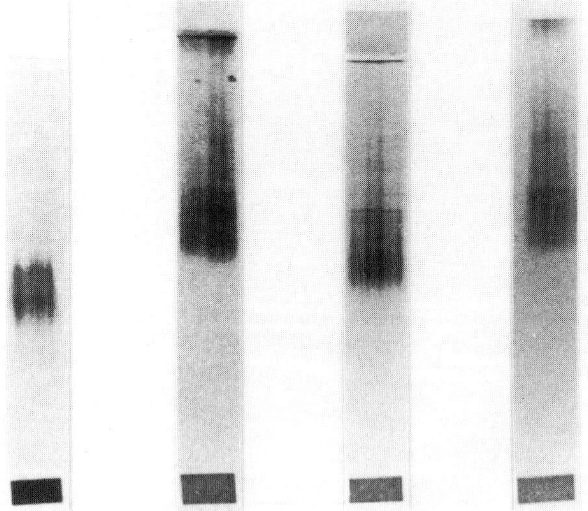

Figure 3: EF pattern of Poly R-480 in Ampholine as influenced by additives. Left to Right: neat gel (no additives), 8M urea, 0.008M CHAPS, 8M urea plus 0.008M CHAPS.

The electrofocusing pattern of Poly R-480 in the buffer mixture was unperturbed by the addition of 8M urea and 0.008M CHAPS, either singly or together (Fig. 2). The pattern of three major bands prevailed in the presence of these hydrogen bond and hydrophobic bond disrupting agents. In contrast, the electrofocusing pattern of Poly R-480 in Ampholine in the presence of these dissociating agents was rather strikingly different from the smear pattern obtained in neat gels without these additives (Fig. 3). Addition of 8M urea or 0.008M CHAPS to the Ampholine gels results in production of a rudimentary banding pattern superimposed upon a polydisperse zonal distribution. Furthermore, the pattern is shifted to the acid region of the gel.

The polydisperse pattern of Poly R-480 in Ampholine gels containing 8M urea and CHAPS was sectioned into quarters and each section was re-focused. All sections generated the _entire_ original pattern. By definition, therefore, the pattern of Poly R-480 in urea and CHAPS is artifactual. Re-focusing of each of two major bands of the dye from buffer gels similarly produced all three original bands, _i.e.,_ the dye focusing pattern in buffers, is also artifactual. Data for these re-focusing experiments is not shown, since re-focused zones were too weakly colored to clearly appear on photographs and solubility limitations of the dye precluded performance of experiments with overloaded gels.

Discussion

Poly R-480 is a polymeric anthraquinone dye with abundant hydrophobic and hydrophilic sites. The repeating backbone unit contains approximately 20 residue percent secondary amino groups and 40 residue percent sulfonate groups, and therefore has an acidic pI. It would be expected that Poly R-480 would be heterogeneous on electrofocusing since it is indeed a copolymer and not a single organic compound.

Polyacrylamide EF of Poly R-480 in simple buffers produces a simple banding pattern in the acidic region of the gradient, and this pattern is unperturbed by the addition of the dissociating agents 8 M urea and 0.008 M CHAPS. By comparison, EF of Poly R-480 in any of the commercially available SCAMs produces a zonal distribution of the dye at pH 6.3-7.2 (Ampholine, Bio-Lyte, Servalyte) or pH 7.4-7.6 (Pharmalyte). The zonal distribution of Poly R-480 in SCAM gels is perturbed by the addition of 8M urea, 0.008M CHAPS or both dissociating agents together. Following addition of these agents, the dye becomes distributed in the acidic region of the gradient, and a rudimentary banding pattern appears.

Re-focusing of the bands of Poly R-480 produced in buffer gels as well as sections of the zones of Poly R-480 in SCAM gels results in re-establishment of the original patterns in all cases. Thus, by definition of an electrophoretic artifact, both the simple and the complex patterns are artifactual. The key difference between electrofocusing in simple buffer carrier constituents and in SCAMs of any four commercial types is, however, that in the latter case the distribution pattern of Poly R-480 is markedly perturbed by the addition of the dissociating agents urea and CHAPS, while in the former case no significant perturbation of the three-band pattern is evident.

Artifactual distribution of polyelectrolytes in EF in SCAMs has been well documented.[1] In general, artifacts are generated in the acid pH ranges and have been attributed to complexing of low pI carrier ampholytes with polyanions. In view of the preponderance of sulfonate groups in Poly R-480, this polymer would be expected to exist as a polyanion over a rather broad pH range. Based upon observations of others[8], it would be anticipated that Poly R-480 would be distributed artifactually in the acidic region pH 3-5 in SCAM gels. However, contrary to expectation, in SCAM gels containing no additives, Poly R-480 is distributed in the pH 6-7 range in all examples with the exception of Pharmalyte in which the zone of distribution is clearly in the alkaline pH 7.4-7.6 range.

Addition of the hydrogen bond disrupting agent 8M urea and the hydrophobic bond breaking agent 0.008M CHAPS, either singly or together, shifts the dye distribution in SCAM gels to the expected pH 3-6 range. Re-focusing experiments of these new polydisperse zones in the acidic region clearly indicate that they are artifactual.

The presence of artifactual zonal distribution of Poly R-480 in neat SCAM gels could be explained by interaction of the dye with ampholytes in the pH 6-7 range. The perturbation of the pattern in the presence of the dissociating agents for hydrogen bonds and hydrophobic bonds suggests that such bonds are operative in giving rise to the original zonal dye distribution pattern. The development of a <u>new</u> artifactual dye distribution in the more acidic region of the gradient in the <u>presence</u> of these dissociating agents renders unlikely the possibility that hydrogen bonds and hydrophobic bonds are operative under these conditions. It is inferred that electrostatic interactions between SCAMs and the dye must be operative in producing the artifactual polydisperse patterns in SCAMs containing urea and CHAPS.
These experiments demonstrate no binding of Poly R-480 in the neutral region in buffer EF gels. Furthermore, the three-band pattern of the dye in neat gels is maintained in the presence of 8 M urea and 0.008 M CHAPS, indicating that hydrogen bonding and hydrophobic bonding are not playing a significant role in establishing the pattern in neat gels. The imperturbability of the pattern in the presence of both of these agents again suggests that electrostatic interactions between buffer carrier constituents and the dye are operative in producing the observed dye pattern.

These findings contradict present dogma that 8 M urea and/or detergents will disrupt all interactions between macromolecules or dyes and SCAMs and that the electric field in EF is always of capable of dissociating electrostatic interactions in the steady state. They similarly caution one not to view zones in EF gels formed in either SCAMs or simple buffers as being necessarily

494

artifact free, or worse, to consider EF zones as real without zonal re-focusing or other critical experimentation.

References

1. Gianazza, E., Righetti, P.G.: Biochim. Biophys. Acta 540: 357-364 (1978).

2. Nguyen, N.Y., Chrambach, A.: Electrophoresis 1: 14-22 (1980).

3. Prestidge, R.L., Hearn, M.T.W.: Anal. Biochem. 97: 95-102 (1979).

4. Cuono, C.B., Chapo, G.A.: Electrophoresis 3: 65-75 (1982).

5. An der Lan, B., Chrambach, A.: In: Gel Electrophoresis of Proteins, Hames and Rickwood, Eds. IRL Press, London 1981, pp. 157-187.

6. Cuono, C.B., Chapo, G.A., Chrambach, A., Hjelmeland, L.: Submitted for publication.

7. Dawson, D.J.: Aldrichimica Acta 14: 23-29 (1981).

8. Righetti, P.G., Gianazza, E., Bosisio, A.B.: Recent Developments in Chromatography and Electrophoresis 10: 89-117 (1980).

PHOTOACOUSTIC MAPPING OF ELECTROPHEROGRAMS

Hans-Peter Köst

Botanisches Institut der Universität München
D-8000 München, FRG

Ulrich Möller and Siegfried Schneider

Institut für Physikalische und Theoretische Chemie
Technische Universität München
D-8046 Garching, FRG

Hans Coufal

IBM Research Laboratory
San Jose, California 95193, USA

Introduction

Densitometry of electropherograms (PAGE-IEF gels) is hampered
by some inherent difficulties: lack of spatial resolving power
of older densitometers with "conventional" light sources and
low signal/noise ratio due to weak bands and/or unsufficient
staining. Another difficulty arises from light scattering of
the carrier (often polyethylene terephthalate, for example
Mylar^R). A number of improvements have been achieved recently
(1,2) by using soft laser scanning densitometers (1). By means
of computer aided densitometry, background corrections can be
performed. In case of an inhomogenous background however, the
results may be misleading. A further improvement concerning
the difficulties arising from weak bands and nontransparent
carriers is made possible by photoacoustic (PA) spectroscopy,
a new method which furnishes information about absorption and
relaxation processes of samples of different types (3-5). The
heat generated by radiationless deactivation processes causes
pressure oscillations in the small gas volume of the PA cell

which follow the chopping frequency of the light used. The emitted sound is recorded by means of a sensitive microphone coupled to the gas in the cell (6). A PA spectrum is obtained by recording the amplitude of the pressure oscillations as a function of the wavelength of the exciting light. A PA mapping is achieved by scanning a focused laser beam at constant wavelength across the surface of a PAGE-IEF gel containing either stained or unstained protein bands.

Results

The gels were evaluated with different instruments, see also (9): 1.) Densitograms of stained gels were recorded with the commercial laser densitometer LKB Ultrascan at λ =633 nm (spatial resolution 50 μm). 2.) The PA mapping of stained and unstained gels was performed using the apparatus shown in Fig.1. Light source is an Ar^+-laser tuned to different emitting lines or a tunable dye laser pumped by the Ar^+-laser. The laser beam is intensity-modulated ($f\approx20$ Hz) and used after correct attenuation and focusing ($\varnothing$ = 50-100 μm) to excite the sample being studied. The sample cell (Fig. 2) is designed to hold a substrate of the dimensions 1x5 cm. Scanning is achieved with high resolution by mounting the deflection prism on a micrometer controlled translation stage (Δx $\approx$ 10 μm). For more details see (9,10). 3.) Absorption spectra were recorded from water solutions of pure chromoproteins using a Perkin Elmer model 551 UV-vis spectrophotometer. PA spectra of solutions and electropherograms were taken with a double beam PA spectrometer described in detail in refs. (5) and (9).
For a photoacoustical evaluation, small gels are especially well suited. Furthermore, the application of small gels is time saving: PAGE-IEF on ultrathin 3x4 cm gels can be carried out within about 20 minutes (7,8).- Fig. 3a displays the laser densitometer trace (LKB-Ultrascan) of such a minigel (stained); in Fig. 3b its photoacoustic counterpart is shown (λ =600 nm;

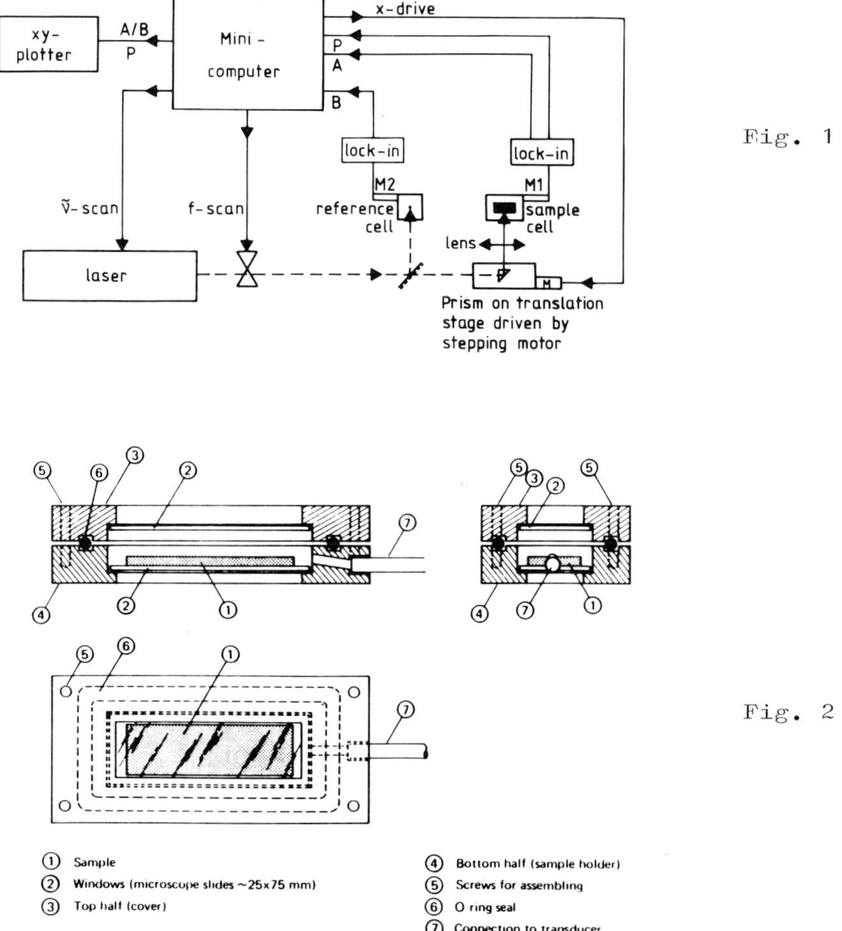

Fig. 1

Schematic of experimental set up for PA mapping. Chopping frequency f$\rightleftharpoons$ 20 Hz. Spatial resolution approximately 50-100 μm.

Fig. 2

Schematic of photoacoustic cell.

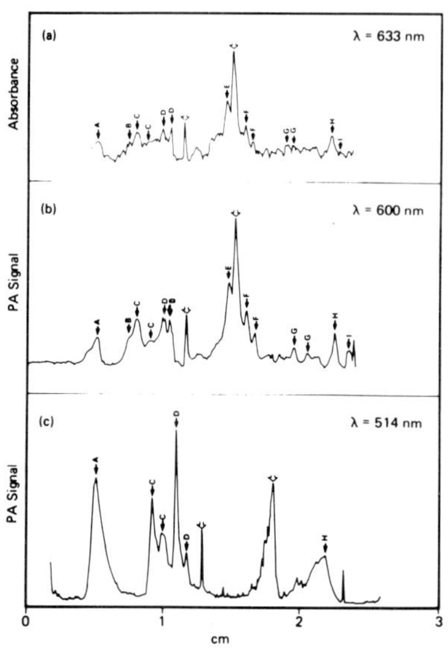

Fig. 3

Densitograms of "minigels" prepared by applying a solution of 1 mg marker protein mix 9 (Serva, Heidelberg, FRG) in 1 ml of 3% Servalyt pH 3-10. Hollow arrows mark artifakts (denatured protein) caused by the application of solution.
A - I: Components of Serva marker mix 9 (see text).
a. Transmission laser densitogram 1 μl marker mix, Serva Blue G stain)
b. PA densitogram of the same electropherogram
c. PA densitogram of unstained gel, 5 μl of marker mix were applied. Only peaks that can be assigned to chromoproteins are observed.

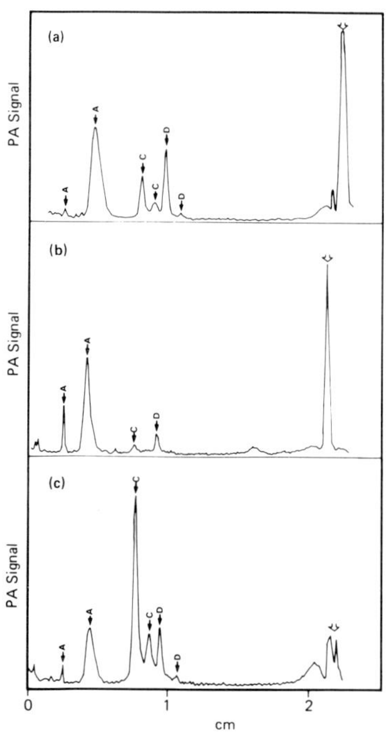

Fig. 5

PA densitogram of "minigels". Protein solutions of 1 mg protein in 1 ml of 3% Servalyte pH 2-11 were applied. For symbols used see text. (a) 0.4 μl of marker mix 9 solution, (b) 0.2 μl of marker solution mixed with 0.2 μl of cytochrome C solution, (c) 0.2 μl of marker solution mixed with 0.2 μl of myoglobin whale.

⌀ = 50 µm). The PA mapping exhibits less background than the laser densitogram, whereas its resolution is at least equal. This is caused by the property of PA spectroscopy to register only light that is absorbed near the surface and converted to heat. The technique is therefore insensitive to light scattering caused by the film which carries the gel (e.g. Mylar[R]). Since both densitograms (Fig. 3a, 3b) are very similar in structure and amplitude ratio, a determination of the amount of protein is possible (compare Fig. 6). In Fig. 3c, a laser scan of the same type of gel, but unstained, is given (λ=514 nm). Only zones comprising the chromoproteins of the marker mix 9 (Serva) used in the experiment cause a PA signal, namely cytochrome C (A), myoglobin (whale) (C), myoglobin (horse) (D), ferritin (H). Each of the four proteins gives rise to several bands. The remaining proteins: ribonuclease (B), conalbumin (E), ß-lactoglobulin (F), bovin albumin (G) and amyloglucosidase (I) are not detected in the unstained sample; their detection would necessitate an UV-excitation source.

Fig. 4 displays six scans of an unstained PAGE-IEF sample at different excitation wavelengths, incident laser power in all scans 4-5 mW. The PA mapping of the substrate does not only locate various protein fractions independently of dying procedures, it allows additionally to identify chromoproteins due to their different absorbance at the exciting wavelength. Whereas e.g. at an exciting wavelength of 570 or 514 nm the intensity of the band of cytochrome C (A) is the highest, the band of myoglobin (horse) (D) dominates for excitation wavelengths $\geqslant$600 nm. Focusing the light at one specific band allows to record the PA spectrum by scanning the frequency of the exciting light. A comparison of the PA spectra of (test) solutions and sample confirm the assignment (for more details see reference 9). Another procedure to identify the chromophores is shown in Fig. 5. Here, to the sample under study (e.g. trace **a**), comparable amounts of one species are added, e.g. cytochrome C (A) in trace b or myoglobin (whale) (C) (trace c). The peak(s) corresponding to the respective protein clearly are enhanced (Fig. 5).

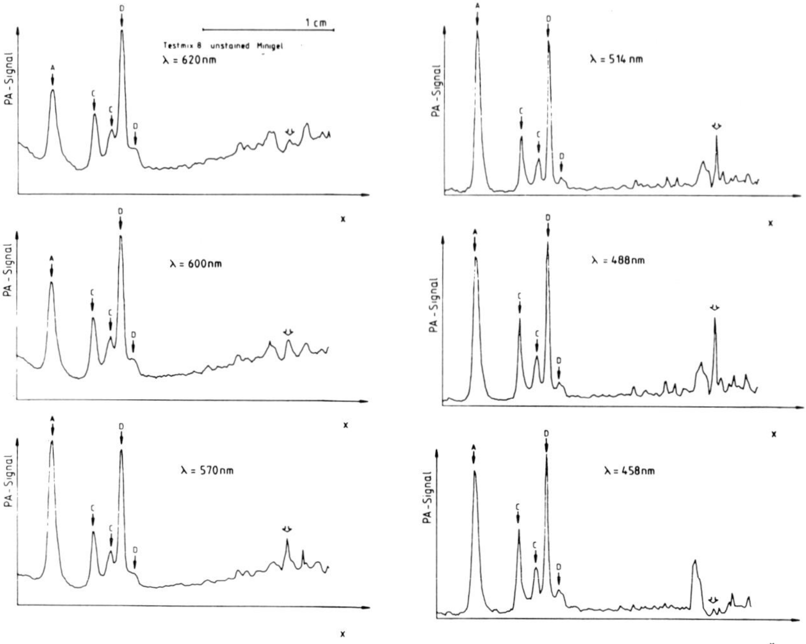

Fig. 4

PA densitogram of unstained gel (compare Fig. 3c) recorded at different wavelenth. Further details see text.

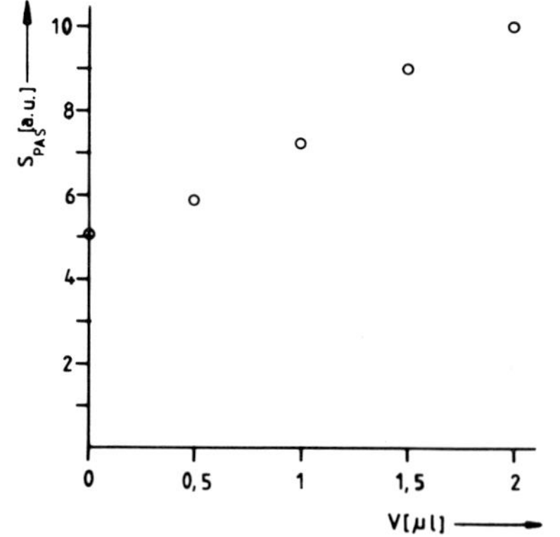

Fig. 6

Photoacoustic response (S_{PAS}) (integral) of band of myoglobin (horse) in arbitrary units as a function of the amount of protein applied.
1 µl = 1 µg protein marker mix 8 (Serva). Further details see text.

If the photoacoustic response is integrated over the whole area
of a protein zone, quantitative measurements can be carried out.
The latter procedure has been performed with a stained gel with
increasing amounts of marker solution applied. The cross section
of the exciting light beam was made large enough to cover the
whole zone. As a consequence, a constant area around the zone
was also illuminated. Since the gel itself was not completely
destained, the surrounding colored area contributed a large
"background" signal, which may be assumed to be roughly constant.
The graph of the PA signal versus the amount of protein (fig.6)
shows a nearly linear dependence, i.e. the PA signal which originates from the stained proteins obviously varies linearily
with the amount of protein in the zone. One prerequisite, as in
conventional densitometry, is that the staining efficiency has
to be equal for all zones; otherwise no quantitative measurements are possible. One virtue of PA mapping lies in the fact
that due to higher sensitivity the staining procedure can be
avoided if one deals with colored proteins.

References

1.   Zeineh, R.A. in Catsimpoolas, N. (Ed.), Electrophoresis '78
     Elsevier-North Holland, Amsterdam 1978, pp 147-151.

2.   Kronberg, H., Zimmer, H.G., Neuhoff, V.: Electrophoresis
     1, 27-32 (1980).

3.   Rosencwaig, A.: Photoacoustics and Photoacoustic Spectros-
     copy, Chemical Analysis, Vol. 57, J. Wiley & Sons, New York
     1980.

4.   Fishman, V.A., Bard A.J.: Anal. Chem. 53, 102-105 (1981).

5.   Schneider, S., Coufal, H.: J. Chem. Phys. 76, 2919-2924
     (1982).

6.   Schneider, S., Möller, U., Köst, H.-P., Coufal, H.:
     J. Photoacoustics 1, 3o9-316 (1982).

7.   Radola, B.J.: Electrophoresis 1, 43-56 (1980).

8.   Kinzkofer, A., Radola, B.J.: Electrophoresis 2, 174-183
     (1981)

9.   Möller, U., Köst, H.-P., Schneider,S., Coufal H.:
     Electrophoresis 4, (1983), in press.

502

10. Möller, U.: PhD thesis, Technische Universität München
     1983.

Acknowledgements

We are indebted to Dr. E. Köst-Reyes and Mrs. S. Nohel for providing us with ultrathin PAGE-IEF gels with focused marker proteins. We wish to thank Mr. K. Sieber from LKB (Munich, FRG)
for recording densitograms, the members of the electronics shop
of our institute for valuable help, Mrs. E. Kudler for skill-
full experimental assistance and Miss E. Benedikt for preparing
the figures. Financial support by the Deutsche Forschungsgemein-
schaft and the Fonds der Chemischen Industrie is gratefully
acknowledged.

ISOTACHOPHORESIS OF SERUM PROTEINS USING AMINO ACIDS AS SPACER ION

Takao Yagi, Kayoko Kojima
Analytical Application Department, Shimadzu Corporation, Kyoto, Japan

Masayo Yagi and Yoshihiro Kajita
Department of Clinical Laboratory Nantan Hospital, Kyoto, Japan

1. Introduction

In the capillary isotachophoresis of serum proteins, a UV absorption detector is often used, because proteins have a strong absorption at 280nm.
Protein fractions separated in the migration tube form discrete zones arranged in the order of the mobilities and travel at a constant speed in the tube, but, since all of these zones have absorption at 280nm, it is difficult to discriminate them with a UV absorption detector. This problem can be solved by the use of spacer ions having no absorption at 280nm. The spacer ions most commonly used in analyses of proteins are carrier ampholytes and a mixture of three amino acids, which are also used for isoelectric focusing. (1, 2)
The isotachopherograms of proteins obtained by the use of a UV absorption detector are too complicated to handle in routine work. There is a demand for a method that permits better fractionation but provides simpler, easier-to-read electropherogram, compared with the electrophoresis that use a cellulose acetate film and gives five fractions.
A trial was made (3) to separate serum proteins by the isotachophoresis that used carrier, and amino acids and aminosulfonic acid as spacer ions.
We used amino acids in the capillary isotachophoresis of serum proteins and obtaind eight fractions. The pattern was so simple that the peak areas could be easily determined by a data processor for chromatography. The quantitative reproducibility was excellent.
This technique was applied to analysis of serum proteins of patients, and gave satisfactorily discriminative data.

2. Material and Methods

2.1 Equipment
Shimadzu Model IP-2A capillary-type isotachophoretic analyzer equipped with a UV absorption detector (280nm) and a potential gradient detector was used. A three-pen recorder was used and the chart speed was set at 40mm/min. Shimadzu

504

CHROMATOPAC C-R1A was used to process the data.

The two-stage migration tube system was adopted: a PTFE tube, 1.0mm I.D. and 80mm long, and an FEP tube, 0.5mm I.D.. and 200mm long, were connected in series. The migration current was set at $150\mu A$ for 14 minutes from the start and at $75\mu A$ afterwords. The thermostatted both was kept at 20°C.

2.2 Reagents

2-(N-morpholino)-ethane-sulfonic acid (MES) purchased from Dojin Chemical Laboratory Ltd, Kumamoto, Japan, 2-amino-2-methyl-1.3-propanediol (Ammediol), ε-aminocaproic acid (EACA) and $Ba(OH)_2$, purchased from Wako Pure Chemical Industries Ltd., Osaka, Japan, and Hydroxyprophyl methylcellulose (HPMC), 15,000 centipoise in viscosity and purchased from Dow Chem. Co., USA, were used to prepare the electrolytes.

Ampholine®, 40% (pH 5~8), 40% (pH 8~9.5), and 20% (pH 9~11), purchased from LKB, Bromma, Sweden was used as the spacer ion. As for amino acids, asparagin (Asp. NH_2), glutamine (Glu. NH_2), glycine (Gly.), valine (Val.), leucin (Leu.), β-alanine (β-Ala.), and tris (hydroxymethyl) methylglycine (Tricine) were purchased from Wako Pure Chemical Industries Ltd, Osaka, Japan.

2.3 Electrolyte System

- Leading electrolyte:　Aqueous solution of 5mM-MES, 10mM-Ammediol, and 0.1% HPMC.
- Terminal electrolyte:　$Ba(OH)_2$ powder was added to the aqueous solution of 5mM-EACA and 10mM-Ammediol, the pH was adjusted to 10.8, and filtered. The filtrate was used as the terminal electrolyte.

2.4 Spacer ion solution

(a):　$180\mu\ell$ of pH 5 ~8 ampholine, $140\mu\ell$ of pH 8~9.5 ampholine, and $300\mu\ell$ of pH 9~11 ampholine were mixed with 4mg each of Gly., Val., and β-Ala., and distilled water was added to make the total volume $10m\ell$.

(b):　8mg each of the seven amino acids, Asp. NH_2, Glu. HN_2, Gly., Val., Leu., β-Ala. and Tricine, were mixed and distilled water was added to make the total volume $10m\ell$.

(c):　The ampholine used for (a) and 4mg each of the amino acids used for (b) were mixed, and distilled water was added to make the total volume $10m\ell$.

2.5 Sample

Normal human serum, IgA myeloma human serum, IgM myeloma human serum, IgG myeloma human serum, and pre- and post- hemodialysis human serum.

2.6   Method

Sample was mixed with spacer ion solution and this mixture solution was injected into the isotachophoretic analyzer, by means of a microlyter syringe. The output signal of the UV absorption detector was recorded as the electropherogram of the protein components. When the spacer ion solution ⓑ was used, the output signal was branched to the data processor to measure the peak areas of the protein fractions. The spacer ions were located from the output signal of the potential gradient detector.

3. Results

3.1   Electropherogram of serum proteins and spacer ions. Fig. 1 shows electrophero-grams of normal serum protein samples containing different spacer ions. Fig. 1 a shows the electropherogram of the sample to which the spacer ion solution ⓐ was added (Delmotte method). About 40 fractions are given, but the electropherogram is so complicated that the cut-and-weight method is the only method usable for quantitative determination, high quantitative accuracy cannot be expected. Fig. 1b shows the electropherogram obtained using the spacer ion solution ⓑ . The sample is bractionated into seven amino acids and the pattern is simple enough to permit easy area measurement. Fig. 1c was obtained using the spacer ion solution ⓒ , to find out which bractions of Fig. 1a correspond to those of Fig. 1b. The numbers at the bottom shows the corresponding fractions of Fig. 1b. The fractions of Fig. 1c were cut and weighed to know the weight (of the chart paper) percentage of the areas corresponding to the fractions of Fig. 1b. The results are Fraction No. 1: 37.6 (37.2), No. 2: 8.7 (9.5), No. 3: 12.6 (12.6), No. 4: 7.3 (6.8), No. 5: 5.2 (6.2), No. 6: 3.5 (4.0), No. 7: 17.6 (18.3), and No. 8: 7.4 (5.5). The parenthesized values are the percentages obtained from Fig, 1b. These two groups of data agree quite well with each other. This shows that the protein components are well separated.

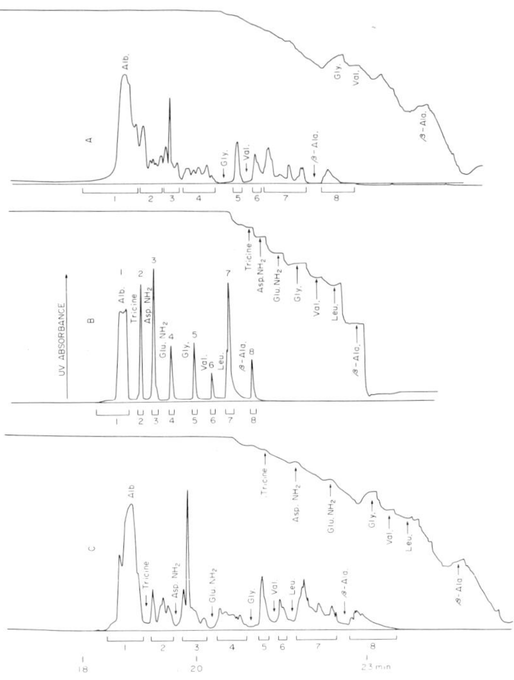

Fig. 1 Isotachopherograms of Normal Human Serum Protein Samples Containing Different
Spacer Ions

A: 1.5μℓ of 1 : 3 mixture of spacer ion solution (a) and serum sample solution was injected.

B: 1μℓ of 1 : 3 mixture of spacer ion solution (b) and serum sample solution was injected.

C: 1.5μℓ of 1 : 3 mixture of spacer ion solution (c) and serum sample solution was injected.

3.2 Effect of Amino Acid Quantity in Serum Sample

The effect of amino acid quantity added to serum sample on the fractionation was
investigated. To 10μℓ of normal human serum, 10 ∼ 70μℓ quantities of the spacer
ion solution (b) were added, and the total volumes were made 100μℓ by adding distilled water. Exactly 3μℓ of these solutions containing different amounts of the
spacer ion were analyzed. The area percentages of the fractions obtained are shown
in Table 1. This data shows that the samples containing the spacer ion solution (b)
of the volume 2 ∼ 6 times that of the serum give smaller deviations from the average
value. Especially, addition of the spacer ion of 2 ∼ 3 times larger volume gives the
minimum scattering of data.

Table 1 Relation Between Spacer Ion Quantity and Peak Area of Protein Fractions.

(%)

Peak No. Space ion ratio	1	2	3	4	5	6	7	8
1	37.63	9.92	12.20	6.94	5.76	3.80	18.28	5.47
2	37.48	9.47	12.54	6.70	5.88	3.89	18.62	5.42
3	37.18	9.46	12.56	6.79	6.01	3.95	18.48	5.55
4	37.27	9.49	12.58	6.62	5.92	4.05	18.63	5.44
5	37.44	9.48	12.85	6.42	5.74	3.96	18.68	5.42
6	37.74	9.24	12.68	6.84	6.02	3.91	18.31	5.63
7	37.67	9.29	13.07	6.85	6.22	3.78	17.68	5.44
AV.	37.49	9.48	12.64	6.74	5.94	3.91	18.38	5.48

AV.: average

3.3 Isotachophoresis of Protein Components

When a protein sample has been separated into the component zones between the leading and the terminal electrolytes, the concentration of the protein in each zone is about ten times higher than that in the original serum. We thought this high concentration can cause precipitation, which may affect the mobility of the zone, and that, hence, a high reproducibility cannot be expected in the isotachophoresis.

To check this presumption, the migration times of protein fraction zones were measured for different amounts of spacer ion. The results are shown in Fig. 2. The ordinate shows the times from the valley of the Fraction 1 to the opexes of the following fraction peaks, while the abscissa shows the mixing ratios of the spacer ion solution. The figure shows that the migration times of the protein fractions are proportional to the amount of the spacer ion added to sample: the migration is constant in speed. Table 1 shows that the protein concentration in the respective zones are independent of the migration time.

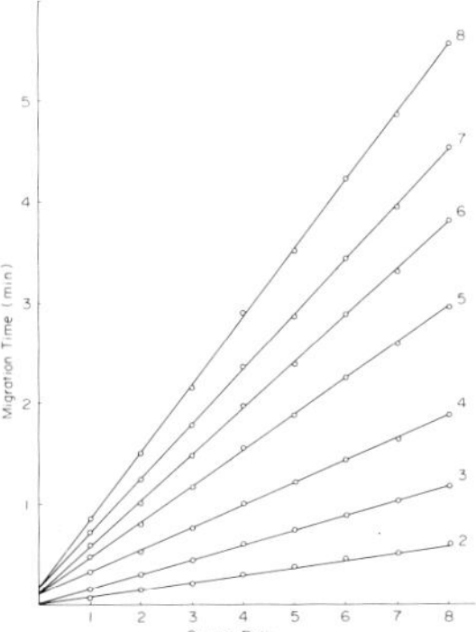

Fig. 2 Relation between Spacer Ion Quantity and Migration Time of Protein Fractions. The numbers at the right end of the curves show the numbers of protein fractions. (10~80μℓ of the spacer ion solution (b) was added to 10μℓ of serum sample solution, and distilled water was also added to make the total volume 100μℓ. Exactly 3μℓ of this mixture solution was injected.

3.4 Reproducibility of Calibration Curve

Fig. 3 shows the calibration curve obtained by injecting samples containing a constant amount of the spacer ion and different amounts of serum. The abscissa shows the real amounts of serum, though it was diluted five times with distilled water to increase the injection volume to obtain higher precision. The area of each protein zone is proportional to the ptotein quantity in a range below $0.8\mu\ell$. The fraction No. 7 has the narrowest linear range. This may be attributable to the fact that, as shown by Figs. 1a and 1c, proteins having a wide range of mobilities are contained in a single zone. The repeatability for one sample was also investigated. To $10\mu\ell$ of normal human serum, $30\mu\ell$ of the spacer ion solution (b) was added, and this mixture was made $100\mu\ell$ by adding distilled water. Exactly $3\mu\ell$ of this solution was analyzed seven times to study the repeatability in percentage of the protein components. The repeatability, expressed in C.V(%) value, were 0.7, 1.9, 1.3, 3.5, 1.1, 3.1, 2.3 and 3.9 for the fractions Nos. 1 ~ 8.

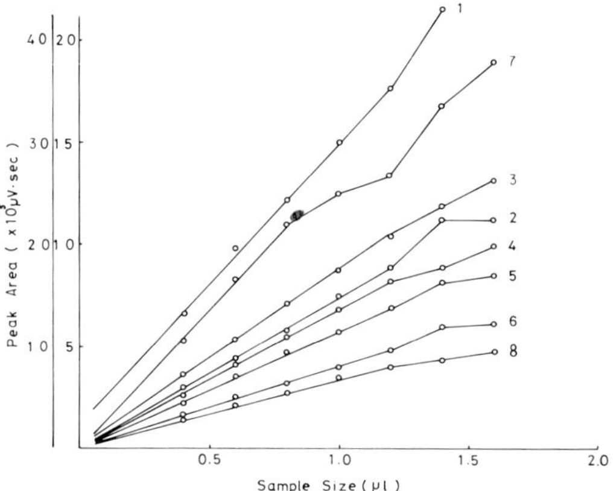

Fig. 3   Calibration cuve for human normal protein

3.5 Application to Practical Sample

Table 2 shows the percentage of protein fractions of abnormal human serum samples. The IgA myeloma serum sample gives larger zones for the fraction Nos. 2 and 3, which correspond to the fast γ position in cellulose acetate film electrophoresis (CAE method). The IgG myeloma serum sample gives larger zones for the fraction No. 8 which corresponds to the slow γ position in the CAE method, or the fraction No. 3 which corresponds to the middle γ position in the CAE method. The IgM myeloma serum sample gives peaks at the fractions Nos. 7 and 8 which correspond to the tailing at the slow γ position in the CAE method, or a peak at the fraction No. 4 which corresponds to the middle γ position in the CAE method. Serum of hemodialysis patients gives larger zones for the fractions Nos. 1 and 2, and smaller zones for the fractions Nos. 7 and 8.

Table 2   Quantitative Determination of Protein Fractions of Various Serum.

(%)

Human Serum	Proteins traction No.							
	1	2	3	4	5	6	7	8
Normal	37.2	9.5	12.6	6.8	6.2	4.0	18.3	5.5
IgA Myeloma (1)	22.2	48.1	7.0	3.8	3.9	1.6	11.9	1.4
IgA Myeloma (2)	26.4	14.3	35.2	5.2	2.9	1.4	13.1	1.4
IgG Myeloma (1)	10.7	4.3	3.8	4.6	1.8	0.6	22.9	51.2
IgG Myeloma (2)	9.2	6.8	55.5	15.7	2.8	0.8	6.1	3.0
IgM Myeloma (1)	21.2	9.2	9.8	5.8	5.7	4.2	31.6	12.3
IgM Myeloma (2)	16.9	6.5	11.0	40.5	7.3	2.2	11.2	4.3
Predialysis	44.9	13.7	13.1	5.8	3.9	2.1	12.3	4.0
Postdialysis	45.4	12.7	11.4	5.8	4.0	2.6	13.5	4.3

4. Discussion

When a spacer ion consisting of ampholine and three amino acids is used in capillary iso-tachophoresis of serum proteins, the electropherogram is too complicated to decipher. When a mixture of tricine, Asp. NH_2, Glu. NH_2, Gly., Val., Leu., and β-Ala. is used as the spacer ion, the serum proteins are separated into eight zones and the pattern is so simple that it is easily processed and gives high quantitative reproducibility.
The optimum volume of the spacer ion solution (b) (8mg/10mℓ) is 2 ~ 3 times that of the serum solution.
Good results are obtainable when the absolute volume of serum solution is below 0.8$\mu\ell$. This follows that, the best result can be obtained when serum sample solution and the spacer ion solution (b) are mixed at a ratio of 1 : 3 and 1$\mu\ell$ of this mixture is analyzed. The analysis is completed in about 20 minutes.
Myeloma serum sample gave considerably different percentages of protein fractions from those of normal serum sample.
The separation of IgA, IgM, and IgG is to be tried in our future study.

5. Reference

1. Delmotte, P.: SCIENCE TOOLS 24, 33–41 (1977).
2. Bours, J., Delmotte, P.,: Reprint from Science Tools. 26, 58–64 (1979).
3. Shimao, K.: Medical Technology 13, 1215–1221 (1979)

ISOTACHOPHORESIS AND ISOELECTRIC FOCUSING OF HUMAN SERUM
PROTEINS

Kiyotsugu Kojima, Takashi Manabe and Tsuneo Okuyama
Department of Chemistry, Faculty of Science, Tokyo Metropolitan University, Tokyo 158, Japan

Introduction

Recently, isotachophoresis has been used to analyze human
fluid proteins (1,2,3,4,5), and when carrier ampholytes were
mixed with the proteins high-resolution was obtained. However,
the mechanisms by which proteins are separated by isotachophoresis in the presence of carrier ampholytes were not investigated. In this report, isotachophoresis of proteins compared
with isoelectric focusing is described.

Materials and Methods

1) Apparatus
    Capillary isotachophoretic analyses were performed on
    a LKB 2127 Tachophor equipped with a PTFE capillary
    (0.5 mm I.D. and 23 cm length) and a Shimadzu IP-2A
    fitted with a PTFE capillary (1.0 mm I.D. and 10 cm
    length connected with 0.5 mm I.D. and 35 cm length).
    Samples were analysed by a UV detector and a thermal
    detector of the LKB 2127 Tachophor, and a UV detector
    and potential gradient detector of the Shimadzu IP-2A.
    The Shimadzu IP-2A was partly reconstructed as described
    later.

Electrophoresis '83

512

2) Leading and Terminating Electrolyte
    The leading electrolyte was 0.01M Cl$^-$ pH 9.0 adjusted
    with Ammediol. This contained 0.1% (w/v) Brij 35
    (non-ionic detergent) for smooth flow of the electrolyte
    in the capillary. The terminating electrolyte was 0.01M
    ε-Aminocaproic acid and 0.01M Ammediol brought up to
    pH 10.8 with Ba(OH)$_2$.

3) Carrier Ampholyte
    Ampholine (pH 3.5-10, LKB) solution (0.2 µl of 10% (w/v))
    was used each time. After injection of Ampholine solu-
    tion into the capillary, 1 µl of human serum was injected.
    Except for Ampholine, other spacers such as amino acids
    were not used.

Results and Discussion

1) Reconstruction of Shimadzu IP-2A
    In the original IP-2A, an N$_2$ gas charging system was used
    both for charging the capillary with electrolytes and
    rinsing the capillary. However, the gas sometimes caused
    bubble to appear in the capillary during isotachophoresis,
    especially at high voltage, we changed the gas charging
    system for the pumping system shown in Fig. 1. By this
    system, high voltages, over 20 kV, could be used without
    any trouble. This pumping system also has the advantage
    of automatic operation of the capillary isotachophoresis
    by addition of an automatic sampler and time controler.

2) Role of Carrier Ampholyte during Isotachophoresis
    Use of adequate amount of carrier ampholyte leads to good
    separation of complex protein mixtures, such as serum,
    in isotachophoresis (2,4,5). However, the role of
    carrier ampholyte during isotachophoresis has not been

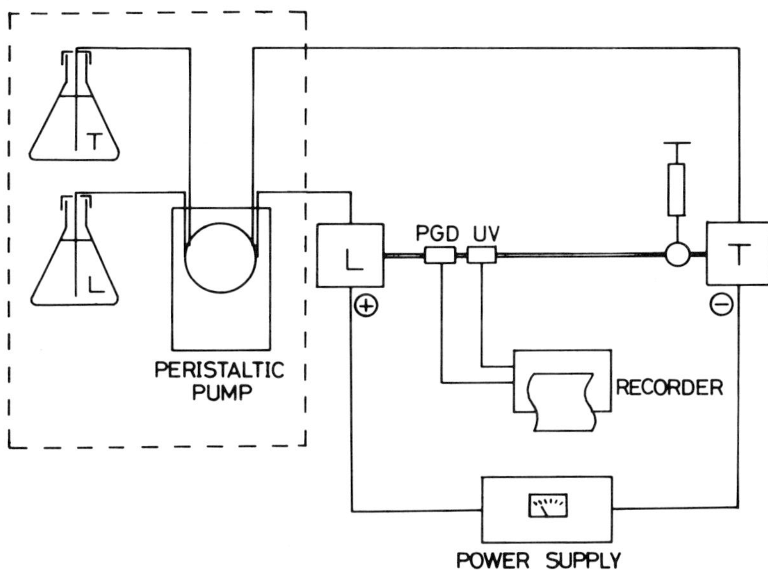

Fig. 1: Block-diagram of Shimadzu IP-2A with reconstructed pumping system. Inside of the dot-line shows the part reconstructed. The leading and terminating electrolytes are sent by a peristaltic pump to the capillary through the reservoirs.

well investigated. One of the reasons of this is that exact identification of the separated peaks is difficult. Therefore, in place of complex protein mixtures we chose six low molecular weight compounds having UV absorbances and pKa values different from on onother: 2-Thiobarbituric acid (pKa =4), Barbital (7.4), Hypoxanthine (8.7), L-Tryptophane (9.4), Uracil (9.5) and Phenol (10.0). As the potential gradient (or thermal gradient) is smooth when carrier ampholyte is used in isotachophoresis, UV absorbance of the samples makes the analysis easy. Fig. 2 shows an isotachopherogram of a mixture of these six compounds when Ampholine (pH 3.5-10 : pH 9-11 = 1:1) was used.

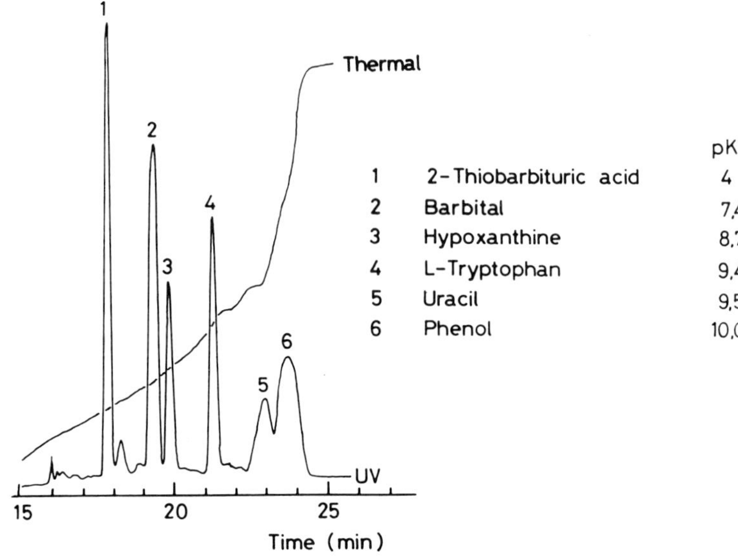

Fig. 2: Isotachopherogram of a mixture of six low molecular weight compounds. After injection of 0.2 µl of 10% (w/v) Ampholine (pH 3.5-10 : pH 9-11 = 1:1), 1 µl of the mixture was injected into the capillary. One µl of the sample contained 0.15 µg of 2-Thiobarbituric acid, 0.51 µg of Barbital, 0.08 µg of Hypoxanthine, 0.19 µg of L-Tryptophan, 0.31 µg of Uracil and 0.69 µg of Phenol. (by LKB 2127 Tachophor)

The compounds were first electrophoresed separately, and then together. The peaks were identified by comparison of isotachopherograms of each compounds electrophoresed independently. As shown in Fig. 2, these compounds were detected in order of their pKa value. This suggests that the mechanism for separating compounds by isotachophoresis using carrier ampholyte seems to be similar to that of isoelectric focusing. In isoelectric focusing, each sample moves to its isoelectric point and finally focused at that point. In isotachophoresis, the samples continue to move until the final analysis. Thus, isotachophoresis using carrier ampholyte appears to be moving isoelectric focusing.

3) Resolution of Peaks Using HPMC in Leading Electorolyte
   In some cases high resolution separation of proteins
   may be obtained by isotachophoresis with the use of
   HPMC (Hydroxypropylmethyl cellulose)(1,4,5). However,
   there seems to be a problem in reproducibility.

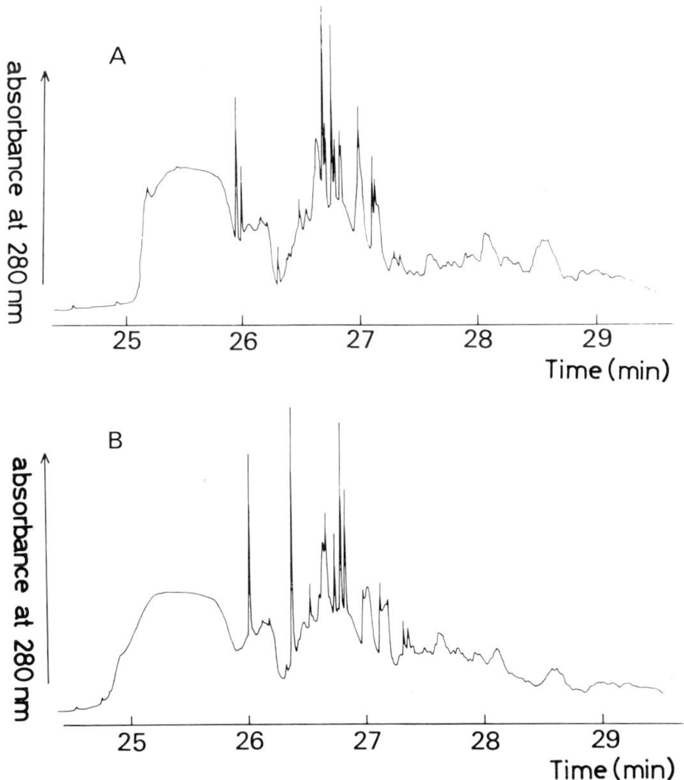

Fig. 3: Two isotachopherograms of the same sample of human
serum proteins run under the same conditions. The apparatus
was a Shimadzu IP-2A. The leading electrolyte contained 0.4%
(w/v) HPMC. The running current was 150 µA for 15 min and
then 75 µA. One µl of human serum was injected after injection of 0.2 µl of 10% (w/v) Ampholine (pH 3.5-10).

516

Fig. 3 shows two isotachopherograms of a human serum
protein sample run under same conditions. The leading
electrolyte contained 0.4% (w/v) HPMC. By use of HPMC,
some sharp peaks were obtained on every run. However,
there seemed to be little reproducibility of these sharp
peaks in position and absorbance as shown in Fig. 3A and
3B. Therefore, we retraced these isotachopherograms
manually omitting the sharp peaks (Fig. 4A and 4B).
The arrows in the figures show the peaks which are
reproducible each of in both A and B. The number of
arrows in the two isotachopherograms was twelve, and at
least these twelve peaks seemed to be reproducible when
HPMC was added to the leading electrolyte.

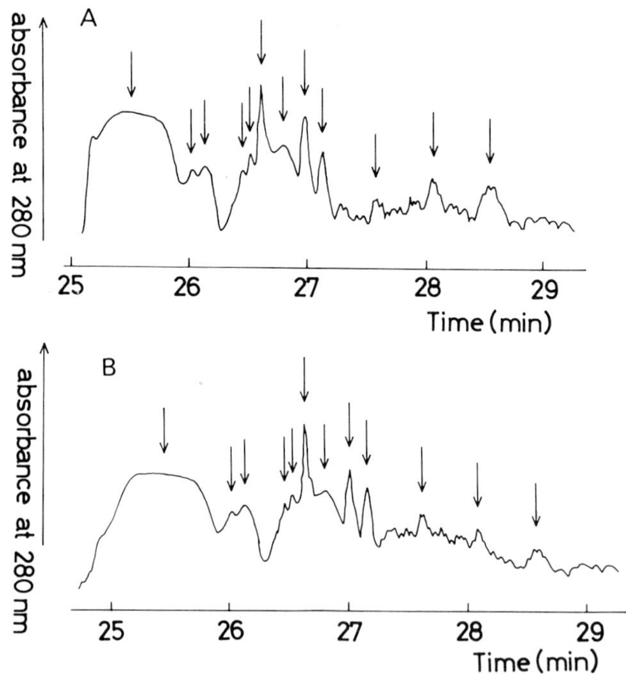

Fig. 4: Retraced isotachopherograms of Fig. 3. A and B
correspond to Fig. 3A and 3B respectively. The arrows show
the peaks which are reproduced in both A and B after retrace.

Apart from the unknown sharp peaks, it is clear that
high-resolution separation of human proteins can be
obtained by addition of HPMC to the leading electrolyte.
Precipitation of proteins in the capillary during isotachophoresis may be on of the causes of the unknown
sharp peaks when using HPMC. With respect to the
separation of low molecular weight compounds described
above, it seems likely that serum proteins may precipitate at their isoelectric point in the capillary.
When small particles of precipitated proteins pass
through the UV detector, sharp peaks appear on the
isotachopherogram.

References

1. Kjellin, K. G.: J. Neurol. Sci. $\underline{26}$, 617-622 (1975)

2. Kopwillem, A., Merriman, W. G., Cuddeback, R. M., Smolka,
    A. J. K., Bier, M.: J. Chromatogr. $\underline{118}$, 35-46 (1976)

3. Levy, L. N., Schoen, T. M.: Neurology, June 1976, Part2,
    62-63 (1976)

4. Delmotte, P.: Electrophoresis '78, ed. Catsimpoolas, N.,
    Elsevier 115-134 (1978)

5. Holloway, C. J., Heil, W., Henkel, E.: Electrophoresis
    '81, Walter de Gruyter 753-766 (1981)

ISOELECTRIC FOCUSING OF CIRCULATING IMMUNE COMPLEXES

Bruno L. Schmidt, Gertrude Steinmetz
Ludwig Boltzmann-Institute of dermato-venerological serumdiagnosis,
Vienna, Austria

Introduction

Soluble immune complexes circulating in the blood stream (CIC) have been
demonstrated in patients with a large variety of autoimmune, infectious,
malignant and other diseases. However, it should be stressed that CIC
represent a normal and effective immunological mechanism for antigenic
clearance. As the term CIC includes a rather complex reaction product
of the interaction of antigen with antibody, some comments about the
nature and isolation of CIC should be given. The nature of a complex
depends upon several factors, including: (1) the number and density of
antigenic determinants within the molecule available for reaction with
antibody, (2) the affinity of the interaction between antigen and the Fab
portion of the antibody molecule, (3) the ratio of antigen to antibody and
the concentration of both, (4) the modifications of immune complexes
occuring after their formation or deposition.
Therefore it is not surprising that more than 50 methods have been
described for detection and also for isolation of CIC. Isolation procedures
can be devided into antigen-nonspecific and antigen-specific methods.
Antigen-nonspecific methods may be based on physical properties (based on
size or solubility changes, e.g. sucrose density gradients, gelfiltration,
ultrafiltration, ultracentrifugation, cryoprecipitation, precipitation
with PEG) or on biological properties of CIC (e.g. Protein A, C_1q,
conglutinin, anti-C_3, anti-C_4, anti-Ig, Fc-receptors on cells (platelets),
C_3b receptors on cells (Raji), antigen specific antibody, DEAE-Cibachrome
Blue). The appropriate isolation procedure has to be selected carefully,
as there are limitations within each method, starting with the class

specific binding of Protein A, the limitations that only complexes
involving the complement system can be isolated with C_1q or anti-C_3
assays and last not least IgG consisting CIC are lost by DEAE-Cibachrome
Blue as these do not bind to the gelmatrix!
Due to the large variety of CIC, it is impossible to even give general
recommendations for IEF of CIC. Therfore focusing was done on 3 typical
examples: (1) IEF of CIC with and without dissociation of antigen and
antibody, (2) IEF of IgM-anti IgG autoantibodies in sera of patients with
infectious diseases, (3) preparative IEF of native undissociated CIC.

Material and Methods

Sample preparation
Crystallized bovine serum albumin (BSA)(Sigma A 7638, St. Louis, USA) was
further purified by gelfiltration over Sephacryl S 200. IgG-Fraction of
rabbit anti BSA was from Dako, Denmark (Z 229, Lot.Nr. 032B). Ten mg BSA
and 2ml anti BSA were incubated at 20°C for 90 minutes and thereafter for
48 hours at 4°C in 0,05M Tris-0,1M NaCl, 0,1% NaN_3, pH 7,5 buffer.

IgG containing soluble CIC were isolated out of sera from patients with
Treponema pallidum infection first by absorption to Protein A Sepharose
CL-4B (Pharmacia, Sweden) according to the recommendations of the
manufacturer. Then CIC were separated from 7S IgG by High Pressure Size
Exclusion Chromatography (HPSEC).

HPSEC was performed at a flow rate of 1ml/min with 2x10cm TSK SW 3000
columns (Toyosoda,Yapan) and with 0,01M phosphate-buffered saline as
eluent. CIC were fractionated with the void volume (2,4min) from IgG
(3,7min).

Gelpreparation
The gels used consisted of 1% agarose IEF (Pharmacia, Sweden), 10% sorbitol
(w/v) and 3% carrier ampholyte as indicated in the legends of the figures.
Gels with a thickness of 0,3mm were casted onto GelBond film with flap

technique (1) or with a casting frame of Pharmacia for 1mm gels. For
comparison polyacrylamid (PAA) gels (5%T, 20%C) were prepared.

Isoelectric focusing

IEF was performed using a Desaga Mediphor chamber with a LKB 2103 power
supply. The coolant was circulating at 15°C through the cooling plate.
Electrofocusing wicks (LKB, Sweden) were used, cut to length with a razor
blade. The anolyte consisted of 0,04M DL-aspartic acid, the catholyte was
0,025M DL-arginine and 0,025M L-lysine (free bases) in 12% ethylendiamine.
10 µl samples were applied with a rubber application band (Serva, Heidelberg, FRG) after 30 minutes prefocusing at 400 volts, and focusing was
started with 500 V, reaching a final voltage of 1200 volts and about
2200 V.h. Staining was done with 0,1% Serva Blau G (Serva, Heidelberg,FRG).
Immunofixation was performed with cellulose acetate membranes (2).

Results and Discussion

Before performing IEF of CIC the question should be answered, if this
method should be used as an analytical one, e.g. to characterize the
constituents, to check the purity of a fraction etc. or as a preparative
one to isolate the native complex. Most work in literature was done with
analytical IEF. The reasons thereof are: (1) CIC can be dissociated
before IEF, avoiding problems with high molecular weight molecules,
(2) well defined PAA gels can be used as a matrix instead of agarose,
(3) surfactants, solubilizing agents as well as denaturing reagents may
be employed avoiding problems with limited solubility and hydrophobic
effects of large aggregates.

IEF of BSA-anti BSA complexes
This   rather small immune complex can be focused in agarose and PAA. In
the most cited method of Maidment and coworkers (3) CIC are precipitated
first with 2,5% polyethylenglykol and then absorbed with Protein A
Sepharose CL-4B. But as might be expected from the title, CIC are not
separated into antigen and antibody by electrochemical forces

522

during IEF, but by dispensing the immune complex coated Protein A-beads
into 0,02M phosphoric acid, which is used as an anolyte. According to the
different nature of CIC some might be separated - through influence of
temperature, solution volume or concentration and nature of ions in
buffer - but for BSA-anti BSA this is not true as can be seen in figure 1:

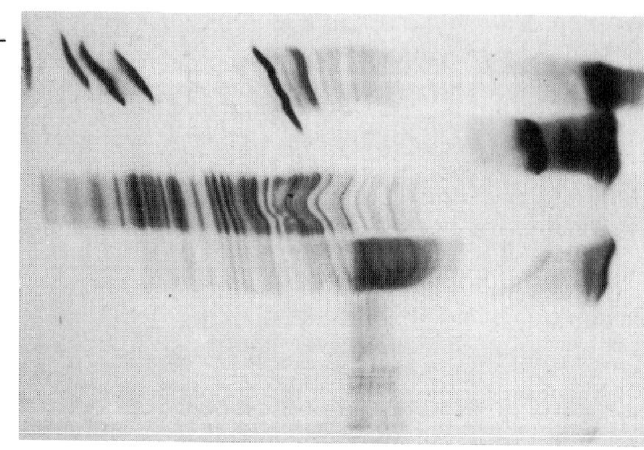

Figure 1
IEF in a 0,3mm layer of 1%agarose IEF (Pharmacia) with 3% carrier
ampholytes (Servalyte T4-9, AG 3-5, AG 5-9, equal parts), 15°C, 2200 V.h.
Samples beginning from the top: Marker mixture 9 (Serva,FRG), BSA,
anti-BSA, BSA-anti BSA, fractions I and II of BSA-anti BSA immune complex
from HPSEC.

The BSA in the complex has a quite different pI than BSA (pI 4,7). IEF was
performed with 0,04M DL-aspartic acid as anolyte and 0,025M DL-arginin and
0,025 M L-lysin in 12% ethylendiamine as catholyt.

IEF of native IgM-anti IgG autoantibodies
During previous work with agarose IEF of immunoglobulins (4) difficulties
were encountered in obtaining sharp focused bands. Compared gels casted
with PAA (5%T, 20%C)showed enhanced sharpness of bands. However, the
practical upper limit of pore size is 5.10^5 in PAA (5), not only due to
restrictive sieving properties of the gel but also due to the greater
hydrophobicity of the crosslinker N,N´methylene-bisacrylamid. Agarose
with its large pore size has considerable advantages.
In figure 2 the separation patterns of an IgM-anti IgG autoantibody
(pI = 8,1) is shown together with contaminating IgG antibodies

resulting from the isolation procedure. The high molecular weight complex
(21S) can be focused as a sharp band. Immunofixation gave a precipitation
with anti human μ-chain specific antiserum. Normal human IgM has a pI
3,5 - 5,0. The specificity of the autoantibody could be examined: Only the
IgG part was reacting with Treponema pallidum whereas the IgM showed no
reactivity with the pathogen (results obtained with ELISA-technique
presented elsewhere).

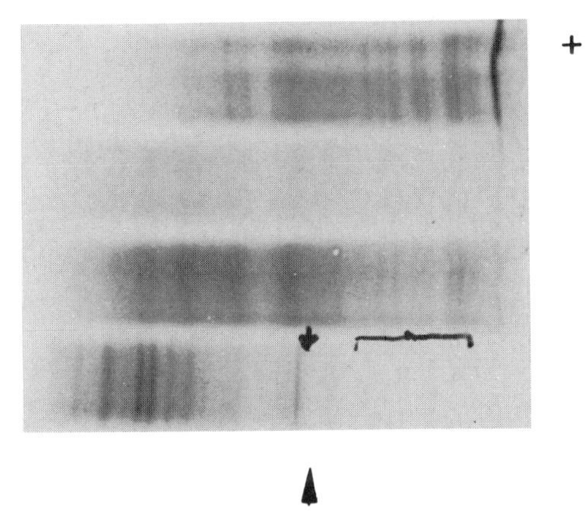

Figure 2
IEF in a 0,3 mm layer of 1% agarose IEF (Pharmacia) with 3% carrier
ampholytes (LKB agarose ampholine 3,5 - 9,5 and Pharmalyte 8 - 10,5; 3+1),
15°C, 2200 V.h. IEF of IgG isolates from patients with infections of
Treponema pallidum. Third sample from top is a pool of IgG, serving as
control.

IEF of native, undissociated CIC
CIC can modify the immune response by influencing lymphocyte traffic and
antigen presentation and by enhancement or suppression of the activity
of helper and suppressor T cells. CIC are thought to cause a direct or
indirect block of specific helper T cells. In order to isolate native,
undissociated CIC free of contaminating proteins, experimental runs with
IEF in agarose were performed (preliminary tests with PAA gels showed
that CIC did not enter the gel). With agarose, migration of CIC can be
achieved, allowing to estimate the pI, however, severe distortion of the

gradient results. Dialysing of the sample caused precipitation. All
tested solubilizing agents (0,1M acetic acid, citrate phosphate pH 3,0,
3M NaSCN, 2% SDS, 2,5M $MgCl_2$ pH 7,0) showed a marked interaction with
CIC resulting in dissociation and interfering with following immunoassays
and also with IEF. It should be mentioned that IEF of large molecules
like CIC is much more sensitive to interfering influences which can be
neglected when focusing is performed with low molecular weight proteins.
Large molecular weight complexes are migrating only slowly, therefore
affording lomg electrophoretic runs and high voltages. Both do increase
cathodic drift in agaroses.available. Thermostating the gel at 10°C
will improve heat transfer, minimize the convection with some sharpening
of the bands, but will also cause surface flooding resulting in distorted
protein patterns and electrical short circuits. On the other hand,
solubility of CIC may be extremely influenced by changing surrounding
conditions, e.g. proteins (colloidal effect) and salts. IEF does cause
such effects. Therefore preparative IEF seems not to be the best method
for isolating native CIC, at least those complexes from patients with
infectious diseases studied.

References

1. Radola, B.J.: Electrophoresis 1, 43-56 (1980)
2. Marschall, M.O.: Clin.Chim.Acta 104,1-9 (1980)
3. Maidment, B.W., Papsidero L.D., Gamarra, M., Nemoto, T., Ming Chu, T.:
   Anal. Biochem. 111, 336-342 (1981)
4. Schmidt, B., Pfeifer, G.: Electrophoresis ´81, R.C. Allen, P. Arnaud
   Eds., Walter de Gruyter, Berlin.New York 1981
5. Rüchel, R.: J. Chromatography 166, 563-575 (1978)

INCREASING RESOLUTION AND IMPROVING REPRODUCIBILITY OF ISOELECTRIC FOCUSING AND 2D-ELECTROPHORESIS BY PERFORMING IEF IN IMMOBILIZED PH GRADIENTS

Angelika Görg, Wilhelm Postel, Johann Weser,
Reiner Westermeier
Lehrstuhl für Allgemeine Lebensmitteltechnologie
der Technischen Universität München,
D-8o5o Freising-Weihenstephan (F.R.G.)

Introduction

Immobilized pH gradients are a new concept in isoelectric focusing, and were introduced at the last Electrophoresis conference in Athens in 1982 (1 - 3). One of the major advantages of the Immobiline system, basing on a patent of Gasparic, Bjellqvist and Rosengren (4) is the possibility to generate pH gradients which are stable with time. Isoelectric focusing with Immobiline ^{R} represents in opposition to the carrier ampholyte system a true equilibrium method without cathodic drift (5). This essential improvement in isoelectric focusing is of fundamental importance for the reproducibility of one- or two-dimensional separations. Another advantage of the Immobiline system is the possibility of generating narrow or ultranarrow pH gradients which fit exactly to the pI's of the protein of interest. Flattening the pH gradient in the desired pH interval increases the resolution of isoelectric focusing and helps to overcome difficult separation problems (1 - 3, 5, 6).

Examples how to optimize the analysis of proteins in forensic haemogenetics and plant genetics are given in the present paper. Also zymogram techniques will be demonstrated. The high reproducibility of isoelectric focusing in immobilized pH gradients is successfully exploited for the reproducibility of two-dimensional electrophoresis.

Material and Methods

Ampholine carrier ampholytes pH 4.o - 6.5 and Immobilines [R] pK 3.6, 4.6, 6.2, 7.o and 9.3, acrylamide, N,N'-methylenebisacrylamide (Bis), ammonium persulfate, N,N,N',N'-tetramethylethylenediamine (TEMED), Repelsilane and Coomassie Brilliant Blue R-25o and G-25o were all from LKB, Bromma, Sweden. All other chemicals were of analytical grade. The IEF experiments were carried out using the LKB Ultrophor apparatus. The second dimension of the high resolution 2D-electrophoresis was run horizontally using the LKB Multiphor and vertically using the LKB vertical electrophoresis unit. For casting the gels, the LKB gradient gel kit with the micro gradient mixer was used.

The pH gradient gels and the pore gradient gels were cast according to Görg et al. (7, 8).

Immobilized pH gradients. For generating pH gradients used for the analysis of serum proteins see refs.: Pi system (9), Gc system (1o) and Tf system (11). The recipes of the Immobiline concentrations for the other pH gradients were calculated after Bjellqvist et al. (12), see also Righetti et al. in this book. Ground seeds or grains were extracted with o,2 M Tris-Gly buffer, pH 8.6. The samples were applied into the slots directly without prerunning. Running conditions: 1o W (max) power; 2o mA (max) current and 25oo V (max) voltage, for 16 hours at 1o $^{\circ}$C. The electrodes were applied directly on the gel surface.

2D electrophoresis: The second dimension was run horizontally in 36o um thin gels (13o x 26o mm) and vertically in 75o μm thin gels (15o x 14o mm). The acrylamide gradient was from 1o % - 22.5 % T with constant 2.5 % C, o,1 % SDS and 375 mM Tris-HCl, pH 8.8. After fixing and staining the IEF strip was washed in water for 1o min and then equilibrated for 3o min in 1 % SDS, o.1 % dithiothreitol and 375 mM Tris-HCl, pH 8.8. The gel strips were transferred to the second dimension according to refs. (7) and (8).

In opposition to the horizontal run, in the vertical run the gel strip has
to be overlaid with agarose (1 %). The horizontal electrophoresis lasted
for 2.5 hours, the vertical one for 5 hours.

Zymograms: Staining for esterase isoenzymes after IEF in immobilized
pH gradients was performed as in conventional carrier ampholyte generated
pH gradients with α-naphthyl-acetate and Fast Blue B. Staining for
trypsin-inhibitors see Ref. (3). Acidic silver stain was in principle
according to Merril et al. (13).

Results and Discussion

Fig. 1 shows the pH gradients, which we have found out to be
optimal for the analysis of inherited heterogeneity in the
human Gc (group specific component, VDBP) system and for the
classification of genetic variants in the human Pi (α_1-anti-
trypsin) and Tf (transferrin) system. The pH gradient from
4.o to 6.5, which is routinely used in forensic haemogenetics,
was subdivided into special pH intervals. We cut and flattened
the pH gradients step by step in order to blow up the patterns
of interest as much as possible. The best pH gradients with
the highest resolution were from pH 4.5 - 4.7, pH 4.9 - 5.2
and pH 5.2 - 5.7 for the Pi, Gc and Tf systems respectively.
With these and slightly modified pH gradients, common and rare
variants of Pi (9), Gc (1o) and Tf (11) were analyzed. For
example the Pi M1 M3 subtype, which is barely separated into
double bands by IEF with conventional carrier ampholytes
pH 4.o - 5.o (9), clearly resolves into two zones in the
immobilized pH gradient from 4.5 to 4.7 (Fig. 1). Because the
optimal tailoring of immobilized pH gradients is only a
question of strategy and calculation of recipes, the Immobiline system offers the possibility to generate very individual
pH gradients, which fit exactly to the proteins of interest.
Also the reproducibility of the results, even in the very
narrow pH ranges with long focusing times, is guaranteed,

528

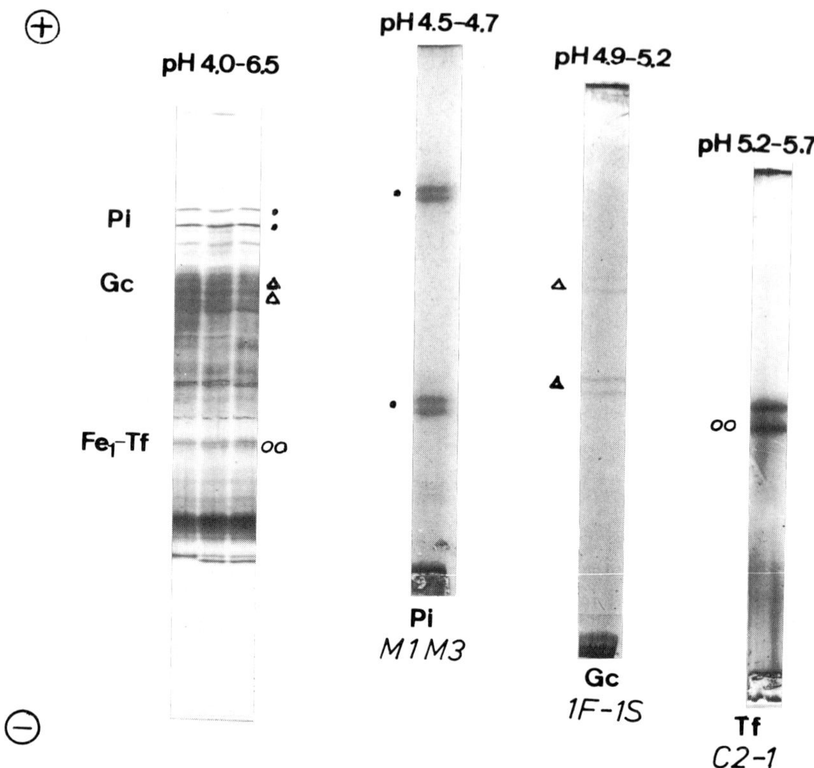

Fig. 1 Selection of pH ranges for optimal resolution of genetic variability. Analysis of three human serum protein polymorphism: Pi, Gc (VDBP) and
Tf. IEF with carrier ampholytes pH 4.o - 6.5. IEF with Immobilines: pH
4.5 - 4.7 (ref 9) pH 4.9 - 5.2 (ref 1o) and pH 5.2 - 5.7 (ref 11). All
photographic reproductions are proportional to the original sizes.

because no pH drift or conductivity gaps are occurring.

IEF has also become a useful tool for plant breeders and gene-
ticists, because different species and varieties have different IEF patterns. For taxonomic studies with IEF, only one
grain or seed kernel is needed. For identifying mutants which
differ only in small changes in few protein species, an IEF
method with very high resolution is required. In first
studies, tailored immobilized pH gradients were successfully
applied for the screening of seed proteins of different pea
varieties (3).

In Fig. 2 the pherograms of IEF in ultranarrow pH gradients of
buffer soluble seed proteins of four different varieties of
soy beans (Glycine max) are demonstrated. The IEF in immobilized pH gradients from 4.4o to 4.85 provides a number of
genetic differences in the band pattern, which have not been
expected in earlier studies (14) with IEF in wide pH ranges
from 3.5 - 1o or in narrow pH gradient from pH 3.5 - 5.o. The
silver stain (Fig. 2A) gives additional information to the
Coomassie staining (Fig. 2B). The trypsin inhibitors (Fig. 2C),
which in conventional IEF have not shown any variety specifity,
display a number of unexpected charge heterogeneities.

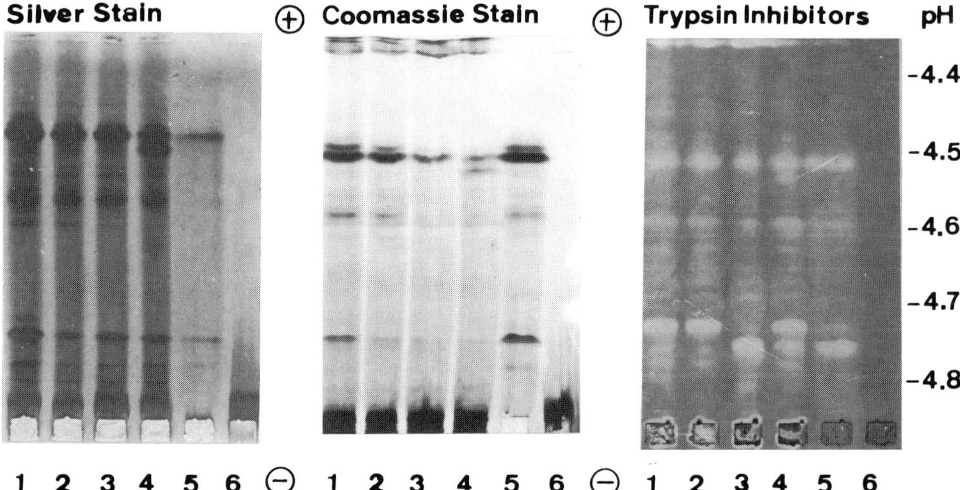

Fig. 2 IEF of soy bean (Glycine max) proteins in immobilized pH gradient
pH 4.4o - 4.85. Samples: (1) - (4) soy bean varieties: (1) Merit,
(2) Cosor, (3) Gieso, (4) Caloria; (5) soy bean trypsin inhibitor isolate,
(6) commercial soy bean protein isolate. IEF overnight, 25oo V, 1o $^{\circ}$C.
pH gradient measured with a surface electrode.

Zymogram stains can be developed in conventional way. Esterase
zymograms from different wheat var. (Triticum aestivum and T.
durum) and rye (Secale cereale) are shown in Fig. 3. Flattening the immobilized pH gradient from pH 5 - 7 (Fig. 3A) to
pH 4.5 - 5.5 (Fig. 3B) demonstrates the improved resolving
power of narrow pH intervals: most of the enzyme bands split

into double bands. T. durum is easily distinguished from
T. aestivum.

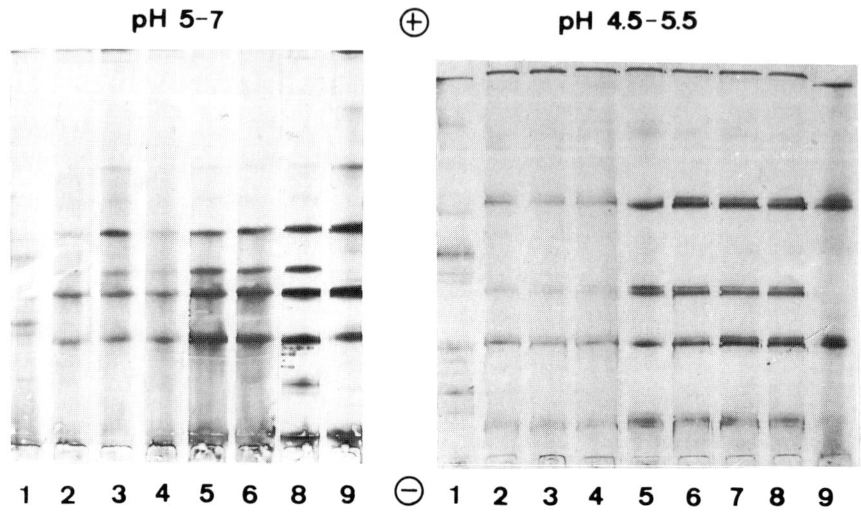

Fig. 3 IEF of wheat and rye esterase isoenzymes in immobilized pH gradients. Resolution was increased by flattening the pH gradient from pH 5 - 7 to pH 4.5 - 5.5. Samples: (1) rye flour, (2) Graham flour, (3) wheat pollard, (4) wheat flour, (5) - (9) wheat varieties: (5) Aureum, (6) Horizont, (7) Arkas, (8) Kolibri, (9) Durum, IEF overnight, 25oo V, 1o °C.

2D-electrophoresis typically exhibit hundreds or even thousands of separated protein spots. Constancy of spot position and intensity in different experiments is a prior condition for any genetic or clinical study. The reproducibility of present day 2D electrophoresis is the result of a combination of different standardizing procedures which have to be carefully followed. One problem is that the first dimension of 2D electrophoresis, the IEF with carrier ampholytes is not a true equilibrium method, therefore the focusing pattern are time dependend. The pH gradients in IEF with carrier ampholytes show a cathodic drift, which results in a net movement of protein zones towards the cathode. This becomes obvious, when the focusing time is long; for example, when the IEF is performed in narrow pH gradients. The stability of immobilized pH gradients offers new possibilities to improve the reproducibility of 2D electrophoresis.

In general, 2D separations with IEF in immobilized pH gradients are performed in the same way as conventional 2D electrophoresis. When pH intervals of the acidic or basic extreme
are used, horizontal streaking of the spots in the second
dimension are observed. There can be two reasons: (1) the IEF
run has not been completed (too much background), (2) the
electroosmosis of immobilized pH gradients becomes more
pronounced and disturbs the migration of proteins from the
first to the second dimension. Point (1) can be solved by
prolonging the focusing time; point (2) is solved by adding
a "starting gel" to the second dimension gel. The starting
gel solution, which is overlaid to the electrophoresis gel
solution, contains 2.5 mM of each Immobiline used in the
first dimension (IEF).

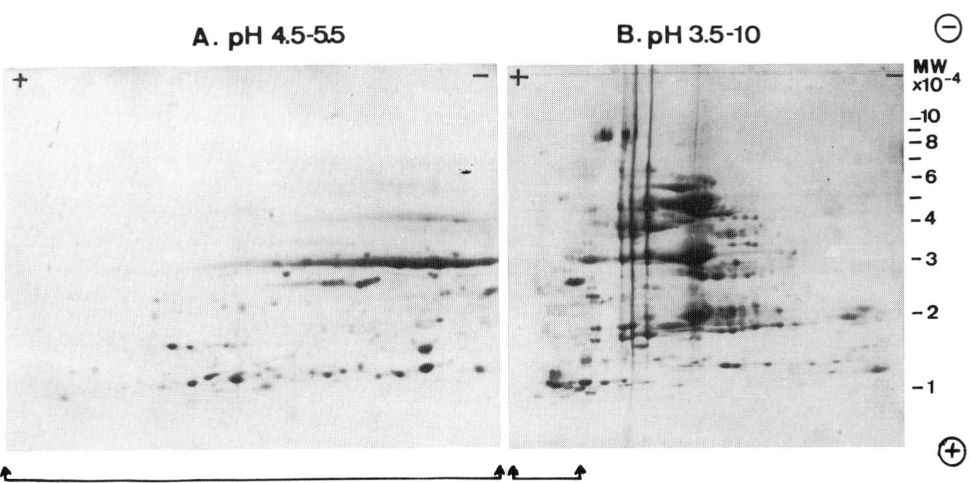

Fig. 4 Horizontal high-resolution 2D electrophoresis of seed proteins of
broad bean (Vicia faba). First dimension: (A) IEF in immobilized pH gradient 4.5 - 5.5, (B) IEF with carrier ampholytes pH 3.5 - 1o. Second
dimension: Ultrathin-layer SDS pore gradient gel electrophoresis. PAG:
T = 1o - 22.5 %, C = 2.5 %. Separation: 2.5 h, 5 °C, max 5o mA, max 3o W,
max 1ooo V. Staining with Coomassie BB R-25o.

An example for the application of IEF in narrow immobilized
pH gradients to high-resolution 2D electrophoresis is demonstrated in figure 4. Broad bean seed proteins are focused in
an Ampholine pH gradient 3.5 - 1o (B) and in an immobilized

pH gradient 4.5 - 5.5 (A), and subsequently separated by SDS pore gradient electrophoresis. Thus, the spot pattern in the section between pH 4.5 and 5.5 is extremely high resolved, when the narrow immobilized pH gradient is used for IEF in the first dimension. - In order to increase the resolution of 2D electrophoresis spot patterns, there is a growing trend towards using longer separation distances compared to those employed nowadays. This means in practice, that the IEF separations will be performed in a flat pH gradient relatively to the absolute migration distances. Because focusing time increases exponentially with the extension of the separation distance, a pH gradient, which is stable with time, is essential. When immobilized pH gradients are used for isoelectric focusing, flattening of a gradient and prolongation of the focusing time are not a limiting factor for the reproducibility of 2D maps.

References

1. Bjellqvist, B., Ek, K., Righetti, P.G., Gianazza, E., Görg, A., Postel, W.: In Electrophoresis '82, Stathakos, D., ed., de Gruyter, Berlin (1983).
2. Righetti, P.G., Gianazza, E., Bjellqvist, B., Ek, K., Görg, A., Westermeier, R.: In Electrophoresis '82, Stathakos, D., ed., de Gruyter, Berlin (1983).
3. Görg, A., Postel, W., Westermeier, R., Bjellqvist, B., Ek, K., Gianazza, E., Righetti, P.G.: In Electrophoresis '82, Stathakos, D., ed., de Gruyter, Berlin (1983).
4. Gasparic, V., Bjellqvist, B. and Rosengren, A., Swedish patent 14o, 49-1 (1975).
5. Bjellqvist, B., Ek, K., Righetti, P.G., Gianazza, E., Görg, A., Westermeier, R., Postel, W.: J. Biochem. Biophys. Methods 6, 317-339 (1982).
6. Görg, A., Postel, W., Westermeier, R., Weser, J.: Science Tools 29, 23-24 (1982).
7. Görg, A., Postel, W., Westermeier, R., Gianazza, E., Righetti, P.G.: J. Biochem. Biophys. Methods 3, 273-284 (198o).
8. Görg, A., Postel, W., Westermeier, R., Righetti, P.G., Ek, K. LKB Application Note 32o (1981).
9. Görg, A., Postel, W., Weser, J., Weidinger, S., Patutschnick, W., Cleve, H.: Electrophoresis 4, 153-157 (1983).
1o. Cleve, H., Patutschnick, W., Postel, W., Weser, J., Görg, A.: Electrophoresis 3, 342-345 (1982).
11. Görg, A., Weser, J., Westermeier, R., Postel, W., Weidinger, S., Patutschnick, W., Cleve, H.: Submitted for publication.
12. Bjellqvist, B., Ek, K.: LKB Application Note 321 (1982).
13. Merril, C.R., Goldman, D., Sedman, S.A., Ebert, M.H.: Science 211, 1437-1438 (1981).
14. Postel, W., Görg, A., Westermeier, R.: Lebensm.-Wiss. u. Technol. 11, 2o2-2o5 (1978).

GENERATION OF HIGHLY-REPRODUCIBLE, EXTENDED pH INTERVALS IN IMMOBILINE GELS

Pier Giorgio Righetti, Elisabetta Gianazza, Giulio Dossi, Fabrizio Celentano
Departments of Biomedical Sciences and Technologies and of Biology, University of Milano, Via Celoria 2 and 26, Milano 20133, Italy

Bengt Bjellqvist, Kristina Ek, Bengt Sahlin and Conny Eklund
LKB Produkter AB, Box 305, S-161 26 Bromma, Sweden

Introduction

Immobilized pH gradients (IPG) represent an entirely new concept in modern separation techniques, rather than being a simple variant of conventional isoelectric focusing (IEF) (1). By grafting the non-amphoteric buffers generating the pH gradient to the anticonvective support (a polyacrylamide matrix) all the troublesome drawbacks and limitations of IEF in amphoteric buffers have been at once eliminated. No more 'cathodic drift', nor salt disturbances, nor near-isoelectric smears or isoelectric precipitation. Moreover the physico-chemical parameters under which the IEF separation is carried out, namely ionic strength (I) and buffering power (β), could be controlled and varied at whim. Such an advanced form of 'pH gradient engineering' is quite unique and cannot be obtained in conventional IEF or in buffer focusing.

IPG's, so far, have been used mostly for generating ultranarrow pH gradients (e.g. 0.1 pH units), where a resolution of $\Delta pI=0.001$ pH could be achieved. There are cases, however, when it might be advantageous to mix two or more buffering Immobilines, in order to cover wider pH intervals, to be used as the first dimension of two-dimensional (2-D) techniques. 2-D maps are most sensitive to disturbances in the first dimension (e.g. cathodic drift, near isoelectric precipitation) which lead to altered or blurred spots in the final 2-D plane. The unsensitivity of grafted pH gradients to such disturbances, and the ease of control of form and width of these gradients, make them the natural choice for this application. There appear to be three ways

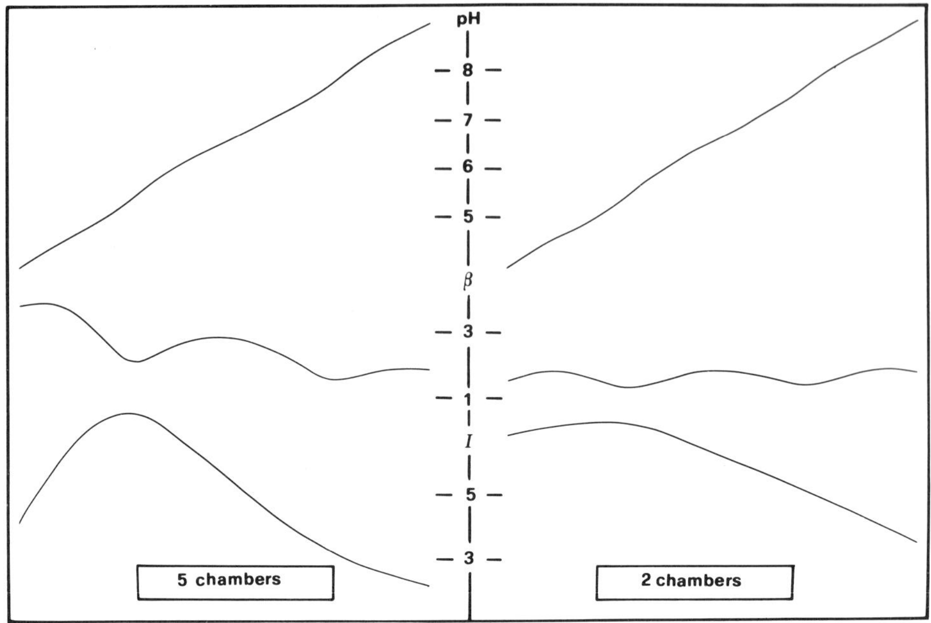

Fig. 1.   Comparison of the properties (pH course, upper; buffering power, middle; ionic strength, lower) of IPG's covering the pH range 4-9. The gradient on the left was eluted from a 5 chamber mixer. Each chamber contained two buffering Immobilines and their titrant(s); the formulation was improved through trial and error. For a volume of 10 ml/chamber the mixture composition was: chamber 1: 225 μl pK 3.6, 225 μl pK 4.6, 207 μl pK 9.3; chamber 2: 250 μl pK 3.6, 275 μl pK 4.6, 275 μl pK 6.2; 228 μl pK 9.3; chamber 3: 205 μl pK 3.6, 250μl pK 6.2, 200 μl pK 7.0; chamber 4: 140 μl pK 3.6, 200 μl pK 7.0, 175 μl pK 8.5; chamber 5: 110 μl pK 3.6, 125 μl pK 8.5, 100 μl pK 9.3. The gradient at the right was from a 2 chamber mixer. The solutions in both vessels contained all the buffering species in identical concentrations and were titrated to the required pH extremes. For 20 ml total volume the mother solution contained: 175 μl pK 3.6, 427 μl pK 4.6, 331 μl pK 6.2, 268 μl pK 7.0, 415 μl pK 8.5, 175 μl pK 9.3; to 10 ml acidic solution, 490 μl pK 0.8 and to 10 ml basic solution 188 μl pK 10 were added. The major differences in performance are that the 5-chamber system does not require titrants (i.e. compounds with pK's at least one pH unit below or above the pH interval), and offers a substantial saving of Immobiline (one third less for the same β_{min}), while the 2-chamber system allows for the easiest experimental approach and is sufficiently resistant to minor mistakes in the preparation of the solutions.

for producing wide (>2 pH units) pH gradients with Immobilines: a) multichamber mixers; b) two-chamber mixers with identical molarities of buffering species and varying concentrations of titrants (3); c) two-chamber mixers containing different amounts of the same Immobiline species. We will briefly review these three approaches, while referring the reader to previ-

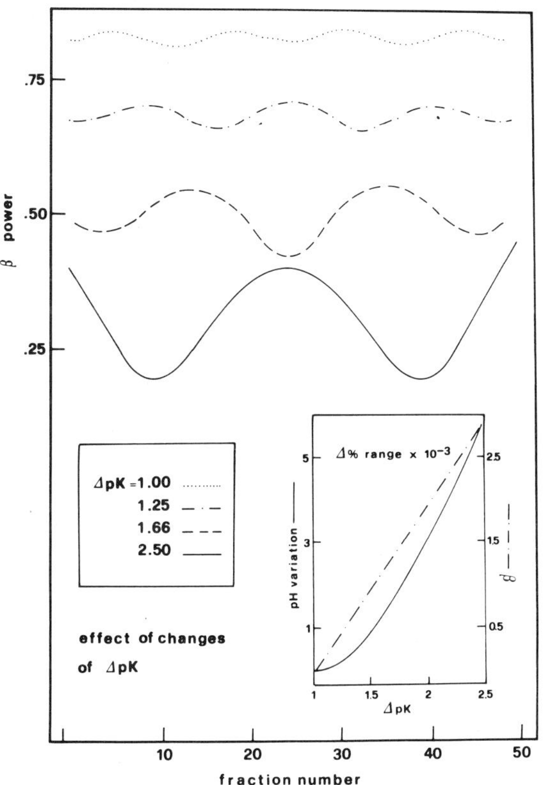

Fig. 2. The effect of changes in the number of buffering species was investigated by simulating the course of β power in cases where either 6, 5, 4 or 3 bases with evenly spaced pK's were used to cover a 5 pH unit interval (whose extremes coincided with the highest and lowest pK's). The deeps between the pK's progressively smoothen and almost level out for ΔpK=1. The course of the pH gradient also becomes more and more linear, as summarized in the insert. For ΔpK=1 the maximal deviation equals 0.0025 pH units; for ΔpK=1.25 it is still as good as 0.0084 pH units; for ΔpK=1.33 it is 4.28/100 of a pH unit while for ΔpK=2.0 the deviation is quite marked, 0.1547 pH units.

ous articles (1-4) for experimental details, beheaded here for lack of space.

Results and Discussion

Multichamber mixers. By using a five-chambered mixing device, containing varying amounts of buffering Immobilines, titrated to appropriate pH's with non buffering species, we have generated a five pH unit interval (pH 4-9), with minimum deviation from linearity (Fig. 1, left). In connection with this, a computer program was developed which would calculate, at any desired pH increment, the accompanying I and β courses (2).

Two-chamber mixers with identical molarities. Since multichamber devices might be quite cumbersome in daily practice, we have reverted to the use of two-vessel mixers, with the proviso that they would contain the same buffering species, in identical concentrations, in both chambers, and would be titrated, with the aid of non buffering Immobilines (fully dissociated

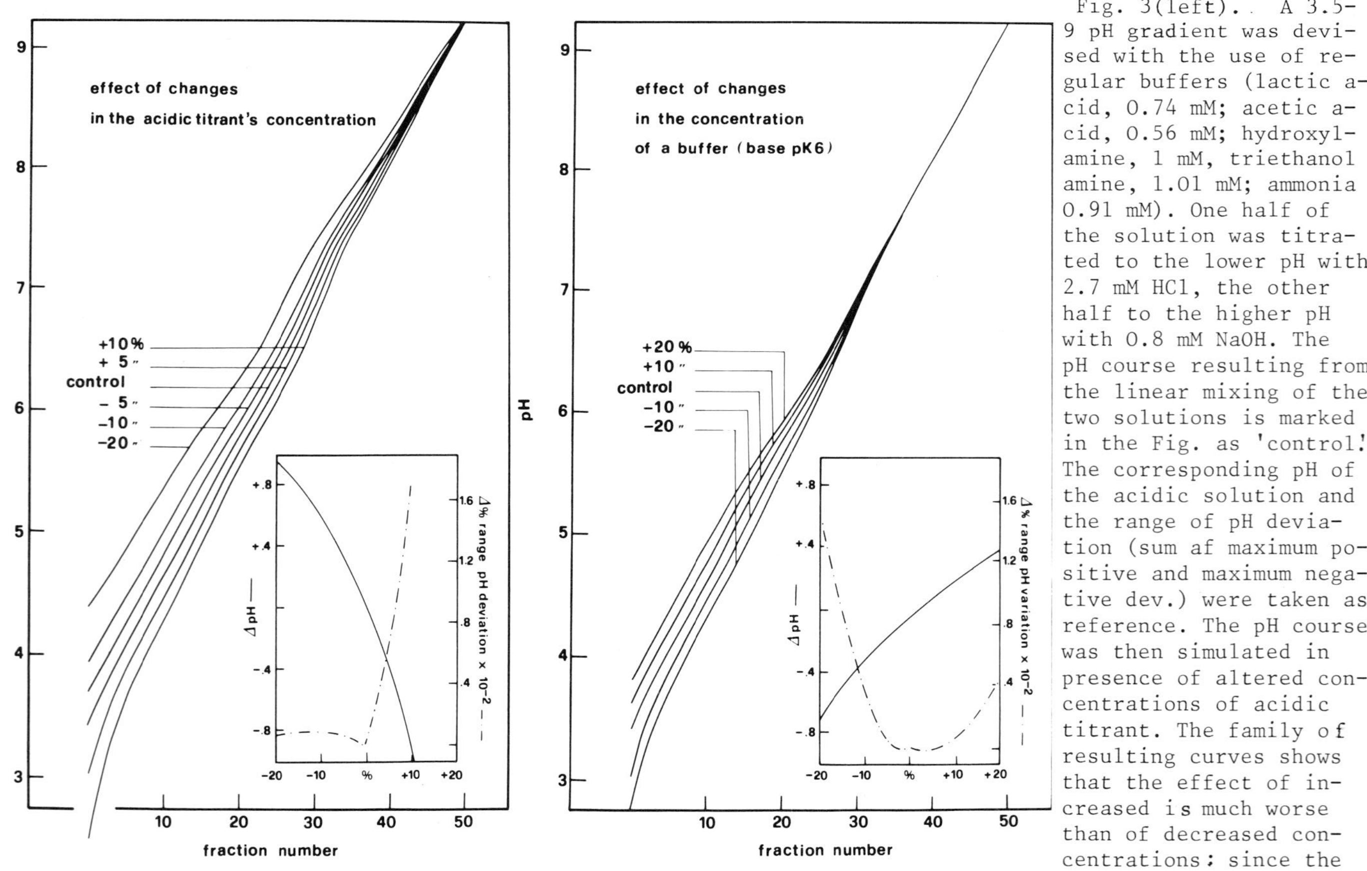

Fig. 3(left). A 3.5-9 pH gradient was devised with the use of regular buffers (lactic acid, 0.74 mM; acetic acid, 0.56 mM; hydroxylamine, 1 mM, triethanol amine, 1.01 mM; ammonia 0.91 mM). One half of the solution was titrated to the lower pH with 2.7 mM HCl, the other half to the higher pH with 0.8 mM NaOH. The pH course resulting from the linear mixing of the two solutions is marked in the Fig. as 'control.' The corresponding pH of the acidic solution and the range of pH deviation (sum af maximum positive and maximum negative dev.) were taken as reference. The pH course was then simulated in presence of altered concentrations of acidic titrant. The family of resulting curves shows that the effect of increased is much worse than of decreased concentrations: since the buffering power of the solution is exceeded, a loss of linearity results (sharp increase of Δ% range deviation, see insert, while in the symmetric case, the gradient only shifts around the alkaline estreme).

Fig. 4 (right). The gradient described in Fig. 3 was tested for its response to alterations in the concen-

(Fig. 4, continued): tration of a buffer (hydroxylamine, pK=6.03). As in the case of Fig. 3, the curves fan out from a single point, but this corresponds in the present case to pH ca. 7.5, i.e. 1.5 pH units above the pK of the relevant base, where its protonation becomes negligible. As this point is midway along the gradient, the linearity of the pH curve decreases **sharply for** any change in the base concentration (see Δ% range pH deviation in the inset) while the shift of the pH of the acidic solution is much lower than in the acid's case (insert).

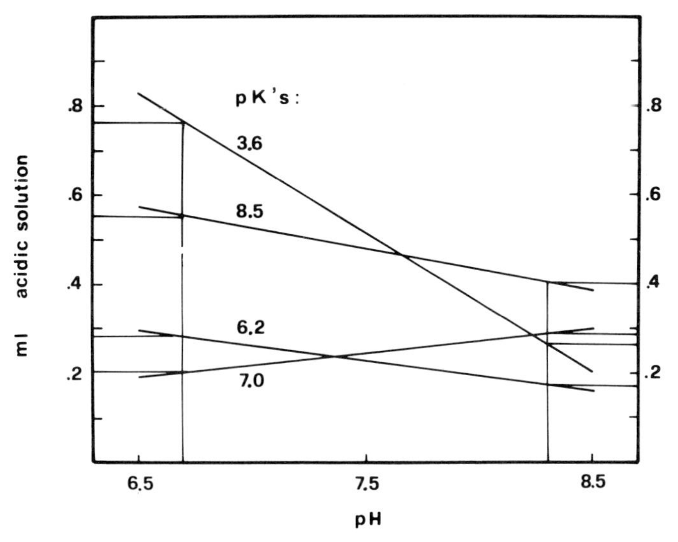

Fig. 5. This is an example of how, from the data of Table I, the solutions' formulation for any narrower gradient centered around the same middle value of pH can be graphycally calculated. Segments are drawn connecting the amounts required for each component in the limiting solutions for the wide interval. The new composition corresponds to the intercept on these segments of vertical lines at the new pH extremes.

in the entire pH interval) to the two extremes of the desired pH range. The results of this approach are seen in Fig. 1 (right): here the *I* and β courses are smoothened since an optimization algorithm has been introduced in our computer program (3).

Computer modelling. Such a sophisticated technique would be misused if one could not control the experimental parameters and the possible sources of errors connected with the dispensing of the Immobiline chemicals. Since a constant β power appears fundamental for generating linear pH gradients with the above approach, we have simulated the behaviour of the β course as a function of ΔpK's of the buffers. As shown in Fig. 2, the smoothest β power is obtained at ΔpK=1; larger or narrower ΔpK's would generate hills and valleys, with concomitant deviations of pH linearity in their proximity (see insert). It is also of interest to simulate errors in the concentrations of titrants or buffering species, in order to evaluate their ef-

TABLE I

PREPARATION OF 2 AND 3 pH UNIT INTERVALS

for 15 ml of acidic solution						μl Immobiline 0.2 M	for 15 ml of basic solution					
pK 3.6	pK 4.6	pK 6.2	pK 7.0	pK 8.5	pK 9.3	pH range (10°C)	pK 3.6	pK 4.6	pK 6.2	pK 7.0	pK 8.5	pK 9.3
394	285	161	–	–	–	3.5 – 5	263	398	589	–	–	–
619	107	473	–	–	–	4 – 6	422	563	299	–	–	782
450	261	540	–	–	–	4.5 – 6.5	–	615	263	255	–	323
69	431	414	–	–	–	5 –7	–	483	276	224	–	328
–	440	347	111	–	–	5.5 – 7.5	340	–	232	281	278	–
441	–	333	203	45	–	6 – 8	293	–	178	333	338	–
824	–	294	194	576	–	6.5 – 8.5	205	–	164	302	387	–
1378	–	–	277	380	863	7 – 9	491	–	–	236	194	557
722	–	–	480	244	375	7.5 – 9.5	221	–	–	998	150	375
350	–	–	325	308	84	8 – 10	81	–	–	293	325	254
619	107	473	–	–	–	4 – 7	323	791	161	287	–	940
799	285	469	150	394	–	5 – 8	199	141	150	394	394	–
728	–	375	86	338	75	6 – 9	225	–	150	420	225	210
558	–	–	389	361	–	7 – 10	92	–	–	333	361	289

Fig. 6. IPG in the pH 7-9 range. (for its composition, see Table I, 8th entry). Samples (50 µg each): 1: conalbumin; 2: hemoglobin; 3: horse heart myoglobin; 4: sperm whale myoglobin; 5: creatine phosphokinase; 6: concanavaline A; Electrophoretic conditions: overnight at 2500 V, 10°C. Staining: Coomassie Brilliant Blue R-250 in acid alcohol solvent.

fects on the pH gradient profile. When a positive or negative error in the concentration of acidic titrant is made, the acid end of the pH gradient is mostly affected (Fig. 3). However a positive error (+10% titrant) has an effect twice as great as a negative error (-20%). When the concentration is altered in one of the buffering species (in this case, a pK 6 base), a similar family of curves, fanning out from a single point, is generated, but having their confluence point at considerably lower pH values (in this case pH 7.5) where the base would be mostly deprotonated (Fig. 4). It should be emphasized here that once these errors, made and revealed by a change in pH of the solutions, are corrected for by titration, the resulting pH gradient is essentially identical to the one obtainable in the absence of such errors.

Two chamber mixers with different molarities. This is the latest approach tried. At present, 10 different, 2 pH unit intervals and four different, 3 pH unit spans have been calculated (see Table I for their composition, given in µl of each Immobiline, in a 0.2M solution, to be added to 15 ml volume in each chamber). One of the advantages of this approach is that, once the molarity courses of the given Immobilines needed to generate a particular pH span are plotted, any narrower pH interval within these outer limits can be easily calculated (Fig. 5). Fig. 6 gives an example of the separations obtainable in a two pH unit interval.

Conclusions. pH gradient engineering has finally become a reality. Separations can now be carried out in a fully defined and controlled chemical milieu (β, I) with a pH gradient tailored to fit exactly any separation problem. The complete reproducibility of pH gradients and abolition of 'cathodic drift' are also particularly attractive in 2-D separations. The Achille's heel of 2-D maps has been in fact the IEF dimension, due to irreproducibility of spot position along the pI axis. This has forced scientists to introduce a locally shifting coordinate system, based on the relative position, in terms of pI and M_r, of a spot in relation to major, dominant spots in the map. With IPG's, 2-D maps can now be interpreted in terms of a rigid coordinate system (manuscript in preparation).

Acknowledgements

Supported by a grant from Progetto Finalizzato 'Salute dell'Uomo', sottoprogetto 'Meccanismi di invecchiamento (CNR, Roma).

References

1. Bjellqvist, B., Ek., K., Righetti, P.G., Gianazza, E., Görg, A., Westermeier, R. and Postel, W.: J. Biochem. Biophys. Methods 6, 317-339 (1982).

2. Dossi, G., Celentano, F., Gianazza, E. and Righetti, P.G.: J. Biochem. Biophys. Methods 7, 123-142 (1983).

3. Gianazza, E., Dossi, G., Celentano, F. and Righetti, P.G.: J. Biochem. Biophys. Methods (1983) in press.

4. Bjellqvist, B. and Ek, K.: LKB Application Note No. 321 (1982).

STUDY OF ISOTACHOPHORESIS CONDITIONS FOR SERUM PROTEIN ANALYSIS

Takashi Hine

Analytical Applications Department, Shimadzu Corporation,
Chofu, Tokyo 182, Japan

Introduction

Capillary isotachophoresis is a useful method for the analysis
of various ion components of smaller or larger molecules. The
analysis of serum protein with capillary isotachophoresis was
reviewed by Delmotte(1,2) or Holloway(3). In the analysis of
serum proteins, they described how the separation of serum
proteins could be tremendously improved by the addition of spacer
ions such as carrier ampholytes or amino acids. However, many
experiments of leading ions and viscous agents on the separation
have not yet been performed.
This report deals with the effects of viscous agents, leading
ions and spacer ions on the separation of serum proteins to
obtain satisfactory and repeatable results, using a data
processor.

Materials and Methods

Analyses were performed with the Shimadzu capillary isotachophoresis IP-2A fitted with two migration tubes of 1.0mm I.D.,
100mm length and of 0.5mm I.D., 300mm length in serial connection. The migration current was kept at 150uA for 17min. and
then decreased to 75uA.
The leading electrolyte system consisted of 4.5mM hydrochloric
acid, lactic acid and MES with ammediol as the counter ion. The
viscous agents such as HPMC(Hydroxy Propyl Methyl Cellulose),

PVA(Poly Vinyl Alcohol), HEC(Hydroxy Ethyl Cellulose) and Triton (Triton X-100) were added to the leading electrolyte. The terminating electrolyte was 10mM 6-amino caproic acid with 10mM ammediol as buffer, corrected to pH10.8 by the addition of barium hydroxide.

For the spacer ions, 0.5ml ampholine pH3.5-10.0, 40%(w/v) solution, was made up to 10ml with distilled water. This solution was used for the spacer solution(A). On the other hand, 0.5ml ampholine pH3.5-10.0, 40%(w/v) solution, 1.0ml amino acid mixture of asparagine, threonine, methionine, glycine, valine, leucine and beta-alanine, 1.0mM solution, and 0.1ml tryptophan as the internal standard, 0.5mM solution, were made up to 10ml with distilled water. This solution was used for the spacer solution(B).

For the samples, the control serum I(Q-pack I) was obtained from Hyland diagnostic Co.(IL., U.S.A.) and the pI marker was obtained from Oriental yeast Co.(Tokyo, Japan). Albumin, transferrin and γ-globulin as the standard sample were obtained from Sigma chemistry Co.(MO., U.S.A.).

These samples were mixed with the spacer solution(A) or (B) at volume ratio 1 to 3 and 1.5ul injected. The samples were detected by UV-signal at 280nm. The isotachopherogram and the peak area of the sample are described by the Shimadzu chromatopac C-R2AX as a data processor.

Results and Discussion

In investigating the effect of viscous agents in the leading electrolyte on the separation of pI marker in the spacer solution(A), to avoid adsorption of the sample on tubes, all migration tubes were changed each time a sample was studied. HPMC, PVA, HEC and Triton as viscous agent were added to the leading electrolyte and the concentration was changed from 0.05% to 0.20% over 0.05% increments. The leading electrolyte was 4.5mM hydrocloric acid corrected to pH9.1 by the addition

of ammediol as counter ion. As shown in Fig.1(a), (c), 0.1%
HPMC and 0.2% PVA resulted in the most specific separation of pI
marker. In addition, HPMC showed a smaller noise level than PVA.
HEC, Triton and non-additive in the leading electrolyte resulted
in the indistinct separation shown in Fig.1(b),(e),(f). However
as shown in Fig.1(d), when all migration tubes were not changed
after using the HPMC in the leading electrolyte and Triton was
added, the pI marker was clearly separated. This showed the
"memory effect" with the HPMC on the migration tubes.(4)

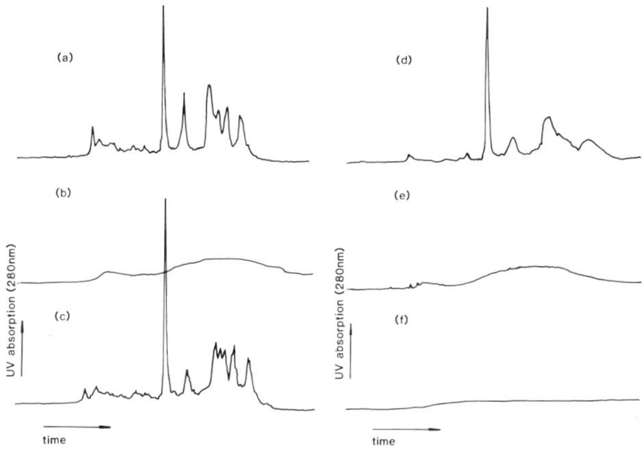

Fig.1 Effect of viscous agents on the separation of pI
       marker in the spacer solution(A).
       (a) 0.1% HPMC, (b) 0.1% HEC, (c) 0.2% PVA,
       (d) 0.2% Triton after using HPMC,
       (e) 0.2% Triton, (f) non-additive

Fig.2 shows the effect of leading ions such as chloride, lactate,
and MES ion on the separation of serum in spacer solution(A).
The leading electrolyte was 4.5mM chloride, lactate or MES ion
with added 0.1% HPMC as viscous agent, corrected to pH9.1 by the
addition of ammediol. The separation of proteins which have
large mobilities, for example albumin, was changed for the better
as the mobility of leading ion became smaller. The MES has the
smallest mobility of the three leading ions carried out with the
exclusion of ion components of larger mobility than albumin.
The MES also obtained the detection of proteins at the highest
sensitivity.
The effect of spacer ion on the reliable separation and repeat-

544

able peak area by using a data processor was investigated. The
leading electrolyte was 4.5mM MES and 10mM ammediol with added

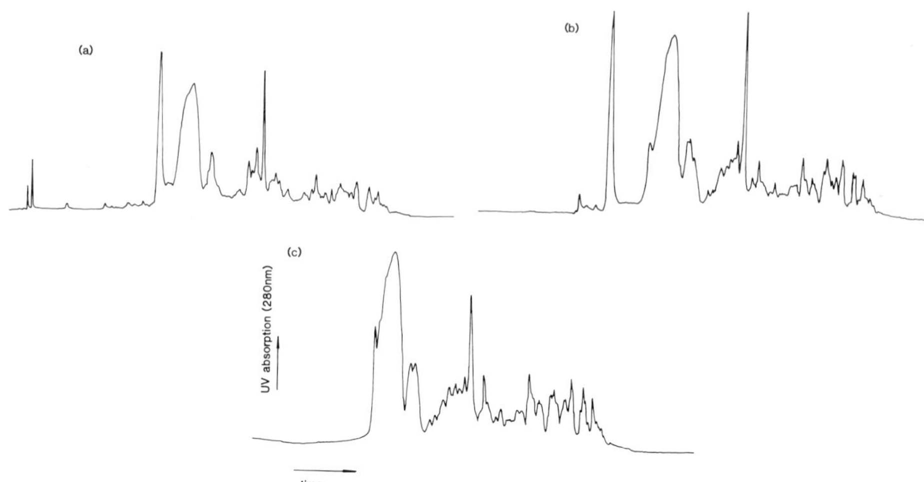

Fig.2 Effect of leading ions on the separation of serum in
       the spacer solution(A).
       (a) chloride ion, (b) lactate ion, (c) MES ion

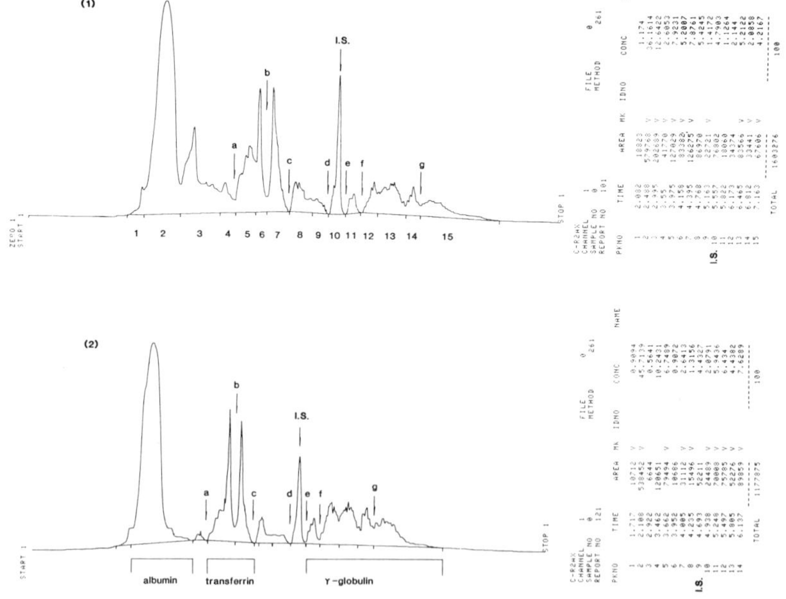

Fig.3 Isotachopherograms of serum and standard protein in
       the spacer solution(B).
       (1) serum, (2) standard protein (15ug albumin, 2.5ug
       transferrin, 5ug γ-globulin)

0.1% HPMC. When the serum was mixed with the spacer solution (A), the peaks of protein were difficult to bring down to the baseline as shown in Fig.2(c). Therefore the data processor could not reliably integrate the peaks and obtain a repeatable peak area. By using the spacer solution(B), amino acids as spacer ion were inserted at the various points(a-g) as shown in Fig.3(1) and the repeatability of peak area was improved. However the number of total peaks decreased when too much amino acids were added to the leading electrolyte.

The repeatability of peak area of serum protein, expressed in C.V(%), was obtained under the optimum condition that the leading electrolyte was 4.5mM MES and 10mM ammediol (about pH9.0) with added 0.1% HPMC and the spacer ion was the spacer solution(B). The C.V(%) values were 5.80,3.26, 2.64, 1.00, 2.53, 2.56, 4.69, 1.67, 7.81, 2.58, 4.28, 3.36, 4.09 and 3.07 for the fractions Nos.1-15 excepting No.10 as shown in Fig.3(1). The calibration curves were linear up to 22.5ug of albumin, up to 7.5ug of γ-globulin and up to 2.5ug of transferrin.

References

1. Delmotte, P.: Science Tools 24, 33-41 (1977)
2. Delmotte, P.: Biochemical and Biological Applications of Isotachophoresis, Elsevier Scientfic Publishing Co., Amsterdam, 259-265 (1980)
3. Holloway, C.J., Heil, W., Henkel, E.: Electrophoresis, Walter de Gruyter & Co., Berlin, New York, 753-765 (1981)
4. Everaerts, F.M., Beckers, J.L., Verheggen, P.E.M.: Isotachophoresis, Elsevier Scientific Publishing Co., Amsterdam, 171-190 (1976)

A NEW PREPARATIVE ISOELECTRIC FOCUSING APPARATUS

N. B. Egen, W. Thormann, G. E. Twitty and M. Bier
Biophysics Laboratory, University of Arizona
Tucson, Arizona, 85721, U. S. A.

Introduction

The high resolution of isoelectric focusing (IEF) makes it
a very attractive process for protein isolation. The RIEF
(Recycling IEF) apparatus was designed for potentially
industrial applications (1,2). The protein solution to be
fractionated is rapidly recycled through a multicompartment
focusing cell, individual components of the mixture migrating
in each pass to their equilibrium position in the pH gradient.
The construction of the RIEF is modular, permitting its
scaling up to any desired capacity. With our laboratory
unit, protein volumes in the range of 200 to 10,000 ml can
be processed in a few hours, best resolution being of the
order of 0.1 pH units (3).

The RIEF apparatus has been applied to a wide range of
proteins submitted to us by various collaborators: monoclonal
antibodies, vaccines, snake venoms, interferon, peptide
hormones, enzymes, etc. In some instances, however, but
minute quantities of sample were available, resulting in an
inefficient use of the RIEF.

This has prompted us to develop a smaller preparative focusing
apparatus. The new device separates a 45 ml sample into 20
fractions. As in the RIEF, the focusing cell is segmented
into subcompartments by means of screens of woven monofilament
nylon. The main difference from the RIEF is that recycling

548

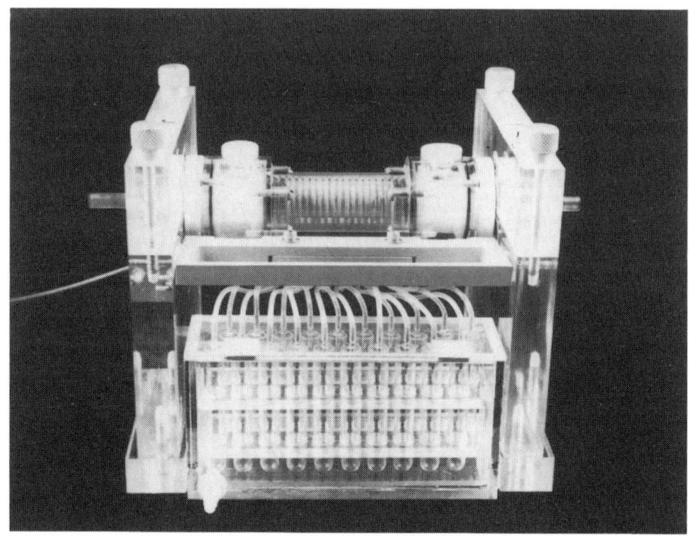

Fig. 1. Photograph of the Rotofor and its fraction collector.
Cooling fluid and electrical cables are not connected.

has been replaced by rotation of the apparatus about its
horizontal axis. Hjerten (4) has demonstrated the usefulness
of rotation for prevention of gravity induced convection in
free-fluid electrophoresis. Hjerten's apparatus was limited
to tubes of narrow diameter. Our monofilament screens
contribute further to fluid stabilization and minimize
electroosmosis. This combination of stabilizing factors has
permitted us to substantially increase the internal diameter.

Apparatus Description

A photograph of our apparatus, named Rotofor, is presented
in Fig. 1. The salient features are:

     i. The cylindrical plexiglass focusing cell comprises
a main focusing body and two detachable electrode compartments.

     ii. Cooling liquid is recirculated through a central
glass tube, extending beyond the full length of the column.

Fig. 2. Focusing chamber of the Rotofor with an expanded view of the nylon screens supported by the cooling tube.

    iii. The apparatus is rotated by means of a synchronous motor and gear assembly at 6.5 rpm;

    iv. Samples are collected by means of a vacuum device.

An expanded view of the apparatus is shown in Fig. 2. The nylon monofilament screens, porosity 6 μm, are reinforced by silicone rubber O-rings. These define 20 subcompartments, each accessed by small inlet and outlet ports, clearly visible in the photograph. The electrode compartments are separated from the focusing subcompartments by ion exchange membranes. The electrolyte filling ports are sealed with hydrophobic membranes, permitting venting of gases formed by electrolysis. The apparatus rotates on metallic gears which also carry electric power to the platinum electrodes.

The apparatus is filled by means of a syringe, through the individual inlet ports. During focusing, all inlet and outlet ports are sealed by strips of adhesive tape. For

collection of fractions at the end of a run, rotation is
stopped. A microswitch automatically aligns the exit ports
over a matching array of syringe needles. These feed to the
collecting test tubes, contained in a vacuum box. The upper
tape is removed and a lever is pressed, forcing the needles
to penetrate the bottom adhesive tape, simultaneously and
rapidly aspirating all twenty samples.

Results

Human hemoglobin variants HbA (pI = 7.2) and HbC (pI = 7.8)
are convenient standards for assessing the resolution in
IEF. These can be easily separated in either the RIEF or
the Rotofor, intermediate fractions usually containing some
minor components of the starting mixture. The Rotofor can
also be utilized as a second step, to further purify individual
fractions of the RIEF apparatus. For example, 250 ml of a
liver homogenate were fractionated in the RIEF, two of its
ten fractions were pooled and further run in the Rotofor,
without addition of carrier ampholyte. Whereas the original
homogenate exhibited a continuum of bands from pH 3.5-9,
most Rotofor fractions contained only one or two components.

References

1. Bier, M. and Egen, N.B. in Electrofocus/78, (Haglung,
H., Westerfeld, J.G., and Ball, J.T., Eds.), Elsevier/North-
Holland, New York 1979, p. 35.

2. Bier, M., Egen, N.B., Allgyer, T.T., Twitty, G.E. and
Mosher, R.A. in Peptides, (Gross, E. and Meienhofer, J.,
Eds.), Pierce 1979, p. 79.

3. Binion, S.B., Rodkey, L.S., Egen, N.B. and Bier, M.,
Electrophoresis 3, 284 (1982).

4. S. Hjerten: Free Zone Electrophoresis, Almqvist &
Wiksells, Uppsala 1967.

IMPROVEMENTS IN PREPARATIVE ISOELECTRIC FOCUSING IN AGAROSE GELS

R. McLachlan

Cancer Institute, Melbourne, Australia, 3000

When analysing IgGs from sera containing ≥ 2 monotypic IgGs it
is useful to eliminate their initial microheterogeneity and an-
alyse spectrotypically pure protein. The proteins are purified
by preparative isoelectric focusing (pIEF) in agarose gels.

Methods

The basic method is as previously described (1). Thinner gels
(0.1 cm) are poured by capillary action on mylar sheets (Gel-
Bond^{TM}, Marine Colloids) using glass templates (11.0 cm long and
variable width). The sample wells (0.05 cm deep, 0.5 cm long)
are cast into the gels 2 cm from the anode. These gels are easy
to make and avoid the need to remove sample slots. Samples, in
2% ampholyte, are pipetted into the sample wells, the electrodes
placed on the gel and the sample focused (LKB Multiphor 2103).

pIEF in urea gels

Gels are composed of 6.5 M urea, 1% agarose (Pharmacia, IEF
grade), 1% non-crosslinked polyacrylamide (ncPA) and 2%
ampholytes pH 3-10 (Pharmacia). Templates used are similar to
pIEF templates except that the sample wells are 0.025 cm deep.
Focusing proceeds at constant power of 0.03 Watt/cm^2 surface
area for up to 120 V.hr/cm gel length.

Samples used in all experiments were human pathological sera,
purified IgGs or Ig heavy and light chains.

Results and Discussion

Resolution

Experiments were performed to observe the effects of different
final field strengths on resolution of proteins. No difference
in resolution was observed between final field strengths of
27.5 and 110 V/cm (Fig.1). The apparent variance with
Svensson's theory (2) (zone width is inversely proportional
to the square root of the final field strength), is because
diffusion occuring while the imprint is taken obscures the
fine resolution acheived by higher field strengths.

Resolution is improved and endosmosis reduced by the addition
of ncPA. Fig.2 shows two analytical IEF gels and Fig.3 shows
imprints of two pIEF gels. Gels in Figs.2A & 3A contain ncPA,
those in Figs.2B & 3B do not. The most noticeable improvement
in both resolution and reduction of endosmosis occurs in the
proteins of highest pIs.

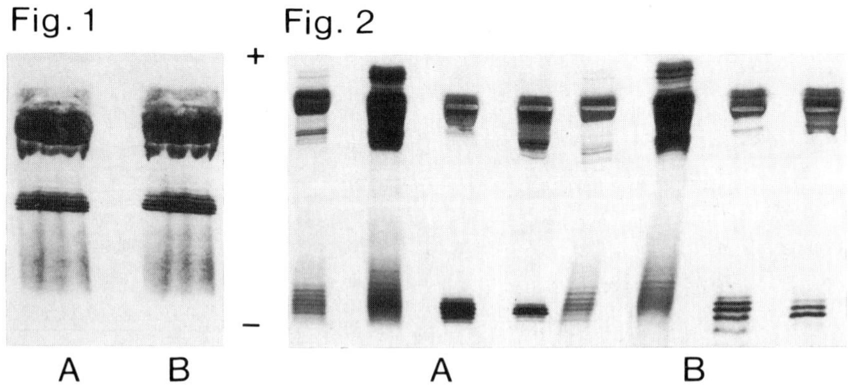

Fig.1. Imprints from pIEF gels run for 40 V.hr/cm final field
strength 55 V/cm, then 10 V.hr/cm at 27.5(A) & 110(B) V/cm.
Sample:- IgG myeloma serum.

Fig.2. Analytical IEF gels with(A) and without(B) ncPA (0.03
Watt/cm^2, 35 V.hr/cm.
Samples:- pathological sera, same samples repeated.

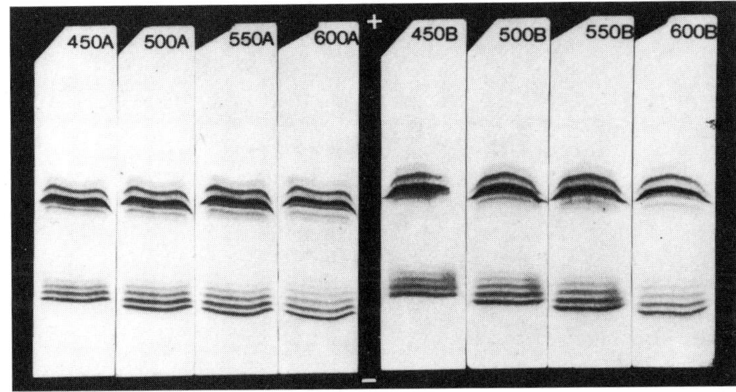

Fig.3.   Imprints from pIEF gels with(A) and without(B) ncPA,
taken at 450-600 V.hr (10 cm electrode gap). Sample:- monotypic
IgG (cathodal) and monotypic free light chains.

Urea gels require much longer V.hr integrals to achieve optimal
resolution. Fig.4 shows a 6.5 M urea, 1% ncPA analytical IEF
gel run for 35, 70, 105 & 140 V.hr/cm. The higher the pI of
the protein, the longer the V.hr integral required, presumably
due to molecular sieving of the denatured protein by the gel.
An imprint taken from a 6.5 M urea pIEF gel, run for 120 V.hr/
cm, is shown in Fig.5.

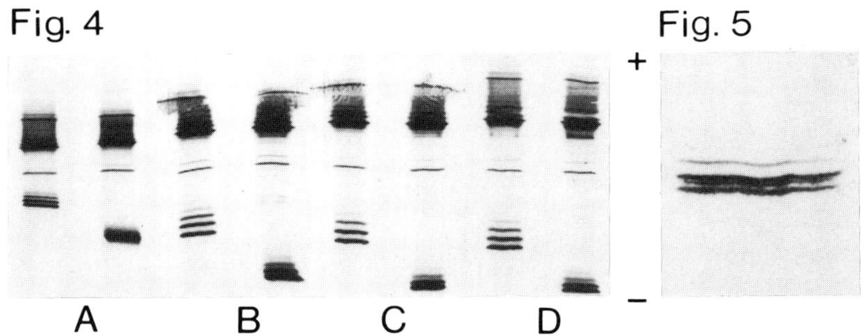

Fig.4.   6.5 M urea analytical IEF gel run for 35(A), 70(B), 105
(C) & 140(D) V.hr/cm. Samples:- two IgG myeloma sera.

Fig.5.   Imprint from a 6.5 M urea pIEF gel (120 V.hr/cm).
Sample:- reduced and alkylated monotypic IgG.

Loading capacity

Loading capacity depends on the amount of protein in each homogeneous band. Monotypic IgGs typically focus as a microhetero-
geneic group of bands. Fig.6 shows a paper imprint of a 9 ml
pIEF gel loaded, tracks A-C, with 67, 200 & 600 µg of monotypic
IgG. The bands in track C contained, from the cathode, 30, 258,
162, 84 & 66 µg of protein. The volume of gel occupied by each
band was approximately 20 µl. Therefore the concentration of
proteins in the two major bands was approximately 13 & 8 µg/µl
focused gel. At 13 µg/µl the gel was overloaded but at 8 µg/
µl it was not. From these and similar experiments, the loading
capacity is approximately 10 µg of protein in a single band/µl
of focused gel. In practice, 2-3 mg of monotypic IgG can be
separated in a 10 ml gel. An equivalent amount of another monotypic IgG or other protein of different pI (eg. Fig.3) may be
separated in the same gel.

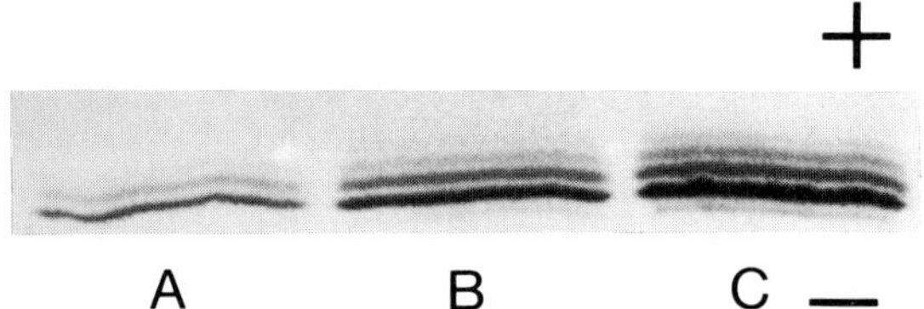

Fig.6. Imprint from pIEF gel (50 V.hr/cm).
Sample:- monotypic IgG, 67(A), 200(B) & 600(C) µg protein.

Overloaded proteins in one part of the gel do not interfere
with the resolution of proteins in other parts. In Fig.1, 3.5
mg of serum protein containing 2.0 mg of albumin and 0.6 mg
of IgG was separated in a 2.5 ml gel. The very large quantity
of albumin did not affect resolution of the monotypic IgG.

References

1. McLachlan, R., & Cornell, F.N.: Electrophoresis '82, Walter
   de Gruyter, Berlin. New York, in press (1983)
2. Svensson, H.: Acta Chem. Scand., 15, 325-341 (1961)

ISOELECTRIC FOCUSING WITHOUT CARRIER AMPHOLYTES

Kazuo Shimao

Department of General Education, Tokyo Medical and Dental University

Ichikawa-shi 272, Japan

Introduction

Isoelectric focusing without carrier ampholytes was discussed by
Rilbe (1). The basic equation of his treatment is not strict, because
the conductance of the equation does not contain terms from diffusional
current. In spite of this, his conclusion applies to experimental results owing to the fact that diffusional flow of electrolytes is very
slow and negligible compared to electrophoretic flow under the usual experimental conditions of electrophoresis. Experiments to prove the above
statement were carried out (2) and the results are shown in this report.

Experimental and results

Typical electrode buffer systems are shown in Table 1, and experimental
procedures in Fig. 1. 1.5 ml of anode buffer (or 1.2 ml of anode buffer
+ 0.3 ml of water) and one drop of pH indicator solution (0.1 % bromphe-
nol blue + 0.1 % methyl red in alcohol) were added to 7 ml of 1.4(or 1.2)
% Agarose IEF solution and the mixture (A) was poured into a frame (B,
100 mm wide x 115 mm long x 1 mm deep) on a GelBond Film (C). Then, the

Table 1. Electrode buffer system.

System No.	1		2		3	
Electrode	Anode	Cathode	Anode	Cathode	Anode	Cathode
HOAc Conc.(M)	0.21	0.07	0.20	0.05	0.25	0.05
Base*Conc.(M)	0.07	0.21	0.05	0.20	0.05	0.25

* Trisaminomethane,Ammediol or imidazole.

556

surface was covered by a glass plate (D) on a slant to which "GelFix for
Covers" (E) had been stuck by a thin water layer. After gelification,
the glass plate was removed and "GelFix for Covers" was peeled off, leav-
ing a wedge-shaped acidic gel in the frame (Fig. 1, step 1). A wedge-
shaped alkaline gel (F) prepared from the cathode buffer was formed on
the acidic gel in the reverse direction (Fig. 1, step 2). Mutual diffusion of electrolytes into acidic and alkaline gel resulted in formation
of uniform concentration gradients of both the weak acid and the weak
base (Fig. 1, step 3). One hour was sufficient for this step, but standing up to 20 hours did not make much difference. The pH indicators gave
a rough estimate of pH gradient in the gel. The sum of concentrations of
the acid and the base was constant throughout the gel, keeping osmotic
flow of water very slow.

Sample solutions were applied in the slots previously formed by putting
filter paper pieces (4mm wide x 6 mm long, Whatman 3MM) vertically on the
gel surface (Fig. 1, step 4). Sample sizes were between 0.2 and 1 µl.
The gel was set on a cooling plate (H), acidic and alkaline ends were
connected to the corresponding electrode buffers (I, I') by filter paper
sheets (4 sheets, 100mm wide x 45mm long, Toyo No.2, J, J'), and the
whole surface was covered by a polythene film (Fig. 1, step 5).
Electrification was carried out at constant current (1.5 to 2 mA per cm
width of the gel) for 120 to 150 min at 6° C.

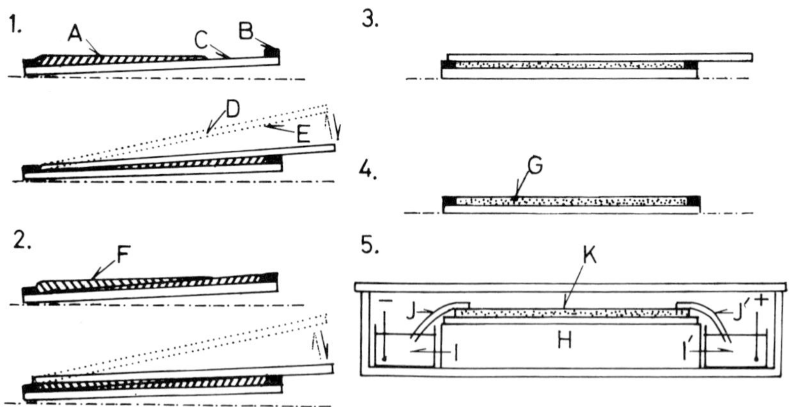

Fig. 1. Experimental procedure.

Protein fractions were stained by Coomassie brilliant blue R-250 (0.1 %)
dissolved in 30 % methanol containing 10 g/dl TCA.

pH of the gel before and after electrification was measured by direct
contact of a combination-type micro glass electrode (Jookoo Ltd., Model
SE-1600 GC) to gel surface.

An example of voltage change during electrification, migration of brom-
phenol blue bound to serum albumin, pH of the gel before and after electrification and Coomassie brilliant blue stained pattern are all shown
in Fig. 2. Some of the serum protein fractions were identified by immunofixation procedure (3) using specific antisera obtained from Hoechst
Japan Ltd.

Patterns obtained by electrode buffer systems of Table 1 are shown in
Fig. 3. These results show clearly that isoelectric focusing without
carrier ampholytes can be realized by the procedure mentioned above.

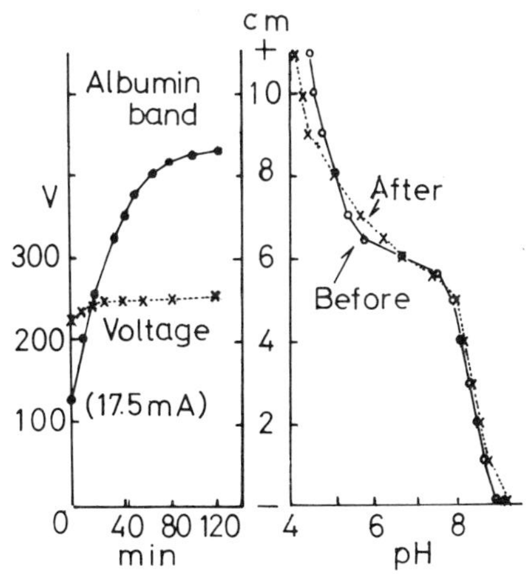

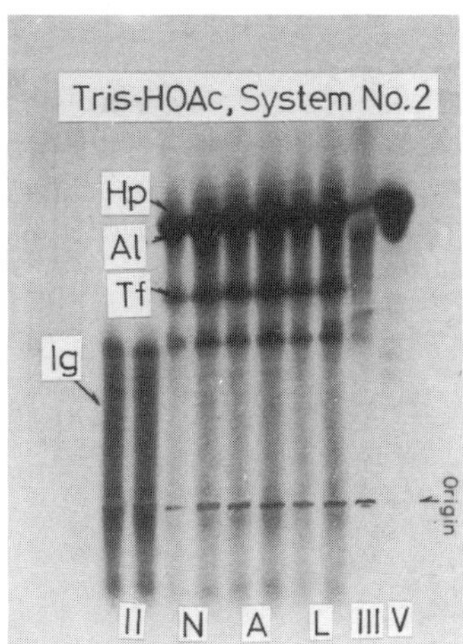

Fig. 2. An example of the course of electrification and protein pattern.

N: ORTHO Normal Control Serum, Lot No. 020V07

A: ORTHO Abnormal Control Serum, Lot No. 3V115

L: ORTHO Elevated Lipids Control Serum, Lot. No. 050V02

II, III, V: Fraction II, III, V (Sigma)

Al: Serum albumin,   Hp: Haptoglobin

Tf: Transferrin,   Ig: Immunoglobulins

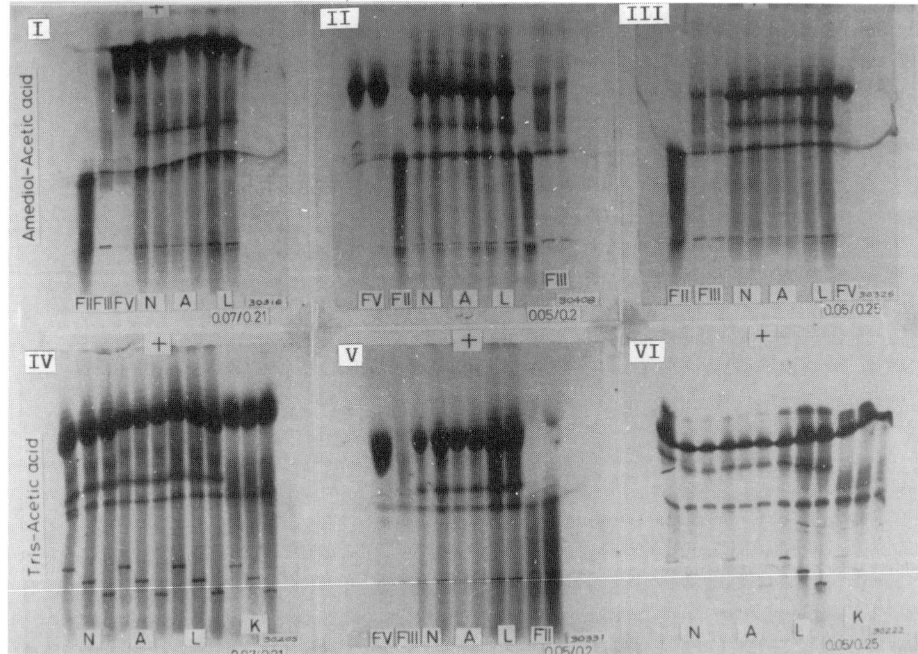

Fig. 3. Typical patterns. Experimental conditions are listed below.

Buffer	Ammediol-Acetic acid			Tris-Acetic acid		
Pattern	I	II	III	IV	V	VI
System No.	1	2	3	1	2	3
Diff.(hr)	1	21	1	1	19	1
pH range	4.4-8.7	4.1-8.9	4.0-9.0	4.3-8.6	4.1-8.9	4.0-8.8
Current(mA)	15	20	15	15	17.5	15
Time (min)	130	120	120	150	120	120

Final pH gradient in the gel seemed to be very sensitive to the initial concentration gradient. The procedure of cencentration gradient formation must be refined before this type of technique can be used widely in routine experiments.

References

1. Rilbe, H.:Electrophoresis, 2, 261-267,268-272 (1981).

2. Shimao, K.: Physico-Chem. Biol. (in Japanese), 26,283 (1982).

3. Nakano, E.: Med. Technol. (in Japanese), 17, 1273-1276 (1979).

ISOELECTRIC FOCUSING WITH IMMOBILIZED PH GRADIENTS FOR THE ANALYSIS OF HUMAN GENETIC SERUM PROTEIN POLYMORPHISMS

Hartwig Cleve, Walburgis Patutschnick, Sebastian Weidinger
Institut für Anthropologie und Humangenetik der Universität München
Richard-Wagner-Strasse 10 I, D-8000 München 2

Wilhelm Postel, Johann Weser, Reiner Westermeier, Angelika Görg
Lehrstuhl für Allgemeine Lebensmitteltechnologie
Technische Universität München, D-8050 Freising-Weihenstephan

A new method for generating stable pH gradients for isoelectric focusing (IEF) has been described recently (1). The method is based on the principle that the buffering groups are covalently linked to the matrix used as anticonvective medium. A set of acryloyl monomers has been developed with different dissociation constants which are called Immobilines. PH gradients are produced by mixing different Immobilines and co-polymerization with acrylamide and N,N'-methylenebisacrylamide. PH gradients can be produced in polyacrylamide gels which are as narrow as 0.01 pH units per cm. These immobilized pH gradients have several attractive features: 1) the pH gradient is stable with time; cathodic drift is completely abolished, 2) they give high resolution with high loading capacity, 3) the conductivity is uniform and conductivity "gaps" are absent, 4) the pH gradient is insensitive to the presence of salts in the sample.

We have recently explored the possibilities of this new technique for the analysis of genetic variability in human serum proteins. We wish to present here, briefly, the results on the polymorphic systems Gc (group-specific component; vitamin D binding protein), Pi (alpha-1-antitrypsin; protease inhibitor), and Tf (transferrin). The details of the procedures used are described elsewhere (2, 3, 4).

For classification of the common and rare phenotypes of the Gc system the
pH range from 4.90 to 5.15 was chosen. The six common Gc subtypes are
shown in fig. 1. With this pH range most rare Gc variants can also be identified. For the analysis of rare anodal Gc variants (Gc 1A and Gc 2A variants) a pH gradient from 4.75 to 5.05 was employed. For cathodal Gc variants (Gc 1C and Gc 2C variants) a slightly more basic pH gradient is recommended. The Gc patterns may be visualized without specific immunofixation by staining with Coomassie Blue only (2).

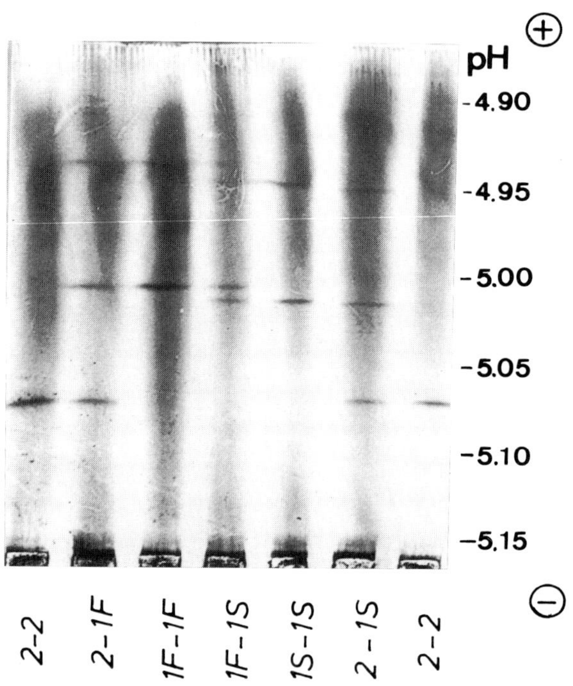

Fig. 1   The six common Gc phenotypes as obtained by isoelectric focusing
with an immobilized pH gradient. PH range was from 4.90 to 5.15.
Staining with Coomassie Blue.

For the Pi types of alpha-1-antitrypsin the so-called M4 and M6 region was
used for classification. The pH range selected for classification was from
4.40 to 4.80. Some of the common Pi M subtypes are demonstrated in fig. 2.
Remarkable is the clear resolution of the subtype M1M3 into a double band
pattern. Also presented are the PiM subtypes M3, M2, M2M3, and M1M2. With
this method not clearly different from M3 is the subtype M4: M1M4 resembles M1M3, M2M4 resembles M2M3.

Also some of the less common Pi variants have been analyzed with Immobiline pH gradients (3). The same pH range may be applied for their demonstration; with a gradient from pH 4.4 to 4.8 the Pi variants L and I as well as V and S were readily resolved (3). A more basic pH range from 4.55 to 5.05 was optimal for the demonstration of PiZ (3). Also for the Pi system the identification of the Pi bands was accomplished by specific immunofixation. Future studies have to address the problem of classification of further PiM subtypes.

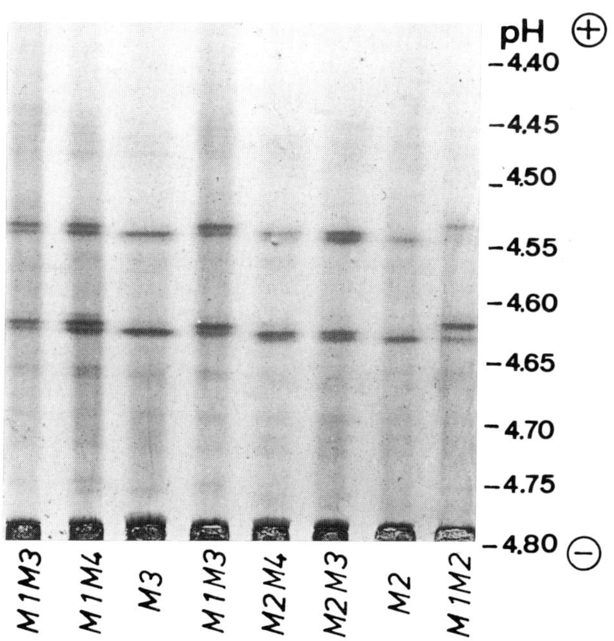

Fig. 2 Classification of PiM subtypes by isoelectric focusing in immobilized pH gradients. The range was from pH 4.40 to 4.80. Illustrated are four common PiM subtypes. PiM4 is not clearly distinguishable from PiM3.

The phenotypes of some of the transferrin C subtypes and some of the transferrin variants are shown in fig. 3. Fig. 3A illustrates four of the common TfC subtypes;TfC 3-1 has a double band pattern. Some of the anodal (TfB) and cathodal (TfD) variants are shown in fig. 3B. The identification of the transferrins by specific immunofixation is given in fig. 3C.

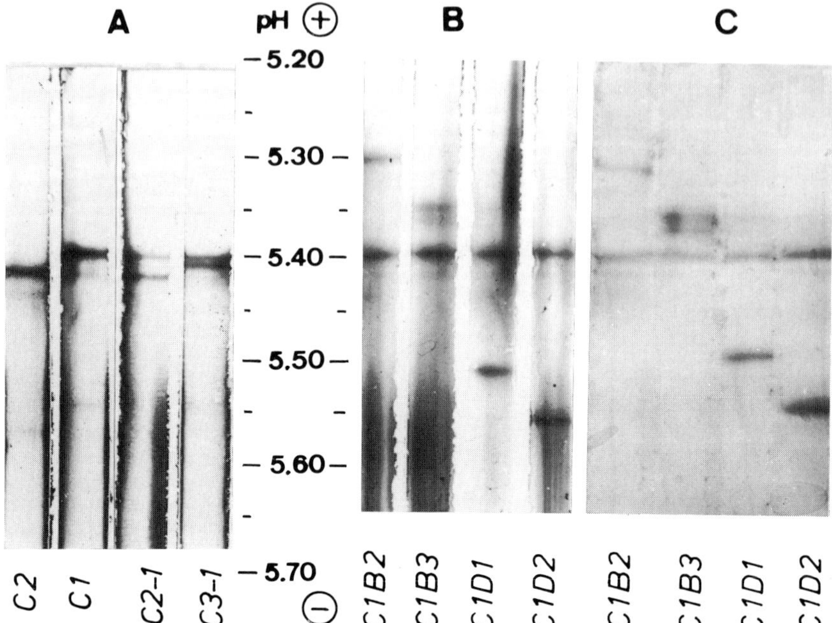

Fig. 3 A: Transferrin C subtypes, B: Transferrin variants - Protein stain with Coomassie Blue, C: Transferrin variants - Immunofixation. PH range from 5.20 to 5.70. Demonstrated is the Fe_1-Tf region.

For the analysis of the transferrin system a pH range from 5.20 to 5.70 was found to be suitable (4). Within this range are located those transferrin molecules which have one of the two iron binding sites saturated with iron.

References

1. Bjellqvist, B., Ek, K., Righetti, P.G., Gianazza, E., Görg, A., Postel, W., Westermeier, R.: J. Biochem. Biophys. Methods 6, 317-339 (1982).

2. Cleve, H., Patutschnick, W., Postel, W., Weser, J., Görg, A.: Electrophoresis 3, 342-345 (1982).

3. Görg, A., Postel, W., Weser, J., Weidinger, S., Patutschnick, W., Cleve, H.: Electrophoresis 4, 153-157(1983).

4. Görg, A., Weser, J., Westermeier, R., Postel, W., Weidinger, S., Patutschnick, W., Cleve, H.: Submitted for Publication.

PREPARATIVE ISOELECTRIC FOCUSING IN IMMOBILIZED pH GRADIENTS

Kristina Ek, Bengt Bjellqvist

LKB Produkter AB, Box 305, S-161 26, Bromma, Sweden

Pier Giorgio Righetti

Chair of Biochemistry, Faculty of Pharmaceutical Sciences and Dept. Biomedical Sciences and Technologies, University of Milano, Via Celoria 2, Milano 20133, Italy

Introduction

There is only a handful of papers on immobilized pH gradients (IPG) (1-3), and yet the technique appears to be one of the most promising innovations of the eighties, just as conventional isoelectric focusing (IEF) (4) and disc electrophoresis (5, 6) were the leading techniques of the two past decades. After describing, in the first paper (1), the general principle and methodology, we directed our efforts at generating wide pH intervals, to be used as the first dimension in two-dimensional maps (2, 3). This approach appears very promising, since IPG's fully overcome the weakest points connected in pattern recognition and spot identification in 2-D maps, namely zone smears due to isoelectric precipitation and irreproducibility of pI position due to the 'cathodic drift' (7). However, we have soon become aware that, coupled to an extraordinary resolving power (of the order of $\Delta pI=$ 0.001 pH units), IPG's appear to offer a very high loading capacity, similar to that commonly found in isotachophoresis (ITP).

In the present report, we should like to investigate some parameters leading to successful preparative runs in IPG's: a) its load capacity; b) its resolution at high protein inputs; c) protein recoveries from these matrices.

Materials and Methods

Immobilines, acrylamide, Bis, Repelsilane, ammonium persulphate, TEMED, a-

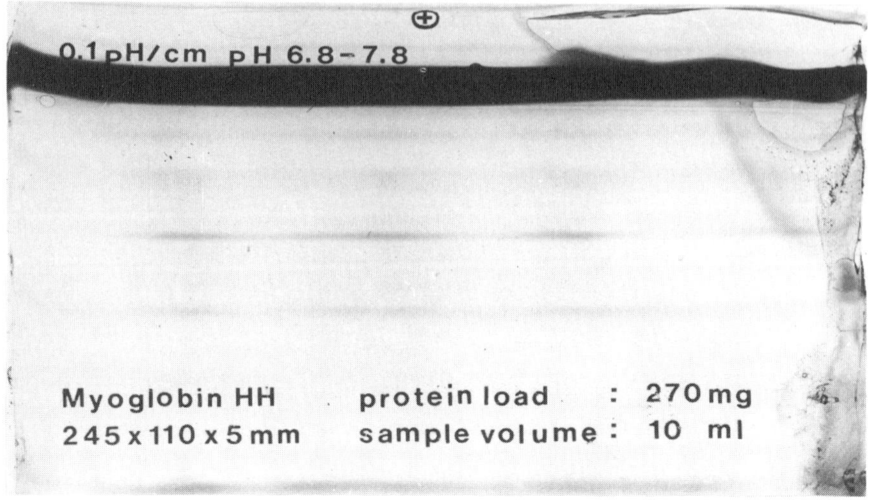

Fig. 1.   Preparative IPG runs. 270 mg horse heart myoglobin were loaded in
a trench in an Immobiline gel, pH 6.8-7.8, of 245 x 110 x 5 mm dimensions.
The buffering capacity was 3 mequiv. L^{-1}(pH unit)$^{-1}$. The run was continued
overnight at 10°C, 2500 V. Sample volume: 10 ml. Load in major band: 170 mg

garose M, HA-ultrogel, Tris and Gel Bond PAG were from LKB Produkter AB.
All proteins were from Sigma, St. Louis, Mo. Casting of Immobiline gels was
done according to ref. (8). For protein recovery from IPG matrices, a 0.8%
agarose gel, in 100 mM Tris-Gly, pH 9.1, is cast in the LKB tray 2117-115
of dimensions 245 x 120 and thickness from 3 to 5 mm. A central gel portion
(2 cm wide) is removed and the trench filled with hydroxyapatite beads, e-
quilibrated in the same buffer. The Immobiline gel strip, with the protein
of interest, is applied 5 mm on the cathodic side of the HA-beads and electrophoretically eluted for 1 hour at 30W. After transferring the HA-gel to
a syringe, the protein is recovered with 5-6 washings of 0.2M phosphate buffer, pH 6.8.

Results and Discussion

Load capacity.   Fig. 1 shows a myoglobin run in a pH 6.8-7.8 Immobiline
gel, 5 mm thick: 270 mg total protein were focused in the system, with no
overloading effects (e.g. droplets, smears, streaking).The main band contained ca. 170 mg protein, the concentration in this zone being of the order of 45 mg protein/ml gel volume. This load limit is common to all protein species we have investigated (see Table I) and appears to be the general

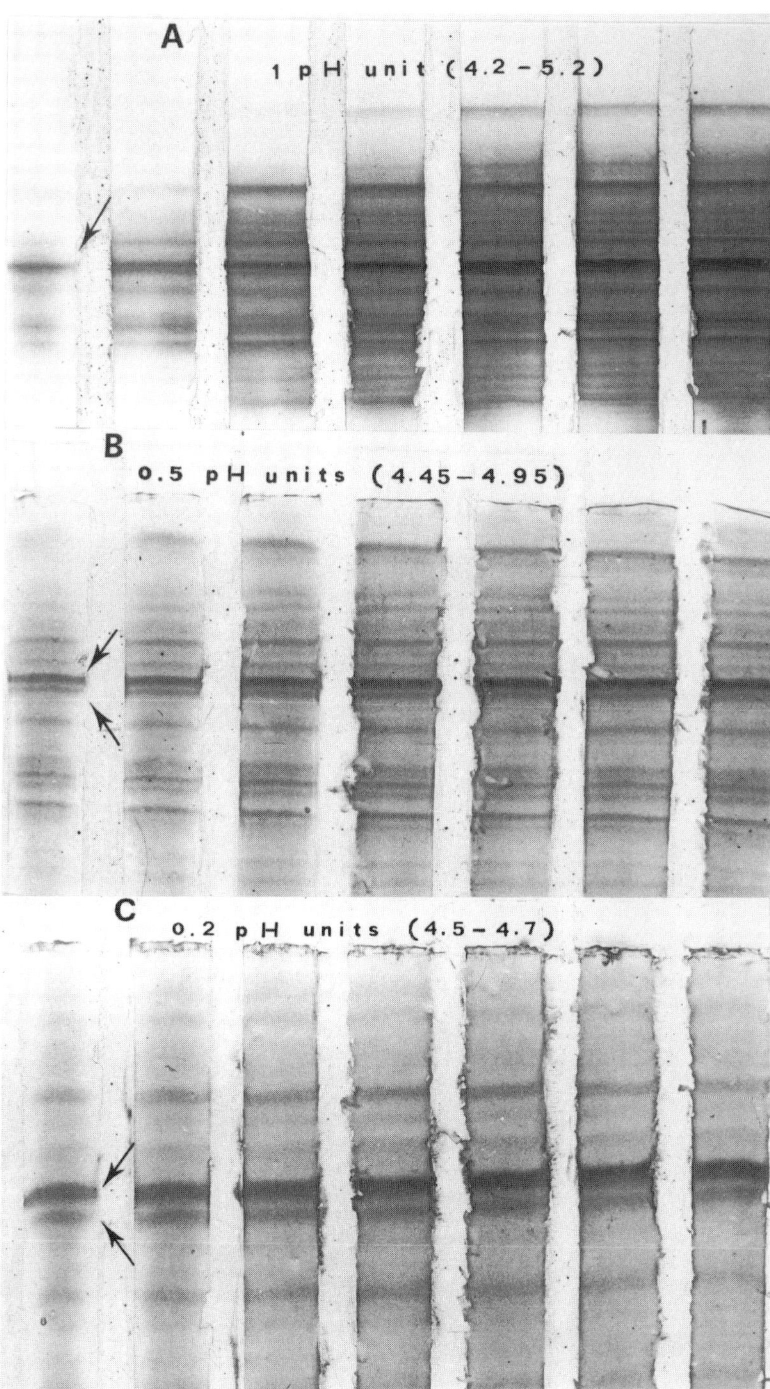

Fig. 2. IEF of ovalbumin in different widths of IPG's. A: pH 4.2-5.2; B: pH 4.45-4.95; C: pH 4.5-4.7. Sample load (from left to right): 125, 250, 500, 750, 1000, 1500, 2000 and 3000 µg/sample track of 10 mm width. Gel thickness: 0.5 mm. Sample application: at the anodic side on filter paper pads (except for 125, 250 and 500 µg, which were applied as droplets on the gel surface). Run: overnight at 2500 V, 10°C. The arrows indicate two protein components whose resolution is increased in progressively shallower pH gradients (B and C). The ΔpI was estimated to to be 0.007 pH units. Staining: Coomassie Brilliant Blue R-250 in acid-alcohol solvent.

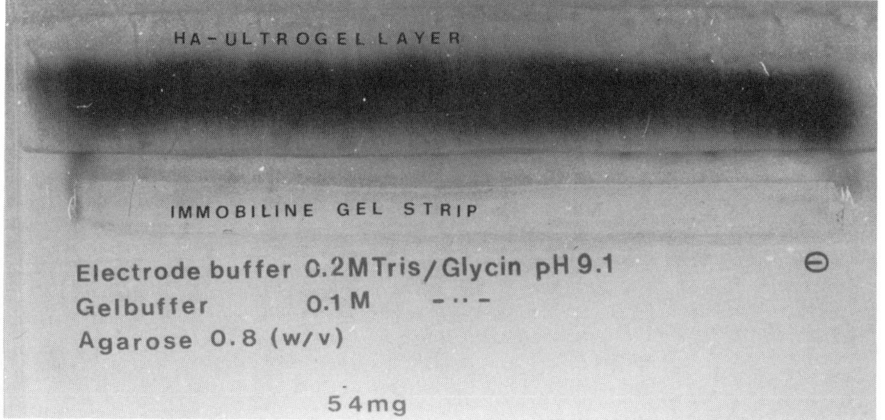

Fig. 3.   Recovery of proteins from Immobiline gels into hydroxyapatite
beads. The IPG strip was 5 mm wide, 11 cm long and 5 mm thick. The 0.8% a-
garose gel, 5 mm thick, was made to contain 100 mM Tris-Gly, pH 9.1. A
trench was cut in the agarose gel to accomodate the IPG strip. Electrical
conditions: 30W constant for 1 h at 10°C with an initial voltage of 420V.

upper load of the system: in any event, this load capacity is greater by one
order of magnitude than in conventional IEF (9) and typical of isotachophoretic runs.

Resolution.   As shown in Fig. 2, even at the maximum permissible loads (45
mg/ml), a high resolving power is still retained, so that a major **ovalbumin**
band could be fully separated from a minor contaminant, the ΔpI between the
two species being 0.007 pH units. This could be obtained by 'pH gradient
engineering': the separation was not operative in a 1 pH unit interval (Fig.
2A), was visible in a 0.5 pH unit span (Fig. 2B) and was complete in a 0.2
pH unit **range** (Fig. 2C). Conventional IEF or buffer focusing cannot possibly allow such a manipulation of pH gradients.

Sample recovery.   Fig. 3 shows the results of a second dimension run into
an HA-gel: it is seen that the Immobiline gel strip, from the IPG run, is
totally devoided of protein, the myoglobin zone being condensed within the
hydroxyapatite gel layer. Table I gives the total protein load (in IPG's)
and relative recovery (from HA-beads) of six different proteins: while the
maximum load ability is centred around 45 mg protein/ml gel volume, recoveries range from a minimum of 76% (bovine serum albumin) up to 98%, typical yields being between 80 and 90% (BSA is known to bind and interact with
a multitude of ligands, including anticonvective matrices).

TABLE I

PROTEIN LOAD AND TOTAL RECOVERY AFTER PREPARATIVE IEF IN IPG'S (1 pH UNIT)

Protein	mg protein/ ml gel volume	total recovery (%)
hemoglobin	43	98
myoglobin	43	98
bovine serum albumin	46	76
carbonic anhydrase	46	81
ovalbumin	46	89
transferrin	46	87

Acknowledgements

P.G.R. is supported by a grant from Consiglio Nazionale delle Ricerche (CNR, Roma), Progetto Finalizzato 'Chimica Fine e Secondaria', Sottoprogetto 'Metodologie di Supporto'.

References

1. Bjellqvist, B., Ek, K., Righetti, P.G., Gianazza, E., Görg, A., Wester-meier, R. and Postel, W.: J. Biochem. Biophys. Methods 6, 317-339(1982).

2. Dossi, G., Celentano, F., Gianazza, E. and Righetti, P.G.: J. Biochem. Biophys. Methods 7, 123-142 (1983)

3. Gianazza, E., Dossi, G., Celentano, F. and Righetti, P.G.: J. Biochem. Biophys. Methods (1983) in press

4. Rilbe, H.: Ann.N. Y. Acad. Sci. 209, 11-22 (1973)

5. Davis, H.: Ann. N. Y. Acad. Sci. 121, 404-427 (1964)

6. Ornstein, L.: Ann. N. Y. Acad. Sci. 121, 321-349 (1964)

7. Righetti, P.G. and Drysdale, J.W.: Ann. N. Y. Acad. Sci. 209, 163-186 (1973)

8. Bjellqvist, B. and Ek, K.: LKB Application Note No. 321 (1982)

9. Radola, B.J.: In: Isoelectric Focusing (Catsimpoolas, N., ed.) Academic Press, New York, pp. 119-168 (1976)

AN INTEGRATED APPROACH TO THE ANALYSIS OF HUMAN HEMOGLOBIN VARIANTS BY COMBINING IEF, FPLC AND ELECTROPHORETIC TITRATION CURVE ANALYSIS

L. Wahlström, V. Nylund, P.E. Burdett, H. Englund
Pharmacia Fine Chemicals AB, Uppsala, Sweden

Introduction

A variety of methods have been used for studying hemoglobins, in particular electrophoretic methods, isoelectric focusing and chromatographic procedures (1,2,3,4). In this presentation an integrated approach has been taken to provide a rapid and easy way of separating the hemoglobin variants both analytically and preparatively. High resolution isoelectric focusing (IEF) has been combined with the Pharmacia Fast Protein Liquid Chromatography system (FPLC). The chromatographic separation procedures were optimised using titration curve analysis.

Experimental

Reagents: Pharmalyte carrier ampholytes 5-8 and 6.5-9, and Polybuffer^{TM} chromatofocusing eluent, were used. All other reagents were of analytical grade.

Chromatography equipment : Pharmacia FPLC system consisting of a Mono P HR 5/20 prepacked chromatofocusing column (0.5 x 20 cm), a Mono S HR 5/5 prepacked cation exchange column (0.5 x 5 cm), a GP-250 Gradient Programmer, a V7 injector valve, a UV-1 monitor fitted with a HR-10 flow cell, filter 280 nm and 405 nm and a Hewlett-Packard integrator. Isoelectric focusing and titration curves: Flat Bed Apparatus (FBE 3000) using Pt/Ti electrodes, ECPS 3000/150, Capillary Casting Mould for titration curves and IEF-gels.

570

Sample treatment-preparation of hemoglobin

1 ml of heparinized whole blood was incubated in 10 ml of 0.15 m NaCl at 37°C for 5 hours to remove an intermediate form of HbA_{1c}, which interferes with the determination of HbA_{1c}. The sample was centrifuged and the plasma-NaCl fraction was removed. The erythrocytes were hemolysed with 4.2 ml of distilled water. 0.7 ml carbon tetrachloride was added, and after mixing the sample was centrifuged. The supernatant (hemolysate) was removed and stored at -20°C, after treatment with CO. The hemoglobins were generous gifts from Dr. Jeppsson of Malmö General Hospital, Sweden.

Titration curve analysis and Isoelectric focusing

1 mm thick polyacrylamide gels were cast using the Pharmacia Capillary Casting Mould. The polyacrylamide gel was T5C3. A sigmoidal pH gradient was prepared (5) using ß-alanine as a chemical spacer. (For 30 ml of gel solution, 0.9 ml of both Pharmalyte 5-8 and 6.5-9 were used, with 700 mg ß-alanine). Gels were cast on GelBond[R] PAG, using ammonium persulphate and TEMED as described previously (6).
Electrode solutions: anode: 0.04 M glutamic acid; cathode: 1 M NaOH
Running conditions, Titration curves:1st dimension IEF-15 W for 1,400 Vh
2nd dimension ELP-1,000 V for 400 Vh
Isoelectric focusing, running conditions: prefocusing, 20 W to 400 Vh
focusing, 20 W to 1,000 Vh then 30 W to 4,200 Vh.

Results.

Cation Exchange Chromatography of HbA_{1c} and HbA.

A purified fraction of hemoglobin A and A_{1c} were analysed by titration curve analysis (Fig 1a). As can be seen the charge differences were maximum at around pH 6. Consequently the Mono S cation exchanger was used, at a pH close to 6. The separation of HbA_{1c} from other hemoglobins is shown in Fig 1b using Mono S ion exchanger. Base line separation was achieved.

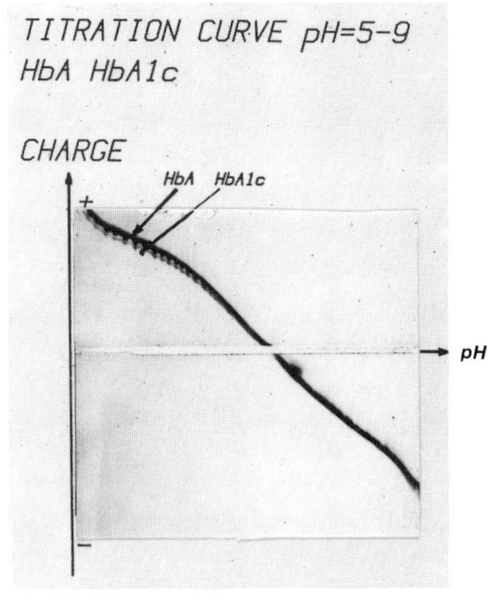

Fig 1a.

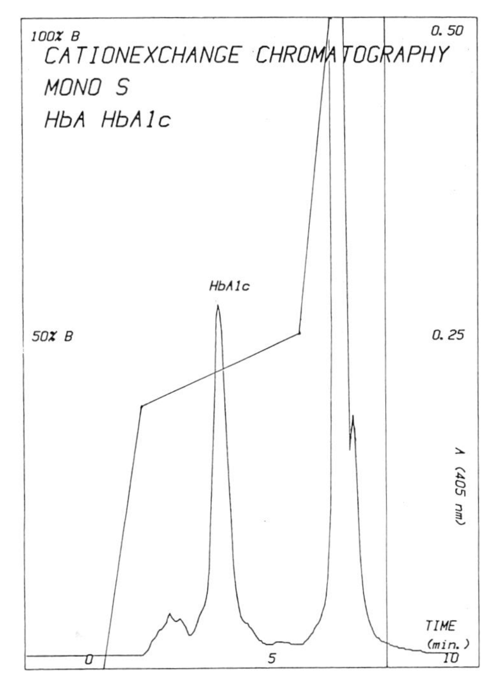

Fig 1b.

Separation of Hemoglobin variants.

From the titration curve analysis of purified hemoglobins A, F and E
(Fig 2a) it was decided that chromatofocusing would provide the most
information. In some cases Mono Q could also provide relevant
information, but this would involve working at a rather high pH.

Fig 2.

TITRATION CURVE pH=5-9
HEMOGLOBIN VARIANTS
HbE   HbA   HbF

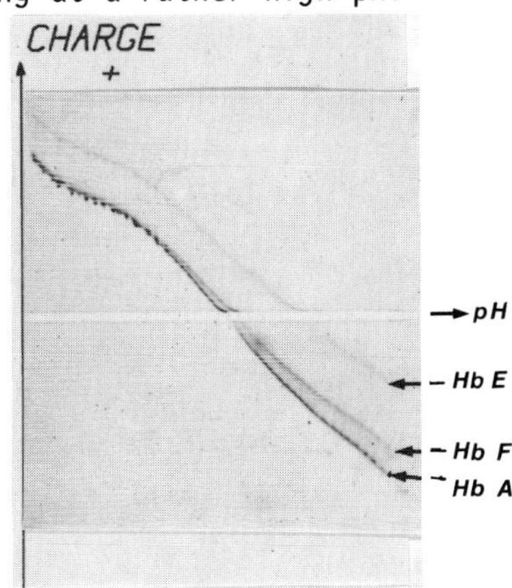

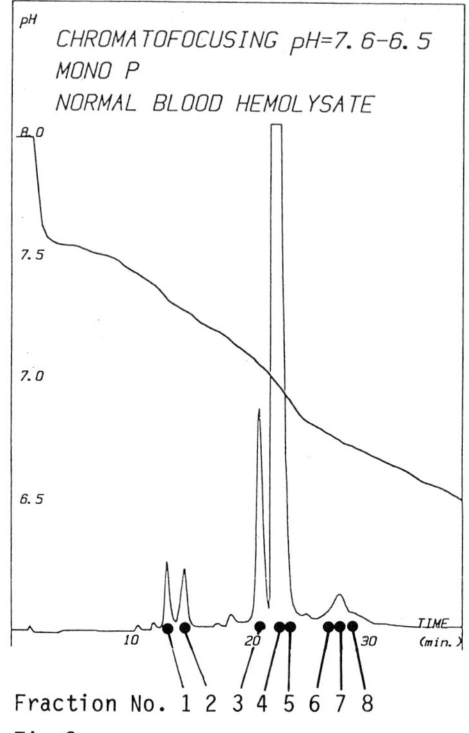

Fraction No. 1 2 3 4 5 6 7 8

Fig 3a.

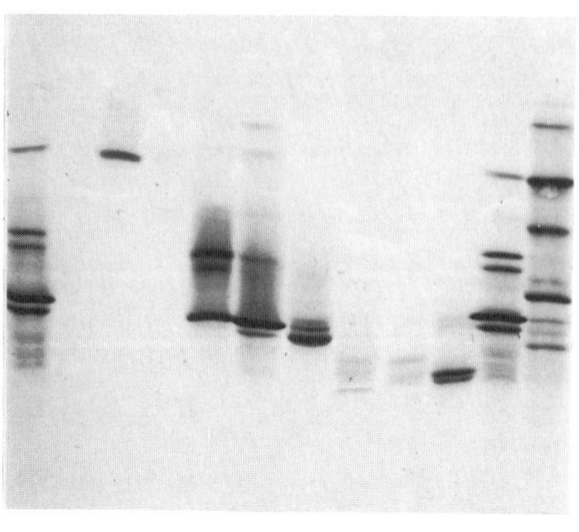

Fig 3b. IEF of fractions recovered from chromatofocusing of normal blood hemolysate (Fig 3a). A, normal blood hemolysate; E, hemolysate E. Fractions 1-8 as in Fig 3a.

Hemoglobin A

In Fig 3a is shown the chromatofocusing pattern of hemoglobin A using the pH range 7.6 to 6.5, whilst in Fig 3b is shown the analysis of these fractions by IEF. Fraction 1 was found to contain essentially pure A_2, fraction 4 contains mostly A_{1c} whilst fractions 5, 6 and 7 probably contains A_{1a}, A_{1b}, and perhaps A_3 respectively. However their precise identity awaits further work.

Footnote:

Cation exchange chromatography Mono S for $HbA-HbA_{1c}$. (Fig 1b).

Buffer A: 0.01 M Malonate/NaOH pH=5.7
Buffer B: 0.01 M Malonate/NaOH pH=5.7 + 0.3 M LiCl
Column: Prepacked HR 5/5 Mono S
Flow rate: 2 ml/mn.

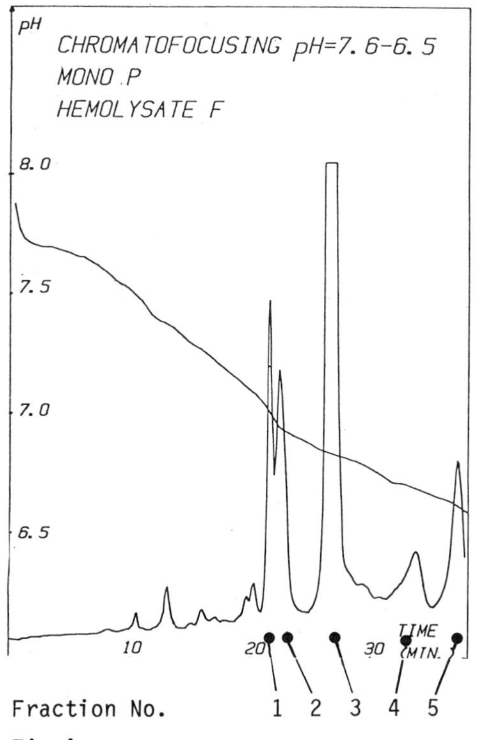

Fraction No.   1 2 3 4 5

Fig 4a.

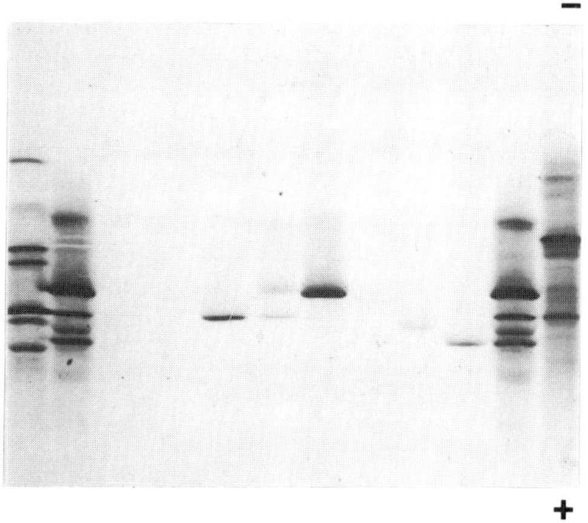

Fig 4b. IEF of fractions recovered
from chomatofocusing of fetal cord
blood. (Fig 4a).
A: normal adult hemolysate
F: cord blood hemolysate
S: hemolysate from heterozygous A+S

Hemoglobin F.

In Fig 4 is shown the separation of hemoglobins from a lysate containing
hemoglobin F by chromatofocusing, together with the corresponding
analysis of the resulting fractions by IEF. Hemoglobin F (fraction 3)
elutes somewhat later than what would be expected from its pI point, a
phenomenon which has also been reported in ordinary ion exchange
chromatography and chromatofocusing with regular Polybuffer exchanger.
Fraction 1 probably contains hemoglobin A, and fraction 5 possibly
contains hemoglobin F_1.

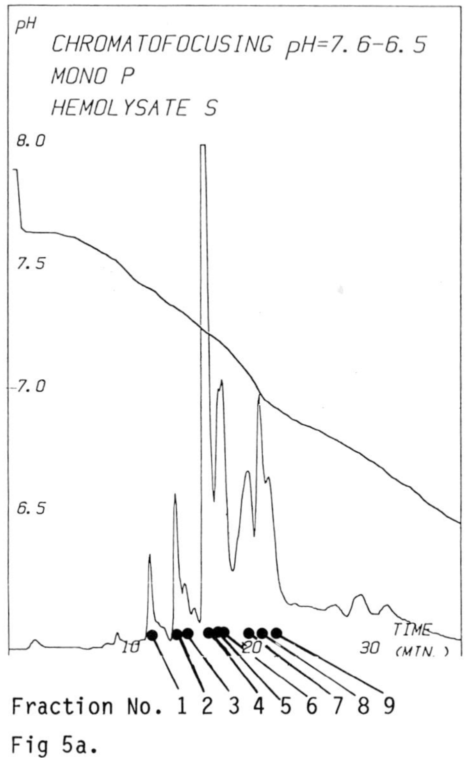

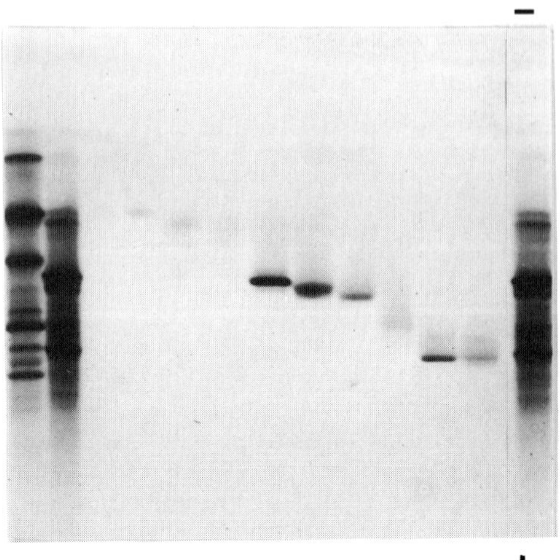

Fig 5b. IEF of fractions recovered from the chromatofocusing of hemolysate from heterozygous A+S (Fig 5a) Fractions 1 to 9 are as defined in Fig 5a. E, hemolysate E; S, hemolysate from heterozygous A+S.

Fraction No. 1 2 3 4 5 6 7 8 9
Fig 5a.

Hemoglobin S

A hemolysate containing hemoglobin S has been analysed by chromato-focusing and IEF (Figures 5a and 5b). Fraction 4 is hemoglobin S, and fraction 8 is HbA. Fraction 1 is probably A_2. The other fractions are as yet unidentified.

Footnote:

Chromatofocusing conditions in the pH range 7.5 to 6.5 (Figs 4-7)
Start buffer: 25 mM triethanolamine/methanesulphonic acid, pH 8.1
Eluent: Polybuffer 96 diluted 1/16-methanesulphonic acid, pH 6.65
Column: Mono P^R
Flow rate   1 ml/min
Monitor   UV-1 set at 280 nm, 0.5 AU F.S.
Chart speed: 0.5 cm/min

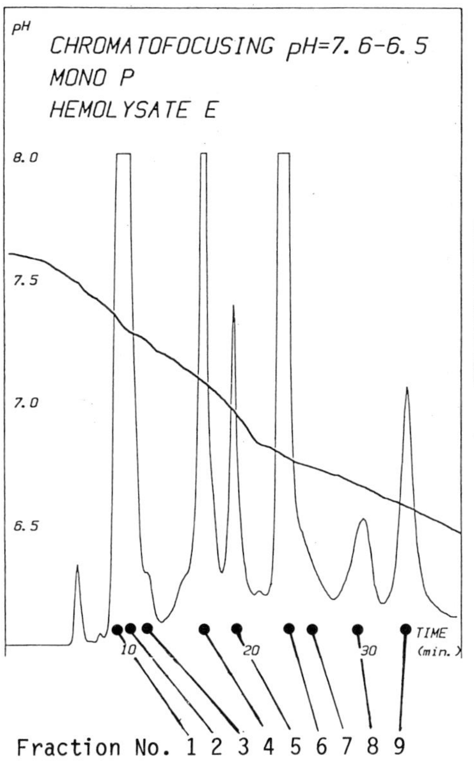

Fig 6a.

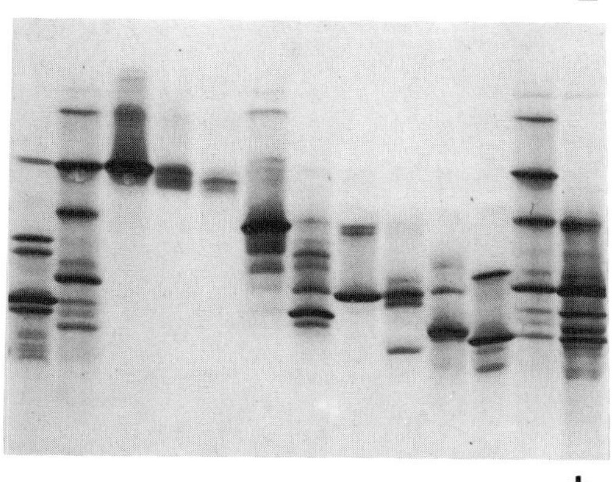

Fig 6b. IEF of fractions recovered
from chromatofocusing (Fig 6a).
A, normal adult hemolysate
E, hemolysate from hemoglobin E
F, cord blood hemolysate

Hemoglobin E

The chromatofocusing pattern of a lysate containing hemoglobin E is shown
in Fig 6 and the corresponding analysis of these fractions by IEF is
shown in Fig 6b. Fractions 1, 5, and 6 can be identified as containing
hemoglobins E, A and F respectively, but the identity of fractions 2, 3,
4, 7, 8 and 9 is less clear. Fraction 9 may be acetylated hemoglobin F.
By combining preparative and analytical procedures in this way, further
identification work can readily by undertaken.

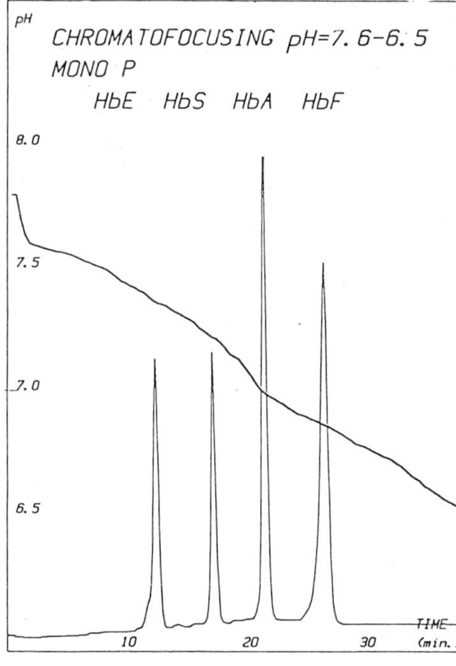

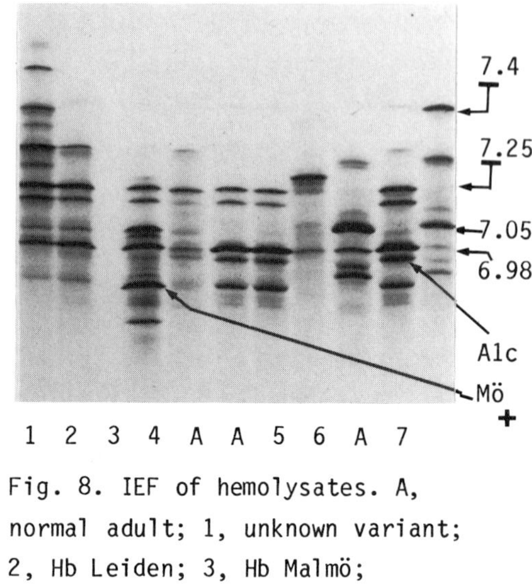

Fig. 8. IEF of hemolysates. A, normal adult; 1, unknown variant; 2, Hb Leiden; 3, Hb Malmö; 4, Hb Uppsala; 5, Heterozygous A+S; 6, Cord blood; 7, Hb E.

Fig 7.

In Fig 7 is shown the separation of a mixture composed of purified hemoglobins E, S, A and F by chromatofocusing. The clear base line separation is obvious.

IEF of a number of hemoglobin variants, (lysates) including an unidentified hemoglobin, is shown in Fig 8. The pI points of some hemoglobins are shown alongside, as follows: HbE and A_2, 7.40; HbS, 7.25; HbF, 7.05; HbA, 6.98; HbA_{1c}, 6.94. The unknown variant (No 1) contains a variant with a very alkaline pI.

Footnote:

Chromatofocusing of hemoglobins A_2 and E in the pH interval 7.6 to 7.0. (Fig 8).

Start buffer: 25 mM triethanolamine/methane sulphonic acid pH 8.1.

Eluent: Polybuffer 96 diluted 1/16-methanolsulphonic acid, at 7.2 (including 10% betaine).

Column: Mono P

Flow rate: 0.5 ml/min

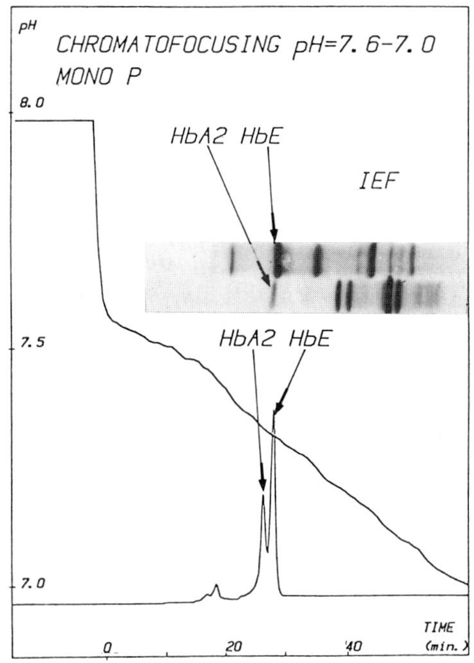

Fig 9.

Separations of Hemoglobins A_2 and E.

By using a shallower pH gradient on chromatofocusing (pH 7.6-7.0) it was possible to separate hemoglobins A_2 and E (Fig 9), illustrating the powerful resolving power of this technique. On IEF, these two hemoglobins are barely resolvable (inset).

Discussion and conclusions

1. Up until now the separation of hemoglobins has rested primarily on slow chromatographic procedures, or unsuitable electrophoretic methods. The use of FPLC and thin layer IEF provides a versatile system that allows for reliable identification and quantification.

2. A chromatographic method for the determination of hemoglobin A_{1c} using Mono S ion exchanger has been investigated. Separation of HbA_{1c} from other hemoglobins was accomplished by gradient elution using a complex salt gradient. The separation time was ten minutes including the time required for column regeneration. The stability and reproducibility of the method were confirmed by a precision study yielding coefficient of variance values better than 2.5%. Unlike other chromatographic methods, no significant temperature influence on the quantitation was observed in the range of 20-30°C. Normal levels of Hb F do not influence the quantitation of HbA_{1c}. The results obtained correlated well (r=0.96) with standard IEF methods used for the determination of HbA_{1c}.

3. Chromatofocusing on Mono P has been shown to be a useful technique for the analysis of hemoglobin variants. The high resolving power and the ability to use narrow pH intervals makes chromatofocusing a suitable method for separation of proteins with small differences in charge. This was confirmed by the separation of HbA_2 and E.

4. Isoelectric focusing using the pH interval described has the ability to resolve essentially all the hemoglobin variants described. In particular, when combined with chromotofocusing, the separation between HbE and HbA_2 was confirmed.

5. The combination of IEF and FPLC makes a powerful system for routine clinical analysis as well as for hemoglobin research.

References

1. Basset, P., Braconnier, F., and Rosa, J., J. Chromatogr. 267-304, _227_ (1982).

2. Andreasen, T.L., Hansen, E., and Olesen, H., Scand. J. Clin. Lab. Invest. 119-122, _42_ (1982).

3. Bunn, H.F., Gabbay, K.H., and Gallop, P.M. Science 21-27, _200_ (1978).

4. Peterson, C.M., Kolan G, Jovanovic, L. et al, Am. J. Obstet. Gynecol. 85-88, _135_ (1979)

5. Isoelectric Focusing, principles and methods. Pharmacia Fine Chemicals, Uppsala, Sweden 1982.

6. GelBond[R] Pag film. Package Insert. Pharmacia Fine Chemicals, Uppsala

ISOELECTRIC FOCUSING OF PLASMA LIPOPROTEINS IN THE DIAGNOSIS
AND PROGNOSIS OF LIVER DISEASE

D. Stathakos
Laboratory for Enzyme Research, Department of Biology, Nuclear
Research Center "Demokritos", Athens, Greece

G. Rekoumis, A. Avgerinos, T. Kanaghinis
Research Unit, "Evangelismos" Hospital, Athens, Greece

Introduction

In a recent report we described the isoelectric fractionation
on polyacrylamide gel of human plasma lipoproteins pre-stained
with tetrazolium blue (1). This method offered an advantage
over the traditional methodology of ultracentrifugal separation
of lipoproteins (2,3) or their electrophoretic fractionation
(2,3,4), as it provided sharp, reproducible patterns of Lp-
subpopulations which - being intensely coloured - could be also
easily quantified densitometrically (1).
So, beyond the suitability of this technique for the study of
Lp molecular subspecies, we investigated its application as an
eventual laboratory tool in the diagnosis and prognosis of liver disease. When applied in typical cases of liver disorders,
the method gave distinct patterns corresponding to each type of
clinical profile (1). Some of these results are reported here.

Materials and Methods

The procedure of collection and pre-staining of samples, the
conditions of isoelectric focusing by the gel-rod technique (5)

and the quantitation of the obtained patterns have been previously described (1). Here it should be mentioned that all
experiments presented in this report were performed in 3%
Ampholine (pH 3.5-10) for optimal resolution.

Results and Discussion

Since in electrofocusing lipoproteins are usually applied
cathodically, it would be useful to describe their resolution
in terms of "high, intermediate and low regions" within the pH
gradient formed, as shown in normal plasma (Fig. 1).

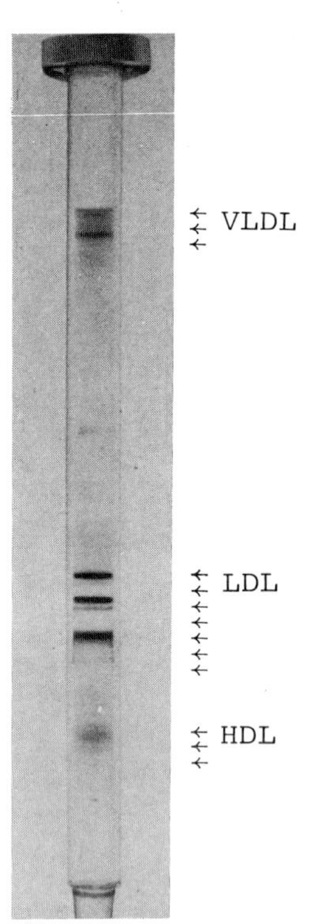

Fig. 1.
Lp pattern of
normal plasma (1)

Nevertheless, we chose to use the terms
VLDL (very-low-density lipoproteins), LDL
(low-density lipoproteins) and HDL (high-
density lipoproteins), respectively, for
reasons of better understanding. (For
identification of the Lp bands in
electrofocusing see 1). There is definitely a general consensus in this usage,
as well as the electrophoretic homologous
terms: pre-β, β, α, especially in the interpretation of abnormal Lp patterns (4).
In comparison to normal plasma the most
striking differences have been observed
in severe obstructive jaundice, where the
LDL region is intensified and densily
populated at the cost of VLDL and HDL
which completely disappear (1).
Three other cases of hepatic lesion are
presented in Fig. 2. Fig. 2A shows the
pattern distribution in a case of advanced hepatoma. The LDL region is enriched with several bands, while a diffuse bilirubin area is noticed at pIs below the original HDL area. At the same

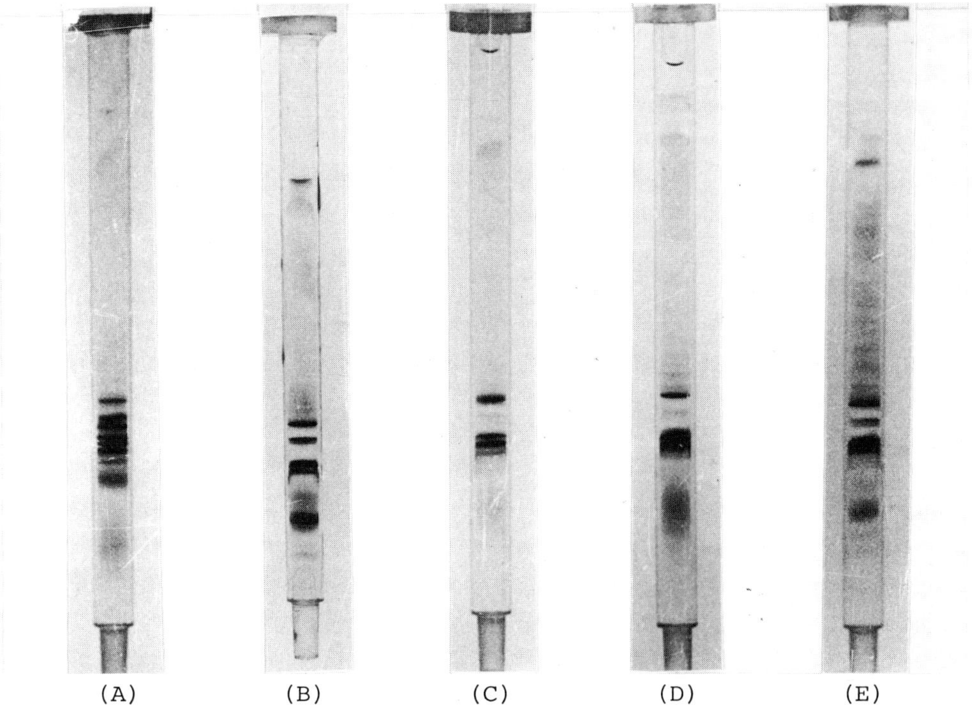

Fig. 2. Lp patterns in liver disease.
(A) Advanced primary liver carcinoma, 40 days from onset of jaundice; (B) cirrhosis of the liver (in hepatic
failure).
Acute viral hepatitis: (C) acute phase; (D) second
week; (E) recovery. Sample applied on gel: 20 μl.
For liver function tests see Table I.

time the colour intensity of the LDL bands is increased sever-
alfold, so that the gel appears "overloaded", a consistent picture with the carcinomas which were investigated. Current "expanded scale" experiments (6) may lead to the isolation and
identification of these densely packed bands.
In cirrhosis (Fig. 2B) a clearly different pattern is presented. In the densely populated LDL region distinct signs of
aggregation are observed under our experimental conditions.

582

Table I

Liver function tests of the cases represented in Fig. 2

Patterns in Fig. 2	Bilirubin mg/%		Alkaline phosphatase U.*	Aspartate amino-transferase I.U.
	total	conjugated		
Hepatoma (A)	26,5	21	90	250
Cirrhosis (B)	2,1	0.6	17	310
Acute viral hepatitis:				
Acute phase(C)	6,5	2,3	32	750
2nd week (D)	9,1	2,1	46	710
Recovery (E)	2,8	0,8	16	210

* King-Armstrong units

The HDL band is dense, but broad and diffuse, an indication of
molecular heterogeneity. This persistent finding might be due
to the presence of nascent HDL newly synthesized in the cirrho-
tic liver (4).
Similar intermediary lipoprotein species, which show a pronounced diversity of pIs, can be observed by this technique
during the course of liver disease. Thus, Fig. 2 (C to E) represents a case of acute viral hepatitis. The first sample (C)
was run at the peak of the disease; it is a typical "only LDL"
pattern, with traces of VLDL barely discernible. Yet, as con-
valescence proceeds, the HDL and VLDL bands gradually reappear
(D). During this period, several Lp bands, which precede the
appearance of VLDL are formed above the LDL region (E); these
bands are eliminated when the pattern returns to normal (Fig.
1). The characterization of these intermediary Lp species (4)
is currently under investigation.
Clearly, the diagnostic potential of this technique remains to
be conclusively evaluated after an adequate number of patients

has been investigated and characteristic "lesion-specific" Lp
patterns have been established.

Acknowledgement: We thank Mr. I. Lagouros for skillful technical assistance.

References

1. Stathakos, D., Rekoumis, G., Avgerinos, A., Kalantzis, N.,
   Kanaghinis, T.: In Electrophoresis '82, Ed. D. Stathakos,
   pp. 665-674, Walter de Gruyter Berlin-New York (1983).

2. Hatch, F.T., Lees, R.S.: Adv. Lipid Res. 6, 1-68 (1968).

3. Day, R.C., Harry, D.S., McIntyre, N.: In "Liver and
   Biliary Disease", Eds. Wright, R., Alberti, K.G.M.M.,
   Karran, S., Millward-Saddler, G.H., pp. 63-82, W.B.
   Saunders Co., London-Philadelphia-Toronto (1979).

4. Sabesin, S.M., Bertram, P.D., Freeman, M.R.: Adv. Int.
   Med. 25, 117-146 (1980).

5. Righetti, P.G., Drysdale, J.W.: J. Chromatogr. 98, 271-
   321 (1974).

6. Stathakos, D., Vellios, A., Koussoulakos, S.: In Electrophoresis '79, Ed. B.J. Radola, pp. 517-528, Walter de
   Gruyter, Berlin-New York (1980).

Determination of Cystathionine in Rat Tissues and its Derivatives in the Urine of a Cystathioninuric Patient
Using Isotachophoresis

Hiroyuki Kodama, Hiroaki Mikasa and Tomiko Ageta
Department of Chemistry, Kochi Medical School
Kochi 781-51 Japan

Introduction

Cystathionine is well known as an important intermediate in
the transsulfuration pathway in mammalian tissues. Especially
high concentrations of L-Cystathionine are found in the
primate brain(1).
Recently, it was reported that experimental cystathioninuria
induced in rats by administration of DL- and L-Propargyl-
glycine was followed by accumulation of large amounts(2,3) of
cystathionine in several tissues.
Cystathionine can be determined according to the general
method of Moore et al(4), and a colorimetric procedure(5).
However, analysis of this compound using an amino acid
analyzer is time-consuming.
The isotachophoretic method(6-9) presented here is the first
description of the use of this techniques for the quantitative determination of cystathionine in biological samples and
has several advantage over those previously described.
It was reported that cystathionine and one of its metabolic
compounds, Perhydro-1,4-thiazepine-3,5-dicarboxylic acid were
excreted in the urine of a cystathioninuric patient(10,11).
The cystathionine excreted in the urine could be easily de-

termined using an amino acid analyzer, but the color value for
the ninhydrin reaction of perhydro-1,4-thiazepine-3,5-dicarbo-
xylic acid due to imino group was too low for amino acid analysis. Therefore, a new method for determining urinary per-
hydro-1,4-thiazepine-3,5-dicarboxylic acid was devised using
isotachophoresis.

Results

Isotachophoretic runs of authentic cystathionine, supernatant
of normal brain homogenate and cystathionine added to the
supernatant are shown in Fig. 1, A-C.
We have ascertained that authentic cystathionine is easily
separated under the following conditions : The leading electrolyte consisted of 0.01M hydrochloric acid and 2-amino-2-
methyl-1,3-propanediol containing 5% polyvinylalcohol(pH8.9).
The terminal electrolyte was 0.01M γ-aminobutyric acid .
The zone from the supernatant of a normal brain homogenate
(Fig. 1, B), which has the same potential gradient as the zone
of authentic cystathionine, was made to overlap by adding
authentic cystathionine to the supernatant from the normal
brain homogenate, resulting in an elongation of the zone of
cystathionine in the supernatant as shown in Fig. 1, C, indicating that the elongated zone contains cystathionine.
The slope of the standard curve drawn by plotting zone length
against different concentrations of authentic cystathionine
was linear between 0 and 50 nmoles.
The recovery of cystathionine by column chromatography was
92-95%.
A comparison of isotachophoresis and an amino acid analysis
to determine cystathionine in several tissues of DL-propargyl-
glycine treated rats is shown in table 1.
These two methods for determination of cystathionine gave
almost the same values. These results indicate the reliability

Comparison of cystathionine content of the tissues of
propargylglycine treated rats determined by isotachophoresis and amino acis analysis.

Samples	Cystathionine content (μmoles/g ')	
	Isotachophoretic analyzer	amino acid analyzer
Liver (5mg)	2.37	2.35
(20mg)	7.05	6.71
Kidney(5mg)	1.82	1.87
(20mg)	5.42	5.29
Brain (5mg)	0.125	0.128
(20mg)	0.411	0.394

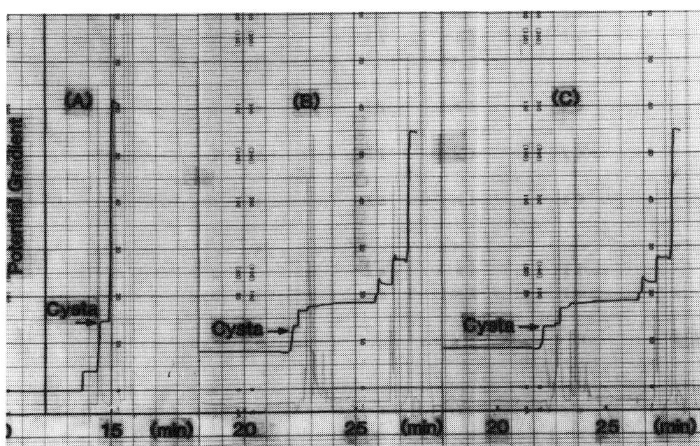

Fig. 1. Isotachophoretic runs of authentic cystathionine
(10 nmoles) (A); brain sample (10 μl) of normal rats (B); and the mixture of authentic cystathionine plus brain sample (C).

of this method for measuring cystathionine from several tissues of rats.

The determination of urinary perhydro-1,4-thiazepine-3,5-dicarboxylic acid of a cystathioninuric patient was carried out under the following conditions : The leading electrolyte consisted of 0.01M hydrochloric acid and β-alanine(pH3.1), the terminating electrolyte was 0.01M glutamic acid. The amount of perhydro-1,4-thiazepine-3,5-dicarboxylic acid in the urine of a cystathioninuric patient was 0.94 $\pm$ 0.021 μmole/(n=6).

The determination of cystathionine and perhydro-1,4-thiazepine-3,5-dicarboxylic acid using an isotachophoretic analyzer presented here is simpler than that using an amino acid analyzer.

References
1. Tallan, H.H., Moore, S. and Stein, W.H. : J. Biol. Chem. 230, 707-716 (1958)
2. Kodama, H., Sasaki, K. and Ageta, T. : Biochem. int. 4, 195-200 (1982)
3. Shinozuka, S., Tanase, S. and Morino, Y. : Eur. J. Biochem. 124, 377-382 (1982)
4. Moore, S., Spackman, D.H. and Stein, W.H. : Analyt. Chem. 30, 1185-1190 (1958)
5. Kashiwamata, S. and Greenberg, D.M. : Biochem. Biophys. Acta 212, 488-500 (1970)
6. Haglund, H. : Sci. Tools 17, 2-12 (1970)
7. Arlinger, L. : Biochem. Biophys. Acta 293, 396-403 (1975)
8. Kodama, H. and Uasa, S. : J. Chromatogr. 163, 300-303 (1979)
9. Kodama, H., Yamamoto, M. and Sasaki, K. : J. Chromatogr. 183, 226-228 (1980)
10. Kodama, H., Yao, K., Kobayashi, K., Hirayama, K., Fujii, Y., Mizuhara, S., Haraguchi, H. and Hirosawa, M. : Physiol. Chem. and Physics 1, 72-76 (1969)
11. Kodama, H., Ishimoto, Y., Shimomura, M., Hirota, T. and Omori, S. : Physiol. Chem. and Physics 7, 147-152 (1975)

AGAROSE ISOELECTROFOCUSING: A PREPARATIVE TECHNIQUE FOR
STUDYING THE IMMUNOSUPPRESSIVE PROPERTY OF α_2MACROGLOBULIN

Ala Alomran, Brian Shenton, David Francis, Peter Veitch,
George Proud, R.M.R. Taylor.
University Department of Surgery
Royal Victoria Infirmary
Newcastle upon Tyne NE1 4LP UK

Introduction

Several authors (1, 2) have suggested that $\alpha_2$macroglobulin
(α_2M) may play an important role in the modulation of immune
response in normal individuals and patients with malignant
disease. α_2M is a major protease inhibitor in human plasma and
it was found that protease binding to α_2M was associated with
an increased ability of α_2M-protease complex (α_2M-P) to
suppress mixed lymphocyte reactivity (MLR) (1). Using the
Isoelectrofocusing (IEF) technique, Ohlsson (3) showed that
α_2M-P and ascitic fluid from patients with acute peritonitis
following appendicitis contained another form of α_2M which was
not found in normal plasma. In this study we have used
analytical and preparative IEF to study and compare the α_2M
pattern of plasma from normal subjects and patients with
advanced malignant disease. Blood samples from 10 normal
subjects and 10 untreated patients with cancer (gastric,
colorectal and breast) were collected in citrate anti-
coagulant in the presence of soybean trypsin inhibitor (SBTI).
Other samples were collected either in lithium heparin tubes
or as serum. Plasma suppressive activity (PSA) of the samples
from the groups were tested using autologous lymphocytes and a
panel of allogeneic lymphocytes stimulated with PPD in the
Tanned Erythrocyte Electrophoretic Mobility (TEEM) test (4).
IEF was performed using 12 x 24cm Agarose EF gel film (LKB)
containing 2% ampholine solution (pH 3.5-10 (LKB)), 10µl 1:6

diluted plasma was used. The condition of the electrophoretic run, staining procedure and immunofixation was
performed as described by application note 317 (LKB).
Crossed immunoelectrofocusing was performed by cutting a
0.6cm thick gel slice, and overlying it onto an agarose plate
containing appropriate amounts of rabbit anti-human plasma
protein and anti-human α_2M (DAKO). Preparative IEF was
performed by applying 200ul of diluted plasma sample (1:6) on
10cm pieces of Whatman paper (1 Chroma). The focused gel was
cut horizontally into a 6mm slice using a metal grid. The
protein content of each gel slice was eluted by ultra-
sonicating the gel for 1-2 minutes. This was followed by
centrifugation. The supernatant was then dialysed against
citrate buffer pH 6.5. Each fraction was then tested for its
suppressive activity against lymphocyte reactivity in the
TEEM test. α_2M from normal plasma and patients with
malignant diseases was prepared as described by Virca (5).
The purity of α_2M was checked by running a sample on 4%
polyacrylamide gel electrophoresis.

Results

The PSA of samples from cancer patients collected by the
three methods showed much higher suppressive activity than
that of normal plasma against lymphocyte response in the TEEM
test. The PSA value (expressed as the number of μl of plasma
to block lymphocyte responses to PPD by 50% in the TEEM test)
was 15 times more suppressive than that of normal plasma
(Table 1). No significant difference was found in the PSA
values of those samples collected in the patient group.
However, the PSA values of normal plasma collected in the
presence of protease inhibitor SBTI was significantly less
than that collected by other methods. Results from
preparative IEF of both normal and cancer patients' plasma

Table 1 - EFFECT OF THE METHOD OF BLOOD COLLECTION ON PSA
Volume of sample (ul) Causing 50% Lymphocyte Inhibition

Groups		SBTI			Li-Heparin			Serum	
	n	$\bar{x}$	SD	n	$\bar{x}$	SD	n	$\bar{x}$	SD
Autologous	10	66.75±8.4		10	35.1±3.0		10	30.8±3.9	
Normals									
Allogeneic	10	82.0 ±3.6		10	38.5±1.5		10	34.7±1.83	
Autologous	12	5.8 ±2.3		14	1.68±0.8		10	1.92±1.2	
Cancer									
Allogeneic	12	3.02 ±1.8		14	1.02±0.5		10	0.87±0.29	

showed that there was a major region of lymphocyte inhibitory activity associated with the α_2M region. However, regions with higher suppressive activity were found in fractions from cancer patients. Pure preparations of α_2M from both normal and cancer patient groups when run on preparative IEF confirmed the previous results. The suppressive activity in this region was due to α_2M, and a small cathodal band of α_2M was shown to retain most of the suppressive activity (Fig. 1). Analysis of the α_2M IEF pattern from both normal and patient groups consistently showed three protein bands - confirmed by immunofixation and crossed immunoelectrofocussing.

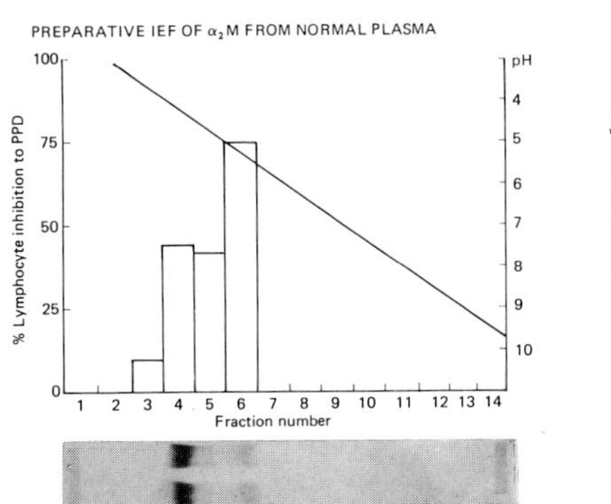

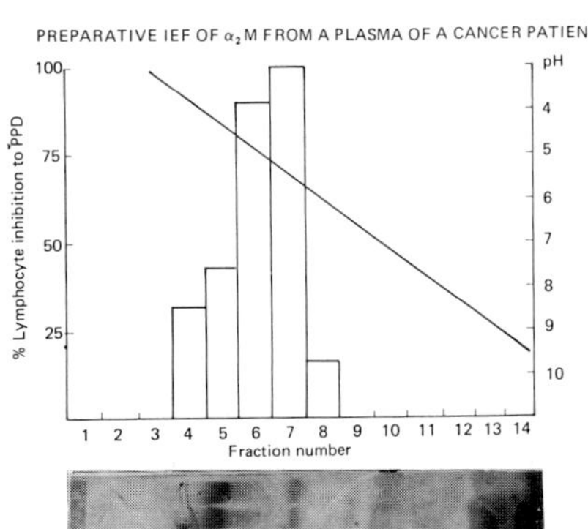

Pure α_2M was focused on a 12 x 24cm Agarose EF gel film containing 2% Ampholine 3.5–10. The initial setting of (100V, 5mA) for 30 minutes was followed by (200V, 10mA) for another 30 minutes and finally (1500V, 15W) for 1 hour. The pH gradient across the gel was determined by using IEF protein markers.

592

Discussion

Our data shows that the immunosuppressive activity of plasma
from patients with malignancy is greater than that from normal
subjects, when either autologous or allogeneic lymphocytes are
used as responding cells. The fact that SBTI collected blood
shows lower PSA than either heparinised blood or serum indicates that spurious proteinase activation during blood
collection may be responsible for elevated PSA. This is not so
prominent in samples collected by the same method from
patients with advanced malignant conditions. This could be
related to high protease levels which are liberated from
rapidly dividing cells in the tumour mass which could saturate
the already existing protease inhibitors. Hubbard (1) has
pointed out the possibility of an immunological control
mechanism produced by α_2M-P complexes. He suggested that
protease binding to α_2M could modulate immunological reaction
as a feedback control to prevent over-immunological reaction
which might harm the body, the association of lymphocyte suppressive activity with α_2M and the reduction of PSA with SBTI
in normal plasma supports the hypothesis proposed by Hubbard.
Our attempt to find any difference with the α_2M IEF pattern
from normal plasma, cancer patient plasma and ascitic fluid
showed no differences. We were, therefore, unable to confirm
the results of Ohlsson (5) on identification of different
forms of α_2M in analytical IEF after α_2M protease binding.

References

1 Hubbard, W.J.: Cell.Immunol. 39, 388 (1978).
2 James, K.: TIBS 2, 43 (1980).
3 Ohlsson, K., Skude, G.: Clinica Chimica Acta 66, 1 (1976).
4 Shenton, B.K., Jenssen, H.L., Werner, H., Field, E.J.: J.
  Imm.Methods 14, 123 (1977).
5 Virca, G.D., et al. Annal Biochem. 89, 274 (1978).

Affinity Electrophoresis

AFFINITY ELECTROPHORESIS WITH
EVALUATION OF GLYCOPROTEIN MICROHETEROGENEITY
AND SCREENING FOR LECTINS

T.C.Bøg-Hansen and J.G.Grudzinskas
The Protein Laboratory, University of Copenhagen, Denmark, and
Department of Obstetrics and Gynaecology, University of Sydney,
Australia

There are other parameters than charge and size to userfor analytical characterization of macromolecules by electrophoresis.
Therefore, the work of S.Nakamura is a cornerstone in the history of analytic electrophoresis, as acknowledged by the historical exhibition arranged at this conference. Nakamura devised
a simple electrophoretic system for the study of interaction
between macromolecules, two-dimensional paper electrophoresis
called 'crossing diagrams' (see for instance ref. 20). The interacting systems studied by Nakamura and coworkers were antigens and antibodies; trypsin and trypsin inhibitors; enzymes and
substrates; glycoproteins and con A; and proteins binding drugs.

In principle there are three approaches for studying interacting
molecules by analytic electrophoresis: to let the interacting
molecules react either before electrophoresis, during electrophoresis, or after electrophoresis. We believe that much information may be gained by letting interaction take place during
electrophoresis, and we have used the term 'affinity electrophoresis' to describe such electrophoretic systems (2). Well known
examples of affinity electrophoresis are electroimmunoprecipita-
tion methods like 'rocket immunoelectrophoresis' (18) and
'crossed immunoelectrophoresis' (8). An interesting example of
a study of an enxyme with its substrate is (23). We have mainly
worked with glycoprotein analysis with lectins (as reviewed in
refs. 4 and 5).

RECENT DEVELOPMENTS IN AFFINITY ELECTROPHORESIS

In addition to a more generalized understanding of the underlying principles of affinity electrophoresis (1, 6, 13, 14) the
recent developments have been a broadening of the experimental
scope, making new goals within easy reach with electrophoresis.
Examples of this is electrophoretic desorption of affinity gels
either by simple electrophoresis (9) or by isoelectric focusing
in granulated gels (21). Two dimensional methods may reveal
the pH-dependance of th dissociation constant (11) in a system
very similar to the original 'crossing diagrams' of Nakamura,
or they may reveal the dissociation constant of proteins or
subclasses of a single protein separated with respect to their
pI values (24).

Other applications of affinity electrophoresis have been shown
at this meeting (16 and 26). The clinically important separation of HbA_1 was shown by Janik and Ambler (15).

GLYCOPROTEINS STUDIED WITH LECTINS

With the 'crossing diagrams', Nakamura studied the variation of
the precipitation of serum glycoproteins with con A, though the
discrimination was limited to few peaks, the differences were
striking from normal serum to serum from patients with beneign
or malignant tumors (20).

With the introduction of agarose as medium for electrophoresis
additional advantages became apparent. One very important
aspect is the non-restrictiveness of the agarose matrix, allowing macromolecular complexes of a molecular weigth exceeding
10^7 to migrate freely. This has proven very important for detection and description of microheterogeneity of glycoproteins.
Probably the most sensitive technique for detection and description of microheterogeneity of glycoproteins is the 'first dimension technique', crossed immunoelectrophoresis with lectin
in the first dimension gel (3). We have used this technique
repeatedly as reviewed in (4, 5). An interesting modification

with two lectins in the first dimension gel has even greater resolution (26). Earlier we evaluated available techniques and discussed problems related to quantitation of lectin-binding and non-binding fractions of soluble proteins (1).

In last year's report we concluded that the microheterogeneity forms of human serum proteins are remarkably constant in normal circulation (12). An earlier study of orosomucoid showed a progressive change of the microheterogeneity pattern towards the less con A-binding orosomucoid subclasses during pregnancy (27, 28). Raynes has recently reported a study of five major serum proteins, the acute phase proteins: orosomucoid (α_1-acid glycoprotein, AGP), ceruloplasmin, α_1-antitrypsin (α_1-protease inhibitor), α_1-antichymotrypsin, and α_2-HS glycoprotein (22). The variation with con A was examined in pregnancy, in response to surgical wounding, during acute septic and aseptic inflammation and the chronic inflammation often associated with rheumatic disease and metastatic cancer, and it was found that the variation in the amounts of components of each acute phase glycoprotein in severe disease and pregnancy was outside the range seen in normal serum samples. The con A non-binding components were generally increased in acute pancreatitis, septicaemia and following major surgery. There may be a link to the severity of the disease or to the size of the acute phase response, and the greatest changes were observed in the pattern of α_1-antichymotrypsin and orosomucoid which are the proteins with the highest carbohydrate content of the five proteins studied by Raynes. These proteins are also the fastest to show plasma concentration increases after a stimulus. Raynes relates the pattern of orosomucoid and α_1-antichymotrypsin, ans well as the lack of significant changes of the patten of α_2-HS glycoprotein to a stimulus which causes a change in the hepatic biosynthetic rate to result in a change of addition of antennary sugars before secretion into the plasma (22).

The changes in the glycosylation pattern in the acute phase response to tissue injury compared with that in chronic inflam-

mation were explained by Raynes by the induction of the glyco-
sylating enzymes (22). Also at the end of an attack in acute
pancreatitis, α_1-antichymotrypsin has been found to change back
towards a higher con A non-binding proportion (7).

With the proteins studied by Raynes (22) and Wells et al. (27,
28) there is a reduction in con A-binding associated with
higher estrogen levels, estrogen possibly affecting protein
biosynthesis by binding to a high affinity receptor at the target surface (10). However, there seems not to be any direct
correlation with the estrogen drop after pregnancy and the
orosomucoid pattern (27, 28) reflecting a complex system of
regulation of synthesis of acute phase proteins and their glycosylation, strengthening the idea of two separate glycosyla-

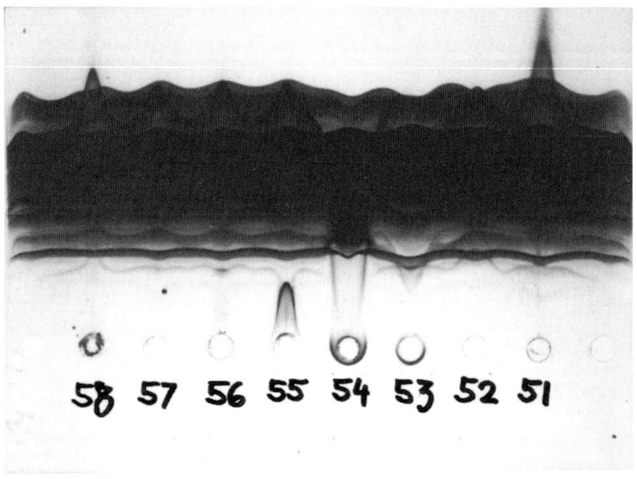

Fig. 1. Line immunoelectrophoresis of human serum proteins with
in situ absorption with plant extracts. Strong reactions are
seen for 53: Albizzia lucida (deflection of the lines) and for
54: Atalya hemiglauca, and 55: Banksia ericifolia (precipitate
formation). Similar strong reactions were observed in other
experiments with extracts from: Caesalpinia gilliesii, Calliste-
mon salignus, Cassia fistula, javanica, and nemophile, Casuarina dislyla, Cissus antartica and hypoplauca, Erythrina indica
and vespertillio, Eucalyptus eximia, grandis, haemastoma, radi-
ata, rossi, rubida, saligna, tenueramis, trabuti, and viminalis,
Indigofera australis, Jacaranda mimosaefolia, Leptospermum
squarrosum, Peltophorum africanum, Rhus toxidendron succadanea,
Schotia affra, brachypetala and lalifolia, and Sophora japonica.

tion systems (19) only one of which being affected by estrogen.

The main question is whether one may be able to deduce a change in synthesis from the changes in carbohydrate microheterogeneity forms, or whether these changes reflect the clearance or an alteration during blood circulation, thus to reflect the function of the glycoprotein. At this point there is no reason to eliminate any explanation of the microheterogeneity of the serum glycoproteins.

The clinical importance of detection of AFP (α-fetoprotein) microheterogeneity forms was outlined in (4, 5), and recently Taketa and Izumi has shown this with special reference to hepatocellular carcinoma (25). Generally it is our approach to assay for microheterogeneity changes. As several important marker proteins in human diseases are glycoproteins, their in-

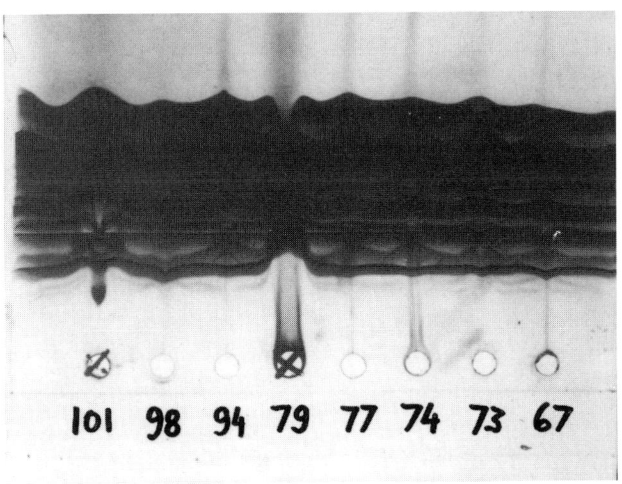

Fig. 2. Line immunoelectrophoresis of human serum proteins with in situ absorption with plant extracts. 79: Cissus antartica and 101: Indigofera australis show strong reaction, and 94: Ficus rubiginosa shows a weak reaction. Similar weak reactions were observed for: Acacia elongata, melanoxylon, mollissima, podalyraefolia, pycnantha, sophorae and stricta, Actinotus helianthi, Callitris rohmboidea, Calodendron capense, Cassia javanica and nodosa, Casuarina littoralis, Clitoria ternatea, Delonix regia, Leptospermum flavescens and Rhus lancea.

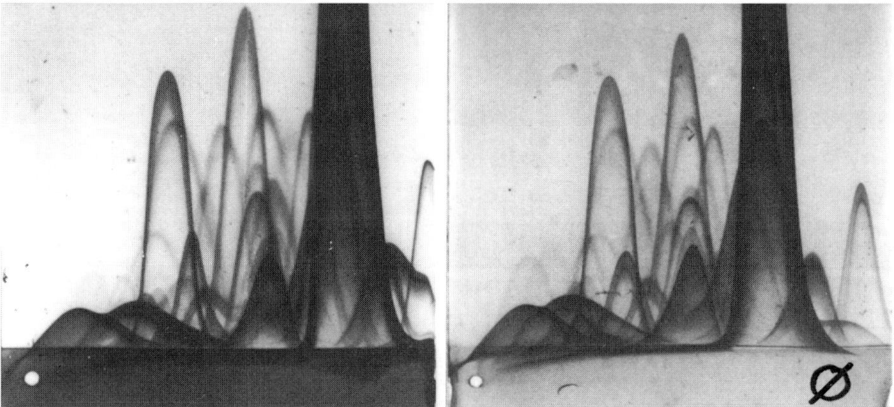

Fig. 3. The first dimension technique, crossed immunoelectrophoresis with plant extract in the first dimension gel. Normal serum proteins are tested with an extract of Banksia ericifolia seeds in the first dimension gel. Ø is the control
experiment without extract in the first dimension gel.

teraction with lectins may be expected to give important clues
to the state of the disease.

SCREENING FOR LECTINS

Presently we are screening a number of Australian seeds for
lectins. The screening is performed by means of electrophoresis and agglutination (details to be reported elsewhere).
Prior to electrophoresis, the dry seeds are crushed in a meat-
grinder, then homogenized with mortar and pestle with a small
volume of buffer (less than 10 ml/g). After centrifugation the
supernatant is tested by electrophoresis.

Electrophoresis is first performed as 'line immunoelectrophoresis with in situ absorption' (17), see Figs. 1 and 2. This
type of electrophoresis is probably the easiest way to test a
large number of lectin extracts for precipitating capacity towards glycoproteins.

Several different reactions are seen in Figs. 1 and 2, and the
specificity of the lectin in each reacting extract is analyzed

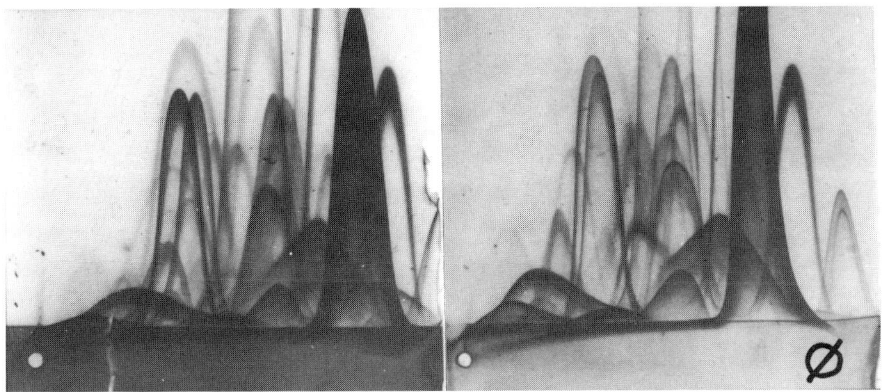

Fig. 4. The first dimension technique as in Fig. 3, but the
serum proteins were from a patient suffering from acute phase
reaction. Otherwise as for Fig. 3.

in the first dimension technique, see Figs. 3 and 4. As mentioned above this analysis is probably the most sensitive. The
search for lectins have revealed some rather discriminating lectins with specificity for acute phase proteins, as shown in Fig.
4. It is our aim to prepare such lectins for detailed description of the carbohydrate changes during biological processes.

References:

1. Andersen, M.M., Hau, J. and Bøg-Hansen, T.C. in Lectins -
    Biology, Biochemistry, Clinical Biochemistry Vol. 2 (Bøg-
    Hansen, T.C., editor) W. de Gruyter, Berlin 1982, pp. 779.

2. Bøg-Hansen, T.C. Anal.Biochem. 56, 480-488 (1973).

3. Bøg-Hansen, T.C., Bjerrum, O.J. and Ramlau, J. Scand.J.
    Immunol. 4 Suppl. 2, 141-147 (1975).

4. Bøg-Hansen, T.C., in Solid Phase Biochemistry. Analytical
    and Synthetic Aspects (Scouten, W.H., editor) J.Wiley and
    Sons, New York in press.

5. Bøg-Hansen, T.C. and Hau, J. J.Chrom.Library 18B, 219-252
    (1981).

6. Bøg-Hansen, T.C. and Takeo, K. Electrophoresis 1, 67-71
    (1980).

7. Bowen, M., Raynes, J.G. and Cooper, E.H. in Lectins - Bio-
    logy, Biochemistry, Clinical Biochemistry Vol. 2 (Bøg-
    Hansen, T.C., editor) W. de Gruyter, Berlin 1982, pp. 403.

8. Clarke, H.G.M. and Freeman, T. Clin.Sci. 35, 403 (1968).

9. Dean, P.D.G., Qadri, F., Jessup, W., Bouriotis, V., Angal, S., Potuzak, H., Leatherbarrow, R.J., Miron, T., George, E. and Morgan, M.R.A. in Affinity Chromatography. Les Colloques de l'Inserm Vol. 86 (Egly, J.M. editor) Inserm, Paris 1979, pp. 321-344.

10. Eisenfeld, A.J., Alen, R., Weinberger, M., Haselbacher, G. and Halpern, K. Science 191, 862 (1976).

11. Ek, K. and Righetti, P.G. Electrophoresis 1, 137-140 (1980).

12. Hinnerfeldt, F., Albrechtsen, J.M. and Bøg-Hansen, T.C. in Electrophoresis '82 (Statachos, D. editor) W. de Gruyter, Berlin 1983.

13. Horejsi, V. J.Chrom. 178, 1-13.

14. Horejsi, V., Ticha, M., Tichy, P. and Holy, A. Anal.Biochem. 125, 358-369 (1982).

15. Janik, B. and Ambler, J. These proceedings.

16. Kasai, K. and Shimura, K. These proceedings.

17. Krøll, J. and Andersen, M.M. J.Immunol.Methods 9, 141-146 (1975).

18. Laurell, C.B. Anal.Biochem. 15, 45 (1966).

19. Mega, T., Lujan, E. and Yoshida, A. J.Biol.Chem. 255, 4057 (1980).

20. Nakamura, S. Cross Electrophoresis. Its Principle and Applications. Igaku Shoin, Tokyo and Elsevier, Amsterdam 1966.

21. Rautenberg, P., Reinwald, E., Kaufmann, E. and Risse, H.J. in Lectins - Biology, Biochemistry, Clinical Biochemistry Vol. 2 (Bøg-Hansen, T.C. editor) W. de Gruyter, Berlin 1982, pp. 619-632.

22. Raynes, J. Biomedicine 36, 77-86 (1982).

23. Szylit, M. Ann.Biol.Clin. 29, 215-227 (1971).

24. Takeo, K., Suzuno, R., Tanaka, T., Fujimoto, M., Kuwahara, A. and Nakamura, K. These Proceedings.

25. Taketa, K. and Izumi, M. Prot.Biol.Fluids 31, in press.

26. Taketa, K. and Izumi, M. These proceedings.

27. Wells, C., Cooper, E., Glass, R.M. and Bøg-Hansen, T.C. Clin.Chem.Acta 109, 59-67 (1981).

28. Wells, C., Cooper, E. and Bøg-Hansen, T.C. in Lectins - Biology, Biochemistry, Clinical Biochemistry Vol. 1 (Bøg-Hansen, T.C. editor) W. de Gruyter, Berlin 1981, pp. 339-346.

ELECTROPHORETIC PROPERTIES OF RCA-2 BINDING GLYCOPROTEINS IN
HUMAN SERA AND ASCITIC FLUID WITH HEPATOCELLULAR CARCINOMA

Masaki Fukushima, Terukatsu Arima, Fumiaki Suwa,
Junichi Watanabe, Norio Koide, and Hideo Nagashima
The First Department of Internal Medicine, Okayama University
Medical School, Okayama 700, Japan

Introduction

The carbohydrate chains of cell surface glycoproteins are considered to change in the course of differentiation. Some
malignant cells have abundant cell surface glycoproteins which
have galactose (Gal) and N-acetyl-galactosamine (GalNAc) residues on their non-reducing end of carbohydrate chains (1).
These glycoproteins may appear in the circulation by being
shed from the surface of malignant cells (2).
To prepare carcinoma associated glycoproteins, we investigated
the glycoproteins having terminal Gal and/or GalNAc residues,
in serum and ascitic fluid. To this purpose, two lectins were
utilized, Ricinus communis agglutinin-1 and 2 (RCA-1, RCA-2),
which are specific for terminal Gal and both Gal and GalNAc
respectively.

Materials and Methods

The total number of sera subjected to hemagglutination inhibition assay was 119, composed of 23 normal, 21 hepatocellular
carcinoma (HCC), 22 liver cirrhosis, 20 chronic hepatitis, 10
acute hepatitis, 23 other gastrointestinal cancer patients. A
large volume of ascitic fluid was obtained from a patient with
HCC. These samples were stored at -20°C until use.
RCA-1, RCA-2, RCA-2 agarose were purchased from E.Y.

Laboratories, antiserum to whole human serum (rabbit) from DAKO, and other media and reagents from other commercial sources. The inhibition assay was carried out in a microtest plate, 96 wells round bottom, as follows. To each 25μl of serum, diluted with gelatin Veronal buffer (GVB^{++}) in series, was added an equal volume of RCA-1 or RCA-2 solution, diluted to have sufficient hemagglutinating strength. After incubation for 10min at room temperature, 25μl of human O-type RBC, $2x10^8$/ml in 0.01M sodium phosphate buffer, pH 7.3(PBS), were added. The mixture was kept for 30 min, at 37°C and then examined for hemagglutination. The serum asialo-glycoprotein (AGP) level was evaluated by the method described by a co-auther Arima (3). It was measured by inhibition of AGP in the sample in binding reaction between rat plasma membrane and ^{125}I-labeled asialo-orosomucoid. The amount of AGP was expressed in terms of nanograms of asialo-orosomucoid equivalents. Ten ml of ascitic fluid (170mg protein) obtained from a patient with HCC were applied to a column (1x14cm) of RCA-2 agarose (4mg/ml of RCA-2) and glycoproteins bound to this column were eluted with 0.3M β-lactose after washing the column with PBS. After the removal of β-lactose, the eluate was applied to a column (1.6x18cm) of Sepharose 4B, conjugated with IgG fraction of anti-whole human serum (WHS), 15mg/ml gel bed, to eliminate serum components. Glycoproteins bound to this column were eluted with 8M urea. In all, 160ml of the ascitic fluid were subjected to these procedures of affinity chromatography. Electrophoretic properties of these glycoproteins, the eluate of RCA-2 agarose column, the effluent and the eluate of anti-WHS-Sepharose 4B column, were studied by SDS-polyacrylamide gel electrophoresis. After the removal of salt, these glycoproteins were heated at 100°C for 3min in 0.025M Tris-HCl, pH6.8, 3% SDS, 5% 2-mercaptoethanol, 10% glycerin and 0.001% Bromophenol Blue, and 50-100μg protein in the solution were subjected to 7.5% gel slabs (14x12cmx0.75mm). The gel slabs were run at 25mA per slab. Gels were stained

Tab. I   RCA-I   hemagglutination inhibition

	Serum dilution, cases(%)			
	< ×8	×8	×16	positive*/total(%)
HCC**	8(38.1)	10(47.6)	3(14.3)	13/21(61.9)
Liver cirrhosis	7(31.8)	10(45.5)	5(22.7)	15/22(68.2)
Chronic hepatitis	12(60.0)	8(40.0)		8/20(40.0)
Acute hepatitis	5(50.0)	5(50.0)		5/10(50.0)
Other cancer	10(43.5)	9(39.1)	4(17.4)	13/23(56.5)
Others	20(87.0)	3(13.0)		3/23(13.0)

*hemagglutination inhibition titer more than 8

** hepatocellular carcinoma

using Coomassie Brilliant Blue R-250 and the molecular weights
of proteins located on the gel were calculated from their
mobility compared to standard proteins (from Pharmacia).

Results

By hemagglutination inhibition assay with lectins, it was
shown that glycoproteins having binding activity to these
lectins increased in liver disease and carcinoma. As shown in
table 1 and 2, when the samples were diluted 16-fold, RCA-1
binding activity was detected in sera of HCC, liver cirrhosis
and other gastrointestinal cancers. A similar result was obtained with RCA-2. RCA-1 or RCA-2 binding activity did not
correlate with serum α-fetoprotein (AFP) level, determined by
a RIA method, nor with serum carcino-embryonic antigen (CEA)
level.
As lectin binding activity was elevated in liver disease and
carcinoma, relationships between the binding activity and
liver function tests were studied. Neither RCA-1 nor RCA-2
binding activity were associated with serum glutamic pyruvic
transaminase (SGPT) level and zinc turbidity test (ZTT), while
the binding activity tended to increase, as cholinesterase
activity or Indocyanine Green clearance rate (K_{ICG})
decreased in chronic hepatitis and liver cirrhosis (Fig.1).

There was no correlation between lectin binding activity and serum AGP level.

Tab.2 RCA-2 hemagglution inhibition

	Serum dilution, cases(%)			
	< ×4	×4	×8	positive*/total(%)
HCC **	11(52.4)	10(47.6)		10/21(47.6)
Liver cirrhosis	13(59.1)	8(36.4)	1 (4.5)	9/22(40.9)
Chronic hepatitis	19(95.0)	1 (5.0)		1/20 (5.0)
Acute hepatitis	9(90.0)	1(10.0)		1/10(10.0)
Other cancer	5(21.7)	15(65.2)	3(13.1)	18/23(78.3)
Others	21(91.3)	2 (8.7)		2/23 (8.7)

*hemagglutination inhibition titer more than 4
**hepatocellular carcinoma

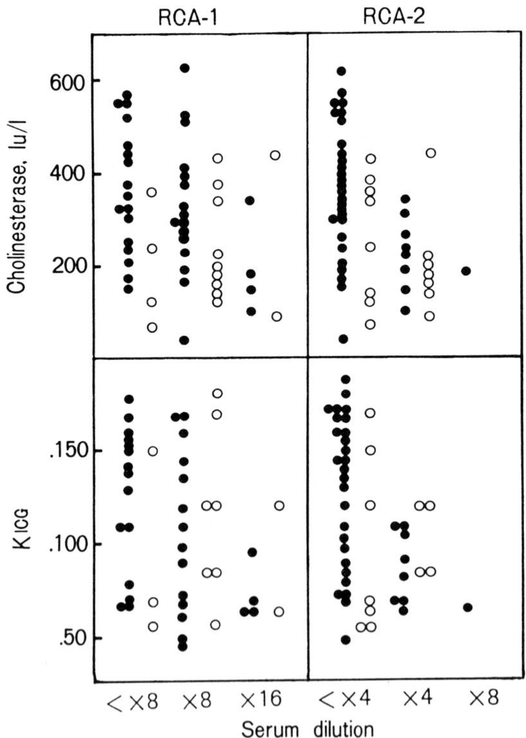

Fig. I Lectin hemagglutination inhibition and liver function tests

RCA-2 binding glycoproteins in ascitic fluid from a patient
with HCC were prepared, with RCA-2 agarose affinity chromatography from which contaminating normal serum components
were absorbed by an anti-WHS-Sepharose 4B column. With 0.3M
β -lactose, 6.5% of the proteins in the fluid were eluted
from a RCA-2 column, which had absorbed RCA-2 binding
glycoproteins in 10ml of ascitic fluid. Similarly 92mg of
RCA-2 binding glycoproteins were prepared from 160ml of
ascitic fluid, and were applied to an anti-WHS-Sepharose 4B
column. Finally 3.6mg of glycoproteins were prepared, which
were equivalent to 0.13% of the original proteins in the
ascitic fluid.

These glycoproteins were subjected to SDS-polyacrylamide gel
electrophoresis. As a result, a main band at 48K daltons of
molecular size as well as several numbers of minor bands were
observed (Fig.2).

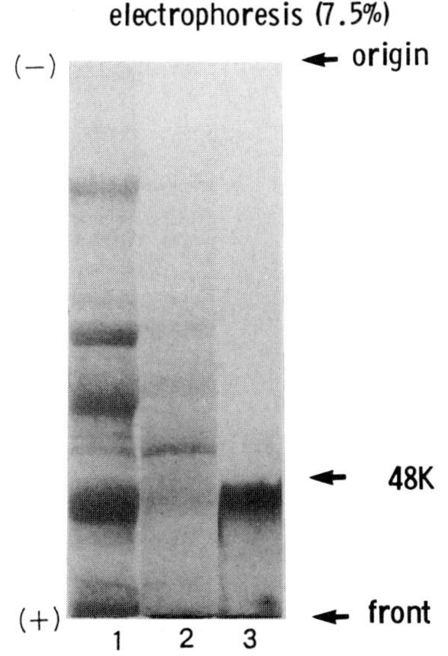

Fig.2 SDS-polyacrylamide gel
electrophoresis (7.5%)

1. Glycoproteins eluted from RCA-2 agarose
   column

2. Eluate of anti-whole human serum-
   Sepharose 4B column

3. Effluent of anti-whole human serum-
   Sepharose 4B column

Discussion

Cellular transformation results in changes of the carbohydrate chains of cell surface glycoproteins and glycoprotein
synthesis by the cell. These glycoproteins may appear in the
circulation by being secreted like AFP, or being shed from the
tumor cell surface. In similar fashion, levels of circulating
glycoproteins are often increased in patients with carcinoma.
These glycoproteins have binding activity to RCA-1, and are
abundant in sialic acid, and can be distinguished from acute
phase reactant glycoproteins associated with malignancy, CEA,
and AFP (4,5,6). In our hemagglutination inhibition assay,
RCA-1 or RCA-2 binding activity increased in liver disease and
gastrointestinal cancers including HCC, and correlated partially with K_{ICG} and cholinesterase activity, but not with AFP,
CEA, SGPT, and ZTT. Liver cell membrane was identified as the
major locus of binding for circulating AGP (7). While in liver
disease, serum AGP level was elevated, especially in liver cirrhosis, and correlated partially with K_{ICG} (8,9). In our
study lectin binding activity did not correlate with serum AGP
level, presumably because of the terminal sialic acid content
and complexity of the structures of carbohydrate chains (10).
Judging from the observations described above, carcinoma associated glycoproteins may participate in the elevation mechanism
of lectin binding activity in HCC, in addition to liver dysfunction, and may be expected to exist in malignant ascitic
fluid. We attempted to prepare such glycoproteins using RCA-2
agarose and anti-WHS-Sepharose 4B affinity chromatography. The
SDS-polyacrylamide gel electrophoretic pattern of these glycoproteins showed a main band at 48K daltons and several minor
bands. Some of them may originate from cancer cells. Further
investigation are in progress.

References

1. Koyama, K., Nundelman, E.: Cancer Res. 39, 3677-3678 (1979)

2. Kapeller, M., Gal-Oz, N.: Exptl. Cell Res. 79, 152-158 (1973)

3. Arima, T., Motoyama, Y.: Gastroenterol. Jpn. 11, 300-306 (1976)

4. Lipton, A., Harvey, H.A.: Cancer 43, 1766-1771 (1979)

5. Harvey, H.A., Lipton, A.: Cancer 47, 324-327 (1981)

6. Bolmer, S.D., Davidson, E.A.: Biochemistry 20, 1047-1054 (1981)

7. Pricer, W.E.Jr., Ashwell, G.: J. Biol. Chem. 246, 4825-4833 (1971)

8. Marshall, J.S., Green, A.M.: J. Clin. Invest. 54, 555-562 (1974)

9. Arima, T.: Gastroenterol. Jpn. 14, 349-352 (1979)

10. Baenziger, J.U., Fiete, D.: J. Biol. Chem. 254, 9795-9799 (1979)

SEPARATION AND IDENTIFICATION OF DIFFERENT MOLECULAR SPECIES
OF HUMAN α-FETOPROTEIN BY DOUBLE AFFINITY ELECTROPHORESIS WITH
CONCANAVALIN A AND <u>LENS</u> <u>CULINARIS</u> HEMAGGLUTININ A

Kazuhisa Taketa and Masaki Izumi
Health Research Center, Kagawa University, Takamatsu 760,
Japan

Hidekazu Nakabayasni and Jiro Sato
Division of Pathology, Cancer Institute, Okayama University
Medical School, Okayama 700, Japan

Introduction

Human α-fetoprotein (AFP) in body fluids consists of varying
proportions of distinct molecular species of AFP with different affinities to lectins. With concanavalin A (con-A) and
<u>Lens</u> <u>culinaris</u> hemagglutinin A (LcH-A) as affinity ligands,
four major molecular species of AFP (AFP_1, AFP_2, AFP_3 and AFP_4)
have been separated by affinity chromatography in combination
with affinity electrophoresis and characterized by determination of dissociation constants of AFP-lectin complexes
(1,2). AFP_1 has no affinity to either con-A or LcH-A, AFP_2
has an affinity to con-A but a weak or no affinity to LcH-A,
AFP_3 has affinities to both con-A and LcH-A, and AFP has no
affinity to con-A but a definite affinity to LcH-A.
The proportion of con-A- or LcH-reactive AFP in body fluids is
known to vary under several pathophysiological conditions
(3,4). It is, therefore, of some interest to determine the
proportion of individual molecular species of AFP in the
conditions associated with AFP production.
In this report, we describe a simple and quantitative method
for separation of the four major molecular species of AFP by a

pair of double affinity electrophoreses using con-A and LcH-A
gel-strips linked together in alternative orders. The results
were compared with those obtained by affinity column chromatographies with similar lectins.

Materials and Methods

As sources of AFP, frozen samples of sera or ascites fluids
from patients with hepatocellular carcinomas, spent culture
media of human hepatocellular carcinoma (HuH-7) and hepato-
blastoma (HuH-6 Cl-5) cell lines (5), sera from cases of
fulminant hepatitis, saved for us by Dr. H. Endo (Kurashiki-
chuo Hospital, Kurashiki), and of ovarian tumor (embryonal
carcinoma, Group C, or yolk sac tumor), kindly provided by Dr.
R. Ohkawa (Nippon Medical School, Tokyo), an amniotic fluid at
term and a cord serum (human α-1-fetoprotein standard, DAKO-
immunoglobulins Lab., Copenhagen) were employed.
Crossed immuno-affino-electrophoresis of B∅g-Hansen et al.(6)
was performed as reported earlier (1,2) with 1% Litex agarose
(LSB, Litex, Glostrup) gels 1.0 mm thick and prepared in a
phenobarbital buffer (pH 8.6, μ=0.025). In double affinity
electrophoresis, separate agarose gels with con-A (2.0 mg/ml,
Pharmacia Fine Chem., Uppsala) and LcH-A (1.0 mg/ml, E-Y Lab.,
San Mateo) were serially linked as illustrated in Fig. 1 for
the first dimension electrophoresis, which was run at 20- 25°C
and at 10 V/cm until bromophenol blue (BPB) moved exactly 8 cm
for the gel with a junction of two lectin gels set at 3.5 cm
from the origin. Diluted samples (20-30 ng of AFP) with the
phenobarbital buffer containing BPB were applied to gel slots
and detected after electrophoresis with rabbit immunoglobulins
to human AFP (10 μl/ml gel, DAKO-immunoglobulins) by the
protein A-peroxidase method of Taketa et al. (1). Human serum
albumin was always present or included as a reference and
detected with its antibody. Mobilities of AFP peaks (R) were
expressed as relative to the migration of albumin peak

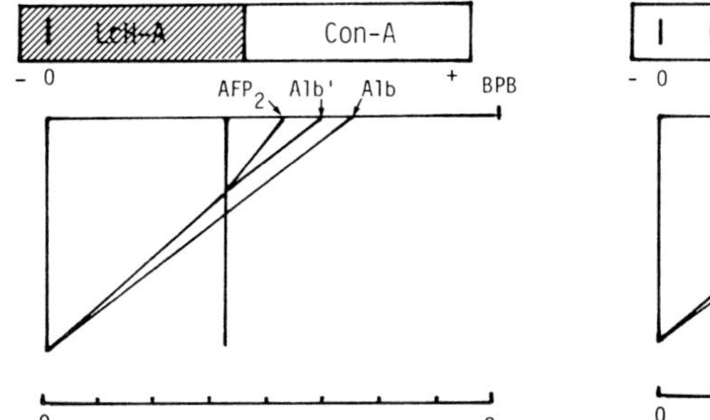
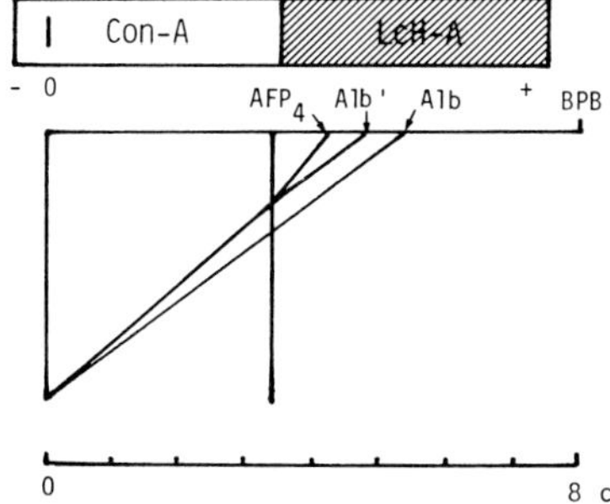

Fig. 1. Schematical representation of a pair of double affinity electrophoreses of AFP with the LcH-A-con-A (left panel) and the con-A-LcH-A (right panel) gels. The agarose gel areas shown by giving the names of lectins are those before electrophoresis and the vertical lines represent the junctions after electrophoresis. Alb, the distance of human serum albumin-BPB migrated from the origin; and Alb', the theoretical distance of Alb-BPB migrated from the junction.

(actually albumin-BPB) (2). Distribution of AFP among the major molecular species was estimated by measurement of the peak areas of immunoprecipitates on enlarged photos. AFP was also fractionated by affinity column chromatographies with Con A-Sepharose (Pharmacia Fine Chem.) and/or LcH-Gel (E-Y Lab.) as described previously (2).

Results and Discussion

The results of double affinity electrophoresis on samples with known molecular species of AFP are shown in Fig. 2. A spent medium of HuH-7, containing AFP_1, AFP_3 and AFP_4, gave two peaks in the LcH-A-con-A gel system and three peaks in the con-A-LcH-A system (Fig. 2A). The most anodic peak in either sequence of lectin gels was regarded as AFP_1, because it did not show any affinity to either con-A or LcH-A. The cathodic peak in the LcH-A-con-A gel was considered as AFP_3 plus AFP_4 since its mobility in the first interacting LcH-A gel corresponded to that of AFP_3 or AFP_4, and thus it did not

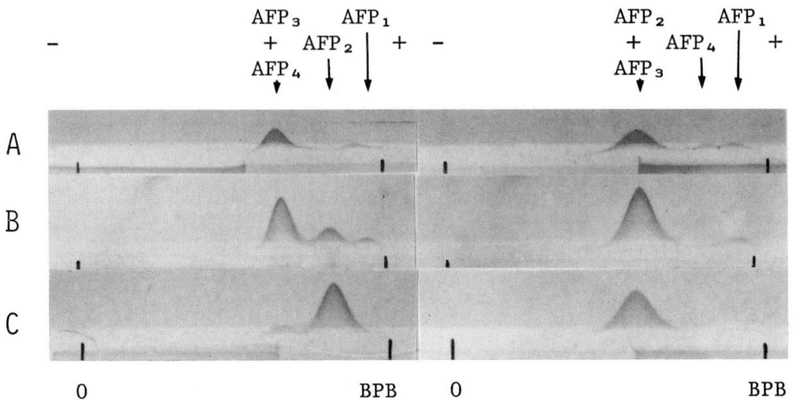

Fig. 2. Immunoprecipitate peaks of AFP in a pair of double affinity electrophoreses with con-A and LcH-A. The LcH-A-con-A gel series are shown in the left panel and the con-A-LcH-A gel series in the right panel. A, spent medium of HuH-7; B, serum from a case of hepatocellular carcinoma; and C, cord serum. 'O' denotes the origin.

enter the second con-A gel. AFP_2, if present, should form an intermediate peak in this system as its mobility would be reduced only after entering the subsequent con-A gel. Similarly, the small intermediate peak found in the con-A-LcH-A gel system should be AFP_4. A serum from hepatocellular carcinoma, having AFP_1, AFP_2 and AFP_3, gave three peaks in the LcH-A-con-A gel, corresponding to AFP_1, AFP_2 and AFP_3 from the anode, and two peaks in the con-A-LcH-A gel, the slower peak being composed of AFP_2 and AFP_3 and the fast peak being AFP_1 (Fig. 2B). In cord serum with a predominant AFP_2 and a trace of AFP_3, the intermediate peak of AFP_2 in the LcH-A-con-A gel was pronounced and a mixed single peak of AFP_2 and AFP_3 was formed in the con-A-LcH-A gel (Fig. 2C).

As may be seen in the above three illustrated cases, the proprotion of the major molecular species of AFP can be readily obtained by subtracting the known area of a peak with single molecular species in one of the paired double affinity systems (AFP_2 in the LcH-A-con-A gel or AFP_4 in the con-A-LcH-A gel) from the area of the peak with two mixed AFP's in the other affinity series (AFP_2 + AFP_3 in the con-A-LcH-A gel or AFP_3 +

AFP$_4$ in the LcH-A-con-A gel) provided that the total peak
areas in each series are made equal.
The mobilities of AFP$_2$ and AFP$_4$ for single lectins in the
second gels in the double affinity electrophoresis can be
estimated for further identification by calculation with the
following formulae:

$$R(AFP_2) = (AFP_2 - 3.2) \times 0.84/(Alb \times 0.84 - 3.2) \ (cm/cm) \quad (1)$$
$$R(AFP_4) = (AFP_4 - 3.4) \times 0.86/(Alb \times 0.86 - 3.4) \ (cm/cm) \quad (2)$$

where Formulae 1 and 2 are for the LcH-A-con-A gel and the
con-A-LcH-A gel, respectively, and the symbols are: R, the
mobility of AFP in the presence of 2.0 mg/ml con-A for AFP$_2$ or
1.0 mg/ml LcH-A for AFP$_4$; Alb $\times$ 0.84, for calculation of Alb'
from the R of AFP$_2$ in the LcH-A gel of the LcH-A-con-A gel
(Fig. 1); Alb $\times$ 0.86, for calculation of Alb' from the R of
AFP$_4$ in the con-A gel of the con-A-LcH-A gel (Fig. 1); 3.2,
the distance from the origin to the junction for LcH-A gel
after electrophoresis (reduced from 3.5 cm to 3.2 cm as con-A
diffused out during electrophoresis); and 3.4, the distance
from the origin to the junction for con-A gel after electrophoresis (reduced from 3.5 cm to 3.4 cm as LcH-A diffused out
during electrophoresis). Reference values of R obtained with
single lectins under the present assay conditions are given in
Table 1 for identification of AFP$_1$, AFP$_2$, AFP$_3$ and AFP$_4$.

Table 1. R values for major molecular species of AFP

Lectins (mg/ml)	R (M + SD)			
	AFP$_1$	AFP$_2$	AFP$_3$	AFP$_4$*
Con-A (2.0)	0.86 ± 0.01 N=8	0.59 ± 0.02 N=8	0.59 ± 0.02 N=7	0.86 ± 0.03 N=2
LcH-A (1.0)	0.87 ± 0.01 N=3	0.83 ± 0.03 N=7	0.60 ± 0.05 N=7	0.65 ± 0.04 N=3

N, the number of determinations; *, from hepatocellular
carcinoma and yolk sac tumor; otherwise from hepatocellular
carcinoma.

616

The distribution of AFP among the four molecular species
determined by the double affinity electrophoreis is presented
in Table 2 by listing the values as per cent of the total peak
area of AFP, and the results are compared with those obtained
by affinity chromatographies with Con A-Sepharose and LcH-Gel
(or Con A-Sepharose chromatography in combination with affinity electrophoresis with LcH-A). The results obtained by the
double affinity electrophoresis agreed well with those obtained by other methods or compatible with the binding ratios
determined by other workers (3,4) with similar individual
lectins. Thus, AFP_2 was a predominant molecular species of
the AFP found in cord serum, amniotic fluid at term or serum
from fulminant hepatitis, AFP_3 was a major but variable
component of the AFP in hepatocellular carcinoma, and AFP_4 was

Table 2. Per cent distribution of major molecular species of
AFP from different sources

Sources of AFP	Procedure	AFP_1	AFP_2	AFP_3	AFP_4
Hepatocellular carcinoma:					
Case HR (ascites)	E-E	18	33	49	0
	C-E	20	28	51	1
	C-C	21	35	44	0
Case NT (ascites)	E-E	7	58	35	0
Case KM (serum)	E-E	8	25	67	0
Case TK (serum)	E-E	7	0	93	0
HuH-7 (medium)	E-E	10	0	79	11
	C-E	7	7	74	12
Hepatoblastoma:					
HuH-6 Cl-5 (medium)	C-E*	54	8	7	31
Yolk sac tumor:					
Case YO (serum)	E-E	4	0	50	46
	C-E	4	0	41	55
Fulminant hepatitis:					
Case MI (serum)	E-E	5	59	36	0
Others:					
Cord serum	E-E	0	91	9	0
Amniotic fluid (at	E-E	0	92	8	0
term)	C-E	0	90	10	0

*, a reduced mobility of AFP in the absence of lectins was
found and this did not permit us to separate AFP's by double
affinity electrophoresis. E, affinity electrophoresis; C,
affinity column chromatography; and -, sequence of procedure.

characteristic of the AFP produced by yolk sac tumor. Although further heterogeneity exists in each molecular species of AFP separated in the present study, as revealed by the splitting of a peak in prolonged electrophoresis with the same lectin or in electrophoresis with other lectins (7,8), the double affinity elecrophoresis with con-A and LcH-A is a simple and quantitative technique for one-step separation of the major molecular species of AFP and may prove to be a clinically valuable tool for differential diagnosis of AFP-producing pathological conditions. The method is also applicable to other combinations of lectins with different affinities to AFP or other glycoproteins. This would provide a useful inform-ation for the fractionation of oligosaccharides by serial lectin-affinity chromatography (9), which should be preferably directed to the analysis on the isolated single species of glycoprotein with respect to the lectin reactivity.

APPENDIX: A SUGGESTED SYSTEM FOR IDENTIFICATION OF DIFFERENT MOLECULAR SPECIES OF AFP DUE TO SUGAR CHAIN HETEROGENEITY

The number of different molecular species of AFP, which can be separated by differential affinities to lectins, has been increasing as the additional lectins, such as erythroaggluti-nating phytohemagglutinin (PHA-E) (8), are used as separation media. This necessitated us to introduce a new system of nomenclature for identification of AFP heterogeneity as revealed by different affinities to lectins.
The rule is to affix '0' or '1' to AFP, depending on the affinity to lectins, in the order of interacting lectins employed; '0' denoting no or negligible affinity and '1' a definite affinity under the defined assay conditions. Low affinity species may be differentiated by employing other symbols, 1', 1'' and so forth, in place of 1. The names of lectins should be given, at least at first mention, in parentheses in the order corresponding to the preceding number. Examples are given below.

Numbering system	Suggested new system
AFP_1 :	AFP_{00} or AFP-00 (con-A, LcH-A)
AFP_2 :	AFP_{10} or AFP-10 (con-A, LcH-A)
AFP_3 :	AFP_{11} or AFP-11 (con-A, LcH-A)
AFP_4 :	AFP_{01} or AFP-01 (con-A, LcH-A)

The AFP_4 with affinity to PHA-E (8) may be expressed as AFP_{011} or AFP-011 (con-A, LcH-A, PHA-E).

This work was supported by Grant-in-Aid from the Ministry of Education, Science and Culture (57570836). We thank Mrs. Eriko Ichikawa for her excellent technical assistance.

References

1. Taketa, K., Toguchi, E., Izumi, M., Takeo, K.: in Electrophoresis'82, D. Stathakos, ed., Walter de Gruyter, Berlin, in press.

2. Taketa, K., Izumi, M., Ichikawa, E.: Ann. New York Acad. Sci. in press.

3. Bręborowicz, J., Mackiewicz, A., Bręborowicz, D.: Scand. J. Immunol. 14, 15-20 (1981).

4. Miyazaki, J., Endo, Y., Oda, T.: Acta Hepatol. Jpn. 22, 1559-1568 (1981).

5. Nakabayashi, K., Taketa, K., Miyano, K., Yamane, T., Sato, J.: Cancer Res. 42, 3858-3863 (1982).

6. Bøg-Hansen, T.C., Bjerrum, O.J., Ramlau, J.: Scand. J. Immunol. 4, Suppl. 2, 141-147 (1975).

7. Toftager-Larsen, K.: in Lectins, Biology, Biochemistry and Clinical Biochemistry, Vol. 2, T.C. Bøg-Hansen, ed., Walter de Gruyter, Berlin, (1982), pp. 433-444.

8. Taketa, K., Izumi, M.: in Protides of the Biological Fluids, Colloquium 31, H. Peeters, ed., Pergamon Press, Oxford, in press.

9. Cummings, R.D., Kornfeld, S.: J. Biol. Chem. 257, 11235-11240 (1982).

AFFINOPHORESIS OF TRYPSINS

Electrophoresis of Trypsins in the Presence of a Soluble Polyelectrolyte Bearing Affinity Ligand

Kiyohito Shimura and Ken-ichi Kasai
Department of Biological Chemistry, Faculty of Pharmaceutical
Sciences, Teikyo University, Sagamiko, Kanagawa, Japan 199-01

Introduction

Most of electrophoretic separations are based on the
difference of the intrinsic electric charges of individual
substances.   The electrophoresis of complexes with some ions,
however, is often effective; e.g.,   neutral sugars with borate
and polypeptides with dodecylsulfate.   Affinophoresis is also
based on the complex formation of a certain protein with
affinophore (1).   Affinophore means a soluble macromolecule

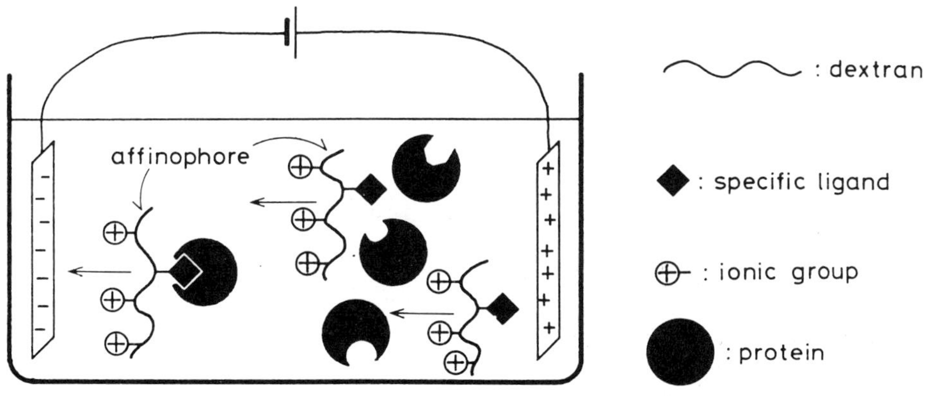

fig. 1   The principle of affinophoresis

which has not only specific affinity ligands but also many electric charges, either positive or negative.   It forms a complex with a certain protein owing to the specific affinity and changes the electrophoretic mobility of the protein.   If the electrophoretic mobility of the complex is much differed from other substances, then a specific separation can be achieved.

Methods

1) Preparation of an affinophore for trypsin (1)

Dextran (average mol. weight 10,500) was dissolved in dimethylsulfoxide and its hydroxyl groups were converted to alkoxides with methylsulfinyl carbanion and followed by the alkylation with diethylaminoethylbromide and 6-chloroacetamidohexanoic acid. m-Aminobenzamidine, a competitive inhibitor for trypsin, was coupled to the carboxyl groups of the dextran derivative by using water-soluble

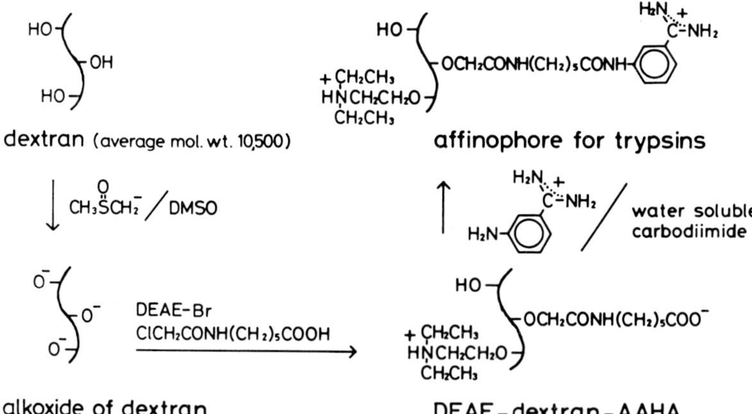

Fig. 2   Preparation of the affinophore
DEAE: diethylaminoethyl, AAHA: 6-acetamidohexanoic acid

carbodiimide and the product was used as an affinophore for trypsins. The affinophore is a polycation in a solution of neutral to acidic pH owing to the positive charges of the diethylaminoethyl group and the amidino group of the m-aminobenzamidine portion. The m-aminobenzamidine portion of the affinophore was expected to function as an affinity ligand for trypsin (2). The ligand was spaced out from backbone of the dextran with the distance of 9 atom lengths to eliminate steric hindrance by the backbone.

The content of the substituents in the affinophore was given in Table 1 with methods utilized for the determination. The affinophore inhibited the activities of bovine and Streptomyces erythreus trypsin on benzoyl-L-arginine p-nitroanilide.

Substituent	Content (μmol/mg)	Method
Diethylaminoethyl group	0.94	Titration
6-Acetamidohexanoic acid	0.64	Amino acid analysis
m-Aminobenzamidine	0.63	Spectrophotometry

Table 1   Content of the substituents in the affinophore

2) Affinophoresis

Agarose gel was formed by spreading 1.3 ml of a hot solution on a glass plate (slide glass for microscopy, 76 x 26 x 1.3 mm). The solution comprised 0.8% (w/v) of agarose, 0.1 M of sodium phosphate, pH 7.0, with (0.1 mg/ml) or without the affinophore. A sample solution (2 μl), which contained 4 μg of protein and 40 mM of sodium phosphate with (4 μg) or without the affinophore, was applied in the middle of the gel plate. After electrophoresis, proteins were visualized by

fluorescamine as fluorescent bands.   Coomassie Brilliant Blue
R 250 and Amido Black 10B were not suitable because both
simultaneously stained the affinophore.

Results and Discussion

Affinophoresis of <u>Streptomyces</u> <u>erythreus</u> trypsin (SE-trypsin)

SE-trypsin (3) which is produced by a species of streptomyces,
is an anionic protein (pI=4.0) in contrast with the well-known
trypsin of bovine pancreas (pI=10.1-10.8).   As shown in lane
7 of Fig. 3, SE-trypsin ordinarily moves toward cathode.   SE-
trypsin can be inactivated by the reaction with L-1-chloro-3-
tosylamido-7-amino-2-heptanone (TLCK), an active site
modification reagent for trypsin-type proteases.   Thus TLCK-
treated SE-trypsin was used as a control protein to see the
role of the active site in the affinophoresis.

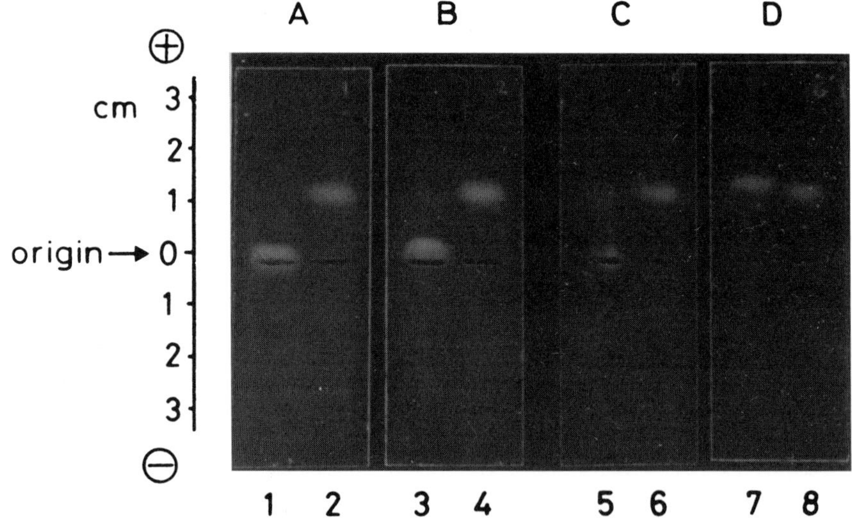

Fig. 3   Affinophoresis of SE-trypsin.   Electrophoresis
of SE-trypsin (lane 1,3,5,7) and TLCK-treated SE-trypsin
(lane 2,4,6,8) was carried out for 1.5 hr, at a constant
current of 4 mA/plate.   Plate A and B contained the
affinophore (0.1 mg/ml) in the gel.   Samples loaded on
lane 1, 2, 5 and 6 contained the affinophore (4 μg in 2
μl).

When the affinophore was included throughout the gel, apparent electrophoretic mobility of SE-trypsin toward anode was greatly reduced (see lane 1 and 3 of Fig. 3), while that of TLCK treated SE-trypsin scarcely changed (see lane 2 and 4 of Fig. 3). The decrease of electrophoretic mobility of SE-trypsin is related to the intactness of the active site. Thus, SE-trypsin must have bound to the benzamidine portion of the affinophore by its active site. The net negative charge of the complex must have become smaller than that of SE-trypsin. Therefore we were able to confirm the affinophoresis really occurred.

When electrophoresis of a mixture of SE-trypsin and the affinophore was carried out in agarose gel which did not contain the affinophore, SE-trypsin did not form a definite band (see lane 5 of Fig. 3). This result is interpreted as follows. Free SE-trypsin and free affinophore in the sample zone will migrate toward opposite directions, anode and cathode, respectively. Trypsin molecules initially located in the anodic side of the sample zone can escape early from the zone of the affinophore and will migrate toward anode with its intrinsic migration rate. However, those located in the cathodic side of the sample zone will escape from the affinophore much later. The delay of the time when the migration with the intrinsic electrophoretic mobility begins resulted in the formation of the indefinite band seen in lane 5 of Fig. 3. This interpretation supposes a rapid equilibrium in the complex formation. This is probably the case for the complex, because the affinophore showed rapid and constant inhibition for trypsin. This is the reason why an equilibrium concentration of free affinophore must be maintained to continue affinophoresis.

At pH 4, isoelectric point of SE-trypsin, feature of the affinophoresis was more evident (1). In the presence of the affinophore, SE-trypsin rapidly migrated toward cathode, the same direction as the affinophore moves. On the other hand, without the affinophore, SE-trypsin remained at the origin.

624

Discrimination between trypsin and chymotrypsin by the affinophore.

Electrophoresis of bovine pancreatic trypsin and chymotrypsin was carried out in the presence of the affinophore along with SE-trypsin (Fig. 4). Electrophoretic mobility of chymotrypsin was scarcely changed (lane 6 and 8 of Fig. 4), while that of trypsin toward cathode obviously increased (lane 2 and 4 of Fig. 4). In spite of the structural resemblance of the two proteins, the affinophore discriminated the specificity determining site of trypsin from that of chymotrypsin. The result also indicates that the trypsin-affinophore complex was formed through the specific interaction between the benzamidine portion of the affinophore and the active site of trypsin.

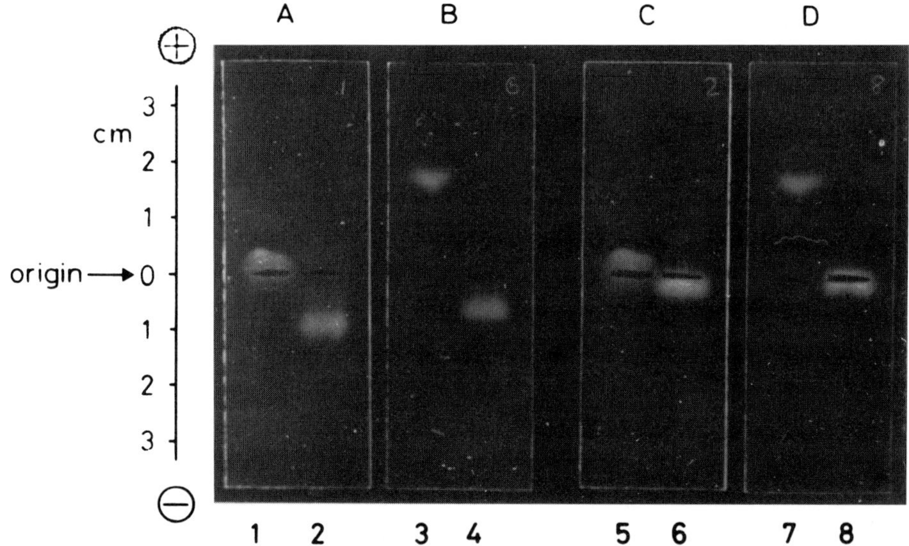

Fig. 4   Affinophoresis of bovine trypsin and chymotrypsin. Electrophoresis of bovine trypsin (lane 2 and 4), bovine chymotrypsin (lane 6 and 8) and SE-trypsin (lane 1,3,5,7) was carried out for 1 hr, at a constant current of 7.3 mA/plate. Plate A and C contained the affinophore (0.1 mg/ml) in the gel. Sample loaded on plate A and C contained the affinophore.

Comparison with conventional affinity techniques and the future of affinophoresis

The most striking difference between affinophoresis and conventional affinity techniques for separation of biological molecules, affinity chromatography (4) and affinity electrophoresis (5,6), is that the latter two require gel matrices to immobilize affinity ligands. Although, in the present work, the affinophoresis of trypsin was carried out in agarose gel, what was expected for the gel was only anti-convectional function. Affinophoresis is basically possible in a free solution. This distinctive feature may be particularly advantageous for the separation of insoluble particles like cells and subcellular organelles, because gel matrices are not suitable for them. Affinophores which bind to specific surface receptors will modify the surface electric charges of the particles and enable us to separate them electrophoretically on the basis of specific affinity.

References

1. Shimura, K., Kasai, K.: J. Biochem. $\underline{92}$, 1615-1622 (1982).

2. Mares-Guia, M., Shaw, E.: J. Biol. Chem. $\underline{240}$, 1579-1585 (1965).

3. Yoshida, N., Sasaki, A., Inoue, H.: FEBS Lett. $\underline{15}$, 129-132 (1971).

4. Cuatrecasas, P., Wilchek, M., Anfinsen, C.B.: Proc. Natl. Acad. Sci. U.S.A. $\underline{61}$, 636-643 (1968).

5. Takeo, K., Nakamura, S.: Arch. Biochem. Biophys. $\underline{153}$, 1-7 (1972).

6. Hořejší, V., Kocourek, J.; Biochim. Biophys. Acta $\underline{336}$, 338-343 (1974).

HETEROGENEITY OF RABBIT ANTI-DNP ANTIBODY STUDIED BY TWO-DIMENSIONAL AFFINITY ELECTROPHORESIS

Kazusuke Takeo, Ryosuke Suzuno, Tatehiko Tanaka,
Masanori Fujimoto, Akira Kuwahara, and Kazuyuki Nakamura
Department of Biochemistry, Yamaguchi University School of Medicine
J-755 Ube, Japan

Introduction

The two-dimensional electrophoresis technique measuring differences in charged state and biological activity is effective when separating heterogeneous protein mixtures of equal molecular weight, such as immunoglobulins. We developed a two-dimensional affinity electrophoresis system the first direction of which is isoelectric focusing(IEF) and the second direction affinity electrophoresis(AEP)(1). Using homologous (Dnp) and heterologous hapten (Tnp) as affinity ligands for AEP, rabbit anti-Dnp antibodies were separated. They consisted of several groups of monoclonal IgG groups. Dissociation constants of the individual IgG spots for Dnp-hapten and for Tnp-hapten were calculated.

Methods

Rabbits were immunized with Dnp-human-γ-globulin (5mg as an emulsion of an equal volume mixture of CFA) by injection into the footpads, followed by intracutaneous injections into the dorsum at monthly intervals. The animals were bled 10 days after the sexth boost. The anti-Dnp antibodies were purified from plasma by ammonium sulfate fractionation

followed by affinity chromatography with a Dnp-lysine-
Sepharose column (2). The antibody preparations obtained
were composed of IgG species in verified by tests by SDS-
disc electrophoresis, Ouchterlony procedure, and immunoelectrophoresis.

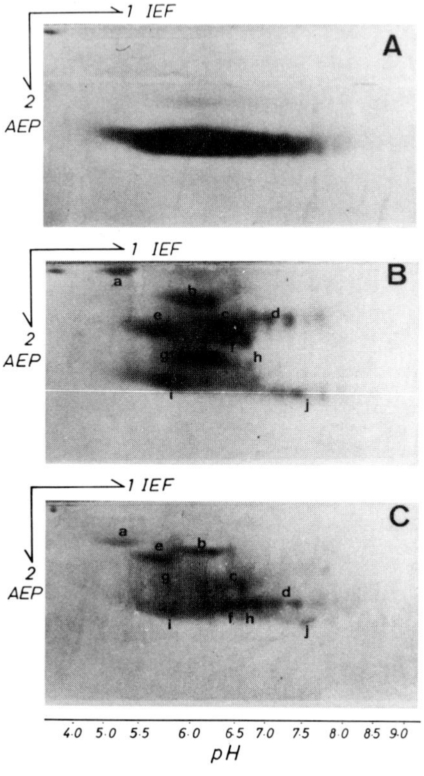

Fig. 1. Two-dimensional affinity electrophoresis patterns of rabbit anti-Dnp antibody.

In the first dimension, IEF was carried out (4) with 5% polyacrylamide
gel (PAG) (5 x 11.4 x 0.1 cm) containing 6M urea, 10% sucrose and an
eight fold diluted Pharmalyte of an equal volume mixture of the stock
solutions having a range of pH 4.0 to 6.0 and 6.0 to 9.0. The purified
antibody (470 ug) was applied in 3 mm width. A 2 mm gel strip from the
finished IEF gel was applied to the gel for the second dimension, AEP,
carried out by using the acidic buffer system of Reisfeld et al. (10)
with 5% PAG (13.2 x 10 x 0.2 cm) containing no affinity ligand (Fig.
1A), 25.1 uM Dnp-PA (Fig. 1B), and 25.1 uM Tnp-PA (Fig. 1C) until the
tracking methylene blue (MB) band migrated about 10 cm through the gel.
Protein was visualized under silver staining (11).

The water soluble Dnp-polyacrylamide conjugate (Dnp-PA) and
its heterologous analogue (Tnp-PA) were prepared by coupling
dinitrofluorobenzene or trinitrobenzene sulfonate with a non-
cross-linked acrylamide allylamine (10:1 w/w) copolymer (3).
The relative migration distance (Rm) is expressed as a ratio
of the migrating distance of the protein to that of the
tracking methylene blue (MB) band.

Results

Fig. 1A shows a two-dimensional electrophoresis pattern in
the absence of the affinity ligand. The anti-Dnp-antibody
migrated as a broad line in the pI range between 5.0 and
8.0. The Rm value was 0.33 at pI 5.3 and 0.38 at pI 7.4.
Fig. 1B shows a two-dimensional electrophoresis pattern in
the presence of Dnp-PA (25.1 uM). The antibody was separated into more than 40 distinct spots. The Rm values were
distributed in a range from 0.025 to 0.37, and tentatively
grouped from a - j, as shown in the figure. All groups were
composed of several spots having almost identical Rm values,
except for group a which was diffusely stained. Fig. 1C
shows the analogous electrophoresis pattern obtained with
25.1 uM Tnp-PA. The antibody separation resembled a fin-
gerprint. While the pattern was different from Fig. 1B, the
spots composing every group migrated the same distance. This
indicates that each group of IgG may be a monoclonal antibody. If this is the case, the anti-Dnp antibody is composed of at least nine distinct monoclonal antibodies.
Groups h and i have weak affinity for both affinity ligands.
In the other eight groups, only group e had a stronger
affinity for Tnp-hapten than the Dnp-hapten. As seen in
Figs. 1B and 1C, the pI fractions still contain several
antibody fractions having different affinities for the
hapten. Hence, the binding constants obtained under IEF (5.

630

6, 7) only give the average for the heterogeneous antibody.

When the dissociation constant(Kd) was calculated by the
original affinity equation (8,9), the Kd of group a for the
Dnp-hapten was 1.99 x 10^{-6}M, whereas the value for the Tnp-
hapten was 1.33 x 10^{-5}M. On the other hand, the Kd value of
group e for the Dnp-hapten was 3.14 x 10^{-5}M, and the value
for the Tnp-hapten was 2.01 x 10^{-5}M. Hence, group a antibody had an affinity for the homologous hapten that was 7
times stronger than for the heterologous hapten, while group
e antibody had a 1.6 times stronger affinity for the heterologous hapten.

References

1. Takeo, K., Suzuno, R., Fujimoto, M., Tanaka, T., and Kuwahara, A.:
    Electrophoresis '82, Walter de Gruyter, Berlin - New York, 277-283
    (1982).

2. Eisen, H. N., Gray, W., Little, J. R. and Simms, E. S.: Meth.
    Immun. Immunochem. 1, 351-360 (1967).

3. Nakamura, K., Kuwahara, A., Ogata, H. and Takeo, K.: J. Chromatogr.
    192, 351-362 (1980).

4. Awden, Z. L., Williamson, A. R. and Askonas, B. A.: Nature 219, 66-
    67 (1968).

5. Hoffman, D. R., Grossberg, A. L. and Ressman, D.: J. Immunol. 108,
    18-25 (1972).

6. Cisar, J., Kabat, E. A., Dornaer, M. M. and Liao, J.: J. exp. Med.
    142, 435-459 (1975).

7. Congy, N. and Mihaesco, C.: Immunology 35, 307-315 (1978).

8. Takeo, K. and Nakamura, S.: Arch. Biochem. Biophys. 153, 1-7 (1972).

9. Takeo, K. and Kabat, E. A.: J. Immunol. 121, 2305-2310 (1978).

10. Reisfeld, R. A., Lewis, U. T. and Williams, E. E.: Nature 201, 281
    (1962).

11. Merril, C. R., Goldman, D., Sedman, S. A. and Ebert, M. H.: Science
    211, 1437-1438 (1981).

MEASUREMENT OF GLYCOSYLATED HEMOGLOBIN (HBA$_1$) ON CELLULOSE ACETATE
MEMBRANES IN A MOBILE AFFINITY ELECTROPHORESIS SYSTEM

Borek Janik

Gelman Sciences Inc., Ann Arbor, MI 48106, U.S.A.

Jeffrey Ambler

Clincal Chemistry Dept., University Hospital, Nottingham NG7 2UH, U.K.

The measurement of HbA$_1$ has generated a lot of excitement in recent years with the realization of its great potential in monitoring diabetic control and hence, lessening the sequelae of diabetes. The methods for measurement of HbA$_1$ generally require its separation from the nonglycosylated hemoglobins. The short ion exchange column procedure, although technically simple, have several drawbacks like temperature dependence of the results. Isoelectric focusing and HPLC, despite their obvious advantage of being able to measure HbA$_{1c}$ directly, are not suitable for routine applications. The colorimetric method has been much criticized because of the uncertain nature of the reaction and failure to compensate for all the unknown variables.

All attempts to use standard electrophoretic techniques for separating HbA$_1$ and HbA inevitably failed due to only minute charge differences in these two hemoglobin species. The use of agar and the principles of electrochromatography represented a major breakthrough.(1) In order to overcome certain disadvantages of agar media, we developed a procedure using cellulose acetate membranes and a novel principle of mobile affinity electrophoresis.(2) We are presenting the results obtained with a commercially available Glyco-Phore Test which is based on the above principle.

Procedure

Glyco-Phore buffer in conjunction with Mylar^{TM}-supported cellulosic membranes (Super Sepraphore) and standard electrophoretic equipment (all Gelman Sciences) and the procedure recommended by the manufacturer were used. To establish the effects of temperature, the buffer, chamber and media were equilibrated to the desired value and kept at it during the entire run.

Results

The normal values obtained with the Glyco-Phore procedure on 100 individuals with no confirmed diabetes agree well with the generally accepted normal HbA_1 values (Table I).

TABLE I. NORMAL VALUES OF HbA_1 OBTAINED BY THE GLYCO-PHORE PROCEDURE

	% HbA_1
Mean	6.1
Median	5.9
5th-95th Percentile Range	4.6-7.9
10th-90th Percentile Range	5.0-7.1
25th-75th Percentile Range	5.4-6.7

Unlike the mini-column results, those obtained with the Glyco-Phore test were found to be independent of temperature within the range which could possibly be expected in laboratories (Table II).

TABLE II. EFFECTS OF TEMPERATURE ON HbA_1 VALUES DETERMINED BY GLYCO-PHORE TEST

HBA_1	Temperature (^{O}C)			
	14	16	23	30
X ± SD		5.8+0.4	5.8+0.4	
CV (%)		7.1	7.1	
n		4	16	
X ± SD	16.7+0.4		16.8+0.5	16.4+0.7
CV (%)	2.1		2.9	4.2
n	8		24	8

The concentration of HbA_1 was measured in clinical samples with normal and elevated HbA_1 levels by Glyco-Phore Test and three commercial procedures: Corning (electrophoresis), Isolab (mini-column) and Bio Rad (mini-column). Each sample was run in duplicate with all procedures. The correlation parameters are summarized in Table III.

TABLE III. CORRELATION OF GLYCO-PHORE TEST WITH OTHER COMMERCIAL HbA_1 TESTS

	Gelman Glyco-Phore Vs.:		
	Corning	Isolab	Bio Rad
Correl. Coef.	0.917	0.856	0.890
Slope	1.011	0.919	0.734
y-Intercept (% HbA_1)	0.48	1.73	3.24
n	64	54	64

The Bio Rad results correlated rather poorly not only with the Glyco-Phore test but also with the Corning and Isolab tests. Obviously, differences do exist between the results obtained by methods which are based on the same principle (e.g., Isolab vs. Bio Rad).

The Glyco-Phore test results were found independent of the electrophoretic chamber (all Gelman Sciences) if the conditions recommended by the manufacturer were used (Table IV). One membrane per chamber with eight samples per membrane was used in all cases.

TABLE IV. RESULTS OBTAINED BY GLYCO-PHORE TEST IN THREE DIFFERENT ELECTROPHORETIC CHAMBERS

HbA_1	Sepratek	Semi-Micro II	Semi-Micro
$\bar{X} \pm$ S.D. (% HbA_1)	6.6 ± 0.3	6.7 ± 0.3	6.5 ± 0.5
CV (%)	4.3	4.5	8.3

Occasionally, distortions of the HbA band were observed (staple effect). These were caused by nonuniform distribution of the sample at the point of application. Higher proportion of the hemolysing solution (e.g., $>$ 450 uL per 50 uL of washed red cells in samples with high Hb) and longer depression of the applicator key (30 sec) generally helped to remedy the problem.

634

Conclusions

The Glyco-Phore test was found technically simple to perform and suitable for both high and low volume routine use (e.g., up to 24 samples
can be run simultaneously in Semi Micro II chamber). Absence of temperature effects and overall economy of the test were considered as
commendable advantages.

References

1. Allen, R.C., Stastny, M., Hallett, D., and Simmons, M.A.; in
    Electrophoresis 1979, B.J. Radola, Ed., Walter de Gruyter, New
    York, NY, 1980, pp. 663-670.
2. Ambler, J., Janik, B., and Walker, G.; Clin. Chem. 29, 340-343
    (1983).

IMMUNOCHEMICAL INVESTIGATIONS OF $Rh_o(D)$ ACTIVITY
DETECTED IN BAND 3 OF RED CELL MEMBRANE

Tsuyoshi Yokoi, Mineo Iwasa, Kaoru Sagisaka
Department of Legal Medicine, Gifu University
School of Medicine, Gifu 500, Japan

Introduction

Characterization of Rh-Hr antigens has been frustrated by the
inability to solubilize red cell membrane without loss of antigenicity. In a previous paper(1), we described the isolation
and purification of $Rh_o(D)$ antigen which is located in Band 3
possessing no sialoglycoproteins. In this study, the immune
complex of anti-D and D active Band 3 which had been prepared
from deoxycholic acid(DOC) solubilized stroma was investigated
by some electrophoretic methods.

Materials and Methods

Preparation of Band 3 and complex of anti-D and Band 3. Stroma
prepared by the method of Fairbanks et al.(2) was solubilized
by 1% DOC and applied onto Sepharose 4B and 6B, followed by
fractionation with reduced thiol Sepharose 4B(2). Band 3 prepared by this method was incubated with half a volume of human
anti-D(Ortho Co.) for 30 min to 3 hr at 37°C. The resulting
immune complex was soluble in a 0.2% DOC solution. This complex
was separated by Sepharose 4B gel filtration.
Antisera. Rabbit anti-Band 3 was prepared by the method of
Fukuda et al.(3). Anti-IgG was also raised in rabbits.
Immunoassay. The Ouchterlony test, immunoelectrophoresis(IEP),
SDS polyacrylamide gel disc electrophoresis (SDS-PGDE) and
crossed-IEP were carried out as reported by Iwasa et al.(4).
The D activity of the fractions of the chromatographies was

determined as described previously (2).

Results

Chromatographies of the complex of Band 3 and anti-D on Sepharose 4B. Figs.1-1,1-2 and 1-3 show elution profiles of the complexes prepared by incubation for 30 min, 1 and 3 hr under the same conditions. The peak in Fig.1-2 shifted to the left in Fig.1-3 with longer incubation time. The molecular weight of the complex prepared by incubation for 1 hr was about 500,000 to 700,000 increasing to about 1,000,000 after incubation for 3 hr. The resulting complex was not prepared from red cell components other than Band 3, and D negative red cells produced no complex at all.

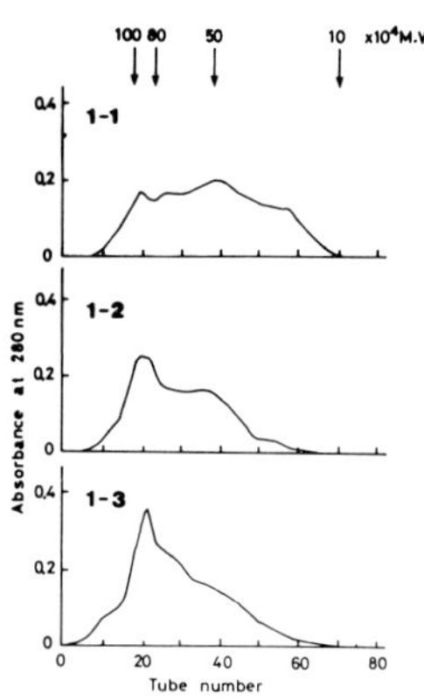

Fig.1, Chromatographies of the complex on Sepharose 4B.
The complex was applied onto Sepharose 4B and eluted with 10 mM Tris buffer pH 7.5 containing 0.2% DOC. Incubation times of the mixture were 30 min(1-1), 1 hr(1-2) and 3 hr(1-3), respectively.

Immunoassay and SDS-PGDE analyses. The fractions containing immune complex reacted definitely with anti-Band 3 and anti-IgG in Ouchterlony tests. In Fig.2, the complex(B) gave a single precipitation line with anti-human IgG at the γ-globulin region on IEP. The complex agglutinated papain-treated D positive red cells, the potency of which corresponded to 8-fold diluted anti-D (original titer 1:512), but did not agglutinate D negative red cells. In crossed-IEP, the mixture gave two peaks and the complex gave a single peak with anti-Band 3 (Figs.3 and 4).

Only a single peak was observed for the mixture prepared from Band 3 of D negative red cells. SDS-PGDE of the complex showed a pattern consisting of fragments of IgG and of Band 3 which were not stained with PAS reagent (Fig.5).

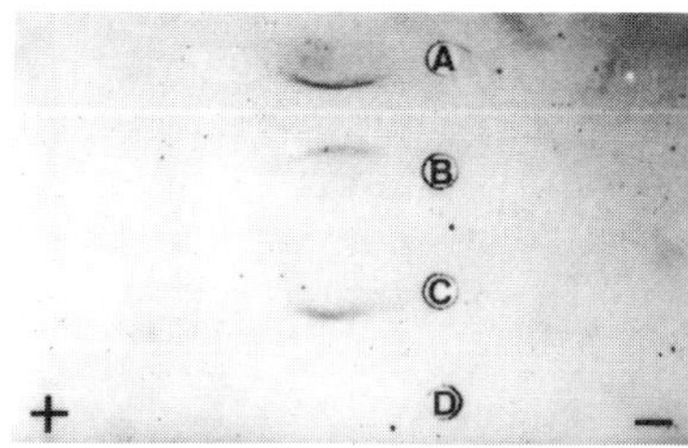

Fig.2, IEP patterns of Band 3 and the complex with anti-human IgG. A and C, human IgG; B, complex; D, Band 3.

Fig.3, Crossed-IEP of the complex with anti-Band 3. IEP in the first dimension was run at 10 mA for 1 hr. The second run was carried out on agarose gel containing anti-Band 3 at 5 mA for 12 hr.

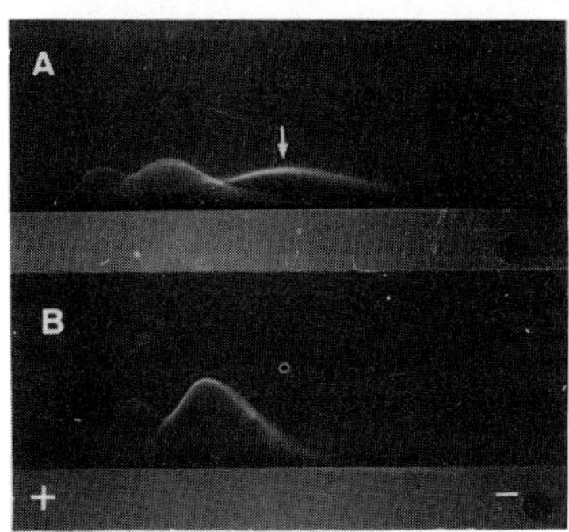

Fig.4, Crossed-IEP of the mixture of stroma and anti-D with anti-Band 3. A. Stroma was reacted with half a volume of saline. The precipitation line indicated by an arrow was supposed to be complex of Band 3 and anti-D. Electrophoresis was done as Fig.3.

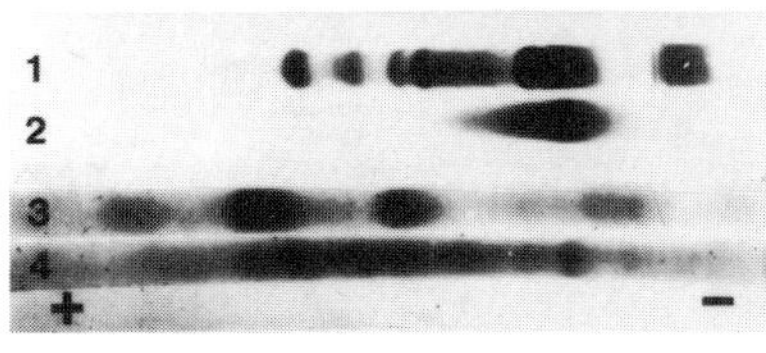

Fig.5, SDS-PGDE patterns. Track 1, human stroma; 2, Band 3; 3, human IgG;4, complex.

Discussion

The IEP and crossed-IEP applied to the final fractions demon-srrated that the D antigen located in Band 3 reacted specifically with anti-D. The shift of the first peak to the left according to the time of incubation indicates that aggregation of the immune complex progressed during the longer incubation time. The complex was stable in 0.2% DOC solution. As the hydrophilic portions of membrane proteins were in contact with the aqueous phase(5), the reactivity of the D antigen with anti-D was maintained in the DOC solution. The anti-D in the complex retained its agglutinin activity weakly.

Recently Plapp et al.(6) have reported that D antigen is present in not only D positive but also in D negative red cells. That the complex could not be produced from Band 3 of D negative red cells in the present study argues against this view. The intensity of the bands from IgG fragments in SDS-PGDE was scarcely reproduced. This suggests that the D antigen in Band 3 reacted with anti-D IgG at multiple rather than integral ratios. D antigen was detected in both fractions separated with reduced thiol Sepharose 4B. The fraction absorbing the resin, which was only one-fifth of Band 3, was known to have disulfide bonds (7). It is clear therefore that the disulfide bonds in Band 3 and sialoglycoproteins are not essential for D antigenicity. Which components in Band 3 account for D antigenicity awaits further studies.

References

1. Yokoi,T.,Iwasa,M.,Sagisaka,K.: Tohoku J.exp.Med. in press.
2. Fairbanks,G.,Steck,T.L.,Wallach,D.F.H.: Biochemistry 10,2606 -2624 (1971).
3. Fukuda,M.,Eshdat,Y.,Tarone,G.,Marchesi,V.T.: J.Biol.Chem. 253,2419-2428 (1978).
4. Iwasa,M.,Yokoi,T.,Kanemitsu,K.,Sagisaka,K.: Tohoku J.exp. Med. 136,303-312 (1982).
5. Helenius,A.,Simons,K.: Biochem.Biophys.Acta 415,29-79 (1975).
6. Plapp,F.V.,Kowalski,M.M.,Tilzer,L.,Brown,P.J.,Evans,J., Chiga,M.:Proc.Natl.Acad.Sci.USA 76,2964-2968 (1979).
7. Lutz,H.U.,Daniken,A.,Sememza,G.,Bachi,T.:Biochem.Biophys. Acta 552,262-280 (1979).

The Heterogeneous Reaction of Seminal Acid Phosphatase to Canavalia gladiata DC Lectin

Mineo Iwasa, Tsuyoshi Yokoi and Kaoru Sagisaka
Department of Legal Medicine, Gifu University School of
Medicine, Gifu 500, Japan

Introduction

Affinity electrophoresis has been employed to analyze interaction between lectins and carbohydrates (Takeo & Nakamura 1972; Horejsi et al. 1977), and the number of binding sites of glycoproteins to lectin has been analyzed by crossed immuno-affinoelectrophoresis (Bøg-Hansen & Brogren 1975). Using modified procedures, we investigated the reactive heterogeneity of acid phosphatase (AcP) of seminal plasma (SP) to Canavalia gladiata DC lectin (Can G).

Materials and Methods

Can G-SP complex. Can G and SP were mixed, and the resulting complex was separated by centrifugation. After being washed adequately with cold saline, the complex was suspended in 0.15 M phosphate buffered saline pH 7.2 containing 0.02 M α-methyl-D-mannoside. After centrifugation, an aliquot of the supernatant was fractionated on a Sephadex G-200 column. AcP activity of the effluents was assayed using p-nitrophenyl phosphate as the substrate.
Prostate sections and antisera. A prostate frozen at -25°C was cut into 5μ thick sections by a cryostat. AcP activity of the

sections was examined by the method of Grogg and Pearse (1952). Anti-Can G and -SP were raised in rabbits and the anti-Can G was labeled with fluorescein isothiocyanate (FITC).

Immuno-affinoelectrophoresis (IAEP). Purified Can G was coupled to agarose gel by the cyanogen bromide activation method. For IAEP with anti-SP, three types of intermediate gels were prepared: a reference gel, a gel containing Can G, and agarose gel-bound Can G at a concentration of 0.1 mg/cm^2. The precipitation line due to AcP in each IAEP was visualized by the azo-dye method using α-naphtyl phosphoric acid and diazonium o-dianisidine.

Results

AcP activity of Can G-SP complex. Can G gave a strong precipitation line possessing AcP activity against SP in the Ouchterlony test. Gel filtration of SP and of solubilized complex of Can G and SP on Sephadex G-200 gave two peaks with AcP activity, the profiles of which were quite similar (Fig.1).

IAEP. To examine the heterogeneity of AcP to Can G, IAEP was carried out on SP (Figs.2-A) and the first and second fraction (Figs.2-B and -C) from the gel filtration on Sephadex G-200 using anti-SP. SP and the second fraction from Sephadex G-200 gel filtration gave a large, definite precipitation line (Fig. 2-1). The height of the AcP lines due to SP and the second fraction was reduced but one due to the first fraction was not reduced in the gel containing free Can G (Fig.2-2, indicated by an arrow).

Inhibition test of AcP with Can G by histochemical method. AcP activity of the prostate sections was examined before and after treatment with Can G. Controls were without substrate or tartaric acid as an enzyme inhibitor. Tartaric acid inhibited AcP activity of the sections but Can G did not. However, Can G bound to the sections could be detected with FITC-labeled anti-Can G.

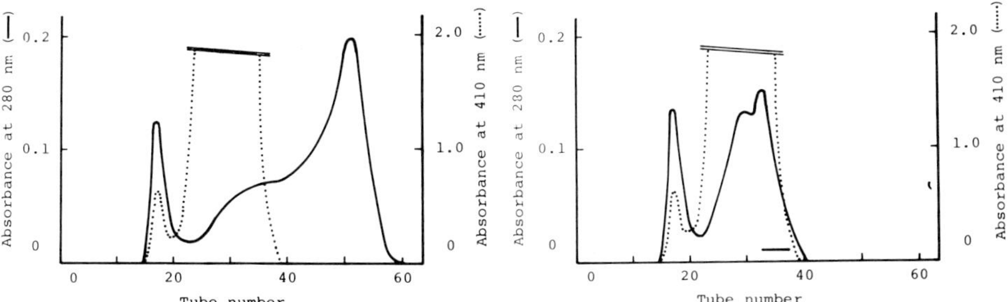

Fig. 1. Fractionation of SP (left) and solubilized complex of Can G and SP (right) on Sephadex G-200 (2.5 X
40 cm). SP and complex gave two peaks with AcP activity
of which profiles were quite similar.

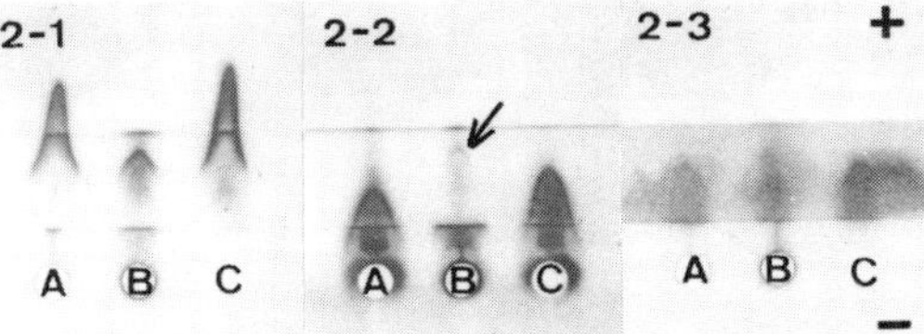

Figs. 2-1, 2-2 and 2-3. IAEP on SP(A), the first (B)
and second fractions (C) using anti-SP. Three types of
intermediate gels were prepared; a reference gel (Fig.
2-1), a gel containing free Can G (Fig.2-2) and gel-bound
Can G (Fig.2-3). The precipitation line due to AcP in
each IAEP was visualized by the azo-dye method.

Discussion

The heterogeneity of AcP of SP to Can G was examined. Our
previous study revealed that Can G had an affinity for AcP of
SP (Iwasa and Sagisaka 1983). In the present study, we found
that Can G produced a precipitation line possessing AcP activity against SP in the Ouchterlony test, indicating that the AcP
of the complex retained partial activity. Fractionation of the
solubilized complex of Can G and SP on Sephadex G-200 suggested
that the reaction of Can G and AcP was reversible. In IAEP,
glycoproteins with no binding site to the lectin are not retained in the intermediate gels containing either free or im-

mobilized lectin. In the present study, the first fraction from
Sephadex G-200 gel filtration was not retained in the intermediate gel containing free Can G (Fig.2-2). This phenomenon
was not observed in the gel containing immobilized Can G (Fig.
2-3). However, the second fraction was retained in both intermediate gels containing free and immobilized Can G. These findings suggested that the first fraction did not react with Can G
but the second one did. The number of binding site of the second fraction to Can G could not be confirmed in this study.
Histochemical study of the prostate sections indicated that the
binding sites of AcP of SP to Can G differed from those to the
substrate and that those binding sites were different.
In conclusion, AcP of SP seems to have a heterogeneity recognizable by gel filtration and reaction with Can G.

References

1. Bøg-Hansen,T.C.,Brogren,C.-H.:Scand.J.Immunol.4,Suppl.2,
   135-139 (1975).

2. Grogg,E.,Pearse,A.G.E.:Brit.J.exp.Path.33,567-576 (1952).

3. Horejsi,V.,Ticha,M.,Kocourek,J.:Biochim.Biophys.Acta 499,
   290-300 (1977).

4. Iwasa,M.,Sagisaka,K.: In Lectins-Biology,Biochemistry,
   Clinical Biochemistry Vol.2, ed. by Bøg-Hansen,T.C., Walter
   de Gruyter, Berlin,New york, pp. 371-377 (1983).

4. Takeo,K.,Nakamura,S.:Arch.Biochem.Biophys. 153,1-7 (1972).

EFFECT OF METAL IONS ON THE INTERACTION BETWEEN CONCANAVALIN
A AND CARBOHYDRATE STUDIED BY AFFINITY ELECTROPHORESIS

Akira Kuwahara, Masanori Fujimoto, Ryosuke Suzuno and
Kazusuke Takeo
Department of Biochemistry, Yamaguchi University School of
Medicine
J-755 Ube, Japan

Introduction

Polyacrylamide gel disc electrophoresis of concanavalin A
(Con A) has been performed, mainly, using the discontinuous
buffer system described by Reisfeld et al.(1). While a
discontinuous buffer system is superior to a continuous
buffer system to obtain a sharp band, it is impossible to
perform electrophoresis at any constant concentration of
metal ions. We described previously (2) a continuous buffer
system for polyacrylamide gel disc electrophoresis over a
broad pH range. Using this buffer system, we have investigated the effect of metal ions on the interactions between
dextran and Con A and succinyl-Con A (s-Con A).

Method

Electrophoresis was carried out at 20°C with a thermostatic
electrophoresis apparatus (3). Electrophoresis gels, 5 cm
in height, were prepared using 5.0% polyacrylamide gel (PAG)
for Con A and 6.9% PAG for s-Con A by photopolymerization
(2). For determination of dissociation constants (Kd)(4, 5),
one set of electrophoresis was run with 12 tubes, composed

of duplicated gels of 6 different concentrations of dextran
(0, 0.04, 0.08, 0.12, 0.16, and 0.20%). The relative migration distance (Rm) was expreassed as the ratio of the
migrating distance of the protein to that of the indicator
dye, methylene blue (MB) or bromophenol blue (BPB). Demetal-
lized Con A (apo-Con A) was prepared by preparative affinity
electrophoresis (6). Commercial Con A was dialyzed overnight against 20 mM EDTA at pH 5.0, and then applied to the
preparative electrophoresis (7) with 5.0% PAG containing

Fig. 1.   Effect of metal ions on Con A - dextran
interaction at pH 4.3.

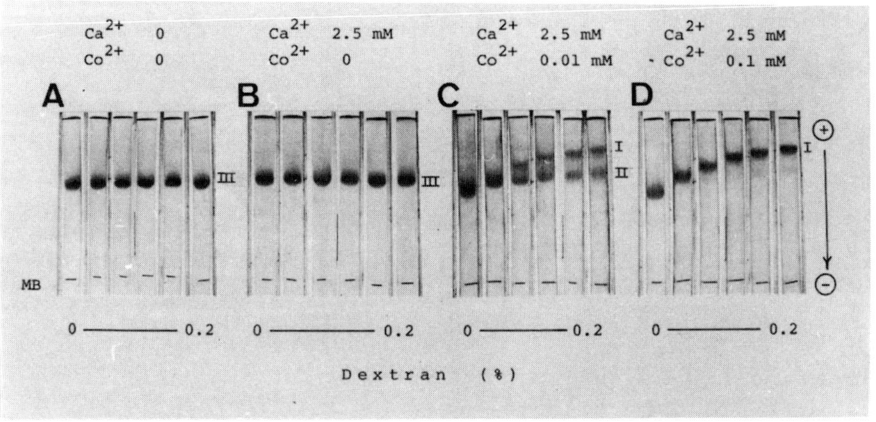

0.5% dextran using the buffer system described by Reisfeld
et al.(1). The Con A sample separated into 3 bands depending on the affinity to dextran. The fastest band was extracted. This band showed no affinity to dextran as seen in
Fig. 1A. We have used this fraction as apo-Con A. Apo-s-
Con A was prepared using the same method as apo-Con A from
commercial preparation.

Results

Con A exists electrophoretically in 3 molecular forms: Type
I, native and fully active, type II, partially active, and
type III, apo-Con A or inactive. At pH 4.3, type III Con A
was not converted to type II or I even in the presence of
2.5 mM Ca^{2+}ion as seen Figs 1A and 1B. When 0.01 mM Co^{2+}ion
was added to the 2.5 mM Ca^{2+} ion solution (Fig 1C), type III
was converted to type II and I. In the presence of 0.1 mM

Fig. 2.   Effect of 2.5 mM Ca^{2+} ion on succinyl-Con A -
          dextran interaction at varying pH values.

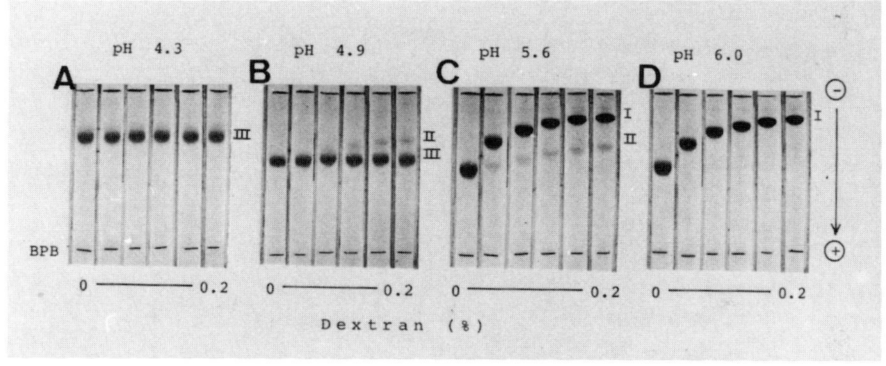

Co^{2+}ion (Fig. 1D), Con A converted completely to type I.
Other transition metal ions such as Mn^{2+} and Ni^{2+}ions have
the same effect as Co^{2+}ion. Mg^{2+}ion can not replace the
transition metal ions. Kd values of type I or II for dextran are independent of metal ions. At pH 4.3 and 20°C, Kd
values were 0.29 mM for type I and 1.1 mM for type II Con A.

Mobilities of three types of Con A are different. Type II
(Rm = 0.42) migrated about 10% faster than type III (Rm =

0.38) and type I (Rm = 0.46) 10% faster than type III. This enhancement of mobility appears to be due to binding of metal ions. From these results, it can be postulated that type III is a metal free form, while type II binds one molecule of metal ion, and type I two molecules of metal ions per subunit. S.Con A also exists in three molecular forms as does Con A. At pH 4.3, it showed no affinity for dextran in the presence of 2.5 mM Ca^{2+} ion. In the presence of the transition metal ion together with Ca^{2+} ion, it recovers its affinity for dextran. When pH of the medium increased in the presence of 2.5 mM Ca^{2+} ion, s-Con A converted to the active forms as seen in Fig. 2. At pH 4.9 (Fig. 2B), weak band of type II appeared, and at pH 5.6 (Fig. 2C) type I predominated. At pH 6.0 (Fig. 2D) or 6.8 (Fig. 2E), all s-Con A converted to type I. Kd values of type I or II are not changed by increasing pH values. These values were to 0.25 - 0.27 mM for type I and 1.1 - 1.2 mM for type II s-Con A.

References

1. Reisfeld, R. A., Lewis, U. T. and Williams, E. E.: Nature 201, 281 (1962)

2. Takeo, K., Fujimoto, M., and Kuwahara, A.: Lectins 3, 397-404 (1982).

3. Takeo, K., Fujimoto, M., Kuwahara, A., Suzuno, R., and Namakura, K.: Electrophoresis '81, Walter de Gruyter, Berlin - New York 33-40, (1981).

4. Takeo, K. and Nakamura, S.: Arch. Biochem. Biophys. 153, 1 - 7 (1972)

5. Takeo, K. and Kabat, E. A.: J. Immunol. 121, 2305-2310 (1978).

6. Takeo, K., Suzuno, R., Fujimoto, M., and Kuwahara, A.: Bull. Yamaguchi Med. Sch. 27, 85-98 (1980)

7. Takeo, K., Suzuno, R., Fujimoto, M.,Kuwahara, A., Ogata, H., and Nakamura, S.: Bull. Yamaguchi Med. Sch., 23, 165-183 (1976).

Application

BINDING SPECIFICITY OF TRANSFERRIN RECEPTORS ON CULTURED
TUMOR CELLS

Yoichi Urushizaki, Yutaka Kohgo, Yoshiro Niitzu,
Ichiro Urushizaki
Department of Internal Medicine (Section 4), Sapporo Medical
College, Sapporo, Japan

Introduction

Transferrin is a serum glycoprotein which binds two atoms of
iron to its specific binding sites. The main role of
transferrin is to deliver inorganic iron to the erythron and
placenta through its specific receptor on the cell surface.
Recently transferrin receptors have been found also in
several established cell lines (1). However, the binding
specificity of transferrin receptors for heterogenous
transferrin molecules still remains controversial. In this
paper, a receptor assay using ^{125}I-labeled transferrin was
performed on various cultured cell lines, and the
interraction of transferrin receptor and transferrin with
different iron and sialic acid content is presented.

Materials and Methods

Human cultured cell lines
Eight kinds of human cultured cell lines were used: K562
originally established from chronic myelogenous leukemia, HL
60 from promyelocytic leukemia, U 937 from histiocytic
lymphoma, Chang liver cell from hepatocyte, C-Hc 4 from
primary hepatoma, M 7609 from colon cancer, HGC-25 from
pancreas cancer and HMV from malignant melanoma. These cell

lines were maintained in RPMI 1640 medium supplemented with
10% fetal calf serum.

Human transferrin and its iodination
According to the method of Okada (2), human diferric
transferrin was isolated from iron saturated human plasma by
35 - 65% ammonium sulfate precipitation, DEAE cellulose ion
exchange column chromatography and high liquid chromatography.
The final purification was performed by running preparative
polyacrylamide gel electrophoresis. The diferric transferrin
was labeled with sodium iodide (^{125}I) by the chloramine T
method of Hunter (3). The specific activity of transferrin
was 4.71 x 10^6 cpm/pmole. Apotransferrin was prepared by
dialyzing the diferric transferrin against 10mM citrate
buffer, pH 5.0. Asialo-diferric transferrin was obtained
after incubation between the diferric transferrin solution
and 0.1 u/ml neuraminidase at 37°C for 2 hours.

Transferrin receptor assay of cultured cell lines using
^{125}I-labeled diferric transferrin
Washed cells were suspended in RPMI 1640 medium containing 2%
bovine serum albumin. 1.3 x 10^5 cpm of ^{125}I-transferrin
and 8 kinds of diluted unlabeled transferrin were used per
assay. After 10^6 cells were incubated with ^{125}I-transferrin
at 37°C for 30 min, the incubation was centrifuged at 2000 rpm
for 5 min. The cell pellet was assayed for radio-activity by
a gamma counter, and the amount of transferrin bound per 10^6
cells was calculated. The binding characteristics were
examined by Scatchard plot analysis.

Transferrin receptor assay using ^{125}I-labeled transferrin
at different iron saturations
^{125}I-labeled transferrin at different iron saturations was
prepared in RPMI 1640 medium which was preincubated with

^{125}I-labeled apotransferrin and chelators. Solution A was preincubated with ^{125}I-diferric transferrin only, solution B with ^{125}I-apotransferrin only, solution C with ^{125}I-apotransferrin and 100uM tropolone, solution D with ^{125}I-apotransferrin and 100uM desferrioxamine. Transferrin receptor assay was performed using these solutions.

6M urea gel electrophoresis

According to the method of Makey and Seal (4), 6M urea gel electrophoresis was performed on the above four solutions containing ^{125}I-labeled transferrin at different iron saturations. Distribution of four molecular forms of transferrin was examined and the iron saturation of these solutions was calculated.

Results and Discussion

Specific binding of diferric transferrin to K 562 cells
Total binding of transferrin was examined by measuring the amount of ^{125}I-transferrin bound to K 562 cells. To vary the concentration of unlabeled transferrin from 1 nM to 20nM, a total binding curve was obtained. Low-affinity binding, determined from the slope of the total binding curve taken between transferrin concentrations of 0.1 and 1uM and extrapolated to 0, was subtracted from the total binding curve to estimate the amount of specific high-affinity binding. (Fig.1) Scatchard plot analysis was performed upon the data of specific high-affinity binding. Plots of bound transferrin/free transferrin versus the concentration of bound transferrin were linear, thus indicating a single high-affinity binding site. The number of transferrin receptors of K562 cells were calculated as 5.10 x 10^5 per cell and their association constant (Ka) was estimated as 1.98 x $10^8 M^{-1}$. (Fig.2)

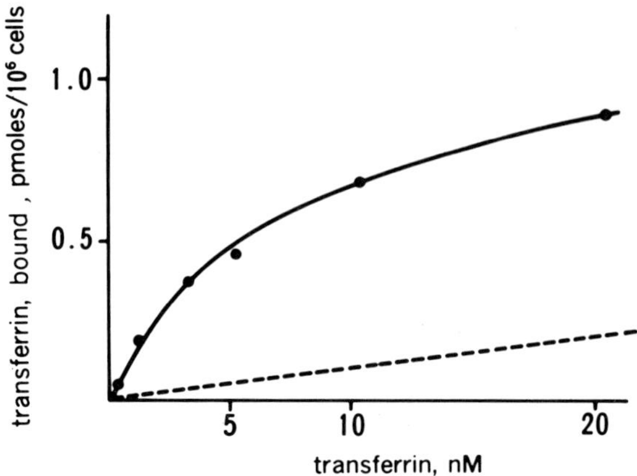

Fig. 1 Binding curves of diferric transferrin to K 562
cells

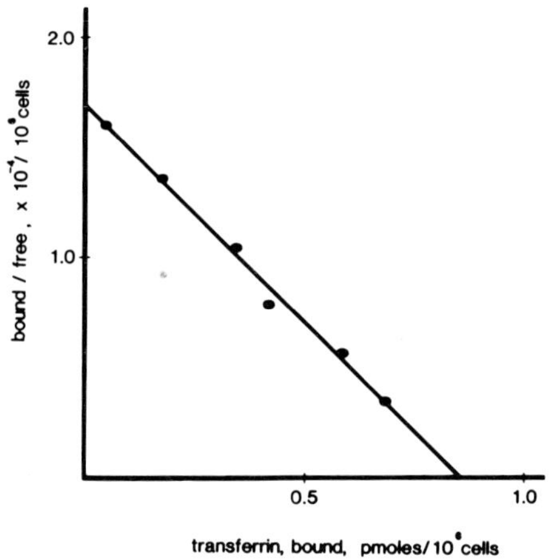

Fig. 2 Scatchard plot analysis of transferrin binding to
K 562 cells

The transferrin receptor assay of eight human cultured cell
lines
Likewise the transferrin receptor assay using Scatchard
plot analysis was performed on eight cell lines. As already
reported by some investigators using monoclonal antibody, all
eight cell lines have abundant receptors for transferrin.
Transferrin binding sites per cell were between 0.37 ± 0.22 x

10^5 and $4.96^{\pm}0.61 \times 10^5$, and the association constant
(Ka) was between $1.91 \pm 0.46 \times 10^8$ and $2.58^{\pm}0.80 \times 10^8 M^{-1}$
as shown in Table 1. In contrast, peripheral blood cells of
normal persons did not have any measurable receptors.

Binding specificity of transferrin receptors for transferrin
at different iron content
This transferrin receptor assay was based on the binding of
diferric transferrin (iron saturated form) to transferrin
receptors. Serum transferrin consists of four molecular
forms (diferric, N-site bound and C-site bound monoferric and
iron free apotransferrin.)

Table 1   Transferrin receptors on the surface of various
human established cell lines

established cell lines	orgin of cells	(number of experiment)	transferrin receptor	
			number/cell	Ka (L/M)
K 562	leukemia (CML)	(8)	$4.96 \pm 0.61 \times 10^5$	$1.91 \pm 0.48 \times 10^8$
HL-60	leukemia (APL)	(2)	3.96×10^5	2.22×10^8
U-937	histiocytic lymphoma	(4)	$0.67 \pm 0.07 \times 10^5$	$2.23 \pm 0.66 \times 10^8$
Chang liver cell	liver cell	(8)	$0.98 \pm 0.40 \times 10^5$	$2.19 \pm 0.28 \times 10^8$
CHc-4	primary hepatoma	(2)	2.81×10^5	2.72×10^8
M-7609	carcinoma of the colon	(4)	$1.23 \pm 0.41 \times 10^5$	$2.06 \pm 0.21 \times 10^8$
HGC-25	pancreas carcinoma	(2)	0.37×10^5	2.58×10^8
HMV	melanoma	(2)	0.95×10^5	2.29×10^8

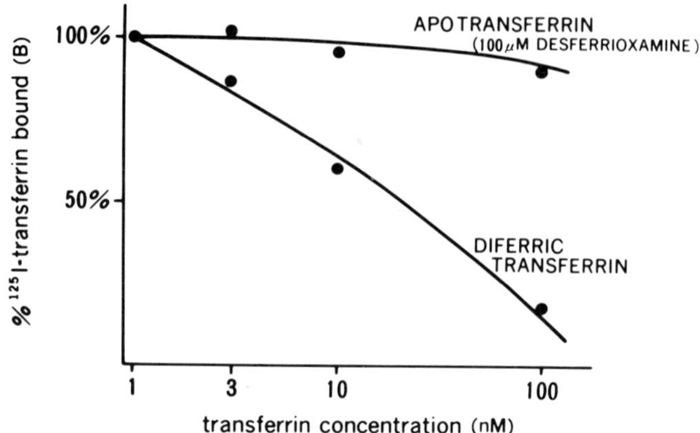

Fig.3　The effect of increasing concentrations of unlabeled apotransferrin and diferric transferrin on the binding of ^{125}I-labeled diferric transferrin

The binding specificity of the receptors for these different transferrins was not clearly demonstrated.　The competitive binding assay using diferric and apotransferrin against ^{125}I-labeled diferric transferrin in K562 cells is shown in Fig. 3.　To avoid binding of free iron to apotransferrin, 100 uM desferrioxamine was added to the incubation medium. Only cold diferric transferrin, not apotransferrin, was able to compete against ^{125}I-labeled diferric transferrin. Furthermore, as shown in Fig. 4, we prepared RPMI 1640 media which contained ^{125}I-labeled transferrins at different iron saturations.　Iron saturation was calculated according to the relative proportions of the four forms of transferrin. Solution A and B contained mainly diferric transferrin at 96% and 82% iron saturation.　On the other hand, solution C contained mainly two forms of monoferric transferrin and apotransferrin at 25% saturation, solution D contained mainly apotransferrin at 2% saturation.　The Scatchard plot analyses

of these transferrin preparations are shown in Fig. 5.
Diferric transferrin and 82% saturated transferrin
specifically bound transferrin receptors on K562 cells.
However, 25% and 2% saturated transferrin, which were mainly
monoferric transferrin and apotransferrin, did not show any
significant binding to transferrin receptors.

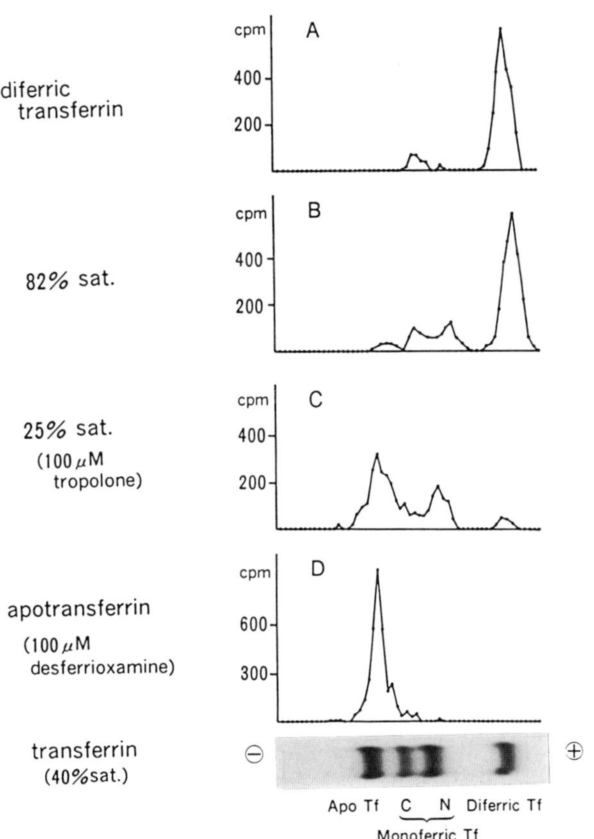

Fig. 4   6 M urea gel electrophoretic patterns of
^{125}I-labeled transferrin at different iron
content

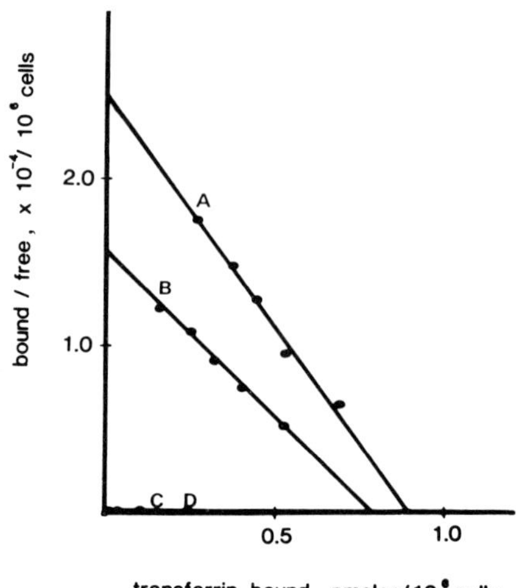

Fig. 5   Scatchard plot of binding of transferrin at
         different iron saturation

Binding specificity of transferrin receptors against sialo-
and asialo-diferric transferrin
Asialo-diferric transferrin was obtained after the
neuraminidase treatment of the diferric transferrin. By
analysis with isoelectric focusing, asialo-diferric
transferrin comprised more basic components than
sialo-diferric transferrin. The Scatchard plot analysis
showed no difference in transferrin binding between sialo-
and asialo-diferric transferrin for transferrin receptors.

It was shown recently that transferrin receptors are found in
abundance not only on human reticulocytes (5) and placental
microvillous membranes (6), but also on the cell surface of
various cultured cell lines. Much of this work on

transferrin receptors was carried out using monoclonal
antibodies against them (7) and, actually, inhibition of cell
growth was reported at 42/6.(8). However, the specificity of
transferrin receptors for heterogenous transferrin molecules
is not well understood. On the basis of our studies, it may
be concluded that the binding specificity of transferrin
receptors is restricted to diferric transferrin, and that the
sialic acid content of diferric transferrin is not related to
its binding specificity. This suggests that the role of
transferrin receptors may be the internalization of iron into
the dividing cells.

References

1. Hamilton, T.A.,Wada, H.G., Sussman, H.H.: Proc. Natl.
    Acad. Sci. 76, 6406-6410 (1979).

2. Okada, S., Jarvis, B., Brown, E.B.: J. Lab. Clin. Med.
    93, 189-198 (1979).

3. Hunter W.M.: Handbook of experimental immunology, 2nd ed.
    1-36, Blackwell Scientific, Oxford (1973).

4. Makey, D.G., Seal, U.S.: Biochim. Biophys. Acta 453,
    250-256 (1976).

5. Seligman, P.A., Schleicher, R.B., Allen, R.H.: J. Biol.
    Chem. 254, 9943-9946 (1979).

6. Liebman, A., Aisen, P.: Biochemistry 16, 1268-1272 (1977).

7. Sutherland, R., Delia, D., Schneider, C., Newman, R.,
    Kemshead, J., Greaves, M.: Proc. Natl. Acad. Sci. 78,
    4515-4519 (1981).

8. Trowbridge, L.S., Lopez, F.: Proc. Natl. Acad. Sci. 79,
    1175-1179 (1982).

EXISTENCE OF MYOGLOBIN IN CHICKEN GIZZARD MUSCLE AND COMPARISON OF MYOGLOBIN IN DYSTROPHIC AND NORMAL CHICKENS

Hisaomi Kawai, Kozo Imai, Kenjiro Masuda, Hiroshi Nishino, Toshio Inui, Yasunori Yoshida and Kanae Kusaka
First Department of Internal Medicine, University of Tokushima School of Medicine, Tokushima, Japan

Kazuo Miyoshi
Okinaka Memorial Institute for Medical Research, Tokyo, Japan

Introduction

With respect to the pathogenesis of human muscular dystrophy, we have been studying human myoglobin(Mb) in parallel with the creatine kinase(CK) isozyme pattern for many years(1-4, 6-9). In the course of studies in humans, we found Mb and muscle type CK in striated muscles, but not in smooth muscles as reported previously(3, 4).
This paper reports studies of the Mb and isozyme pattern of CK in chicken gizzard muscle, which is regarded as smooth muscle. Studies of Mb in dystrophic chickens are also reported.

Materials and Methods

Skeletal muscle, gizzard and intestine were obtained from normal chickens (line-412) and dystrophic chickens(line-413) immediately after decapitation. Portions of skeletal and gizzard muscles used for Mb isolation and study of the CK isozyme pattern were frozen at -20°C. Portions used for light microscopy were frozen in isopentane cooled in dry ice-acetone and sectioned(8-10 μ thickness) for histochemistry. Remaining portions of tissues were fixed in 2.8% glutaraldehyde, post-fixed in 1% OsO_4, dehydrated in acetone, and processed for electron microscopy.

Mb of skeletal muscles of normal and dystrophic chickens was isolated by our modification of Singer's method(5), Sephadex G-75 column chromatography (2x110 cm, 0.02 M phosphate buffer, pH 8.5) and then DEAE cellulose column chromatography(3x5 cm, 0.01 M phosphate buffer, pH 7.2). Heme-proteins similar to Mb were isolated from normal and dystrophic chicken gizzard muscle in the same way. Mb was identified by spectrophotometry, thin layer polyacrylamide gel electrophoresis(Tris-EDTA-borate buffer, pH 8.5), and double immunodiffusion(Ouchterlony method) using rabbit antiserum against Mb of normal chicken skeletal muscle. This antiserum was produced in rabbits immunized with Mb isolated from skeletal muscle of normal chickens.
The Mb content was measured by single radial immunodiffusion(4).
The CK activity was determined by the cysteine-stimulated inorganic phosphoric acid method(10). The isozyme pattern of CK was examined on a zymogram by reduction of the tetrazolium salt(11).

Results

1) Histological findings of gizzard muscle. Gizzard muscle, which is generally considered to be smooth muscle, was examined by light and electron microscopy. Fig. 1 shows sections of gizzard muscle(A, C) and

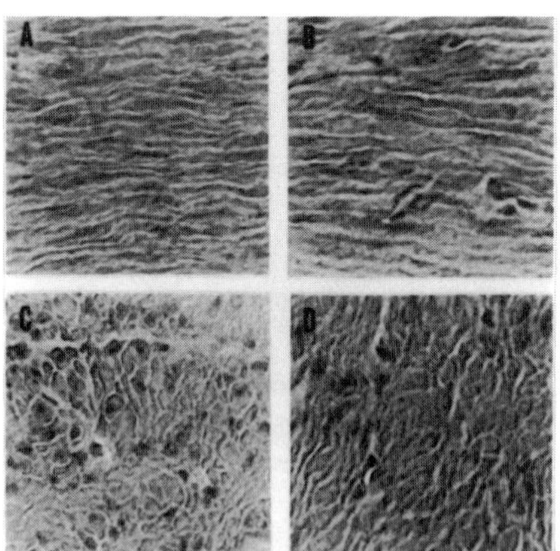

Fig. 1. Histology of gizzard muscle(A,C) and smooth muscle of the intestine(B,D) of normal chicken(A,B: longitudinal sections; C,D:cross sections)

smooth muscle of the intestine(B, D) stained with hematoxylin-eosin.
The longitudinal section of gizzard muscle(A) shows small muscle fibers
of about 3 μ diameter, densely arranged in parallel. They have a central nucleus but no striation. These features are similar to those of
smooth muscle cells of the intestine(B). In cross section(C), gizzard
muscle is seen to be composed of small, densely packed cells which are
similar in nuclear and cellular morphology to smooth muscle cells of
the intestine.
Fig. 2 shows an electron micrograph of several compactly arranged muscle cells of the gizzard. As seen clearly in the central cell, there
is a central nucleus, with diffuse myofilaments in the cytoplasm, among
which there are scattered dense bodies. Attachment plaques are also
observed along the cell membrane. These features are characteristic
of smooth muscle cells, and are also observed in cells of the intestine.

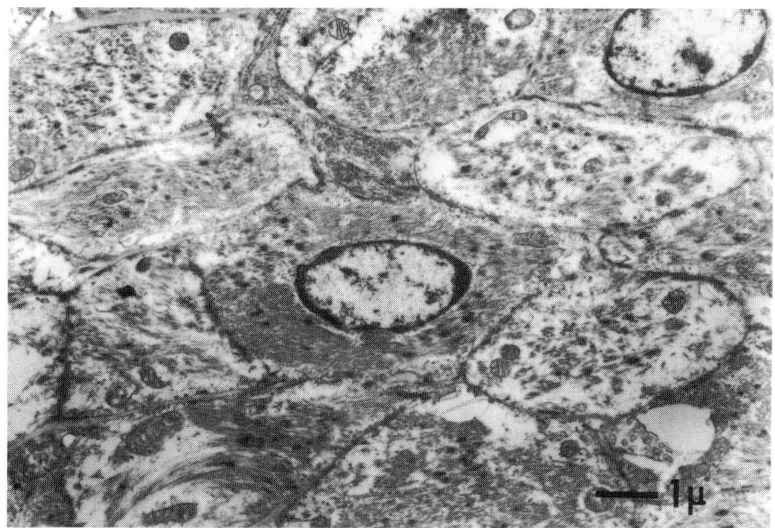

Fig. 2. Electron micrograph of gizzard muscle

2) CK isozyme of chicken gizzard muscle. Fig. 3 shows the CK isozyme pat-
    terns. The CK of skeletal muscle consisted of 78.1% of the muscle
    type(MM) and 13.0% of the brain type(BB), with a trace amount of the
    MB type. In contrast, that of the intestine was entirely brain type
    (BB). The enzyme in gizzard muscle was also entirely BB type, i.e.:
    gizzard muscle has the CK isozyme pattern of smooth muscle.

662

The CK activity was 260 μ moles Pi/min/g in skeletal muscle, 300 in smooth muscle of the intestine, and 100 in gizzard muscle.

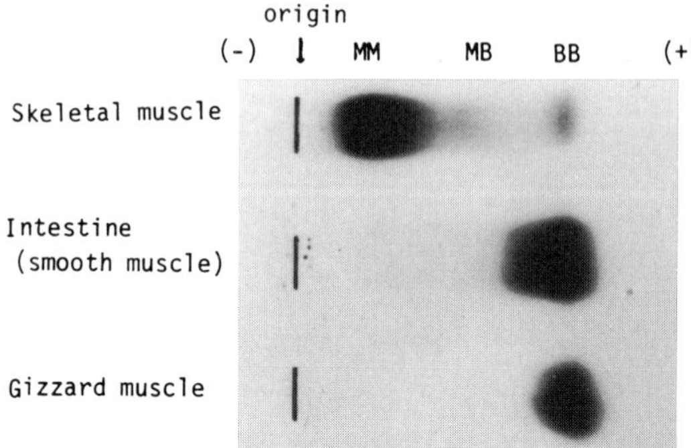

Fig. 3. Creatine kinase isozyme pattern

3) Mb in gizzard. Fig. 4 shows the absorption curves of heme-proteins isolated from normal chicken skeletal muscle and gizzard muscle by the

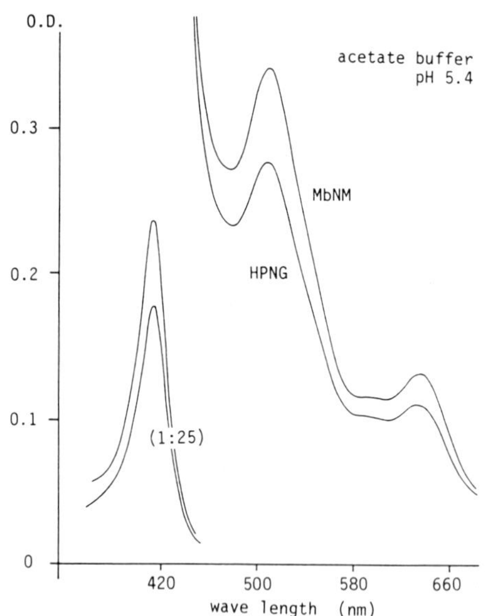

Fig. 4. Absorption curves of Mb from normal chicken skeletal muscle(MbNM) and a heme protein, probably Mb, from normal chicken gizzard muscle(HPNG)

method described above. The solvent was acetate buffer, pH 5.4. Hemes
were converted to the met-form with potassium ferricyanide. Mb of
skeletal muscle showed absorption maxima at 412 nm, 510 nm and 636 nm,
and a plateau due to Mb at 580-610 nm. Heme-proteins isolated from
gizzard muscle gave a similar absorption profile. This indicated that
gizzard muscle contains Mb that is spectrophotometrically identical
with Mb in skeletal muscle.

4) Electrophoresis of Mb. Fig. 5 shows the electrophoretic patterns of
    Mb of normal chicken skeletal muscle and gizzard muscle on polyacryl-
    amide gel. As controls, chicken hemoglobin(Hb) and human Mb were also
    subjected to electrophoresis. All proteins were in the met-form and
    stained with benzidine. The skeletal and gizzard muscle Mb's of normal
    chicken both consisted of four subcomponents which had the same rela-
    tive mobilities. The ratios of the subcomponents also appeared the
    same. Chicken Hb had only two subcomponents with different mobilities.
    Human Mb had four subcomponents, but their mobilities were different
    from those of the subcomponents of chicken Mb.

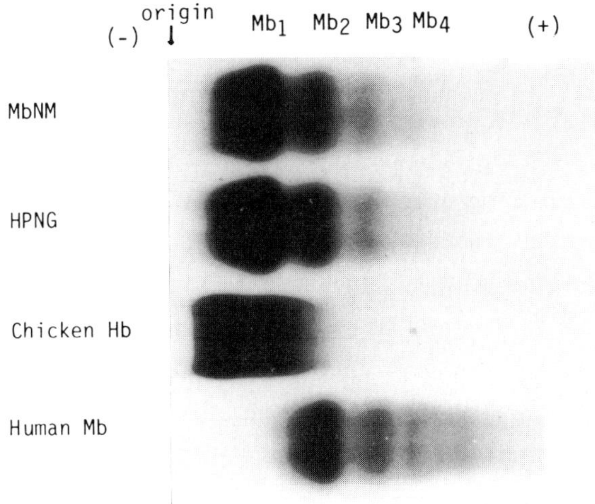

Fig. 5. Polyacrylamide gel electrophoresis of
Mb from normal chicken skeletal muscle(MbNM)
and a heme protein from normal chicken gizzard
muscle(HPNG). 1mA/cm 2.5 h, benzidine stain

664

5) Immunochemical properties of Mb. Fig. 6 shows the results obtained on
an Ouchterlony plate with rabbit antiserum against skeletal muscle Mb
of normal chicken in the center well. Mb of normal chicken skeletal
muscle(MbNM) and Mb of gizzard muscle(HPNG) both formed a precipitin
line, and the two lines fused completely. This suggests that skeletal
muscle Mb and gizzard muscle heme-protein are immunochemically identical. Chicken Hb(Hb) and the supernatant of intestinal muscle homo-
genate(Sup) formed no precipitin line.
The absorption spectra and results of electrophoresis and immunological
studies show that chicken gizzard muscle contains Mb which is identical
with skeletal muscle Mb.

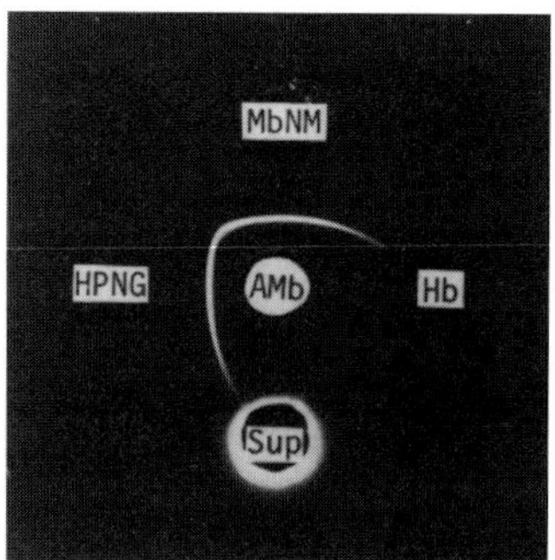

MbNM : myoglobin from normal chicken
       skeletal muscle
HPNG : a heme protein from normal
       chicken gizzard muscle
Sup : supernatant of the homogenate
       of chicken smooth muscle
Hb   : chicken hemoglobin
AMb : rabbit antiserum against Mb
       from normal chicken skeletal
       muscle

Fig. 6. Immunodiffusion of Mb from normal chicken skeletal
muscle and a heme protein from normal chicken gizzard muscle

6) Mb content. The Mb contents of skeletal and gizzard muscle were measured by single radial immunodiffusion. The average Mb content of three
chickens was 0.02 mg/g wet weight in white muscle of the breast and
2 mg/g wet weight in red muscle of the leg. No Mb was detected in
smooth muscle of the intestine, but in gizzard muscle the average content was 6 mg/g wet weight(about 300 times that in white muscle and
3 times that in red muscle).

7) Mb in dystrophic chicken. Mb of the skeletal and gizzard muscles of
dystrophic chickens were studied in the same way as those of normal
chickens. The absorption curves, the electrophoretic mobilities and
ratio of subcomponents, and the immunochemical properties of Mb of
skeletal muscle and gizzard muscle of dystrophic chickens were all
identical with those of normal chickens.

Discussion

As described in the text, chicken gizzard muscle, which is similar morphologically to smooth muscle by light and electron microscopy, was found to contain a considerable amount of Mb that is identical with Mb in skeletal muscle. Moreover, the isozyme pattern of CK in the gizzard was BB type. Table 1 summarizes these results.

There are a few reports of spectrophotometric, or immunochemical studies on Mb in chicken gizzard(12-14). We demonstrated the presence of Mb by three different methods: absorption spectrometry, electrophoresis and double immunodiffusion. In addition, we found a relation between Mb and muscle type CK. No previous reports referred to CK in the gizzard.

In studies on human Mb, we showed that Mb is present only in striated muscles, not in smooth muscles, and that Mb co-exists with muscle type CK in humans(3, 4). It is noteworthy, however, that gizzard muscle, which appears histologically to be smooth muscle and has brain type CK, contained a considerable amount of Mb. This finding prompted us to examine the intracellular localization of Mb in gizzard muscle, and studies on this problem are now under way in this laboratory. The significance of Mb in the gizzard, which is embryologically part of the digestive tract, remains to be clarified. However, aerobic metabolism may be active in the gizzard muscle unlike in other smooth muscles, because the gizzard muscle contracts strongly and repeatedly. In this work we also confirmed that Mb of dystrophic chickens is similar to that of normal chickens, as pointed out elsewhere(15). In this regard, we previously showed that Mb in patients with Duchenne muscular dystrophy is similar to normal Mb(6).

Table 1. Comparison of chicken gizzard muscle, skeletal muscle and smooth muscle

	Histology	Mb	CK
Skeletal muscle	M	(+)~(++)	MM
Smooth muscle	S	(-)	BB
Gizzard muscle	S	(++++)	BB

M:striated muscle type; S:smooth muscle type
(+)~(++++):grade of Mb content
MM:muscle type; BB:brain type

References

1. Kawai, H.: Shikoku Acta Med. $\underline{24}$, 703 (1968)
2. Nakano, M.: Shikoku Acta Med. $\underline{24}$, 715 (1968)
3. Miyoshi, K., Kawai, H., Hiasa, M., Iwasa, M.: Structure and Function of Muscle(Sakai, T., Endo, M., Sugita, H. ed), First Edition, Igaku-Shoin, Tokyo 1977
4. Hiasa, M.: Shikoku Acta Med. $\underline{31}$, 193 (1975)
5. Singer, K., Angelopoulos, B., Ramot, B.: Blood $\underline{10}$, 979 (1975)
6. Miyoshi, K., Saito, S., Kawai, H., Kondo, A., Iwasa, M., Hayashi, T., Yagita, M.: J. Lab. Clin. Med. $\underline{92}$, 341 (1975)
7. Miyoshi, K., Toyota, R., Tomino, H., Yoshimatsu, A., Saijo, K., Katsunuma, N., Fujino, A.: Saishin-Igaku $\underline{20}$, 202 (1965)
8. Matsuoka, M.: Shikoku Acta Med. $\underline{26}$, 70 (1970)
9. Miyoshi, K., Oshima, I., Ohto, Y., Matsuoka, M., Kawai, H., Toyota, R.: Tokushima J. Exp. Med. $\underline{21}$, 75 (1974)
10. Okinaka, S., Sugita, H., Momoi, H., Toyokura, Y., Watanabe, T., Ebashi, S.: J. Lab. Clin. Med. $\underline{64}$, 299 (1964)
11. Tomino, H.: Shikoku Acta Med. $\underline{24}$, 693 (1968)
12. Gröschel-Stewart, U., Jaroschik, U., Schwalm, H.: Experientia $\underline{27}$, 512 (1971)
13. Blessing, M.H., Müller, G.: Comp. Biochem. Physiol. $\underline{47A}$, 535 (1974)
14. Nishida, J.: Jap. J. Vet. Sci. $\underline{38}$, 299 (1976)
15. Goldbloom, D.E., Brown, W.D.: Arch. Biochem. Biophys. $\underline{147}$, 367 (1971)

CHARACTERIZATION OF FERRITIN MONOMER AND DIMER

Yoshiaki Makino, Ikue Kanamaru, Ichiyo Tei and Kunio Konno
Department of Biochemistry, School of Medicine, Showa
University
1-5-8 Hatanodai, Shinagawa-ku, Tokyo 142, Japan

Introduction

Ferritin is an iron-storage protein consisting of an outer
protein shell of molecular weight 450,000 with a central
cavity which contains a variable amount of iron that is
deposited as a microcrystalline ferric hydroxidephosphate
complex (1). Not all ferritin molecules in a tissue are
equally loaded with iron and within any one specimen molecules with iron content varying from 0 (apoferritin) to 4,500
atoms may be identified (2,3). The apoferritin component is
composed of 24 subunits of 2 types, each type having a molecular weight of approx. 19,000 (L subunit) and 21,000 (H subunit) (4). These subunits combine in different proportions
in a presently unknown fashion into the 24 subunit hetero-
polymers, which generate families of isoferritin. Although
most tissues have characteristic isoferritin populations,
these patterns vary markedly during development and in
diseases such as iron overload and cancer. Isoferritins
differ in surface charge and can thus be separated by electrophoresis (5) or isoelectrofocusing (6,7). The various
ferritins from different sources all show at least three bands
of different mobilities when examined by polyacrylamide gel
electrophoresis or starch gel electrophoresis (8,9). These
bands are monomer, dimer and oligomer ferritin. The nature
and origins of ferritin dimer and oligomer have not previously
been studied.

We separated the ferritin monomer and dimer by preparative disc polyacrylamide gel electrophoresis and in this paper describe different characteristics of these ferritin molecules in solution, in vitro.

Materials and Methods

Horse spleen ferritin was purchased from Miles Corp. and Sigma chemical company. Rat liver ferritin was purified by the method of Drysdale and Munro (10). Radioactive iron (^{59}Fe) was employed in the form of ferric chloride procured from New England Nuclear, England. Anti-horse spleen ferritin serum was prepared by injecting a rabbit with 10 mg of the purified ferritin in complete Freund's ajuvant. The anti-ferritin serum drawn 45 days later. Apoferritin was prepared from ferritin by reduction with thioglycollic acid in acetate buffer, pH 5.5, followed by extensive dialysis against several changes of 0.01M phosphate buffer, pH 7.0. Iron was estimated as the ferrous dipyrydyl complex (11). Amino acid analysis was performed on a Hitachi amino acid analyzer (model 835) after hydrolysis for 20 hr at 110°C in 6N HCl. An LKB 2117 Multiphor apparatus was used for preparative flat-bed isoelectric focusing. The beds were formed on glass trays supplied in an LKB 2117-501 Ampholine electrofocusing Kit for granulated gels. Electrofocusing was under constsnt power with the total applied voltage limited to 300V. Isoferritin analysis was subjected to isoelectric focusing in thin layers (110mm x 125mm x 0.2mm) of 1% Agarose and 2% Ampholite; Ampholine, pH4-6 (LKB, Sweden); Pharmalite, pH 4-6.5 (Pharmacia, Sweden). Separation of ferritin dimer and monomer were subjected to polyacrylamide gel electrophoresis as discribed by Ornstain (12). The ratios of ferritin monomer to dimer were measured with a densitometer (Cosmo F-808). Ferritin samples were monitored by electron microscopy according to the procedure of Daivid and Easterbrook (13). Samples were nega-

tively stained with uranyl acetate and viewed in a Hitachi
model H58-1 transmission electron microscope. Protein was
measured by the method of Lowry et al. (14).

Results and Discussion

Rat liver ferritin and horse spleen ferritin all showed at
last three bands (monomer, dimer, oligomer) when examined by
polyacrylamide gel electrophoresis. The dimer and monomer
content of ferritin show smoll differences depending on commercial source or purified lot. In this experiment, the
proportions of monomer to dimer plus oligomer were 85-95:15-5.
Preparations of monomer and dimer were fractionated by preparative disc electrophoresis on 5.25% polyacrylamide gel
(tube size; 2 x 8cm). The proteins were extracted from the
gel as follows : the brown monomer and dimer ferritin bands
were cut from the gel; the proteins were eluted electrophoretically into a dialysis membrane in 25mM Tris-glycine, pH 8.6,
at 100V for 5-10 hr. The purities of the monomer and dimer

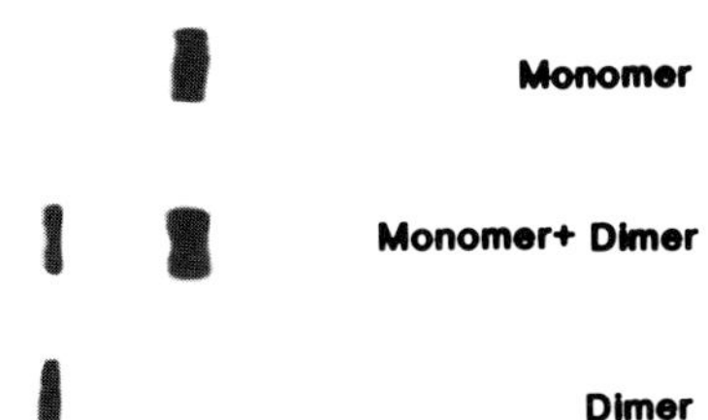

Figure 1. 5.25% polyacrylamide gel electrophoresis of
ferritin monomer and dimer

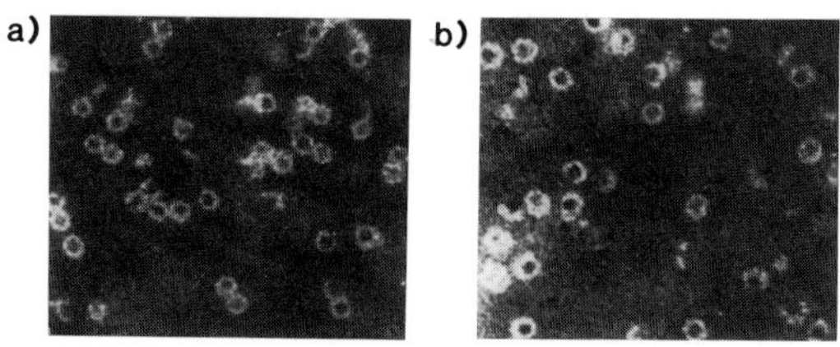

Figure 2. Electron micrographs of ferritin monomer and dimer
a:dimer ferritin, b:monomer ferritin

Table 1. Amino acid composition of ferritin monomer
and dimer (Results expressed as mole per cent of
amino acid/ Protein)

Amino acid	Monomer	Dimer
Aspartate	10.71	10.33
Threonine	4.31	3.99
Serine	6.24	6.35
Glutamate	15.89	15.73
Proline	2.04	3.52
Glycine	7.19	7.22
Alanine	9.28	9.05
Valine	4.24	4.12
1/2 Cystine	0.30	0.71
Methionine	0.77	1.39
Isoleucine	2.40	2.24
Leucine	15.00	14.64
Tyrosine	2.75	2.88
Phenylalanine	4.06	4.18
Lysine	5.36	5.07
Histidine	2.89	2.71
Arginine	5.57	5.43

were checked by both analytical polyacrylamide gel electrophoresis and electron microscopy. Figures 1, and 2 show
results of tests of sample compounds obtained by electrophoresis and checked both methods. Analytical polyacrylamide
gel electrophoresis and electron microscopy both indicated
that the solution of monomer was free of dimer. However, the
dimer solution was contaminated by 5-10% of monomer ferritin.

The electron micrograph showed a small amount of monomer
molecules in the dimer sample (Figure 2a).
The results of amino acid analysis of monomer and dimer ferritin from horse spleen are shown in Table 1. The content of
proline, of 1/2cystine and of methionine were all significantly different in the monomer and dimer ferritin. To study
the iron and dimer content of the isoferritins, horse spleen
isoferritin were separated by preparative flat-bed isoelectric
focusing procedure into six fractions (isoelectric points; pI
3.97, 4.01, 4.10, 4.19, 4.28, 4.31). The iron/protein ratio
was highest in the isoferritin with pI 4.19 and gradually
diminished for the successive pI values: 4.28, 4.10, 4.31,4.01
and 3.97 (Figure 3, right). Measurement of the dimer concentration of isoferritin fractions by analytical polyacrylamide
gel electrophoresis revealed marked differences in dimer
content throughout each isoferritin series (Figure 3, left).
Dimer ferritin content in isoferritin fractions of pI 3.97 and

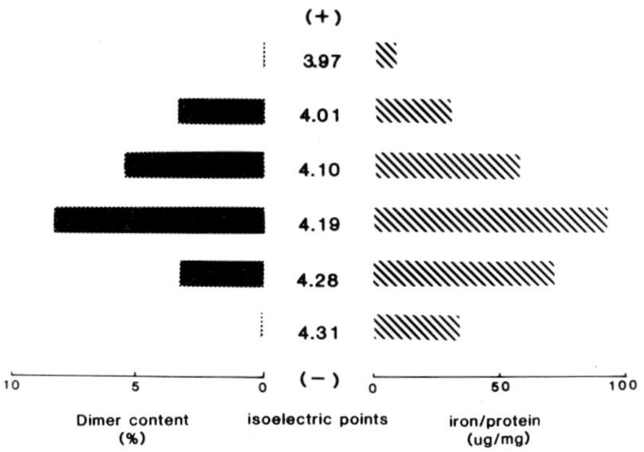

Figure 3. Iron and dimer ferritin content of isoferritin.
Isoferritin was separated by preparative flat-bed isoelectric
focusing

672

4.31 was almost nil; almost all of the dimer ferritin was in
the isoferritin between pI 4.01 and 4.28. In particular, the
highest dimer content (8%) was in pI 4.19 isoferritin. The
patterns of isoelectric focusing on agarose gel were similar
to those described above (Figure not shown).
To study uptake of ^{59}Fe into monomer and dimer, ferritin was
incubated with 0.15M KCl containing 0.64m mole/l of $KHCO_3$ and
^{59}Fe at 37°C. The solution was then added to anti-horse
ferritin serum and was stored overnight at 4°C. The mixture
was centrifuged, the precipitate was suspended in cold 0.9%
NaCl and centrifuged again. This operation was carried out
three times, after which the precipitate was washed with cold
0.9% NaCl. Increasing the concentration of ferritin increased
the rate of ^{59}Fe uptake (Figure not shown). Uptake of ^{59}Fe
was determined for monomer and dimer ferritins of the same
protein (Figure 4). Uptake of ^{59}Fe into dimer ferritin was
consistently lower than that into monomer ferritin (1/4-1/5).

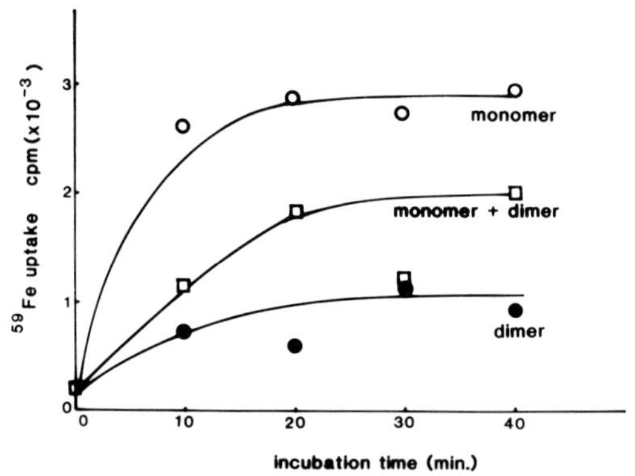

Figure 4. ^{59}Fe uptake into ferritin monomer and dimer

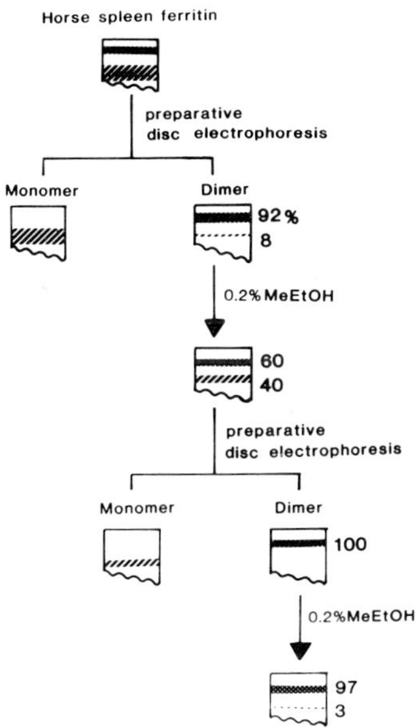

Figure 5. Cleavage of ferritin dimer by mercaptoethanol

The reagents most effective in cleaving the dimer to monomer
were mercaptoethanol and dithiothreitol, which cleave disulfide bonds (15). The effectiveness of sulfhydryl reagent was
dependent on its concentration and condition. When dimer
ferritin was treated with 1.0% mercaptoethanol at 37°C for
1 hr, iron was released and protein was partially degenerated.
However, these phenomenon were not observed in 0.2% mercaptoethanol treatment at 37°C even after 21 hr. The results of
cleavage of dimer into monomer by 0.2% mercaptoethanol are
shown in Figure 5. When dimer ferritin was obtained by pre-
-arative polyacrylamide electrophoresis after 0.2% mercaptoethanol treatment, the dimer decreased to 60%. When this

ferritin dimer was separated by re-electrophoresis, it could
not be cleaved to monomer by re-treatment with 0.2% mercaptoethanol. However, heating to 70°C for 15 min (purification
step) cleaved 45% of the dimer to monomer. These results
suggest that part of the dimer was aggregated by formation of
interchain bridges other than disulfide. The final question
is whether the dimer and oligomer are entirely artefacts.
A conclusive answer can not yet be given to this question.
However, the dimer content in purified ferritin was not much
different than in a purified lot. It is possible that monomer
formation from dimer occurs during the process of purification
of ferritin. It is conceivable that ferritin molecules in
vivo exist as both monomeric and dimeric form in cells.

References

1. Harrison, P.M., Hoare, R.J., Hoy, T.G., Macara, I.G.: in
    iron in Biochemistry and Medicine (Jacobs, A., Worwood, M.
    eds.) pp73-114, Academic Press, New york 1974
2. Harrison, P.M.: Semin. Haematol. 14, 55-70 (1977)
3. Fineberg, R.A., Greenberg, D.M.: J. Biol. Chem. 214, 107-
    113 (1955)
4. Drysdale, J.W.: Ciba Found. Symp. 51, 41-67 (1977)
5. Bomford, A., Lis, Y., Mc Farlane, I.G., Williams, R.:
    Biochem. J. 167, 309-312 (1977)
6. Makino, Y., Konno, K.: J. Biochem. (Tokyo) 65, 471-473
    (1969)
7. Drysdale, J.W.: Biochem. Biophys. Acta 207, 256-258 (1970)
8. Harrison, P.M., Gregory, D.W.: J. Mol. Biol. 14, 626 (1965)
9. Kopp,R., Vogt, A., Maass, G.: Nature 198, 892 (1963)
10. Drysdale, J.W., Munro, H.M.: Biochem. J. 95, 851-858(1965)
11. Drysdale, J.W., Munro, H.M.: J. Biol. Chem. 241, 3630-
    3637 (1966)
12. Ornstein, L.: Ann. New York Acad. Sci. 121, 321 (1964)
13. David, C., Easterbrook, K.: J. Cell. Biol. 48, 15-28 (1971)
14. Lowry, O.H., Rosebrough, N.J.: J. Biol. Chem. 193, 265
    (1951)
15. Niitsu, Y., Listowsky, I.: Biochemistry 12, 4690-4695
    (1973)

ELECTROPHORETIC ANALYSIS OF LIVER SPECIFIC LIPOPROTEIN (LSP)

Mikio Zeniya, Atsushi Arashiyama, Hideki Ando,
Masaki Nishiyama, Yumiko Fujita, Hiroshi Takahashi,
Masamichi Deura, Yoshikazu Shimizu, Yoshio Aizawa,
Mariko Akiba, Ichiro Asukata, and Haruo Kameda

The First Department of Internal Medicine, Jikei University,
School of Medicine, Tokyo

Introduction

It is well known that Liver specific lipoprotein (LSP, Meyer zum Büschenfelde et al.) plays an important role as an antigenic membrane protein in the pathogenesis of development of chronic liver cell injuries (1). This LSP is one of membrane lipoprotein fractions of supernatant of liver homogenate, and has liver specific antigenicity. Meyer zum Buschenfelde and his collegue demonstrated hepatic inflammation resembling chronic active hepatitis in human could be induced in rabbit by chronic immunization with LSP experimentally (2). One of the present authors also studied about the role of LSP in chronic active hepatitis. He reported that about 40% of patients with chronic active hepatitis showed the positive lymphocyte transformation test to the LSP and sera from chronic active hepatitis can cause liver cell destruction by antibody dependent cell-mediated cytotoxicity (ADCC) mechanism using isolated rat hepatocytes as a target cell, and the activity of sera for this cytotoxicity was diminished after absorption of sera against LSP (3). The nature of LSP is still unknown because it cannot be electrophoresed by the reason of its high molecular weight. In this paper the authors showed a new electrophoretical analysing method of LSP by treating LSP with Deoxycholate (DOC) without any changes in its antigenicity.

Materials and Method

Preparation of LSP. LSP was prepared as LP-1 fraction by the method of McFarlane et al. from a rat liver (Wister, male) (4). That is as follows, liver was perfused with 0.25M sucrose containing 0.01M Tris/HCl (pH 7.4) via portal vein for washing out blood components. Then the liver was resected and a 50% (w/v) homogenate in 0.25M sucrose (pH 8.0, adjusted with NaOH) was prepared using a Potter homogenizer with Teflon pestle, and was centrifuged at 105,000g for 1 hr. The supernatant of liver homogenate was applied to a column of Sephadex G-100 (gel bed: 70x2cm) which had been previously equibrated with Tris/HCl pH 8.0, containing 0.2M NaCl and 1mM disodium EDTA. The column was eluted with the same buffer, elution speed: 10ml/hr, and the protein concentration of the elute was analyzed photometorically (transmission at 280nm). The first peak of the elute was concentrated by ultrafiltration by Emmercible filter (Millipore). This concentrate was applied to a column of Sepharose 6B (gel bed: 80x1.5cm) equibrated and eluted with the same buffer. The first peak was concentrated and designed as LSP. The protein concentration of LSP was determined using the Lowry techniques.

Preparation of antiserum against LSP. Rabbits were injected s.c. with 5mg LSP in Freund's complete adjuvant and repeat injections without adjuvant given 2 weeks after the first injection. Rabbits were bled 1 week after the final injection, and the antisera were absorbed against the rat spleen aceton powder and KSP (supernatant fraction of rat kidney homogenate prepared with the same method as LSP). Before using anti-LSP, IgG fraction was prepared from the anti-LSP sera by ion-exchange chromatography.

Immunoassays. Double-immunodiffusion against LSP was performed in 1% agarose in barbiturate buffer, pH 8.4. Five μl of each LSP solution (1mg/ml) and of antisera were added, then incubated for 24 hr at 37°C.

Electrophoretic analysis. Before electrophoretic analysis the LSP was dialyzed with 0.01M Tris/HCl buffer (pH 8.0) containing 1% Deoxycholate (DOC) or 1% Triton X-100 for 24 hr. at 4°C. Polyacrylamide gradient gel (4-30% continuous gradient) slabs were prepared. Electrophoresis was performed at 150V constant power for 16 hr. in 0.09M Tris/0.08M Borate buffer containing 0.0025M disodium EDTA (pH 8.4) and stained with Coomassie brilliant blue. Sodium dodecylsulphate polyacrylamide gel electrophoresis (SDS-PAGE) was performed by a modified Laemmli method using 12.6%T, 2.6%C gel rods (0.6x6.5cm) in 0.05M Tris/0.38M

Glycine buffer containing 0.1% SDS (pH 8.3) (5). Samples containing 1.0mg/ml were incubated in a 0.01M Tris/HCl buffer pH 8.0 containing 1% SDS, 5% β-mercaptethanol and 0.001M EDTA for 5 min. at 100°C and then electrophoresed at 120V (5).

Isoelectrofocusing (IEF) was performed with 1mm thick slab containing 1% Agarose IEF (Pharmacia), 12% sorbitol and 5% Pharmalyte (Pharmacia), using 0.2M L-Histidine as anode and 1M NaOH as cathode, at 1500V max. for 2 hr. Polyacrylamide slab gel were also used for IEF analysis in some experiments, at that time, anode: 0.04M DL-Asparagic acid, cathode: 1M NaOH were used. After phoresis all gels were stained with Coomassie brilliant blue.

Results

The antiserum (IgG fraction) gave only a single line on immunodiffusion against LSP. Against dialyzed LSPs the antiserum also gave a single line which was distinctly fused with undialysed LSP (Fig. 1). No antigenic differences were observed between DOC-dialyzed LSP and Triton X-100 dialyzed LSP in immunodiffusion test.

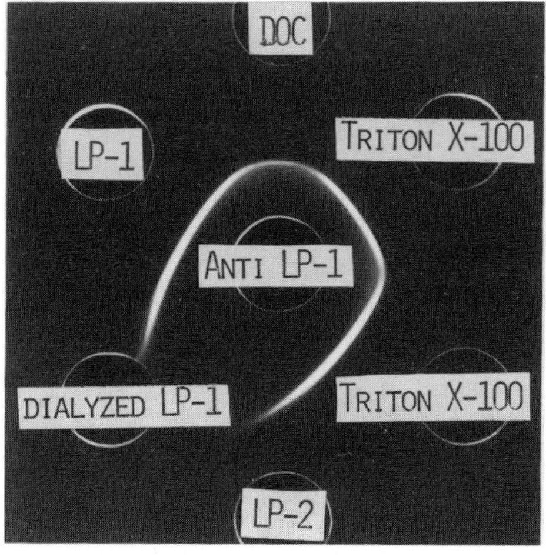

Fig. 1. Immunodiffusion using rabbit antiserum against LSP and dialyzed LSP.

PAGE of LSP revealed that only the lipoprotein precipitates were observed at the site of application and no bands appeared. After dialyzing, LSP could be electrophoresed successfully and seven clear bands were observed from MW 500,000 to 150,000, but this electrophoresis was not complete, some parts of LSP remained still at the top of the gel (Fig. 2).

PAGE of LP-1 (4 ~ 30 %)

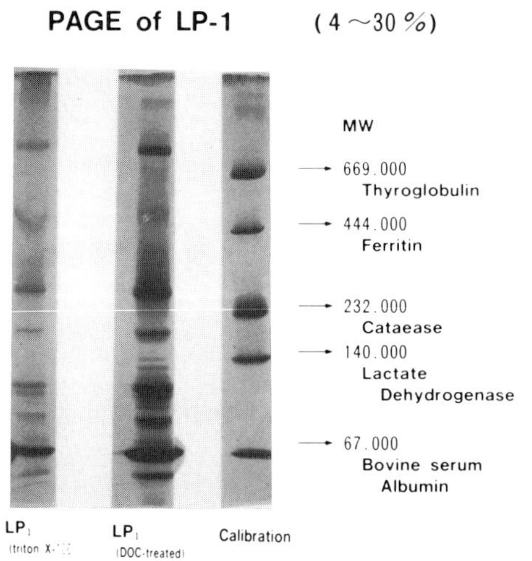

Fig. 2. PAGE of LSP. Seven main bands and seven sub-bands were observed clearly in DOC-dialyzed LSP.

The electrophoretical condition was better in DOC-dialyzed LSP than in Triton X-100 dialyzed LSP.

In SDS-PAGE, the electrophoretical patterns of both DOC- and Triton X-100-dialyzed LSP were simmilar as shown in Fig. 3 Clear 14 main bands were observed from MW 170,000 to 18,000.

In analysis by IEF, DOC-dialyzed LSP was focused from PI 7.35 to PI 9.59 mainly, and made clear 10 bands. IEF was unsuccessful in using Triton X-100 dialyzed LSP because of much tailing from the initial application part of the sample.

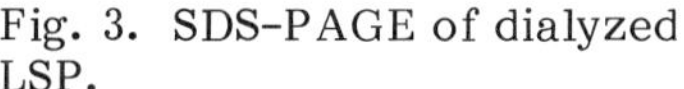

Fig. 3. SDS–PAGE of dialyzed LSP.

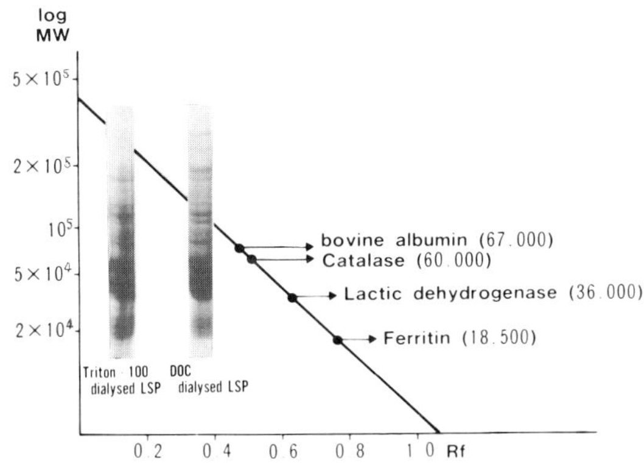

Fig. 4. IEF of DOC-dialyzed LSP.

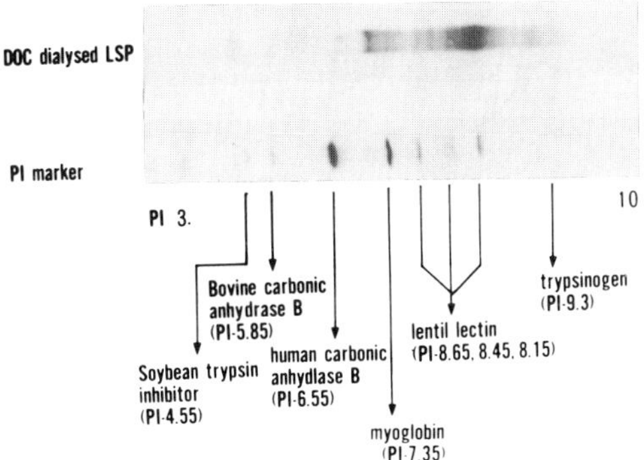

Discussion

It has been demonstrated that the sera of patients with chronic active hepatitis contain autoantibodies against liver membrane antigens, one of which could be absorbed by LSP purified on Sepharose 6B. To clarify the main role of LSP in immune response of chronic liver cell destruction, it is necessary to elucidate the nature of LSP.

The present studies confirm LSP can be analyzed electrophoretically by dialysing with DOC or Triton X-100 without any changes of antigenicity of LSP.

It was shown that electrophoretical conditions were much better in DOC–dialyzed LSP than Triton X–100 dialyzed LSP. In IEF, LSP was well separated and the components of LSP were clearly definite. Further purification of LSP must be carried out by IEF.

In this study it is clear that there are many components in LSP. For clarifying the immune reactions of chronic active hepatitis the heterogenicity of LSP must be concerned.

References

1. Meyer zum Büschenfelde, K.H. and Miescher, P.A.: Clin. exp. Immunol. 10, 89-102 (1972)

2. Meyer zum Büschenfelde, K.H., Kössling, F.K. and Miescher, P.A.: Clin. exp. Immunol. 10, 99-108 (1972)

3. Zeniya, M.: Jikeikai Med. J. 28, 311-330 (1981)

4. McFarlane, G., Barbara, M., Wojcicka, G., Zucker, M., Eddleston, A.L.W.F. and Williams, R.: Clin. exp. Immunol. 27, 381-390 (1977)

5. Laemmli, U.K. and Favre, M.: J. Mol. Biol. 80,575-599 (1973)

OPTIMIZING ELECTROPHORETIC METHODS FOR ANALYSIS OF GENETIC
ORIGINS IN GRAIN AND FOOD SAMPLES

Colin Wrigley, William Campbell, Diana du Cros
CSIRO Wheat Research Unit
North Ryde, N.S.W. 2113, Australia

Joel Margolis
Gradient Laboratories
Pyrmont, N.S.W. 2009, Australia

Agricultural and food products often need to be analysed to
determine the genotypic source of the product or ingredients.
Answers are required to such questions as: Is this barley
seed true-to label?   What variety is this delivery of wheat
grain?   What type of flour is this? Is this spaghetti made
exclusively from durum wheat?   Does this cheese or sausage
contain vegetable protein?   What is the species of this fish
fillet?   Are these rogue plants wild oats, fatuoid oats, or
just minor variants of the correct oat cultivar?   This paper
sets out to describe the use of protein electrophoresis in
routine analysis for these purposes and in dealing with cases
where modifications are needed to make the necessary
distinctions.

Proteins as Information-Carrying Documents

Electrophoretic analysis of protein composition has proved a
valuable means of providing answers to these questions,
because of the fidelity of the transcription-translation
process in reflecting genetic information in protein
characteristics.   Zuckerkandl and Pauling (1) rate proteins

as tertiary semantides (sense-carrying molecules), whose structure indicates the identity of an organism and even its history. The reliability of this information ranks below that of mRNA and DNA (secondary and primary semantides, respectively), but well above that of episemantic molecules (the products of enzymes) or of asemantic molecules (not specifically produced by the organism). This characteristic of proteins, their accessibility and the power of gel electrophoresis to resolve proteins provide a good combination for characterization of genotype in routine analysis or in studies of taxonomic relationships.

Within this general approach, there is considerable scope for varying procedural details to ensure that information of the type needed is provided. This information may lie at any of several taxonomic levels, e.g. genus (irrespective of species), species, variety (cultivar) or biotype. There are four main stages at which variations may be introduced: when choosing the tissue for extraction, the extracting solvent (and thus the class of protein for electrophoresis), the electrophoretic method (cathodic or anodic, isoelectric focusing) and the staining procedure (for all proteins, for carbohydrate or for an enzyme).

Analysis of Cereals

The range of variations can be illustrated by analyses of cereals such as wheat, barley and oats. Anodic electro-phoresis of the leaf proteins (2), for example, is best suited to comparisons at the genus level, since similar electro-phoretic patterns are obtained for different species of the same genus (Table 1). By way of contrast, information about genus identity based on the gliadins of the endosperm is much more difficult to interpret, due to differences in gliadin composition at the species and varietal levels (Table 1).

Table 1. Suitability of different classes of cereal proteins
for comparisons at various taxonomic levels

Protein Class	Comparison of electrophoregrams for individuals at the respective taxonomic levels			
	Genus	Species	Cultivar	Biotype
Leaf proteins	different	similar	similar	similar
Grain albumins	too different	different	similar	similar
Grain prolamins	too different	too different	different	some different
Grain glutelins	too different	too different	different	some different

Examination of Table 1 shows that study of the water-soluble
grain proteins is best suited to comparisons at the species
level.   A practical example of this requirement is the
analysis of pasta products, such as spaghetti and macaroni,
which should be made from semolina of durum wheat (_Triticum
turgidum_ group _durum_).   Admixture of common wheat (_Triticum
aestivum_), not permitted in some countries, can be detected by
water extraction of the pasta followed by electrophoresis and
protein detection (3,4,5).

The most frequent requirement of such techniques is
distinction between varieties of a cereal, sometimes to check
the identity of seed for sowing, or alternatively to establish
the quality type of harvested grain.   Gel electrophoresis of
the prolamin proteins of the grain (Table 1) has become almost
universally adopted for this purpose in grain-growing
countries (5).   In Australia, gradient gel electrophoresis at
pH3 has been adopted (6) as a standard procedure for
establishing varietal identity in commercial samples of cereal

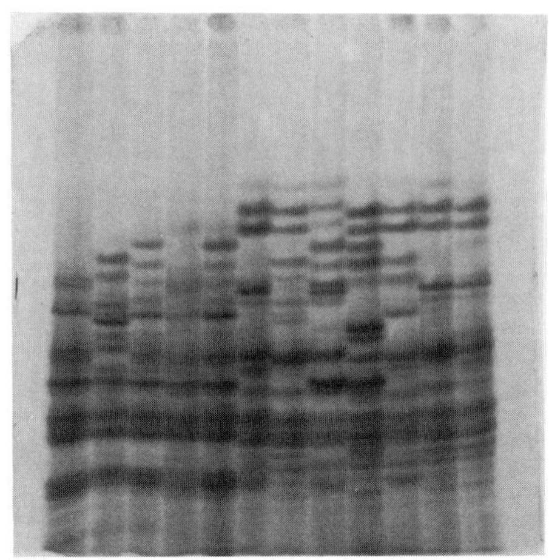

Figure 1.   Cathodic gel electrophoresis (pH 3) of gliadin
proteins extracted with 6% urea from grain of the varieties
(left to right): Duramba A, Duramba B, Durati, Dural,
Kamileroi (durum wheats), Timgalen, Millewa, Eagle A, Halberd,
Tincurrin, Condor A and Banks (hexaploid wheats). The gel is
a gradient of 3% polyacrylamide (at the top) to 13%, supplied
by Gradient Laboratories (5,7).

grain, particularly wheat (Figure 1).   The procedure has been
simplified so that inexperienced operators can obtain good
results with a minimum of laboratory facilities.   These
simplifications include the use of preformed gradient gels (to
avoid the need for such operators to make gels and to provide
automatic sharpening of zones), a simple extraction procedure
(about one hour in 6% urea solution for meal, flour or crushed
grain), and a protein-staining method that does not require a
destaining step (7).

This system has now proved itself for the segregation of
wheats of different qualities in several harvests, and has
been accepted in court as forensic evidence.   In legal cases
particularly, it is important to obtain statistical
information about homogeneity or about the confidence limits
in the proportions of individual components of a mixture
(5,7).   We have found it necessary to identify over one

hundred grains one-by-one in a few cases of this type (5).

Special Cases of Varietal Identification

There have been a few varieties for which electrophoresis of gliadins has provided too much information. The durum wheat Duramba, for example, exists in at least four electrophoretically distinguishable biotypes, two of which (A and B) appear in Figure 1. Such instances have not proved to occur frequently. They make the task of identification slightly more complicated, but by no means impossible. In addition, the Duramba biotypes have proved useful in studying the relationship between proteins and grain quality (8).

On the other hand, there are cases where combinations of closely related varieties are difficult to distinguish by the standard electrophoretic procedure. When identification of quality type is the main aim of analysis, it may not be necessary to make such distinctions. However in other cases, variations to the routine procedure have had to be introduced to permit distinction. Such a case was the combination of wheats Condor and its backcross CSP44, which were electrophoretically similar but different in quality type. The simple expedient of adding mercaptoethanol (1%) to the normal extractant (6% urea) dissolved a further set of proteins that permitted these wheats to be distinguished. In other cases, the solution to this problem has been to change to a completely different system for extraction and electrophoresis. For another combination of similar varieties (Condor, Egret and Oxley), good distinction could be made by esterase staining after gel isoelectric focusing of the grain water-soluble proteins. In addition, gel electrophoresis in sodium dodecyl sulfate (SDS) has served well as an alternative procedure (5,7). Since the need for this type of analysis has arisen on a routine basis, this procedure has also been simplified by designing a preformed SDS gel, containing the gel buffer and a gradient of acrylamide concentration. In addition, it is formed as a thin (1 mm) gel, fixed to plastic film, and is loaded with a thin plastic slot-former. Before use, a layer of concentrating gel is applied and the electrode vessels are filled with the electrode buffer. This system now introduces the opportunity of providing preformed SDS gradient gels for the full range of applications.

Extraction with SDS and a reducing agent also offers the best combination for characterizing protein composition after denaturation by heating or baking. The use of SDS is preferable in the case of several other grains such as oats and rape-seed. SDS gel electrophoresis has proved itself particularly useful in making distinctions within the genus Avena (oats), both for identifying cultivated varieties and for distinguishing them from rogue plants (fatuoids), which appear during pure-seed propagation, and from wild-oats.

Non-Cereal Products

The principles enunciated above for characterizing cereals may also be applied to the analysis of many other products. Important examples include identification of meat and fish species (after filleting) (9) and checking for the presence of vegetable protein to cheese and meat products (10). Standard electrophoretic procedures are becoming established for these and other purposes, but on occasions the routine method is inadequate. Such cases may be handled by modifying aspects of the procedure, as exemplified for cereal grains above. Fish species, for example, may be identified by conventional gel electrophoresis or gel isoelectric focusing (9). Figure 2 shows how another form of electrophoretic analysis can pick out different aspects of protein composition to assist in distinguishing between fish species not readily handled by

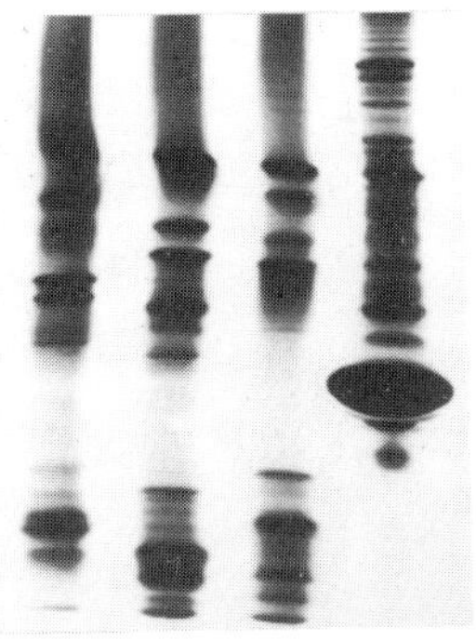

Figure 2.   Anodic gel electrophoresis (pH 8.3) of proteins
from fish fillets on a HX (3-40%) gradient gel after 1700
volt hours.   Species (left to right):   flathead, red fish,
bream (and human serum).

standard methods.   The gel used is available preformed. It
is a gradient gel in which the concentration of polyacrylamide
rises to a high level (40% T, 12% C) (11).   Because of the
resulting small pore size, the composition of smaller proteins
can be determined better than with other procedures, thus
providing a new dimension for characterization and permitting
distinction between closely related fish species.

Future Developments

Despite the considerable use of electrophoretic analyses for
these purposes, attempts to automate such procedures have been
largely unsuccessful: gel electrophoresis is still a
reasonably labour-intensive analysis, even using preformed
gels.   Alternative approaches may thus become more attractive
for these analyses.   Such possibilities (12,13) include
immunological procedures and high-performance liquid
chromatography (HPLC).   For both procedures, the methodology
is highly developed for rapid analysis of many samples.   In
one case, antisera must be raised having the appropriate
specificity.   For the other, the new generation of HPLC

columns, suited to protein fractionation, must be examined. Then an adequate comparison with electrophoretic methods will indicate which type of analysis will best suit future requirements.

References

1. Zuckerkandl, E., Pauling, L.: J. Theor. Biol. 8, 357-366 (1965).

2. Wrigley, C.W., Webster, H.L.: Aust. J. Biol. Sci. 19, 895-901 (1966).

3. Feillet, P., Kobrehel, K.: Ann. Technol. Agric. 21, 17-24 (1972).

4. Cubadda, R., Resmini, P.: Proc. Fifth World Cereal Bread Congr., Dresden 6: 159-166 (1970).

5. Wrigley, C.W., Autran, J.C., Bushuk, W.: Adv. Cereal Sci. Technol. 5, 211-259 (1982).

6. Anon: Method E2, Official Testing Methods, Cereal Division, Royal Australian Chemical Institute, Melbourne (1981).

7. du Cros, D.L., Lawrence, G.J., Miskelly, D.M., Wrigley, C.W.: Technical Publication No. 7, CSIRO Wheat Research Unit, Sydney (1980).

8. du Cros, D.L., Wrigley, C.W., Hare, R.A.: Aust. J. Agric. Res. 33, 429-442 (1982).

9. Lundstrom, R.C.: J. Assoc. Off. Anal. Chem. 63, 69-73 (1980).

10. Armstrong, D.J., Richert, S.H., Riemann, S.M.: J. Fd Technol. 17, 327-337 (1982).

11. Campbell, W.P., Wrigley, C.W., Margolis, J.: Anal. Biochem. 129, 31-36 (1983).

12. Hitchcock, C.H.S., Bailey, F.J., Crimes, A.A., Deen, D.A.G., Davis, P.J.: J. Sci. Fd Agric. 32, 157-165 (1981).

13. Griffiths, N.M., Billington, M.J., Griffiths, W.: J. Ass. Public Analysts 19, 113-119 (1981).

APOPROTEIN A-I, A-II, AND HIGH DENSITY LIPOPROTEIN CHOLESTEROL IN DIABETES MELLITUS

Eiji Nakano, Kenji Ozawa, Kazunari Kumasaka, Kinya Kawano,
Toshio Tsuchiya, Kooichi Watanabe
Department of Clinical Pathology and Department of Internal
Medicine, Nihon Univeristy, School of Medicine
Tokyo 173, Japan

Introduction

Abnormalities of lipid metabolism are common in diabetes
mellitus, the most common being an elevation of serum
triglyceride and very low density lipoprotein (VLDL), but
there is no general agreement whether these changes are
significant as atherosclerotic factors. Recently high
density lipoprotein (HDL) has been thought to be a protective
factor in coronary heart disase. However according to
published reports (1, 2, 3) in diabetes, HDL cholesterol
(HDL-C) level varies and there are a few reports(4) on
apoprotein A-I (A-I) and apoprotein A-II (A-II) which are the
main apoprotein of HDL. In the present study, A-I, A-II and
HDL-C levels were determined in non-insulin dependent diabetic
patients and differences of these levels between normolipidemic
and hyperlipidemic patients were evaluated. Furthermore the
correlations between A-I, A-II and HDL-C were investigated.

Materials

One hundred and forty four patients treated in Nihon
University Hospital (mean age 59.3) were divided into four
groups as follows: no lipidemia 56 cases, hypertriglyceridemia

(>130 mg/dl) without hypercholesterolemia 45
cases, hypercholesteroemia (>250 mg/dl) with or without
hypertriglyceridemia 16 cases, and presence of extra pre-beta
band (5) on electrophoretic pattern 11 cases. As a control
group, 35 normal subjects (mean age 43.8) were examined.
Blood samples from patients and normal subjects were drawn in
the morning before breakfast.

Methods

Serum triglyceride and cholesterol were measured by enzymatic
methods. Lipoprotein electrophoresis was carried out with
cellulose-acetate membrane using ozonized Schiff staining
(6). A-I and A-II levels were measured by single radial
immunodiffusion method with specific anti-A-I and anti-A-II
sera (The kit was obtained from Daiichi Kagaku Co., Japan).
The coefficient of variation of this method was 4% in our
study. HDL-C was determined by phosphotungstate-Mg method.
All of these examinations were performed within 12 hours from
venopuncture.

Results

1) A-I, A-II and HDL-C levels from normal control and
    diabetics are summarized in Table 1.
    In the diabetic patients, the A-I level of hypercholeste-
    rolemic group was slightly higher than hypertriglyceride-
    mic group, but this difference was not significant by
    T-test analysis (p<0.025). Other diabetic groups showed
    values similar to that of normal controls (Fig. 1).
2) For the A-II level, there were no significant differences
    between normal control and each diabetic group, but the
    hypercholesterolemic group showed a significant icrease

(36.4$^{\pm}$5.9 mg/dl) over the hypertriglyceridemic group
(29.0$^{\pm}$6.0 mg/dl) (p<0.025) (Fig.2).

3) The A-I / A-II ratio was no different than those of the
controls diabetic groups (Fig.3).

Table 1. A-I, A-II and HDL-C in diabetes mellitus (mg/dl)

	N	A-I		A-II		HDL-C	
		Mean	SD	Mean	SD	Mean	SD
Normal control	35	135.6	22.4	32.1	5.1	52.5	15.7
Normolipidemia	56	126.9	24.1	30.3	6.0	46.2	11.5
Hyper-TG.*	45	125.0	25.0	29.0	6.0	38.9	13.0
Hyper-Cho.*	16	149.0	30.2	36.4	5.9	47.0	16.0
Extra pre-beta	11	132.8	25.2	31.3	7.3	39.0	9.6

* Hyper-TG.= Hypertriglyceridemia (higher than 130 mg/dl)
  Hyper-Cho.= Hypercholesterolemia (higher than 250 mg/dl)

Fig. 1 A-I in diabetics   Fig. 2 A-II in diabetics

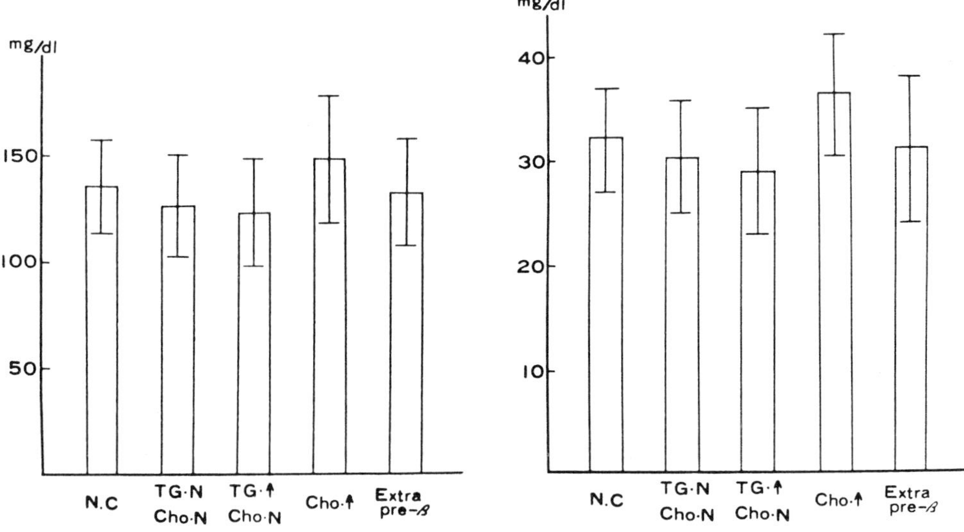

4) The HDL-C level of hypertriglyceridemic group was
35.7$^{\pm}$ 5.9mg/dl and significantly lower than that of

692

the controls (52.5$\pm$15.7 mg/dl) (p<0.025). The other
diabetic groups did not differ in HDL-C level from the
controls (Fig. 4).

5) Correlations between A-I and HDL-C varied among the
    diabetic groups (Table 2). In the controls, there was a
    good linear relationship between A-I and HDL-C levels

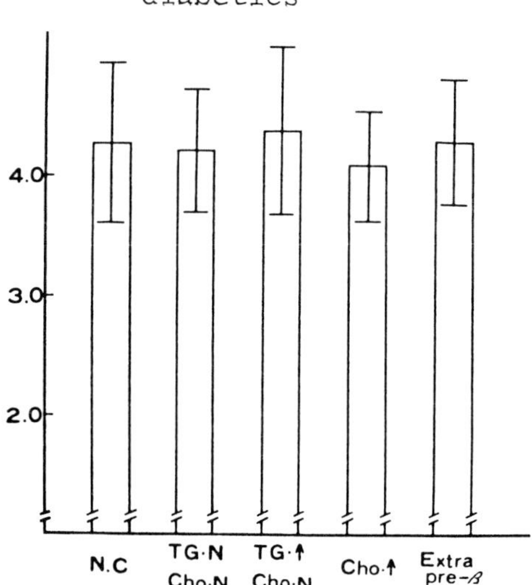

Fig, 3 A-I/A-II ratio in
       diabetics

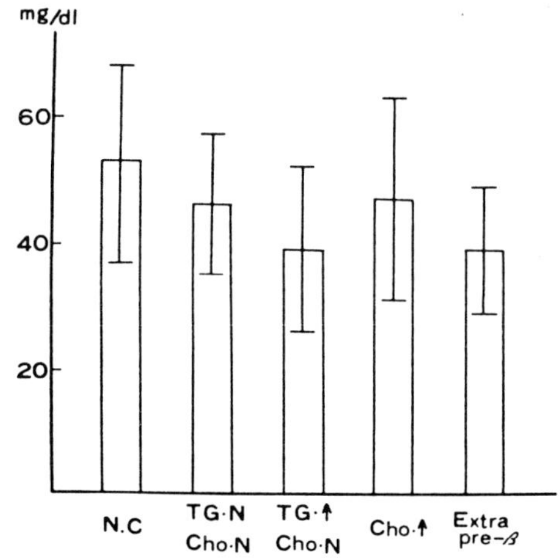

Fig. 4 HDL-C in diabetics

Table 2. Correlation between A-I,
A-II and HDL-C (r value)

	A-I & HDL-C	A-II & HDL-C
Control	0.790	0.479
Normolipidemia	0.671	0.637
Hyper-TG.	0.537	0.398
Hyper-Cho.	0.698	0.492
Extra pre-beta	0.740	0.562

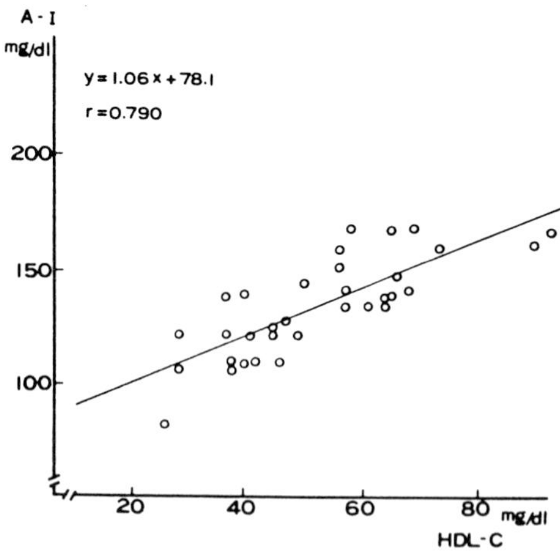

Fig. 5 Correlation between
       A-I and HDL-C in control

(r=0.790, Fig. 5). The correlation in normolipidemic patients was weak (r=0.671, Fig. 6), and in hypertriglyceridemic patients it was rather poor (r=0.536, Fig. 7).

Fig. 6 Correlation between A-I and HDL-C in normo-lipidemic diabetics

Fig.7 Correlation between A-I and HDL-C in hyper-lipidemic diabetics

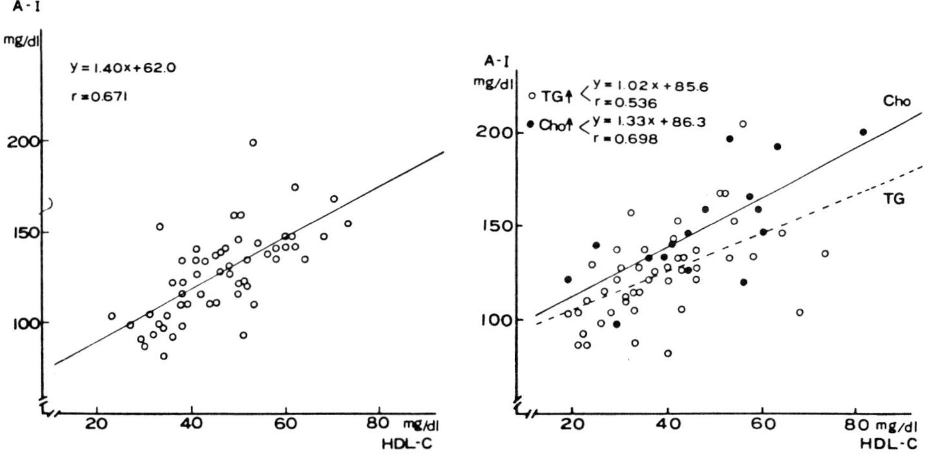

6) Relatively good correlation was observed between A-II and HDL-C in normolipidemic patients, but for normal controls and other diabetic groups did not show good relationship (Table 2).

Discussion

The mechanism of hypertriglyceridemia in diabetics had been clarified by Nikkila et al (7). By kinetic study using isotopics, they concluded that hypertriglyceridemia in diabetics was caused by over production of VLDL in liver, probably due to several factors.
In our study, A-II level of diabetics with hypertriglyceridemia was lower than diabetics with hypercholesterolemia, therefore

A-II level is thought to be related to serum lipid level. It was reported by Biesbroeck et al (4) that A-II level was low in diabetic females, but we did not examine the difference between sexes.

HDL-C level in hypertriglyceridemic group was lower than in normal control as reported by Schonfeld (8), who also described that HDL-C in diabetic patients with type IV hyperlipoproteinemia was lower than those of non-diabetic type IV hyperlipoproteinemia.

A noteworthy result of our study is that the correlation between A-I and HDL-C in the hypertriglyceridemic group was lower than that of other diabetic groups.

Biesbroeck et al described the abnormal composition of HDL in non-insulin dependent diabetics with hypertriglyceridemia, and suggested the possibility of reduced cholesterol in HDL particles (4). The result of our study suggests that the composition of HDL in the hypertriglyceridemic group is not as homogenous as that of normal subjects and other diabetic groups.

References

1. Lopes Virella, M.F.L., Stone, P.G., Colwell, J.A.:
    Diabetrogia 13, 285-291 (1977).
2. Simpson, R.W., Carter, R.D., Moore, R.A., Penfold,
    W.A.F.: Diabetrogia 18 35-40 (1980).
3. Durrington, P.N.: Clin. Chim. Acta 104, 11-23 (1980).
4. Biesbroeck, R.C., Albers, J.J., Wahl., P.W., Weinberg,
    C.R., Basset, M.L., Bierman, E.L.: Diabetes 31 126-131
    (1982).
5. Dahlen, G., Ericson, C., Furberg, C., Lundkvist, L.,
    Svardsudd, K.: Acta Med. Scand, (Suppl.). 531 1 (1972).
6. Hasunuma, S., Nakano, E., Sakurabayashi, I.: Seibutsu-
    Butsuri-Kagaku 16, 15-18 (1971) (in Japanase).
7. Nikkila, E.A., Kekki, M.: Metabolism 22, 1-22 (1973).
8. Schonfeld, J., Birge, C., Miller, J.P., Kesseler, G.,
    Santiago, J: diabetes 23, 827-834 (1974).

PURIFICATION AND PROPERTIES OF APOLIPOPROTEIN A-I AND A-II
FROM HUMAN SERUM HIGH DENSITY LIPOPROTEIN (HDL)

Yasuhiro Nomata, Akira, Matsukawa, Kenichi Hiraoka
Eiken Chemical Co., Ltd., Tokyo Japan

Ikunosuke Sakurabayashi, Yasuo Iimura, Tadashi Kawai
Dept. of Clinical Pathology, Jichi Medical School, Tochigi,
Japan

Introduction

HDL has attracted increasing interest in recent years following the finding of an inverse correlation between coronary
heart disease and serum HDL lelves (1). HDL consists of approximately 50% lipid and 50% protein: the protein moiety
contains two major polypeptides,apolipoprotein A-I (apoA-I),
apoA-II and several minor components. Precise quantitation
of the apoprotein is required in order to assess their role
in the structure and function of HDL as well as its
abnormality. Approaches to apoprotein quantitation include
column chromatography and protein determination of isolated
fractions or gel electrophoresis or isoelectrofocusing
coupled with quantitative densitometry. A more practical
approach is by immunoassay, which eliminats the tedious task
of separating the individial proteins from the complex
protein mixture. We therefore undertook to purify apoA-I and
apoA-II from human serum HDL, immunized with these proteins,
and obtained monospecific antisera reacting with apoA-I or
apoA-II. Furthermore, we developed a quantitative
immunoassay for apoA-I and apoA-II in human serum by single
radial immunodiffusion assay (RID).

Materials and Method

1) Isolation of apoA-I and apoA-II: HDL_3 (density 1.125-1.210) was isolated from fasting plasma of normolipidemic donors by sequential ultracentrifugation, fractionated on sepharose 6B column chromatography, and delipidated with ether-methanol (2:1 v/v) under $-20^{\circ}C$. The delipidated preparation was chromatographed on DEAE Sepharose CL-6B equilibrated with 0.03M Tris/HCl buffer pH 8.0 containing 6M urea, and then eluted with a linear gradient of NaCl from 0 to 0.125M containing of the above buffer to yield crude apoA-I and apoA-II fractions. These crude apoA-I and apoA-II fractions were further purified by gel filtration on Sephadex G-100 (apoA-I) or Sephadex G-75 (apoA-II) equilibrated with 0.03M Tris/HCl buffer pH 8.3 containing 6M urea. Fractions corresponding to apoA-I and apoA-II were extensively dialyzed against 10mM NH_4HCO_3 buffer pH 8.0 containing 1mM EDTA and lyophilized. Lyophilized protein was stored at $-20^{\circ}C$. In some cases, the final liophilization step was omitted and stored at $-80^{\circ}C$. The purity was checked by SDS-PAGE(2), UREA-PAGE(3), Ouchterlony method and amino acid analysis. Protein concentration was measured by the method of Lowry et al(4) or absorption of 280nm: 1.0mg of apoA-I and apoA-II estimated by the Lowry method was equivalent to 0.96mg (apoA-I) and 0.98mg (apoA-II) calculated from amino acid analysis.
Physicochemical properties of these apolipoproteins was measured by HPLC. A molecular sieve type column, Toyosoda TSK-GEL G 3,000 SW (0.75 x 60 cm) was used with a Hitachi 635 HPLC System. Up to $50\mu l$ of protein solution was loaded on the column. The flow rate was 1 ml/min using 0.05M phosphate buffer pH 7.2 containing 0.25M NaCl. Ultraviolet absorption was monitored at 280nm.

2) Iodination of apoA-II: ApoA-II was iodinated by the lactoperoxidase method(5). The labeled peptide (Specific

activity 36–58mCi/mg) was separated from unreacted ^{125}I
by gel filtration on Sephadex G-75. Iodinated apoA-II
was stored at 4°C or frozen in aliquots at -20°C.
The iodinated tracer was used for a maximum of 4 weeks.

3) Preparation of Antisera: New Zealand white rabbit was
immunized with purified apoA-I (1 mg) or purified apoA-II
(1 mg) mixed with complete Freund's adjuvant and injected
subcutaneously into all four legs. Booster injections
were given in the back once a week for 7 weeks. Blood
was taken from a marginal ear vein at one week after the
last injection. Specificity of antiserum was checked by
Oucterlony's method, immunoelectrophoresis and double
antibody radioimmunoassay.

4) Single Radial Immunodiffusion assay (RID): The levels of
apoA-I and apoA-II in human serum were measured by the
single radial immunodiffusion assay described by Mancini
et al(6). Agarose gel (1g/100ml) was prepared in 0.1M
Tris/HCl buffer pH 8.3 containing 0.2M NaCl. The
monospecific antiserum against apoA-I or apoA-II was
added to the gel with a final concentration of 1.5% (v/v)
agarose gel. Four ml of the gel mixture was poured into
a plastic vessel (80 x 40 mm). Ten antigen wells of 2.5
mm diameter were punched out of the agarose plate.
Plasma samples were diluted five times with 0.9% NaCl and
5μl of the diluted samples were applied to the wells.
Four standard solutions of apoA-I or apoA-II were applied
in different wells on the same plate. The immunoplate
was incubated for 48 hr at room temperature. The
diameter of ring-shaped immunoprecipitates were meaured
using a measuring projector for immunoanalysis.

5) Double antibody radioimmunoassay: The assay was carried
out using a 0.1M borate buffer (pH 8.0) containing 0.5%
bovine serum albumin. Standards or samples were made up
to a total volume of 1.0ml; 0.2ml of appropriatly diluted
I-labeled apoA-II (25,000 cpm) and 0.2ml of antibody

(1:15,000) were added to give approximately 50% binding
of the tracer in the absence of added apoA-II. After
incubation at room temperature for 3.0 hr, 0.1 ml of goat
anti-rabbit IgG and rabbit serum as carrier (1:100) were
added. After a further 3.0 hr incubation at room
temperature, the tubes were centrifuged. The supernatant
were decanted and the precipitates were counted. Results
were expressed as B/Bo.

Results and discussion

1) Purification and properties of apoA-I and apoA-II: Fig.1
shows the results of polyacrylamide gel electrophoresis
in SDS and 6M urea of purified apoA-I and apoA-II. No contaminated protein band was observed. The molecular
weight of apoA-I and apoA-II were estimated to be 27,000
daltons (apoA-I) and 18,000 daltons (apoA-II) by SDS-PAGE.
The yield of apoA-I and apoA-II from one liter of human
serum was about 150mg and 50mg respectively. Table shows
the amino acid composition of purified apoA-I and
apoA-II. These data are compatible with the results
reported by Scanu et al (6) and Lux et al (7).

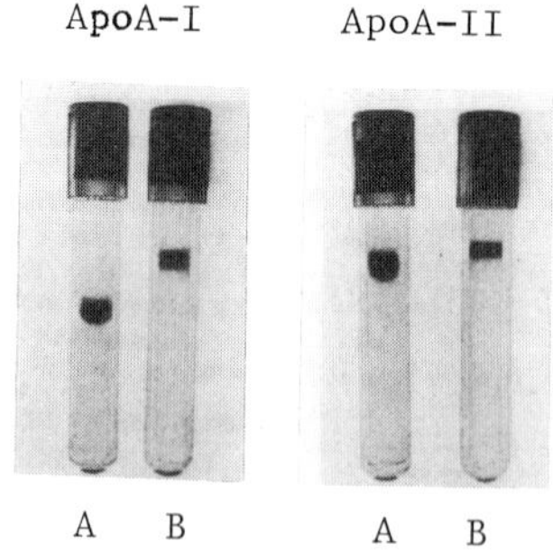

Fig.1 Disc electrophoresis of
purified apoA-I and A-II
A: SDS-PAGE
B: UREA-PAGE

Table. 1 Amino acid composition of apoA-I and apoA-II

(Mol %)

Amino acid	ApoA-I	Ref.(7)	ApoA-II	Ref.(8)
Lys	8.76	8.53	11.60	11.67
His	2.11	2.05	0.144	0
Arg	6.76	6.48	0.22	0
Asp	7.87	8.53	3.96	3.86
Thr	4.19	4.10	7.52	7.82
Ser	5.79	5.69	7.22	7.82
Glu	19.67	19.11	21.01	20.86
Pro	4.69	4.10	5.25	5.22
Gly	4.69	4.10	4.59	3.86
Ala	7.95	7.73	6.69	6.46
Cys[a]	0.03	0	0.80	1.25
Val	5.32	5.23	7.61	7.81
Met	1.24	1.25	1.77	1.25
Ile	0.07	0-0.46	1.37	1.25
Leu	15.60	15.81	10.66	10.43
Tyr	3.17	2.84	4.33	5.22
Phe	2.67	2.39	5.25	5.22

ApoA-I

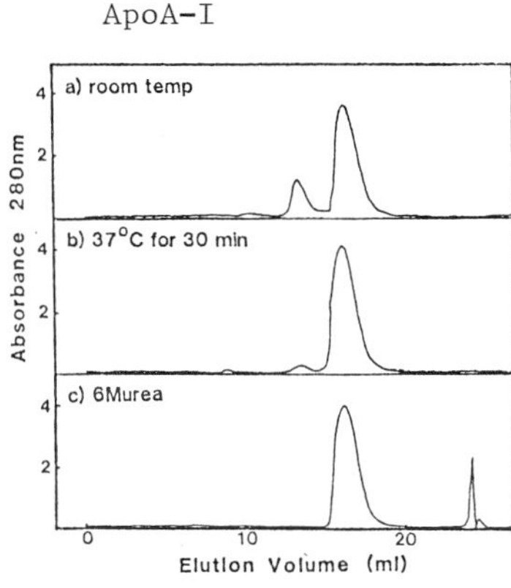

ApoA-II

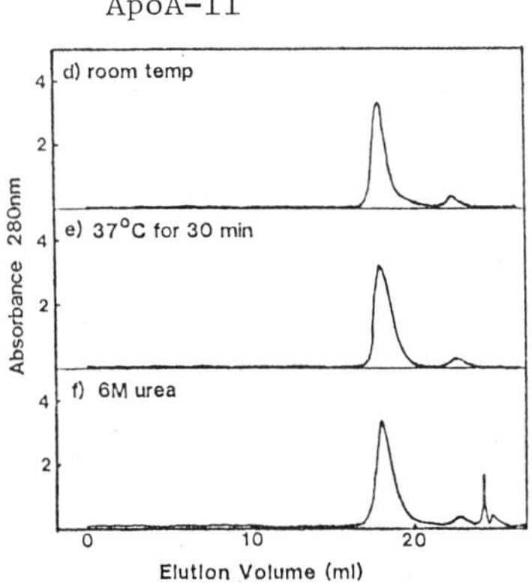

Fig.2 Elution profiles of ApoA-I
and ApoA-II from HPLC column

The molecular forms of apoA-I and apoA-II in aqueous solution were investigated by HPLC. After thawing of apoA-I solution at room temperature, the solution was passed through a column of HPLC at 20°C. Two peaks were observed clearly separated (Fig.2-a). The two peaks of apoA-I corresponded to moleclar weights of 170,000 daltons(peak I) and 56,000 daltons (peak II). Considering of these molecular weights, peak I appeared to be oligomer and peak II to be dimer. When the solution was incubated at 37°C for 30 min, peak I decreased (Fig. 2-b), ultimately disappeared after adding 6M urea (Fig.2-c). Two clearly separated peaks were also observed for apoA-II, (Fig.2-d,e,f). Peak I corresponded to 36,000 daltons and peak II to 9,000 daltons. Considering of these molecular weights, peak I appeared to be dimer and peak II to be half molecule. When the solution was incubated at 37°C for 30 min or 6M urea added, peak I and peak II were not changed. When apoA-I and apoA-II, which were treated as mentioned above, were added to an immunoplate, there is no diference of diameter of precipitin rings before or after treatment.

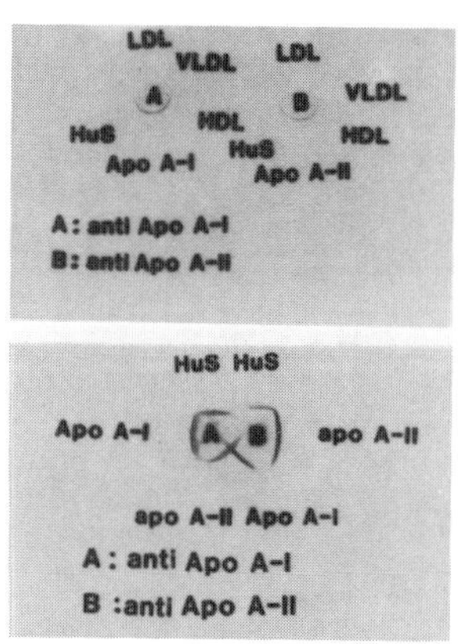

Fig.3 Double immunodiffusion
patterns of anti-apoA-I and
anti-apoA-II

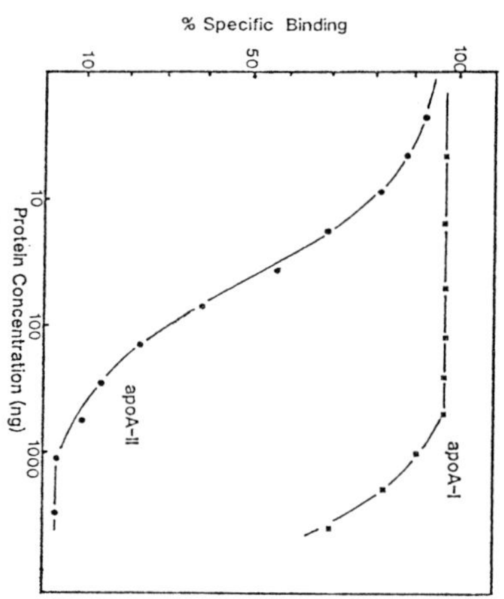

Fig.4 Displacement curves of
apoA-I and apoA-II
The tracer was ^{125}I-labeled
apoA-II and anti-apoA-II was
used

2) Specificity of anti-apoA-I serum and anti-apoA-II serum:
    Fig. 3 shows the results of Ouchterlony's double immuno-
    diffusion test of anti-apoA-I and anti-apoA-II antibodies
    against HDL, LDL (low density lipoprotein), VLDL (very
    low density lipoprotein), HUS (human serum), apoA-I and
    apoA-II. Anti-apoA-I serum formed fused single
    precipitin line with human serum, apoA-I and HDL but
    weakly with VLDL. No precipitin line showed with LDL and
    apoA-II. Similarly, anti-apoA-II serum formed single
    fused precipitin line with human serum, apoA-II and HDL,
    but none with VLDL, LDL and apoA-I. The two precipitin
    lines of anti-apoA-I against apoA-I and anti-apoA-II
    against apoA-II were completely crossed, indicating that
    anti-apoA-I serum and anti-apoA-II serum are monospecific
    and have no cross reactivity with apoA-I and apoA-II,
    respectively.
Cross reactivity of anti apoA-II against apoA-I was inves
tigated further by double antibody radioimmunoassay. The
standard curve with purified apoA-II showed significant

displacement at 1.0ng (Fig 4). The working range of the
assay lay between 1 and 300ng of apoA-II. ApoA-I did not
react and contained approximately 0.3ng of apoA-II per mg
protein (less than 0.05% apoA-II). In a similar study of
cross reactivity of anti-apoA-I with apoA-II, no cross
reactivity was observed (data not shown).

3) Single radial immunodiffusion assay: Preliminary experiment, serum levels of apoA-I and apoA-II of healthy
persons (n=108) were measured by RID assay. Standard
curves of apoA-I and apoA-II are shown in Fig. 5. Each
working ranges of apoA-I and apoA-II lays
5mg/dl-65mg/dl(apoA-I) and 5mg/dl-35mg/dl(apoA-II). In a
random population sample of apparently healthy persons
(n=108, aged 20-60 years), serum levels of apoA-I and
apoA-II were 174$^{\pm}$27mg/dl and 64$^{\pm}$11mg/dl respectively.
Serum apoA-I level was similar to the reports by Fagel et
al (9) and by Børresen et al(10). But apoA-II value was
so high comparing with the reports by Tatu et al (11)and
Albers et al (12).
Since immunological methods are sensitive to the struc-
tual form of the peptide their molecular conformation
may affect quantitation. Purified apoA-I and apoA-II
commonly used as assay
standards, have unique
physicochemical properties
(13-15) which may be
different form those of
apoprotein when these
peptides are complexed
with lipid.
Since, the antigenic sites
probably are modified
during the purification

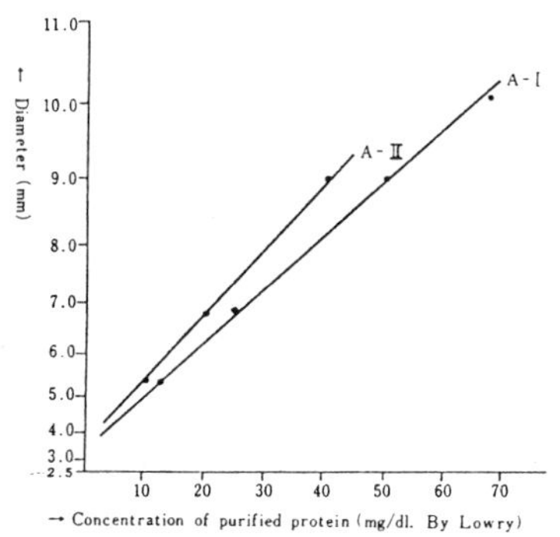

Fig.5 Standard curves of
      apoA-I and apoA-II

step of the apoprotein, the immunoreactivity may be changed.
Wide discrepancies in serum levels of apoA-I and apoA-II have
been reported (16-19), which may be reflected similar problems
in the quantitation of these peptides. Further investigation
is required to solve of these problems.

References

1. Gordon, T., Castelli, W.P.: Amer. J. Med. 62, 707-714
    (1977).
2. Laemmle, V.K.: Nature 227, 680-687 (1970).
3. Kane, T.P.: Anal. Biochem. 53, 350-356 (1973).
4. Lowry, O.H.: J. Biol. Chem. 193, 265-267 (1951).
5. Schonfeld, G., Chen, J.: J. Lipid Res. 18, 645-655 (1977).
6. Mancini, G., Vaerman J.P.: Immunochemistry 2, 235-254
    (1965).
7. Scanu, A.M., Lim, C.T.: J. Biol. Chem. 247, 5872-5846
    (1972)
8. Lux, B.T.: J. Biol. Chem. 247, 7510-7516 (1972).
9. Fagel, G.: Atheroscherosis 36, 67-74 (1980).
10. Borresen, A.L., Berg, K.: Artery 7, 139-148 (1980)
11. Tatu, A., Miettinen, T.: Atherosclerosis 36, 249-259
    (1980)
12. Albers, J.J., Cheung, M.C.: J. Lipid Res. 21, 874-878
    (1980).
13. Vitello, L.R., Scanu, A.M.: Biochemistry 15, 1161-1165
    (1976)
14. Vitello, L.R., Scanu, A.M.: J. Biol. Chem. 251, 1131-1136
    (1976)
15. Swaney, J.B., Brien, K.O.: J. Biol Chem. 253, 7069-7077
    (1978)
16. Karlin, J.B., Juhn, D.J.: J. Lipid Res. 17, 30-37 (1976).
17. Shepherd. J., Packavd, J.C.: Eur. J. Clin. Invest. 8,
    115-120 (1978).
18. Schonfeld, G., Chen, J.: J. Lipid Res. 18, 645-655 (1977).
19. Claus.: Clin. Chem. 29, 1205-1208 (1983).

THE DENATURATION BY LINOLEIC ACID HYDROPEROXIDE
AND AN AUTO-DENATURATION IN HUMAN LIPOPROTEINS.

Masato Ageta, Jiro Nakayama and Seiichiro Yamasaki
Clinical Laboratory & Internal Medicine,
Kurume University School of Medicine,
67 Asahi-Machi, Kurume-City, Fukuoka, 830, Japan.

Introduction

In a lipoprotein particle, cholesterol ester and triglycerides
constitute what has been termed the "lipoprotein core".
Therefore, the lipoprotein core contains a lot of unsaturated
fatty acids (1). Under condition that a lipoprotein particle
coexists with an active oxygen, the active oxygen which is incorporated into the lipoprotein particle can be transferred
into the unsaturated fatty acids portion of the lipoprotein
by a free-radical chain reaction (2). It is possible that an
interacted lipoprotein with active oxygen could induces the
physical denaturation and chemical degradation of lipoprotein
particle (2,3,4). The denatured lipoprotein by lipid peroxides
is said to occur more frequently in low density lipoproteins
(LDL)(2,3,4) which is carrier of plasma lipid peroxide (5).
The experiments reported here demonstrate that other lipoproteins but LDL are also denatured by linoleic acid hydroperoxide. Based on the experiments we propose that the formation of heterogeneous or large molecular size lipoprotein may
result from an interaction with the active oxygen which is
originated in lipid peroxide. We further suggest that a small
quantity of active oxygen can be trigger an auto-denaturation
in lipoprotein particles.

Materials & Methods

Blood samples were collected from healthy male volunteers with normolipidemia and normal lipoprotein electrophoretic patterns. Fresh plasma with the exception of chylomicrons was used in the present study. The plasma was separated into each lipoprotein fraction by the ultracentrifuge (6), i.e., Very Low Density Lipoprotein (VLDL; d=1.006), Low Density Lipoprotein (LDL; d=1.006-1.063), High Density Lipoproteins (HDL$_2$; d=1.063 -1.12, HDL$_3$; d=1.12-1.21). The separated lipoproteins were dialyzed with NaH$_2$PO$_4$ for 24 hr at 4°C.

Preparation of linoleic acid hydroperoxide: Linoleic acid (1 gr) was diluted with 20 ml of 0.02M Tris-buffer containing 0.3 % of bovine serum albumin. Oxidation was allowed to proceed in a loosely covered flask at room temperature by continued stir against the buffer saturated with 85%O$_2$. Hydroperoxide of the linoleic acid was fractionated through gel column by the modified methods of Gardner, et al. (7).

Interaction of fresh plasma with linoleic acid hydroperoxide: Plasma (2 ml) was removed to glass tube, and was pre-incubated for 30 min at 37°C. Several concentrations of linoleic acid hydroperoxide were added into each tube (Table-1). Final volume (2.5 ml) of the mixture was adjusted with 0.02M Tris -buffer. The mixture was incubated for 8 and 12 hours at 37°C with 60 beats/min.

Conc. \ No.	1	2	3	4	5	6	7	8
μmol/ml	0	23.8	40.8	47.5	53.5	58.7	63.4	71.3
MDA nmol/ml	0	5.0	8.6	10.0	11.3	12.4	13.3	15.0

Table 1. Final concentration of linoleic acid hydroperoxide in each tube was shown as a value of linoleic acid (μmol/ml), and as a value of malondialdehyde (MDAnmol/ml) which was measured by the means of thiobarbituric acid reaction (8).

Interaction of isolated lipoprotein with linoleic acid hydroperoxide: Protein concentrations of the isolated lipoproteins were measured by the means of Lowry (9). The isolated lipoprotein was removed into glass tube, and was incubated for

30 min at 37°C. After the pre-incubation, linoleic acid hydroperoxide (14.4 μmol/ml) per 100 mg of lipoprotein-protein was added into tube. The mixture was incubated for 8 hr at 37°C with 60 beats/min.

<u>Auto-denaturation in lipoproteins</u>: The isolated lipoprotein with 10 mM of NaH_2PO_4 was saturated with 85% of oxygen for 1, 5, 10 and 15 min at 37°C. After the oxidation of lipoproteins, the each lipoprotein was incubated for 8 hr at 37°C with 60 beats/min.

<u>Lipid staining pattern & Schlieren pattern of lipoproteins</u>: After the incubation, 0.02% of buthylated hydroxytoluene was added into each tube to stop the peroxidation. Lipid staining patterns of the incubated lipoproteins were determined on 3.5% of polyacrylamide gel-disc electrophoresis which were carried out for 28 min at a constant current of 5 mA. These densitograph were traced by densitometry. Schlieren patterns of LDL and HDL fraction was carried out at d=1.20 in Hitachi-282 analytical ultracentrifuge at a bar angle of 45° and a speed of 53,000 rpm at 20°C. Photograph of the floating lipoproteins was taken at intervals of 3 min.

Results & Discussion

In the incubation study with linoleic acid hydroperoxide for 8 hr using the fresh plasma with the exception of chylomicrons (Fig.1); Low concentrations of the hydroperoxide had a marked effects on the lipid staining patterns of LDL fraction. An unknown band was found out in the area between LDL and HDL_2 at 47.5-53.5 μmol/ml of the hydroperoxide. Both fractions, VLDL and HDL_2 were persistent and resistant to the active oxygen which is originated in linoleic hydroperoxide. Under condition with high concentrations of the hydroperoxide, their modified lipid staining patterns were observed on the electrophoresis. One heterogeneous lipoprotein in the LDL was developed on the Schlieren patterns at 6 and 9 min in the incubation

706

of plasma with 47.5
µmol/ml of the hydro-
peroxide for 8 hr at
37°C (Fig.2). The
heterogeneous LDL
was named as a " λ -
lipoprotein ".
Modification of the
lipid staining
pattern in each lipo-
protein was further
gradual growing by
the long term in-
teraction with
linoleic acid hydro-
peroxide (Fig.3).
The lipid staining
patterns of VLDL, LDL
and HDL₂ were clearly
modified by the lower
concentrations of the
hydroperoxide than
those in the short
term experiment. The
LDL was already dis-
appeared on the

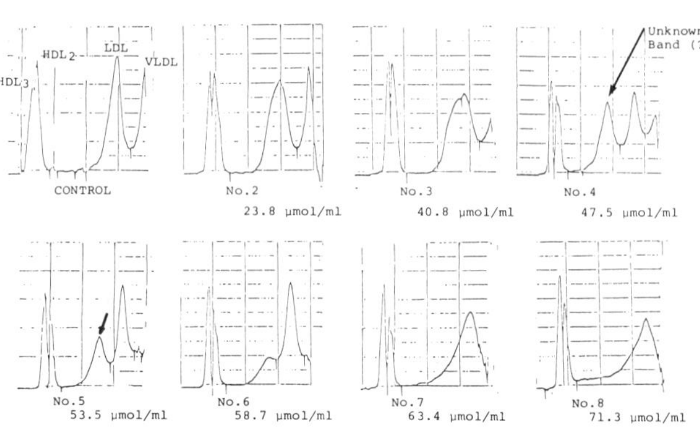

Fig.1. Changes in the lipid staining pattern on
3.5% polyacrylamide gel-disc electrophoresis after
the incubation of fresh plasma with linoleic acid
hydroperoxide for 8 hr at 37°C.

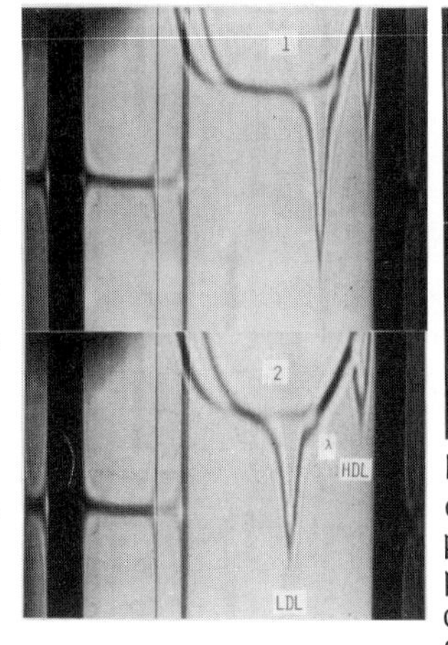

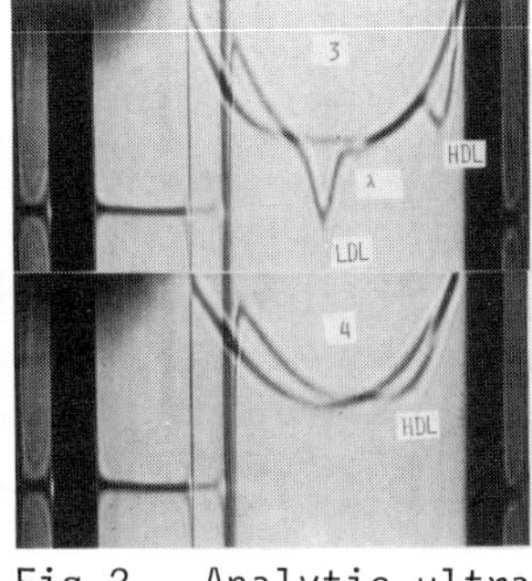

Fig.2. Analytic ultra-
centrifuge Schlieren
patterns of the interacted
plasma with 47.5 µmol/ml
of the hydroperoxide for
8 hr at 37°C.

electrophoresis by low concentraions
of the hydroperoxide (less than 47.5 µmol/ml). Under condition
with a large quantity of the hydroperoxide, the lipid staining
pattern of VLDL was disappeared on the electrophoresis, whereas one of HDL₂ was to be a wide staining pattern. Schlieren
patterns from the sample which was interacted with linoleic
acid hydroperoxide (47.5 µmol/ml) for 12 hr were overt differences from those in the control sample (Fig.4,5). Floating
lipoproteins of the untreated sample with linoleic acid hydroperoxide were still observed on the photograph at 9 min (Fig.5),

though the interacted
plasma lipoproteins with
the hydroperoxide for 12
hr were not observed at 9
min (Fig.4). The phenomenon
was not confirmed in the
short term experiment (Fig.
2). Appearance rates of
the modified LDL with linoleic acid hydroperoxide (
47.5 μmol/ml) which were
inspected on the electro-
phoresis were correlated
with a duration of
the incubation at
37°C (Fig.6). It was
further gradual in-
creased to 100% by
eight hours of the
incubation. It was
concluded that the
modification of plasma
lipoproteins was de-
pendent on the addi-
tional concentrations
of linoleic acid

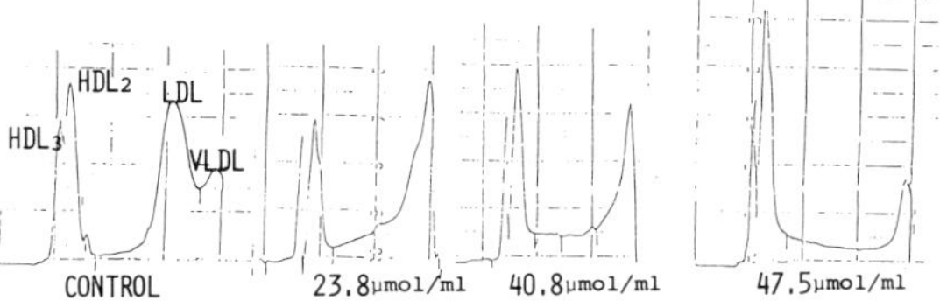

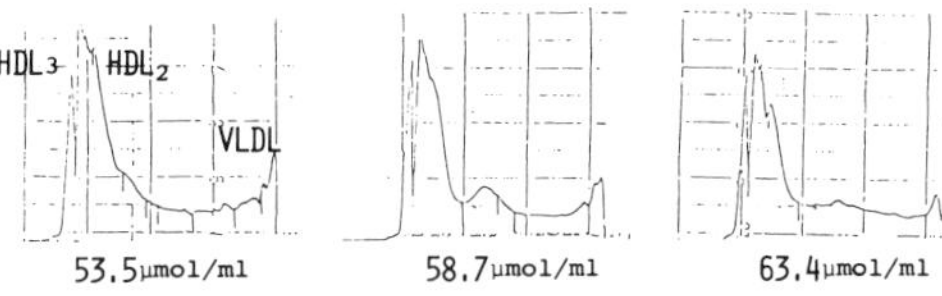

Fig.3. Changes in the lipid staining
pattern of 3.5% polyacrylamide gel-disc
electrophoresis after the incubation of
fresh plasma with linoleic acid hydroperoxide for 12 hr at 37°C.

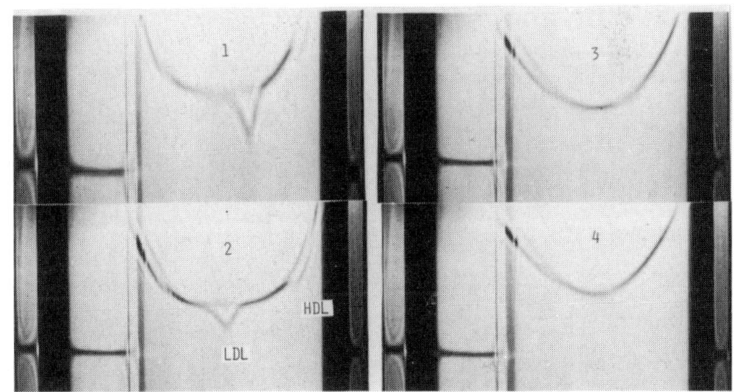

Fig.4. Schlieren patterns of the sample incubated with 47.5 μmol/ml of linoleic acid hydroperoxide for 12 hr at 37°C.

hydroperoxide and the incubated duration with the hydroperoxide
at 37°C. It is possible that an individual lipoprotein may has
a different response or resistance to the active oxygen in
linoleic acid hydroperoxide. Additionally, these results suggest that the interacted lipoproteins with lipid peroxides will
to be lipoproteins having a large molecular size as compared
with that of native lipoproteins. Both subfractions in the
HDL were identified as the resistant lipoproteins to the active
oxygen in linoleic acid hydroperoxide by the experiment using
isolated lipoproteins (Fig.7). Despite the disappearance of

the electrophoretic pat-
terns of VLDL and LDL,
the lipid staining in
HDL_2 and HDL_3 were confirmed on their electro-
phoresis after the incu-
bation with linoleic
acid hydroperoxide for
8 hr. Moreover, the
lipid staining patterns
of the both HDL sub-
fractions were not modi-

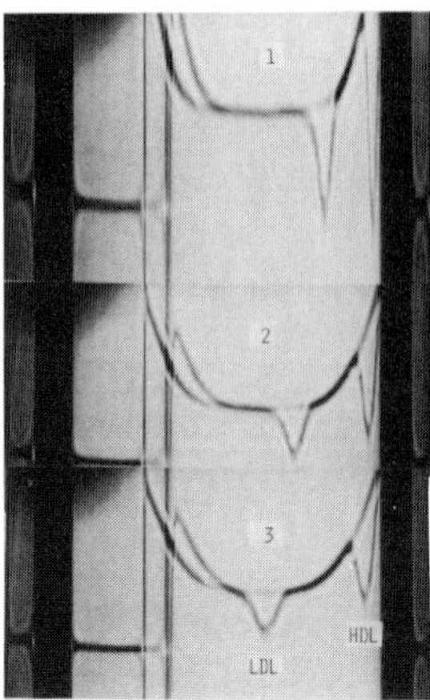
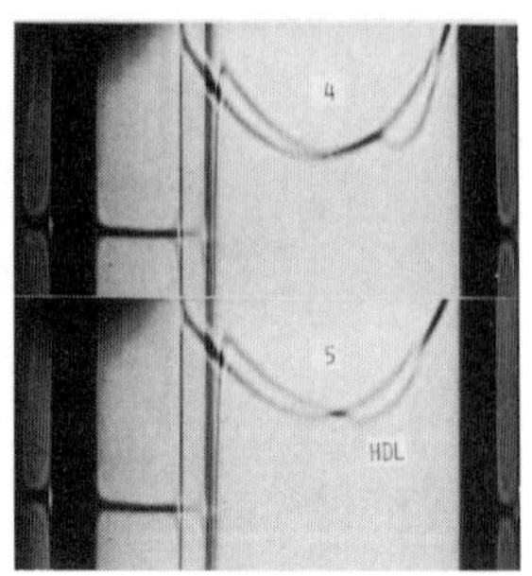

Fig.5. Analytical ultra-centrifuge Schlieren patterns from the sample without linoleic acid hydroperoxide.

fied by the hydroperoxide. A small quantity of active oxygen
triggered the modification of the lipid staining pattern in
each lipoprotein with the exception of HDL_3 (Fig.8). There
were differences of the response to the active oxygen which was
originated in the saturation with $85\%O_2$ for a short time, i.e.
The lipoproteins denatured in order of the sensitivity to the
active oxygen were LDL, VLDL and HDL_2. Mean value of the

plasma MDA in the subjects was
5.34 nmol/ml. The value was
significantly increased to 9.48
nmol/ml by continued stir
against the isolated lipoprotein
saturated with $85\%O_2$ (p<0.001).
High MDA value is more frequently common in the patient with
myocardial infarction or diabetes mellitus is well known(10,11).
The increased MDA value which
was corresponded to 47.5 μmol/ml
of linoleic acid hydroperoxide
(Table 1) will be a physiological value. The lipid
staining patterns in the isolated lipoproteins were easily

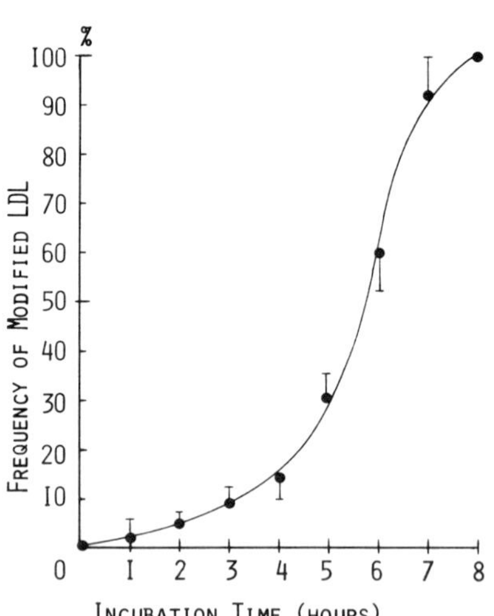

Fig.6. Relationship between the appearance rates of the modified LDL and the incubated duration with linoleic acid hydroperoxide.

modified by a small quantity of the saturated oxygen. It is possible that the incorporated active oxygen into lipoproteins in a state saturated with oxygen can occurs the oxidation of lipids and/or proteins as the composition of lipoproteins (3,4). The studies reported here show that auto-denaturation in human lipoproteins can be developed under a physiological condition rather than a dramatic state for the above reasons. The lipids composition in lipoprotein particles are dependent on the ingestion of daily food. LDL and VLDL have a lot of unsaturated fatty acids in their cores(1,3). The HDL_2 is said to be a variable component in HDL fraction (12). Among the isolated lipoproteins, these different sensitive to oxidation can result from the differences in the lipids component of the lipoproteins, or in the physical state of the lipid phase which is considerably less mobile in LDL (3,13). In the present experiments under condition with a mild quantity of lipid peroxide, LDL was more rapidly denatured and a part of the LDL was formed a hetero-

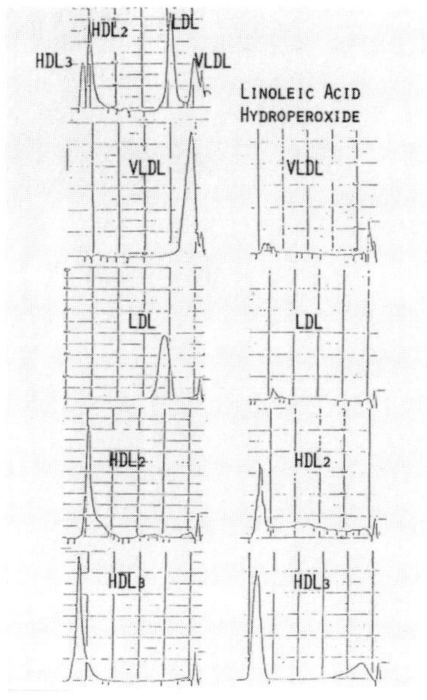

Fig.7. Effects of linoleic acid hydroperoxide on the lipid staining patterns of the isolated lipoproteins.

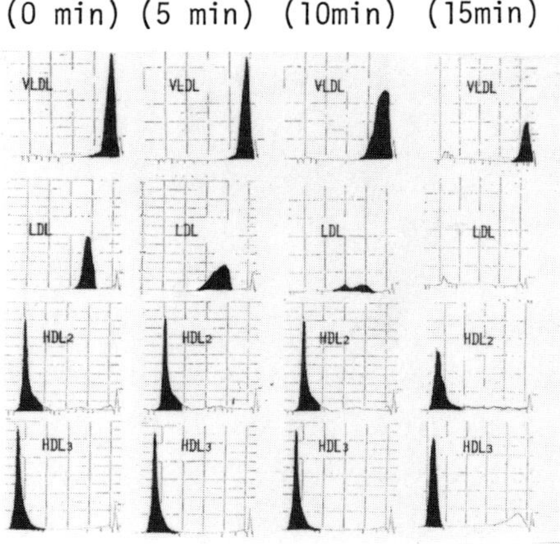

Fig.8. An auto-denaturation in the isolated lipoprotein was examined on the electrophoretic pattern in a state saturated with $85\%O_2$. The LDL was more rapidly oxidized in the early state, whereas HDL_3 was not changed in the lipid staining pattern by the oxidation for 15 min. A short term oxidation with $85\%O_2$ could induces an auto-denaturation in each lipoprotein.

geneous LDL which was named as a λ-lipoprotein. These results suggest that a heterogeneity in lipoprotein fraction may be occured by the interaction with lipid peroxide such as a linoleic acid hydroperoxide. It can be assumed that an abnormal lipoprotein metabolism is occured by the denaturated lipoproteins with active oxygen.

References

1. Spritz N & Mishkel MA.: J. Clin. Invest. 48, 78-86 (1969).

2. Nishida T & Kummerow FA.: J. Lipid Research 1, 450-457 (1960).

3. Schuh J, Fairclough GF,Jr. & Haschemeyer RH.: Proc. Natl. Acad. Sci. USA. 75, 3173-3177 (1978).

4. Fogelman A, et al.: Proc. Natl. Acad. Sci. USA. 77, 2214 -2218 (1980).

5. Szczeklik A & Gryglewski RJ.: Artery 7, 488-495 (1980).

6. Havel RJ, Eder HA & Bragdon JH.: J. Clin. Invest. 34, 1345 -1353 (1955).

7. Gardner HW & Jursinic PA.: Biochim. Biophys. Acta 665, 100 -112 (1981).

8. Yamasaki S.: Japanese J. Clin. Chem. 1, 52-63 (1971), in Japanese.

9. Lowry OH, et al.: J. Biol. Chem. 191, 263-275 (1951).

10. Szczeklik A, et al.: Prostaglandins 22, 795-807 (1981).

11. Yamasaki S. & Ageta M.: Geriat. Med. 19, 396-404 (1981), in Japanese.

12. Anderson DW, et al.: Atherosclerosis 29, 161-179 (1978).

13. Deckelbaum RJ, Shipley GG & Small DM.: J. Biol. Chem. 252, 744-754 (1977).

STUDIES OF IgM-K/IgG MIXED CRYOGLOBULIN

Sumiko Hashimoto, Masakatsu Hashimoto, Kazunari Kumasaka, Kinya Kawano
Department of Clinical Pathology, Nihon University School of Medicine
Tokyo 173, Japan

Introduction

Recently, cryoglobulin has been studied not only for its physico-chemical properties but also as the circulating immunecomplex. Immunochemical analysis of purified cryoglobulins from 80 cases of cryoglobulinemia were performed in our department and classified into three groups (Fig. 1). In this study 72 of the 80 (90%) were classified as mixed type cryoglobulin and 27 of 72 (38%) were shown to have a monoclonal component.

If the immunological reaction plays a main role in forming mixed type cryoglobulin, one of the componens must have antibody activity against the other.

In this study, a mixed type cryoglobulin (IgM-K/polyclonal IgG) was analyzed to determine which one of the component has the antibody-like activity against the other. An attempt was made to analyze the mechanisms of the unique thermodynamic characteristics of the cryoglobulin.

Fig.1 COMPONENTS OF CRYOGLOBULINS

Classification	Components	Case
Simple monoclonal immunoglobulins (Type I)	IgG-K	1
	IgG-L	2
	IgM-K	1
	IgM-L	2
	(BJ-K)	2
Mixed cryoglobulin with a monoclonal component and polyclonal immunoglobulin (Type II)	IgG-K. IgG	2(2)*
	IgG-K. IgG. Clq	1(1)*
	IgG-K. IgM	1(1)*
	IgG-K. IgA	1(1)*
	IgG-L. IgM	6(5)*
	IgG-L. IgM. β-lipo	1(1)*
	IgM-K. IgG	4(1)*
	IgM-K. IgG. $\beta_{-1}c_{A}$	1
	IgM-L. IgG	1
	IgM-K. IgG. IgA	1
Mixed polyclonal cryoglobulin without monoclonal component (Type III)	IgG. IgM	43
	IgG. IgM. $\beta_{-1}c_{A}$	3
	IgG. IgA	5
	IgG. IgA. $\beta_{-1}c_{A}$	1
	IgG. $\beta_{-1}c_{A}$	1

( )* ; M-protein : original serum (−), cryoprecipitate (+).

Methods and Results

712

1) Preparation of the cryoglobulin.

Thrombin was added to plasma obtained by plasmapheresis from a patient with mixed type (IgM-K, IgG) cryoglobulin. After standing at 4°C for 24 hours a cryoprecipitate formed from the serum. The cryoprecipitate was purified by washing 4 times at 4°C with saline. The first and second washings were followed by redissolution of the cryoprecipitate at 37°C.

2) Separation of the cryoglobulin components.

Separation of monoclonal IgM and polyclonal IgG from the cryoprecipitate was performed by gel filtration (Toyopearl HW-55 column) in 1M-propionic acid at 37°C (Fig. 2). The two fractions (Fx.1 and Fx.2) were dialysed against 1/15 M-phosphate buffered saline (pH 6.4) and then concentrated with Centriflo CF-25. Concentrated Fx.1 (6.8mg/ml) and Fx.2 (1.3mg/ml) were analyzed by immunoelectrophoresis (Fig.3). Fx.1 was composed of IgM-K type M-protein and Fx.2 of polyclonal IgG. No cryoprecipitation at 4°C was observed for either of the separate fractions alone.

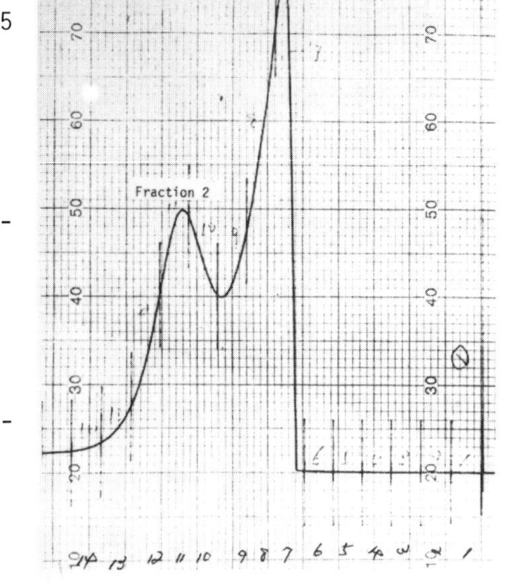

Fig. 2

3) Reactivities of the separated cryoglobulin components.

i) Recombination experiment.

Serial dilutions of Fx.1 and Fx.2 were mixed with each other in 1:1(V/V) ratios, and allowed to stand at 4°C for 18 hours. The mixture consisting of undiluted Fx.1 and Fx.2 formed the largest amount of cryoprecipitate. This optimal mixing condition (6.8:1.3 in W/W) shows that the IgM-M protein and IgG reaction corresponded almost to a 1:1 molar ratio (6.8/900,000:1.3/150,000 = 1:1.15).

ii) Reactivities with normal gamma globulin.

The cold precipitation reaction also occured when IgM M-protein (Fx.1) from the cryoglobulin was added to polyclonal IgG from normal subject

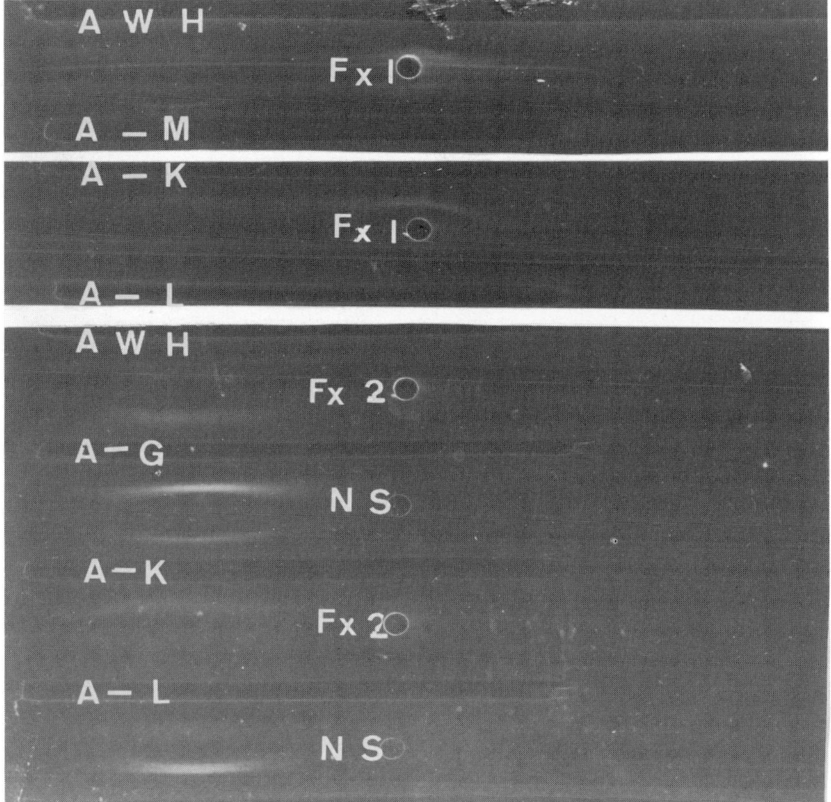

Fig. 3

and the mixture allowed to stand at 4°C. This cryoprecipitate was also redissolved completely on warming to 37°C. On the other hand, addition of normal IgM to the polyclonal IgG (Fx.2) from the cryoglobulin did not form a cryoprecipitate.

4) Reactivities of the treated cryoglobulin components and its fragments. Fabu and (Fcu)$_5$ fragments were prepared by tryptic digestion of Fx.1, according to the method of Tomasi et al and isolated by preparative agar gel electrophoresis. The 7S subunits (IgMs) were prepared by reduction with 0.2M-2-mercaptoethanol and alkylation with 0.01M-iodo-acetamide. Papain digested fragments of normal IgG were prepared by the modified method of Porter and isolated by preparative agar gel electrophoresis. Heat altered IgG was obtained by heating human IgG

(Pentex) saline solution at 62°C for 3 min. Human polyclonal IgG was
also treated with hyaluronidase (bovine testis) or neuraminidase
(CI. perfringens) and the treated IgG was isolated by HPLC (Toyosoda,
G 3000 SW)(Fig.4).

Fig. 4

EFFECT OF THE TREATMENT OF THE EACH CRYOGLOBULIN
COMPONENT ON CRYOPRECIPITATING REACTION

Sample	Treatment	Results of cryoprecipitating reaction		
		Partner	4°C	37°C
IgM-K M-protein (Fx.1)	Trypsin digestion: Fabμ	IgG	(-)	
	(Fcμ)$_5$	IgG	(-)	
	2-ME reduction : IgMs	IgG	(-)	
Polyclonal IgG	Papain digestion : Fab$_r$	Fx.1	(-)	
	Fc$_r$	Fx.1	(+)	(-)
	Heating(62°C,3min)	Fx.1	(+)	(+)
	Hyaluronidase digestion	Fx.1	(+)	(+)
	Neuraminidase digestion	Fx.1	(+)	(+)

Fabu, (Fcu)$_5$ and IgMs obtained from Fx.1 were not able to form cryoprecipitate with polyclonal IgG. Cryoprecipitation occured when the
IgG Fc fragment from polyclonal IgG was added to the monoclonal IgM
obtained from the cryoglobulin, but not when the IgG Fab fragment
from polyclonal IgG was added. This cryoprecipitete was also redissolved on warming at 37°C.
These data suggested that Fx.1 might have antibody activity and the
corresponding antigenic determinant is located on the Fc portion of
the normal IgG molecule.
The reaction of Fx.1 with hyarulonidase-, neuraminidase- and heat-
altered human IgG resulted in formation of irreversible cryoprecipitation which did not redissolve on warming to 37°C.

Discussion

As we described in a previous report, cryoglobulins consist of various
components and can be classified into three groups, according to their
components, as follows; Type 1: Cryoglobulins consisting of monoclonal
immunoglobulin alone, Type 2: Mixed cryoglobulins with a monoclonal

component and Type 3: Mixed cryoglobulins without a monoclonal component.
These data suggested that cryoglobulins are produced by several different
mechanisms.

The cryoglobulin described in this report was a mixed type cryoglobulin
with a monoclonal component (IgM-K). The IgM-K M-protein was able to react
with normal human IgG to form a cryoprecipitate. Therefore, the M-protein
is thought to react as the cryoprecipitating factor in this case.

If the immunological reaction plays a main role in forming mixed type
cryoglobulin, the M-protein must have antibody activity against polyclonal
IgG. The data obtained from the trypsin digestion and 2-ME reduction
experiment of Fx.1, and the papain digestion of Fx.2 are not in contra-
diction to the above supposition.

Since Fx.1 is a monoclonal protein, the antibody activity should be toward
only one antigenic determinant structure. To form a large immune-complex,
the antigen (polyclonal IgG) which reacts with M-protein must have several
identical antigenic determinant sites on its Fc portion. As one explanation for those repeating antigenic structures, the carbohydrate moiety of
the IgG Fc portion is an attractive possibility.

If the cryoprecipitating reaction is a type of the usual antigen-antibody
reactions, the following speculation based on the extremely sensitive
prcipitating reaction to temperature may be made:

(a) The M-protein has an abnormal physico-chemical structure, (b) the
corresponding IgG has an abnormal physico-chemical structure and (c) the
antigenic determinants corresponding to the antibody binding site have
some unique properties. Since the M-protein in this case is able to react
with the polyclonal IgG obtained from normal serum, case (b) appears unlikely.

When the IgG was treated with heat, hyaluronidase and neuraminidase, the
cryoprecipitation reaction became irreversible. This fact suggests that
the antigenic determinant properties of IgG molecule may play an important role in the occurence of the cryoprecipitation. An understanding of
the carbohydrate moiety of IgG Fc may be provide clues to understanding
the unique characteristics of the cryoglobulin.

References

1. Stastny P. et al.: New Eng. J. Med. <u>280</u>, 1376-1381 (1969).
2. Saha A. et al.: Immunochemistry. <u>5</u>, 341-356 (1968).
3. Litman G.W. et al.: Molecular Immunol. <u>17</u>, 337-344 (1980).
4. Lospalluto J.: Am. J. Med. <u>32</u>, 142-147 (1962).
5. Christian C.L. et al.: J. Clin. Inv. <u>42</u>, 823-829 (1963).
6. Melter M. et al.: Am. J. Med. <u>40</u>, 785-793 (1966).
7. Sumiko H.: Physico-Chemical Biology. <u>24</u>, 289-293 (1981).
8. Tomasi J. et al.: Proc. Nat. Acad. Scie. <u>65</u>, 318-322 (1970).

MONOCLONAL GAMMOPATHY ASSOCIATED WITH ANTIBODY ACTIVITY TO
HETROLOGOUS ALPHA-2 MACROGLOBULINS.

Toshikazu Minowa, Kinya Kawano, Kazunari Kumasaka,
Masakatsu Hashimoto, Tatsuyuki Tsuchiya
Department of Clinical Pathology, Nihon University
Itabashi-ku, Tokyo, Japan

Taeko Minami, Yukiro Mutoh
Nakatsu Hospital
Nakatsu, Ohita, Japan

Introduction

Since the first report of Waldenstrom[1], a number of human
murine monoclonal gammopathies associated with various antibody
activities in monoclonal components have been reported.[2-4]
Recently, we have observed a case of lgG-kappa type monoclonal
gammopathy associated with antibody activity to heterologous
alpha-2 macroglobulins. A similar case was reported by
Seligmann et al, where the M-protein was confirmed to react
specifically with horse, rabbit and pig alpha-2 macroglobulin.[5]
The immunochemical properties of M-protein of our case are
described in this report.

Material and methods

The patient was a 63 year-old male, admitted to Nakatsu
Hospital with chief complaint of abdominal pain.
Laboratory examination revealed normal complete blood count,
normal urinalysis and normal blood chemistry, except for
moderate hyperproteinemia (11.0g/dl) with monoclonal peak
(9.8%) in the slow gamma area, and positive HBs and HBe

antigen. M-protein was purified by preparative agar gel
electrophoresis and eluted with saline. The eluate was
concentrated up to 20 to 30 mg/ml of protein concentration
with "Centriflo" system (CF 25, Amicon Co.).
Fab and Fc fragments were prepared by digestion of purified
M-protein with papain for 16 hours at 37°C under the condition described by Porter.[6] The fragments were purified by
preparative agar gel electrophoresis and concentrated with
"Centriflo", as described above. The purity of the Fab and Fc
fragments in eluates were assessed by immunoelectrophoresis and
Ouchterlony methods. Alpha-2 macroglobulin from normal horse
serum was prepared by gel filtration through a Sephadex G-200
column. The macromolecular protein, present in the first peak
consisted mainly of alpha-2 macroglobulin and a trace amount of
lgM.
Immunoelectrophoretic analyses were performed in 1% agar gel
using commercially available anti-human antibodies, arized in
rabbit, goat and horse.
Confirmation of antibody activity to heterologous alpha-2 macroglobulin residing in the Fab fragment of the M-protein was done
by means of passive hemagglutination method, using SRBC sensitized with the Fab fragment of the patient's M-protein.[7]

Results

1) Immunoelectrophoretic analyses
   Immunoelectrophoresis of the patient's serum shows M-bow
   formation in the slow gamma area. In addition, a peculiar
   precipitation arc was also detected at the M-bow site with
   anti-whole human horse serum arrow but not with anti-whole
   human goat serum as shown in Fig.1.

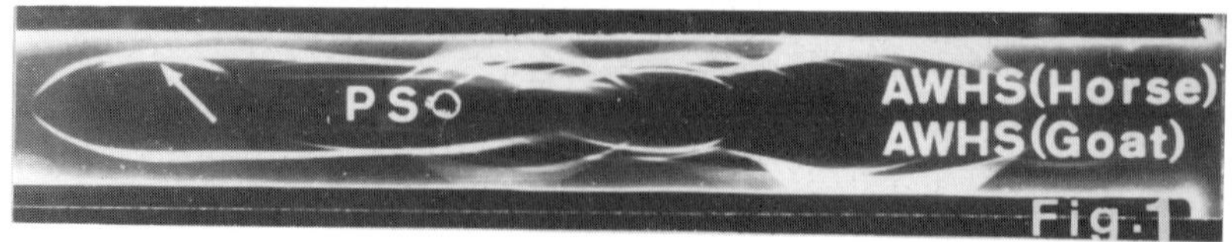

However, immunoelectrophoretic analysis of the serum failed
to define the type of M-protein, because all of the monospecific antisera for human immunoglobulins reacted with the
M-protein, when unpurified antisera were used (Fig.2). The
type of M-protein was defined as lgG-kappa type,(Fig.3)
because peculier precipitation arc was not detected, when
purified (gamma-fraction) monospecific antisera were used.

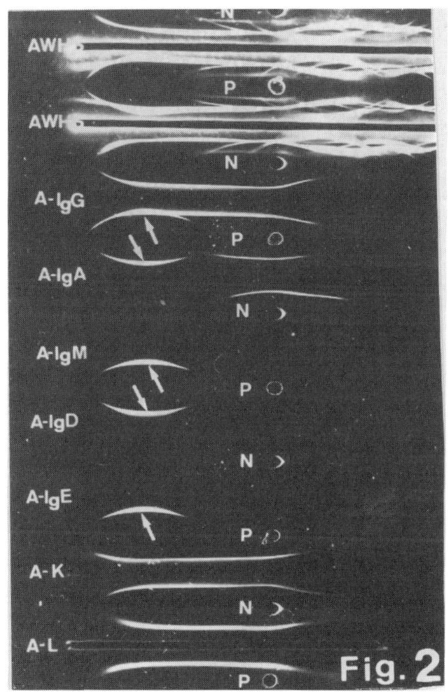

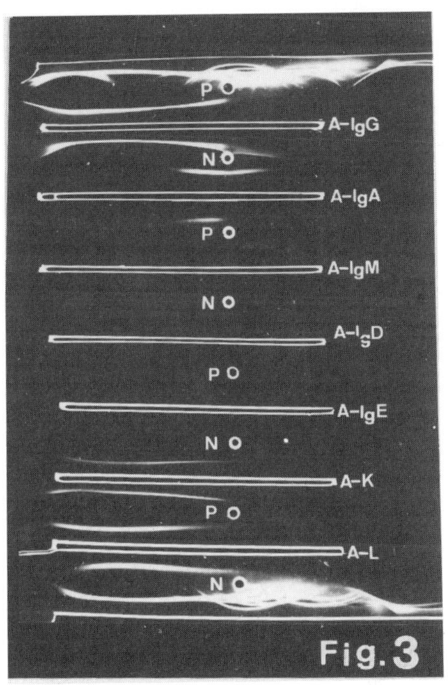

The peculier precipitation arc, identical to that produced
with unpurified antisera, was detected when normal rabbit or
horse sera were added to the antibody trench, instead of
monospecific antisera to human immunoglobulins (Fig.4).

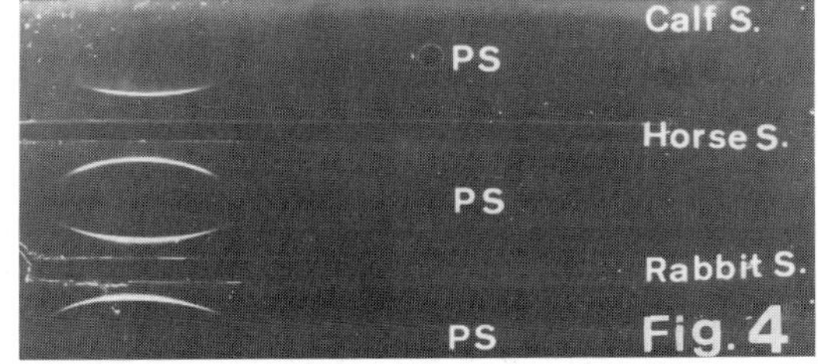

Therefore, we speculted that the M-protein had an antibody activity to heterologous serum protein. In order to clarify the antigenic activity to the M-protein, reversed immunoelectrophoresis (submitting normal horse, rabbit and calf serum to electrophoretic migration and using the patient's serum in the antibody trench) was made as shown in Fig. 5 A single precipitation line was detected between horse or rabbit serum and the patient's serum at the alpha-2 area, as indicated by the arrow, but precipitation line was not formed with calf serum.

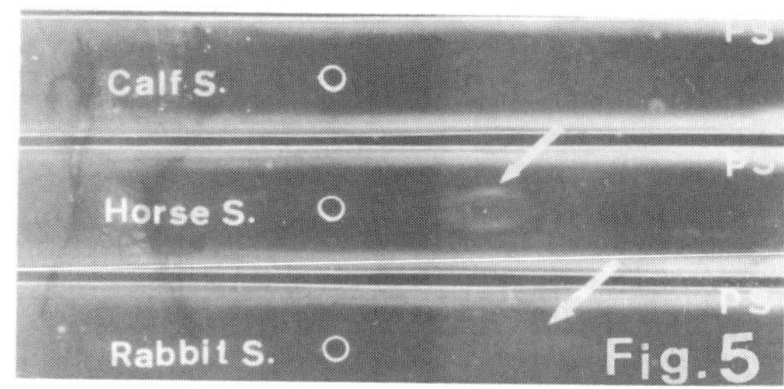

2) Confirmation of the antigen-antibody reaction between alpha-2 marcoglobulin from horse serum and patient's M-protein Macromolecular components from horse serum were prepared by Sephadex G-200 column chromatography and the first peak of the eluate was confimed to consist mainly of alpha-2 macroglobulin and a trace amount of lgM immunoglobulin. It was also confirmed that the macromolecular protein reacting with the patient's

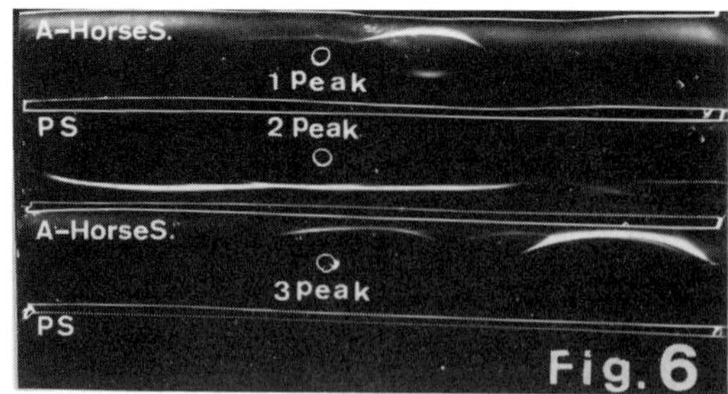

M-protein had alpha-2 mobility on electrophoresis as
shown in Fig. 6 (arrow). Therefore, alpha-2 macroglobulin was considered to have antigenic determinants
corresponding to the antibody activity of the patient's
M-protein. This precipitation reaction was completely
absorbed when the patient's serum was absorbed with
Staphylococcal protein A. Reaction between isolated Fab
fragment from papain digested patient's M-protein and
horse alpha-2 macroglobulin was demonstrated by passive
hemagglutination assay, using SRBC sensitized with the
Fab fragment from the patinet's M-protein, as shown in
Fig.7.

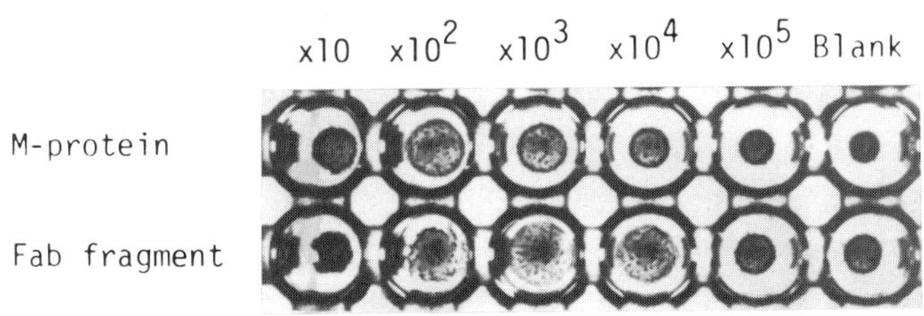

Fig. 7   Pattern of passive haemagglutination.

These findings suggested that the antibody activity to
alpha-2 macroglobulin resides in the Fab fragment of the
patient's M-protein.

Discussion

Once M-protein was thought to be a "dummy: protein with no
antibody activity. However, it has been shown to have the
normal physico-chemical properties of immunoglobulins, but
since the first report of Waldenstrom, a number of M-proteins
exhibiting various antibody activities, such as anti-
streptolysin O, anti-lgG, and cold agglutinin activity etc,
have been documented and now M-protein is thought to have some
sort of antibody activity.
We present here a case of benign monoclonal gammophathy,

whose M-protein was shown to react with hetrologous serum
component by routine immunoelectrophoresis. To our knowledge, this is the second case of M-proteinemia appearing in
the literature with antibody activity to heterologous alpha-2
macroglobulin molecule. A similar case was reported by
Seligmann et al in 1973. In both cases, the heterologous
serum components which reacted with the M-protein were shown
to be horse and rabbit alpha-2 macroglobulin. The activity
was also found to reside in the Fab fragment of the
M-protein, but the intensity of the reaction with horse
alpha-2 macroglobulin was more intense than that with rabbit
alpha-2 macroglobulin. Seligmann attributed the difference
in intensity of the reaction to the slight difference in the
antigenic determinants on each molecule. They also
speculated that the production of the monoclonal antibody to
horse alpha-2 macroglobulin was primed by the injection of
horse anti-tetanus antibody. This speculation was based on
the fact that in addition to the M-protein with anti-alpha-2
macroglobulin activity, the patient's serum was shown to
contain small amounts of precipitating antibodies reacting
with other horse serum protein components. On the other
hand, our case did not have such an anamnestic episode nor
additional antibody activity to the horse serum components,
except for the patient was the employee from the horse race
track. Moreover, double immunodiffusion method of
Ouchterlony revealed that the antigenicity of the horse to
the M-protein was identical to that of the rabbit in both of
the cases. If the antigenicity of the horse and rabbit are
not identical to each other, precipitation on Ouchterlony
method should not show complete identity. Therefore, this
finding is contradictory to the Seligmann's explanation.
At the last annual meeting of the Japanese Society of
Electrophoresis, Dr. Hirai commented, regarding this case
that bacterial infection is more likely to induce the
production of such M-protein associated with an antibody
activity to heterologous protein, and also suggested that if

we make more precise observations of the immunoelectrophoretic patterns, we may encounter more such unusual cases. These cases may be of further clinical significance in that those M-proteins that react with heterologous protein components may alter the results of immunoassay.

Reference

1.   Waldenström, J. et al.: Acta Med. Scand. 176, 619, (1964).

2.   Matzger, H.: Am, J. Med. 47, 837, (1969).

3.   Potter, H.: New Engl. J. Med., 284, 831 (1971).

4.   Seligmann, M. and Brout, J. C., Seminars in Hematol. In press.

5.   Seligmann, M.: J. Immuno., 110, 1, (1973).

6.   Poter, R. R.: Biochem. J., 73, 119 (1959).

7.   Wide, L. and Gemzell, A. C.: Endocrinology, 35, 261, (1960).

RAPID AND SIMPLE PURIFICATION OF HUMAN MITOCHONDRIAL DNA USING AGAROSE GEL ELECTROPHORESIS

Shoji Harada

Institute of Community Medicine, University of Tsukuba,
Sakuramura, Niiharigun, Ibarakiken 305, Japan

Introduction

Genetic polymorphism of human mitochondrial (mt) DNA
sequences has been demonstrated by Case and Wallence (1).
Using a specific radiactive probe, they reported that
restriction endonuclease cleavage patterns of mtDNA verifies
maternal inheritance. These cleavage patterns of DNA are
detected using Southern blotting procedure (2) after slab gel
electrophoresis. This method is complicated and preparation
of the special radioactive probe is time-consuming. DNA
framents after electrophoresis can be identified by
Ethidium-Bromide staining, which is simple, but non-specific
DNA fragments are stained. This means that the circular form
should be highly purified. So far, different methods for
purifing DNA have been reported by many authors (3, 4, 5).
Smith et al. (3) described bouyant equilibrium in CsCl/Et-Br
and fractionation by drop collection. However, such a method
is not suitable for screening large numbers of samples.

Recently, Robert et al. (6) succeeded in achiveing a simple and
rapid isolation of DNA from an agarose gel slice after electrophoresis. Simplicity, speed and yield are important factors
for population genetics involving a large sample numbers. This
paper describes a simple, rapid and improved method for purifying human mtDNA for the study of DNA polymorphisms.

Electrophoresis '83

Materials and Methods

Human placenta was used to obtain mitochondria. After
homogenization of 30 g of the tissue in 0.1 M phosphate buffer
containing 0.25 M saccharose and 1 mM EDTA (pH 7.0), nucleai,
unbroken cells and membrane were removed at 800 g for 30 min.
Mitochondria fraction was obtained by the centrifugation in
the same buffer at 15000 g for 20 min. The fraction
containing mitochondria was purified using percol (Pharmacia
Co. Ltd.).

1) Isolation of crude mtDNA
Purified mitochondria, diluted in the 3 vol of 10mM Tris
buffer (pH 7.9) containing 0.4M NaCl, 2 mM EDTA and 0.5% SDS,
was incubated at 37°C for 60 min. After Phenol-Chloroform
treatment, DNA was precipitated by ethanol. The precipitate
was centrifuged at 15000 g and air-dried. One hundred µl of
Tris-HCl buffer (10 mM Tris and 2 mM EDTA, pH 7.9) were added
to the DNA precipitate followed by agarose gel electrophoresis.

2) Agarose gel electrophoresis
DNA solution was inserted in the slit of agarose gel (0.7%)
and subjected to the electrophoresis based on the submarine
method in Tris-acetate buffer (0.04 M-Tris, 0.02 M-acetate,
2 mM-EDTA, pH 7.9) at 60 mA for 15 hrs.

3) Preparation of pure DNA
After electrophoresis, a small area, 0.5 cm wide, was cut out
from one side of the gel. The sliced gel was immediately
stained by Ethidium-Bromide solution (0.2 mg/ml) as shown in
the Fig. 1. The 16.5 KB DNA fragment was identified under UV
lamp (275 nm). Subsequently, a UV shaped piece of gel zone
0.2 mm wide was cut from the unstained agarose gel. If it
was cut from the agarose, one assumes it was removed.

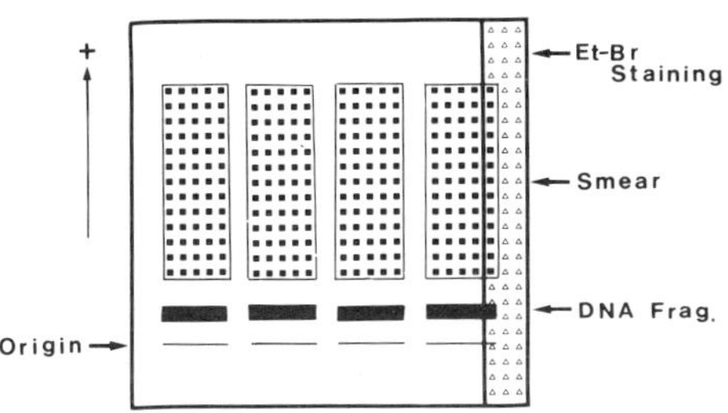

Fig. 1. Illustration of the method for isolating
pure mtDNA from agarose gel after electrophoresis.
DNA Frag indicates pure mtDNA with 16.5 KB.

4) DNA treatment with restriction endonuclease
After a small slice of unstained gel was incubated in an
Eppendorf tube containing the reaction buffer recommended by
the manufacturers for 60 min, the buffer was exchanged with
reaction buffer containing restriction endonuclease and
incubated at 37°C for 90 min. After digestion of mtDNA,
small pieces of gel were again inserted in agarose gel and
electrophoresis was carried out in the same way. DNA
fragments on the gel were stained by Ethidium-Bromide
solution and identified under UV light.

Results and Discussion

The electrophoretic patterns of mtDNA fragments obtained by
the treatment of Eco-RI, Hind-III, Pst-I and Alu-I are shown
in Fig. 2. Sample 6 is untreated pure mtDNA. Sample 5 and 7
show the cleavage patterns of pure mtDNA digested by Eco-RI
and Hind III, respectively. DNA fragment-B possessing 12.5KB

was obtained from mtDNA digested by ECO-RI. This fragment-B
was again digested by Eco-RI, Hind-III Pst-I and Alu-I,
respectively. Sample 1 which was treated by Eco-RI showed
the same numbers of base pairs as fragment-B. On the other
hand, the samples from 2 to 4 were digested into different
fragments. It is noteworthy that digestion of DNA by
restriction endonuclease can be carried out in the agarose
gel without elution. Rober et al. (6) eluted DNA from
disolved agarose gel using glass filter paper. The yield
according to their method was 71-95%. The largest fragment
was recovered at lowest yield. However, the yield in the
present method was 100% in all cases. Therefore, this method
is suitable as a screening test to detect variants in DNA
cleavage patterns in a large sized samples. It should be
enphasized that a small amount of sample can be used and the
procedure is rapid and simple.

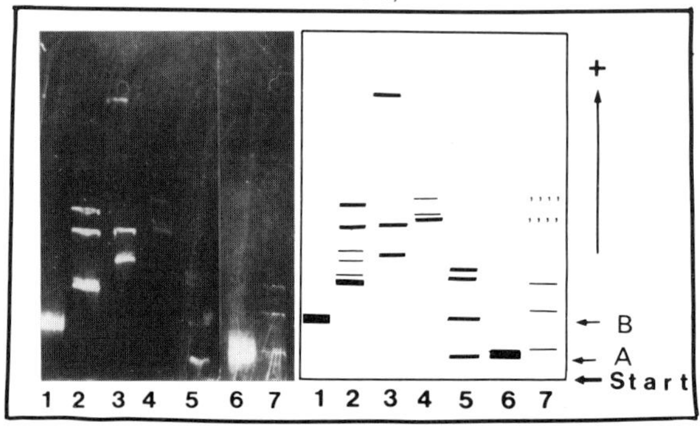

1) Eco R I   2) Hind III   3) Pst I   4) Alu I
5) Eco R I   6) Mt-DNA   7) Hind III

Fig. 2 DNA cleavage patterns on agarose gel
electrophoresis

Acknowledgements

The author is indepted to Dr. Noboru Mesaki, Institute of Clinical Medicine, University of Tsukuba for kindly providing mitochondria samples. Also, he wishes to thank Mrs. Takako Nakamura for technical assistance.

References

1. Case, E.M., Wallance, D.C.: Somat. Cell Genet. $\underline{7}$, 103-108 (1981).
2. Southern, E.M.: J. Mol. Biol. $\underline{98}$, 503-518 (1975).
3. Smith, C.A., Jordan, J.M., Vinograd, J.: J. Mol. Biol. $\underline{59}$ 255-272 (1971).
4. Brawn, W.M., Vinograd, J.: Proc. Natl. Acad. Sci. USA $\underline{71}$, 4617-4621 (1974).
5. Hudson, B., Vinograd, J.: Nature $\underline{216}$, 647-652 (1967).
6. Robert, C.A., Lis, Y.J., Wu, R.: Method in Enzymology 68, Academic Press, New York and London 176-182 (1979).

IMMUNO-ELECTROPHORETIC ANALYSIS OF ACTIVATION OF COMPLEMENT BY CHORIONIC VILLI

Ryoki Ohkawa, Kimiyasu Ohkawa, H.Hatano, M. Nakagawa
Department of Gynecology and Obstetrics Nippon Medical School
1-1-5 Sendagi-cho Bunkyo-ku Tokyo Japan

Introduction

The mechanism of induction of contraction of uterus is not yet unknown.
Aging of chorionic villi (placenta) will induce the contraction of uterus.
As the results the resection of chorionic villi and fetus will occurr.
This will be induction of labour pain, and this mechanism was studied
immunilogicaly.

Methods

1. Complement activity during pregnancy of each trimester was measured.
CH50, Al-CH50 were measured by hemolysis. C3,C3-activator and C4 were
measured by singl-radial-immunodiffusion.

2. Activation of complement by chorionic villi was measured by hemolysis,
and immuno-electrophoresis. The fresh, aged and necrotic chorionic villi
were used. Activation of component of complement by chorionic villi was
determined by immuno-electrophoresis (ph 8.5 veronal buffer including
0.01 mol EDTA). Zymosan and other acid polysaccharides were used as
control.

3 Complement activity in cases of pathological pregnancy and during pregnancy measured in clinical coyrse.

4. C3a-production (anaphylatoxin) by activation of C3 by placenta-extract
was measured by contraction of intestin of guinea pig in Ringer solusion
including atropin.

Results

Complement Profils in Pregnancy

Table. 1

		Cases	CH_{50}	C_3	C_3-Activator	C_4	Al-CH_{50}
Normal		20	40±2.1	66±3.0	18±2	32±3	18±1.1
PREGNANCY	1st	25	52.8±1.5	68±5.2	25±2	42±0.6	21.2±2.0
	2nd	25	56.5±2.5	83.5±2.0	24±1.5	39±2.5	23.4±1.0
	3rd	32	65.5±3	87.6±1.0	25.5±1.2	37±2.0	26.8±1.5
thr.	good	52	53.2±2	70.1±5.1	25.1±6.5	39±7.1	22.2±1.5
ab.	poor	63	38.1±1.5	60±5.2	19±3.1	35±7.2	17.1±1.2
hyt.mole		25	28±2	60±2.1	17±6.1	35±1.5	16.2±1.2
		3	45±2	67±1.5	24±1.1	38±1.1	21.1±2.5
Chor.Ca adenom		5	29±1.1	55.1±1.1	16±1.1	35±1.2	16±1.5
Tiox nmia severe		9	40.5±1.2	65.1±2.1	18.2±2.1	28.1±1.5	17.5±2.1
mild		15	52.1±1.3	22.5±1.2	26.1±3.2	38.2±3.2	23.1±1.8
apoplexia		2	27	49.5	13	24	
ut-pl.			(15)	(41)	(7)	(15)	

In normal pregnancy of each trimester complement activity is enhanced. In cases of poor prognosis and of other abnormal pregnancy activity of complement decreased, as shown in table 1.

Table 2

Anticomplement Activity of Chorion

		CH_{50}	AICH$_{50}$
Pregnancy IX		65.8	26.3
37°36'		63.5	25.1
uterus contraction (−)	IVM chorion (Myom.Op.)	61.0	24.6
	XM chorion Cesarean Sc.	59.5	23.1
uterus contraction (+)	XM chorion	38.5	18.1
	Thr.Ab.	31.5	18.6
Chorion homozinet		30.1	17.1

The fresh chorionic villi without contraction of uterus did not activate the complement, on the other hand aged chorionic villi (after delivery) and necrotic chorionic villi or chorion homozinet activated the complement.

Fig. 1 Activation of complement by chorionic villi

in trough   in trough
anti-C3-activator   anti-C3-

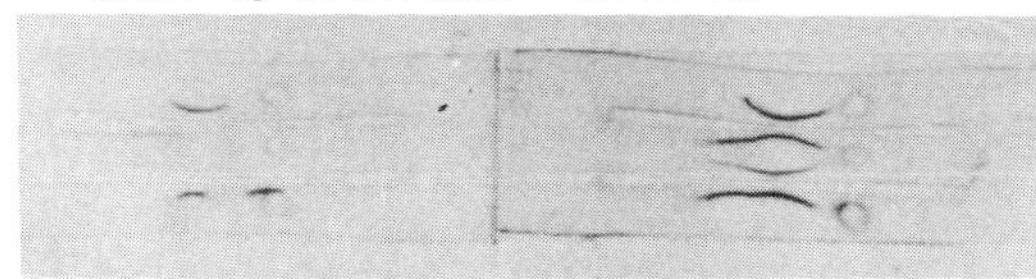

Extract of chorionic villi activate C3-activator, consequently GBG and GGG were found. And C3 convert to C3c and C3d.

Fig. 2 Activation of complement by extract of chorionic villi

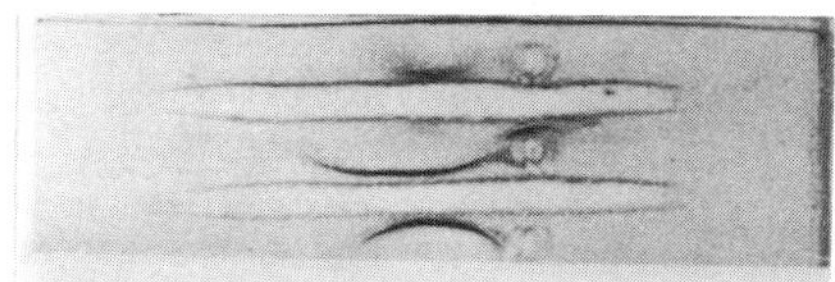

In the immuno-electrophoresis C3 was activated by placenta-extract, and C3a (anaphylatoxin) was found in the zone of r-globulin.

Fig. 3 Contraction of intestin of guinea pig
       by anaphylatoxin

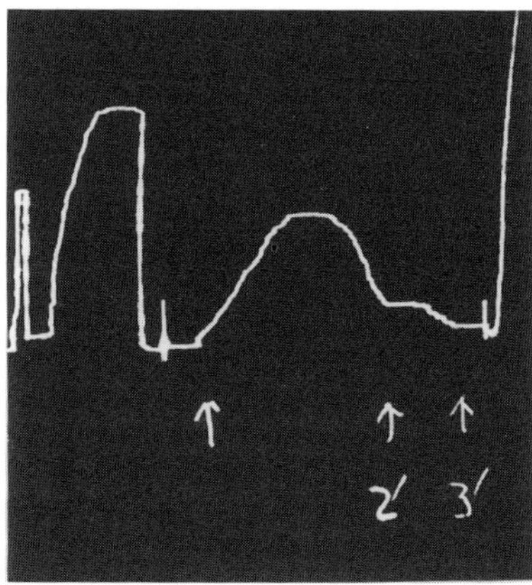

The mixture of serum with extract of chorionic villi causes contraction of intestin guinea pig. This shows that C3a(anaphylatoxin) was produced.

Fig. 4 Complement activation and complement network

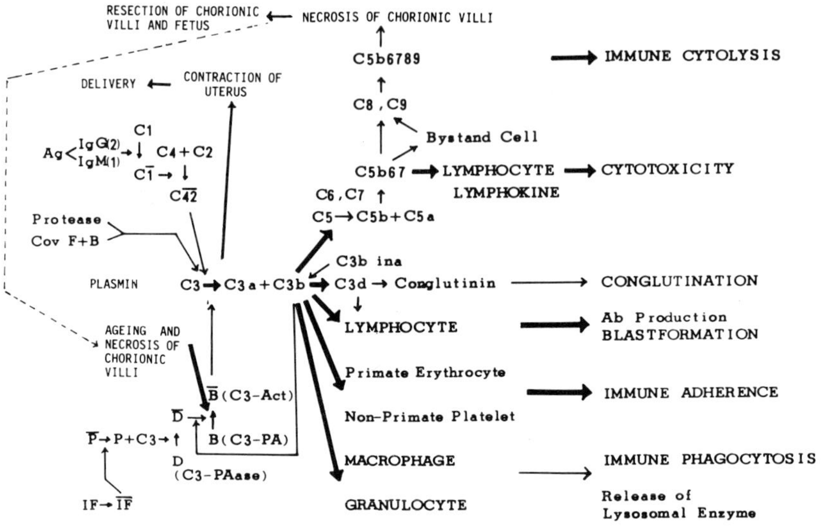

NETWORK OF COMPLEMENT-SYSTEM

In all cases of abnormal pregnancy the complement activity come down owing to activation of extract of chorionic villi. This phenomena is seems to be necessary for the rejection of chorionic villi(abortion or delively) as shown in Fig.4. In immuno-electrophoresis of conversion of C3,,precipitate with anti-C3 appears in the zone of b-globulin and a-globulin. The fresh chorionic villi did not activate complement. The aged and necrotic chorionic villi activate the C3-activator and C3 activated, as the results C3a (anaphylatoxin and C3b ,C3c (C3d) were produced.

Conclusion

The aged and necrotic chorionic villi activate alternative pathway of complement, as the results mechanism of rejection will occurr. C3a (anaphylatoxin) will become trigger of onset of labor.

SENSITIVE AND SIMPLE IDENTIFICATION OF MONOCLONAL PROTEINS IN SERUM
AND URINE BY IMMUNOFIXATION ELECTROPHORESIS (IFE)

Borek Janik

Gelman Sciences Inc., Ann Arbor, MI 48106, U.S.A.

Jeffrey Ambler

Clinical Chemistry Dept., University Hospital, Nottingham NG72UH, U.K.

John T. Whicher

Chemical Pathology Dept., Bristol Royal Infirmary, Bristol BS28HW, U.K.

Nadine Urquhart

Pathology Dept., Shaugnessy Hospital, Vancouver, B.C., Canada

Introduction

Traditionally, monoclonal proteins (M-proteins) are detected by conventional electrophoresis and identified by immunoelectrophoresis
(IEP). When the M-protein is at low concentrations or is masked by
polyclonal immunoglobins (Ig), e.g., an umbrella effect, the identification by IEP might be impossible. IFE proved to be the technique of
choice for identification of M-proteins surpassing IEP in predictivi-
ty, sensitivity, speed, and ease of interpretation.(1-7) IEF includes
two essential steps: electrophoresis, which separates the sample's
proteins, and immunofixation, which anchors one of the proteins in situ
via an immunocomplex. In the procedure described, using cellulosic
media, the simplicity, speed, sensitivity and economy have been maxi-
mized.

736

Procedure

Commercially available immunofixation kit, Ig-FIX Kit, standard electrophoresis equipment, and the procedure recommended by the manufacturer were used (all Gelman Sciences). The antigen-antibody reaction
is accomplished simply by layering the transfer medium (ATS) soaked
with the antiserum over the M-protein-containing zone. The open-pore
structure of Mylar^{TM}-supported cellulosic membranes (Super Sepraphore is used both as the electrophoretic medium and ATS (2)), allows
the individual steps of the procedure to proceed rapidly: penetration
of the antiserum Ig into the electrophoresis medium (fixation time 2
min.), washing out the unreacted proteins, staining and excess stain
removal.

Results

Presence of a "fixation" band indicated positive reaction between the
M-protein and antiserum; identity of the band was determined by the
antiserum with which the M-protein reacted. The "fixation" band typically appeared as a narrow and sharply demarcated band in a position
corresponding to the M-band in the reference track. In extreme cases
of minute concentration differences between monoclonal and polyclonal
Ig (e.g., small IgA or IgM monoclonal gammopathy with normal or increased polyclonal IgG), a lightly stained diffuse band of the polyclonal Ig was seen in addition to the sharp monoclonal fixation band;
at appropriate dilutions, the polyclonal Ig showed a positive reaction
with both light chains.

To establish efficacy of the Ig-Fix procedure, as compared to IEP,
serum samples with high ($>$ 1 g/dL) and low ($<$ 0.5 g/dL) M-proteins
levels were typed. While identical results were obtained at high
M-protein levels, IEP failed to identify 50% heavy chains (HC) and 36%
light chains (LC) in the problematic monoclonal gammopathies (Table I).

TABLE I.   IDENTIFICATION OF M-PROTEINS BY GELMAN Ig-FIX PROCEDURE ON CELLULOSE ACETATE AND BY IEP

Conc. (g/dL)	n	Identified By Ig-FIX		Identified by IEP	
		HC	LC	HC	LC
IgA					
>1.0	4	4	4	4	4
<0.5	4	4	4	3	4
IgG					
>1.0	6	6	6	6	6
<0.5	6	6	6	2	4
IgM					
>1.0	4	4	4	4	4
<0.5	4	4	4	2	1

Ig-Fix correctly typed M-proteins in all "non-problematic" and in nearly all "problematic" monoclonal gammopathies (Table II). Definitive identification of the M-proteins was by a combination of IEP, IFE on agarose gels, staining of bone marrow by peroxidase-antiperoxidase method for Ig and SDS polyacrylamide gel electrophoresis with crossed IEP.

TABLE II.   IDENTIFICATION OF M-PROTEINS IN SERUM BY GELMAN IG-FIX PROCEDURE ON CELLULOSE ACETATE

M-Protein	No. Samples	Note
Correct Identification - Non Problematic Samples		
IgG (K)	2	
IgG (L)	1	
IgA (K)	2	
IgA (L)	1	
IgM (K)	4	Biclonal pattern in one sample
IgM (L)	1	
Correct Identification - Problematic Samples		
IgG (K)	4	
IgG (L)	1	Increased polyclonal Ig's
K	1	In serum & urine; treated LC myeloma
IgA (L)	1	Treated multiple myeloma
IgG & IgA	6	Polyclonal patterns (one with a CRP band)
Failed to Identify - Problematic Samples		
IgG (L)	2	IgG correctly identified
IgA (L)	1	IgA correctly identified; treated myeloma

738

The sample's M-protein is required to be at about 100 mg/dL in serum or 75 mg/dL for BJP in urine. However, we found that positive identification can be achieved at concentrations as low as 15-25 mg/dL. In our experience, no technical problems were encountered as long as the membrane/ATS sandwich was moist and in good contact during the fixation step.

Conclusions

The characteristics of Gelman Ig-FIX Kit procedures are:
o   A fast procedure - results in less than 2 hours.
o   Easy to perform and extremely simple to interpret.
o   Sensitive - identifies M-proteins as low as 15-25 mg/dL.
o   Identifies M-proteins in complex situations, e.g., low concentration of M-protein in the masking presence of polyclonal Ig.
o   Surpasses IEP in all of the above performance characteristics.
o   No equipment is required beyond that of standard protein electrophoresis.

References

1. Cawley, L.P., et al: Clin. Chem. $\underline{22}$, 1262-1268 (1976).

2. Janik, B.: Electroph. Today (Gelman) $\underline{1}$ (No. 1) (1980); $\underline{2}$ (No. 2) (1981).

3. Keshgegian, A.A., and Peiffer, P.: Clin. Chim. Acta $\underline{110}$, 337-340 (1981).

4. Marshall, M.O.: Clin. Chim. Acta $\underline{104}$, 1-9 (1980).

5. O'Reilly, D.S.J., Adjukiewicz, A., and Whicher, J.T.: Clin. Chem. $\underline{27}$, 331-333 (1981).

6. Whicher, J.T., Hawkins, L., and Higginson, J.: J. Clin. Pathol. $\underline{33}$, 779-780 (1980).

7. Whicher, J.T.: Clin. Chem. $\underline{29}$, 402-403 (1983).

FUNCTIONAL HETEROGENEITY OF RAT LIVER ISOFERRITIN FOLLOWING
ON A SINGLE DOSE OF IRON

1. Determination of ferritin and H and L subunits

Chieko Inagaki, Setsuko Shinjo and Yoshio Yoshino
The second Department of Biochemistry, Nippon Medical School,
1-1-5 Sendagi Bunkyo-ku Tokyo, Japan

Introduction

When iron is injected into a rat 20-30% is incorporated into
the liver (1). To investigate the functional characteristics of
rat liver ferritin, we determined the change of ferritin content and of the H and L subunit ratios at 3, 24 and 72 hr after
iron injection.

Material and Methods

Female Wistar rats(BW 250-280g) were used in all experiments.
A solution containing 2 mg carrier iron and $10\mu Ci$ $^{59}FeCl_3$ was
injected intravenously . After 3, 24 and 72 hr the livers were
removed and ferritin was isolated from the liver by the modified
method of Drysdale and Munro (2) . The liver was homogenized
with 4 volumes of distilled water, and heated at 75 C for 15
min. After centrifugation , the supernatant was adjusted to pH
4.8 with acetic acid, the supernatant was dialyzed against distilled water and concentrated by ultrafiltration, and then
purified twice by preparative elctrophoresis. Purity of the
ferritin preparation was confirmed by PAGE, The analysis of
ferritin subunit composition was carried out by SDS-PAGE according to the modified method of Watanabe and Drysdale (3).

The separating gel was 30 cm high, which was higher than those
previously reported (3,4). The solution of purified ferritin
was dissociated by heating at 100 °C for 10 min in 10 % glycerol, 5% 2-mercaptoethanol, 2.3 % SDS and 62.5 mM Tris buffer
pH 6.8. The gel was run at 2 W for 40-48 hr. After staining
with Coomassie Blue R 250, the H and L subunit bands were
quantitatively analyzed by a Chromato Scanner.

Results and discussion

1) A photograph of typical H and L subunits is shown in Fig 1.
Using the higher plates of our method produces more clearly
separated H and L subunit patterns than those of previous
reports (3,4).

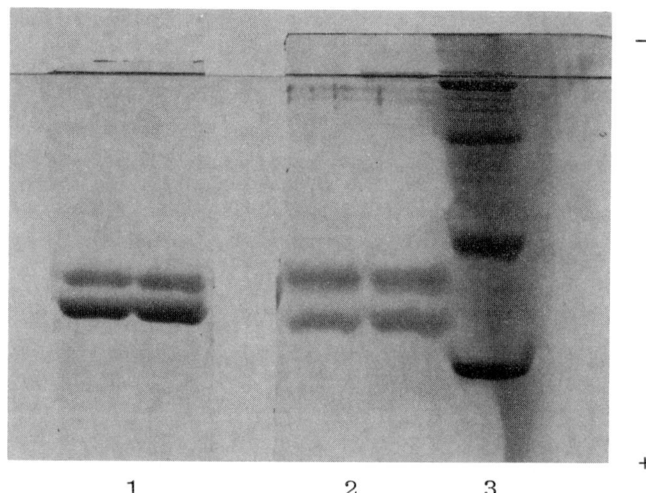

Fig 1. Separation of H and L subunits of rat liver ferritin by
SDS-PAGE : 1. 72 hr, 2. 3 hr, 3. Maker proteins.

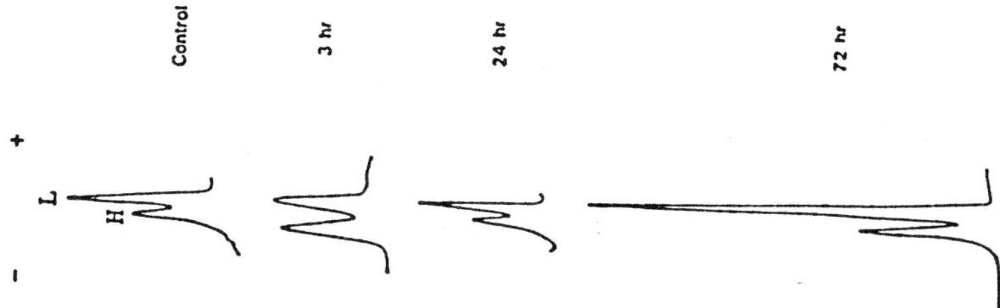

Fig 2. Traces of the densitometric scan of the ferritin subunits preparated by SDS-PAGE. The results show the rat liver
ferritin derived from the iron loaded rats and the control.

The molecular weight of H and L subunits were calculated to be
21,000 and 19,000 respectively from the marker protein patterns
. The patterns of the H and L subunits at 3, 24 and 72 hr
after the iron injection and the control are shown in Fig 2.

2) The cange of ferritin content, H and L subunit ratios of
the total ferritin (original subunit + newly synthesized subunit) and newly synthesized ferritin subunits induced by 2 mg
iron injection at 3, 24 and 72 hr are shown in Table 1.

3) The ferritin content in the liver 3 hr after injection was
three times that of the control level, and decreased to about
twice the control level after 72 hr, In the H and L subunit
ratios of newly synthesized ferritin at 3 and 24 hr after the
iron injection, the H subunit predominated in comparison with
the control, but the ratios reverted to the control level at
72 hr. This indicates that ferritin formed by the induction
by 2 mg iron dose is H rich and thus differs from the control.

Time after ^{59}Fe administration	Ferritin protein /Liver (μg) / (g)	Total subunit (%)		Newly synthesized subunit (%)	
		H	L	H	L
Control	1110	33	67	—	—
3 hr	3055	48	52	62	38
24 hr	2895	46	54	58	42
72 hr	2039	34	66	35	65

Table 1. Effect of single dose of iron on the ampunts of
ferritin protein and subunit ratios. The percentage of H and
L subunits of newly synthesized ferritin at 3 hr was calculated as follows : (3055 · 48/100 - 1110 · 33/100)/ (3055 ·
52/100 - 1110 · 67/100) = 57/43.

Reference

1. Shinjo, S., Inagaki, C., Yoshino, Y., : In press
2. Drysdale, J.W., Munro H.N. : J. Biol. Chem., <u>241</u>, 3630-3637,
1966.
3. Watanabe, N., Drysdale, J.W. : Biochem, Biophys. Res.Commun.
, <u>103</u>, 207-212, 1981.
4. Arosio, P., Adelman, T.G., Drysdale,J. W.: J. Biol. Chem.,
<u>253</u>, 4451-4458, 1978.

Hb F DETERMINATION BY THE AGAR GEL ELECTROPHORESIS

Chang-Se Kim and Man-Hee Cho

Department of Biochemistry, Soonchunhyang University, Medical College,
Onyang PO Box 97, Choong-Nam 331, KOREA

Intrduction

The fetal hemoglobin is present entirely in uterine life (fetus) and present
as less than 1% in normal adult. Hb F is also increased in hemoglobinopathies.
The quantitative methods of Hb F have been used to the alkali denaturation
test (1) which depend upon the alkali resistance of Hb F and the acid
elution test (2) in which hemoglobin other than Hb F is eluted and Hb F
remains in erythrocytes. These methods are either hemoglobins precipitated
unspecifically (3, 4) or uneven distribution of Hb F in each cell (5) respectively. The paper electrophoretic methods (6) of Hb F have developed, at
first, but it was found that the separation between Hb A and F was poor.
The agar gel electrophoresis, however, has had the advantages of the clear-
ness of gel because unadsorbing protein on it due to no interaction of
proteins, and of gelling effect is strong but no sieving. We thought this
advantage of agar gel electrophoresis is good for the determination of Hb F.
And we were chosen the most preferred Hb F determination among various
conditions.

Results

The tris-EDTA-borate buffer in various pH ranges 6.5, 7.0, 7.5, 8.0 and
8.5, and the agar gel concentrations, also, in ranges 1.0, 1.5, 2.0 and
2.5%, were made. Each buffer system which included various agar gel concentrations from 1.0 to 2.5% was runned by electrophoresis.

Electrophoresis '83
© 1984 Walter de Gruyter & Co., Berlin · New York

Figure 1 was schematically drawn the results. As shown in Figure 1
Hb F was separated clearly and definitely at the pH 7.0 and the 2%
agar gel. In the case of discontinued buffer system, in the vessel
barbital buffer, pH 8.6, and in the gel tris-EDTA-borate buffer,
pH 6.5 - 8.5, it could not found the Hb F band on the agar gel.
The agarose gel electrophoresis not agar gel was also found that it was
inappropriate for the Hb F separation. We thought that the branched chain
of agar might possibly be play a role on it.

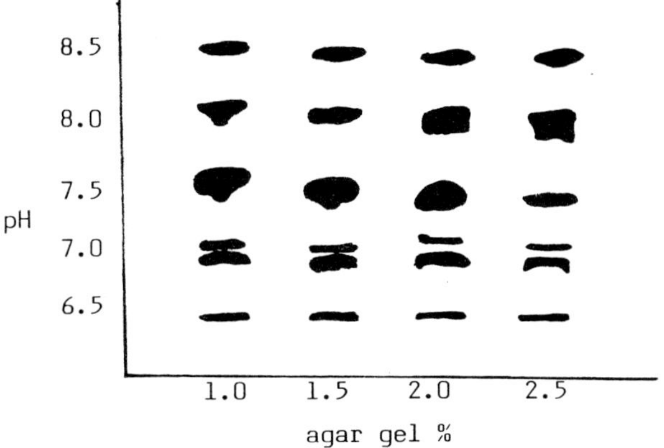

Fig.1. Schematic drawnings of electrophoretic separation
of human hemoglobins at various pH ranges,   6.5
- 8.5, and agar gel concentrations, 1.0% - 2.5%.

References

1. Singer, K., Chernoff, A.I. and Singer, L.: Blood 6, 413 (1951).
2. Shepard, M.K., Wheatherall, D.J. and Conley, C.L.: Johns Hopkins Hosp.
    110, 293 (1962).
3. Itano, H.A.: Adv. Prot. Chem. 12, 215 (1957).
4. Armstrong, D.H., Schroeder, W.A. and Fenninger, W.D.: Blood 22, 554
    (1963).
5. Frandel, S., Reitman, S. and Sonnenwirth, A.C.: Gradwohl's Clinical
    Laboratory Methods and Diagnosis, 7th edition, p 410, The Mosby Co.,
    St Louis 1970.
6. Ivor, S.: Chromatographic and electrophoretic Technique, William
    Heineman, Medical Books Ltd. 1960.

ELECTROPHORETIC ANALYSIS OF HOLSTEIN BEEF CATTLE SERA
AFFECTED WITH LIVER ABCESS

Umetaro Muramatsu

Livestock Sanitory Laboratory, Prefecture of Tochigi.
Hosaku Tomatsuri-cho Utsunomiya-shi, Tochigiken 320, Japan.

Shigenori Ikemoto

Department of Human Biology, Jichi Med. Sch.
Minamikawachi-machi, Kawachi-gun, Tochigi-ken 329-04, Japan.

Kyoichi Tanaka

Veterinary faculty of Azabu University.
Fuchinobe Sagamihara-shi, Kanagawa-ken 229, Japan.

Introduction

The occurrence of liver abcess in intensively or traditionally
fed beef cattle has been reported in many countries(2,3,6).
Similar findings have recently observed in Japan(5,7,9). And
many studies have been reported on the clinico-pathological
changes in the sera of Holstein beef cattle with liver abcess
or dairy cattle(1,4,5,8,9). Some of them include data of
electrophoretic analysis(1,4,9). However, there are no reports
classifying the cellulose acetate membrane electrophoretic
patterns. The present study was conducted to investigate the
serum proteins and to classify electrophoretic patterns
observed in sera of 384 Holstein beef cattles with or without
liver abcess. Several kinds of analyses of the serum protein
are described.

Results

1)   In the sera of Holstein beef cattle with liver abcess, the

     amount of total protein, α-globulin and γ-globulin

     increased and the decrease of alubumin and A/G ratio were

     observed as shown in table 1.

Table 1

Total protein, protein fraction and A/G ratio of fattened steer serum

	Liver abcessed n = 87	Unaffected liver n = 124	Significance
Total protein	7.2 + 0.9 g/dl	6.5 + 0.4 g/dl	II
Alubumin	46.1 + 9.5 % 3.26 + 0.60 g/dl	56.3 + 4.2 % 3.65 + 0.31 g/dl	II II
α - Globulin	13.4 + 2.6 % 0.98 + 0.22 g/dl	12.2 + 1.6 % 0.79 + 0.10 g/dl	I II
β - Globulin	12.7 + 2.2 % 0.92 + 0.21 g/dl	12.4 + 1.8 % 0.80 + 0.13 g/dl	NS II
γ - Globulin	28.0 + 8.8 % 2.07 + 0.88 g/dl	19.3 + 4.0 % 1.26 + 0.31 g/dl	II II
A/G ratio	0.91 + 0.34	1.31 + 0.22	II

I and II : mean significant differences at 5% and 1% level, respectively.

NS : No significant difference

2) <u>Cellulose acetate membrane electrophoretic patterns were divided into 5 types based on the shape of the γ-globulin region</u>, i.e., AA-type: marked increase of γ_1- and γ_2- globulin was observed; A-type: marked increase of γ_1- globulin relative to γ_2 and γ_3 ; B-type: marked increase of γ_2-globulin relative to γ_1 and γ_3 ; C-type: marked increase of γ_3-globulin relative to γ_1 and γ_2 ; D-type: no remarkable increase of γ-globulin. A-type was divided into 12 sub-types, tentatively named A-1 to A-12. Similarly, B-type was divided into 11 subtypes and C-type into 4. Representative patterns of each type are shown in Fig.1.

3) The occurrence of each type in the sera of Holstein beef cattle are shown in table 2. AA-type and C-type were encountered only in the sera of Holstein beef cattle with liver abcess. However, the authors have observed some patterns resembling AA-type in the sera of dairy cattle suffered with liver fluke. <u>The C-type is considered to be the specific pattern to liver abcess.</u>

4) According to the data from immunoelectrophoretic analysis of the sera of Holstein beef cattle with liver abcess (Fig.2), there do not appear to be any precipitin bands specific to liver abcess.

5) Qantitative analysis was made of the sera of Holstein beef cattle with or without liver abcess, and showed a significant increase of Ig G_2 in the group with liver abcess.

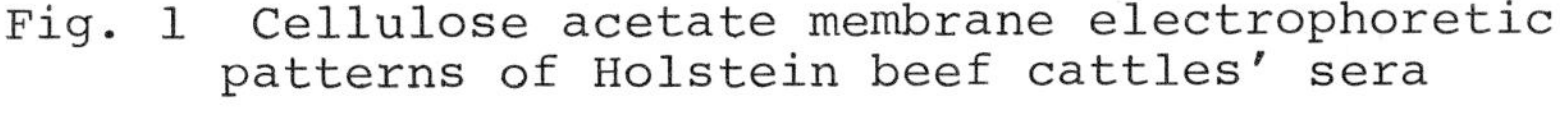

Fig. 1 Cellulose acetate membrane electrophoretic patterns of Holstein beef cattles' sera

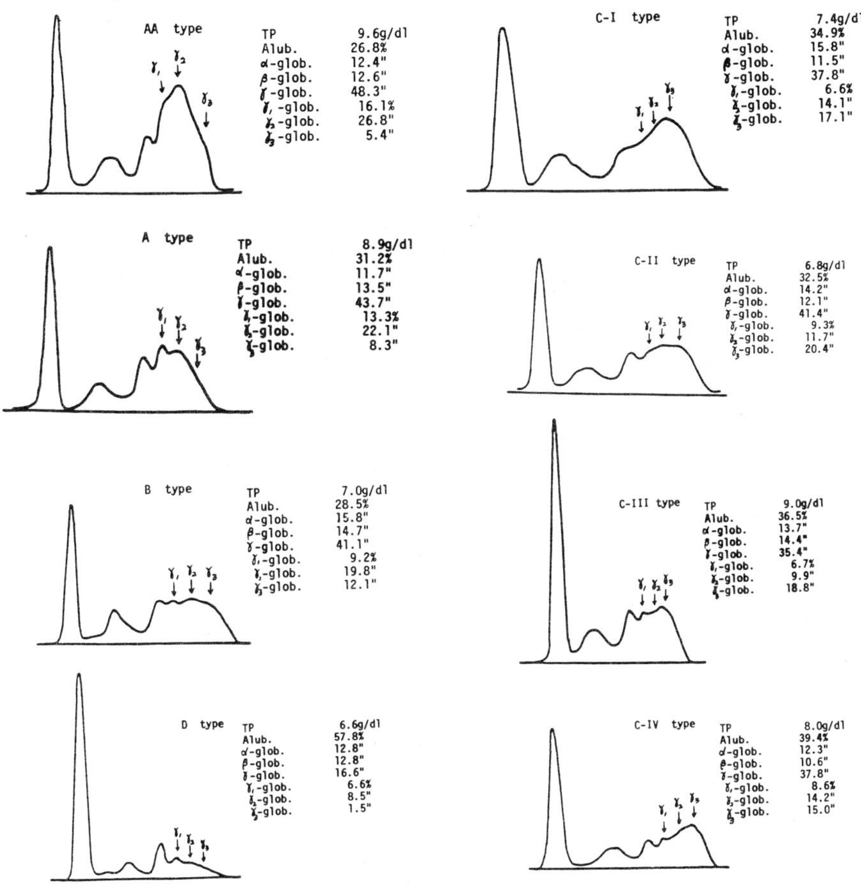

Table 2

Types of electrophoretic patterns of fattened steer sera

Source of serum tested \ Types of electrophoretic pattern	AA	A	B	C	D	Total
Fattened steers affected with liver abcess	2(1.2)	91(56.2)	46(28.4)	8(4.9)	15(9.3)	162(100)
" liver abcess and lung abcess		1	1	1		3
" liver abcess and hepatitis		2	1		1	4
" lung abcess		4			2	6
" lung abcess and hepatitis		1	3		1	5
" hepatitis		4	2			6
" liver fluke		2				2
" abcess of other tissues or organs		1			1	2
Total of affected steers	2(1.2)	106(55.8)	53(27.9)	9(4.7)	20(10.5)	190(100)
Unaffected steers		74(38.1)	13(6.7)		107(55.2)	194(100)
Total	2(0.5)	180(46.9)	66(17.2)	9(2.3)	127(33.1)	384(100)

750

Fig. 2   Immunoelectrophoretic patterns of bovine sera
         affected with liver abcess.

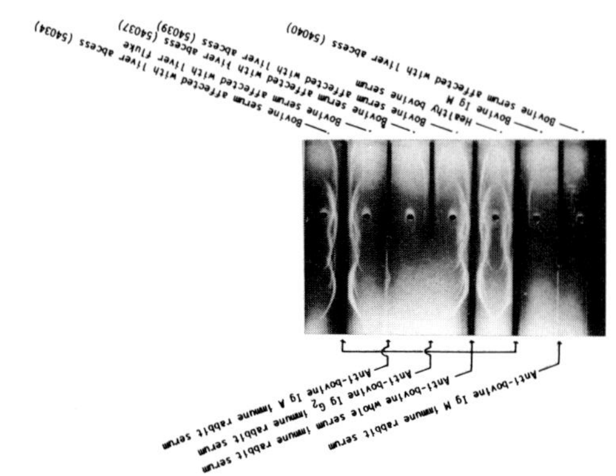

References

1. Goma,K., Muramatsu,U., Takahashi,M., Nakane,H., Yazawa,T.,
    Onoguchi,K., Morishige,N.: Jui-chikusan-shinpo No.667,
    25-29 (1977)(in Japnese).

2. Jensen,R., Dean,H.M., Cooper,L.J., Miller,V.A., Graham,W.R.
    : Am. J. Vet. Res. 15, 202-216 (1954)

3. Jonsen,G., Liberg,P.: Acta vet. Scand. 15, 264-273 (1974).

4. Kimata,H., Tsunekane,T., Sugawa,Y., Watanabe,T.
    Kurabayashi,Y.: Bull. Coll. Agr. & Vet. Med. Nihon Univ.
    No.28, 42-54 (1971) [in Japanese].

5. Muramatsu,U., Takahashi,M., Shibata,H., Kurosaki,H.,
    Matsukura,F., Ogata,T., Saito,Z., Watanabe,K.: Ann. Rep.
    Livestock Sanitory Lab., Pref. Tochigi No.12, 22-26
    (1977)[in Japanese].

6. Rowland,A.C.: Vet Rec. 78, 713-716 (1966).

7. Shimada,Y.: Juikai No.97, 24-34 (1971) [in Japanese].

8. Shimada,Y., Fujioka,S., Mizouchi,T., Asao,T.:
    Chikusannokenkyu 24, 1207-1211 (1970)[in Japanese].

9. Tamate,H.: Tohoku J. agr. Res. 23, 184- 195 (1973)

10. Yoshida,Y.: Kachiku-shinryo No. 229, 27-40 (1982)[in
    Japanese].

LATEX AGGULUTINATION TEST FOR IMMUNOGLOBULIN-ANALYSIS IN CSF

Sachio Ikawa, Kiyoko Ikeda
Department of Laboratory Medicine. Jikei University School of
Medicine, Tokyo, Japan.

Introduction

Routine determination of immunoglobulins in CSF has been
rather difficult because of their low concentration. Recently
direct estimation became possible using a Latex immunoassay
and also a new automatic analyser for this test was developed.
It is suggested that this method may be useful in the clinical
laboratory.

Procedure

The principle of the assay, one that is fully automated, is
to measure immunoglobulins in CSF by an initial change in
absorbance due to the turbidity created by the antigen-
antibody reaction. Latex suspension coated with purified
anti-human immunoglobulin taken from rabbit serum, following
Fritz et al's method (1), was obtained from Eiken Chemical
Co. Delta absorbance was measured at 585nm between 35 and
135 seconds after mixing of sample (10-100µl) and reagent
(400µl) at 37°C using a LA system (Analytical Instrument Co.).

Results

Fig. 1 illustrates a typical calibration curve for IgM,

752

Table 1   Reproducibility (within run)

	IgG		IgA		IgM		
N	10	10	10	10	10	10	
M	1.521	4.923	0.739	2.715	0.915	3.074	mg/dl
SD	0.036	0.0742	0.039	0.0303	0.0347	0.049	
CV	2.4	1.5	5.3	1.1	3.8	1.6	%

which encompassed a concentration range from 0 to 7 mg/dl.
CV of within-run reproducibility of the assay was between
1.1-5.3 % (Table 1). The levels of sensitivity when 100µl
of sample were used, were for IgA 1.0 mg/dl, IgG 1.0, and
IgM 5.0. Upper limits in this situation were for IgA 13 mg/dl,
IgG 18, IgM 15. The samples with higher concentrations were
also analyzed by a single radial immunodiffusion method (LC
Partigen, Hoechst Co.). The values obtained by these different
methods had a high positive correlation coefficient with each
other. Throughoutput is 160 specimens/hr. Fig. 2 illustrates
the frequency distributions of IgA, IgM and IgG in CSF obtained
in our laboratory. The frequency drops rapidly as the concentration increases and showed striking
differences in the levels and
compositions of immunoglobulins
among diseases.
The results from CSF specimens
obtained on the medical and
surgical ward of our hospital
are shown in table 2. Very
high CSF immunoglobulin
concentrations were observed
in subarachnoid hemorrhage,
presumably as a result of
spillage of plasma proteins
into the CSF. Increase in
CSF proteins in disease may
occur in the following cases:

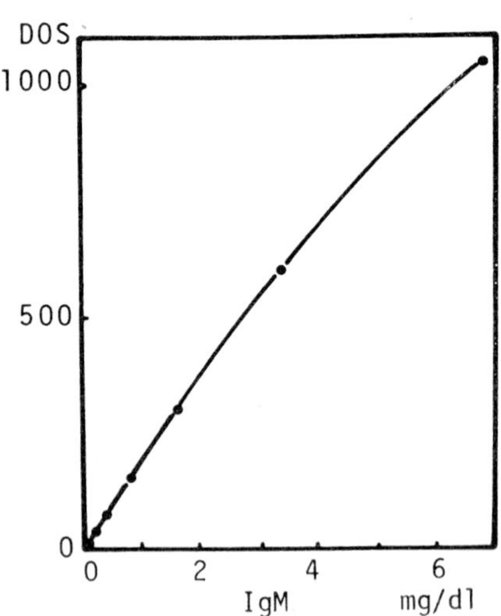

Fig. 1 Standard curve for IgM

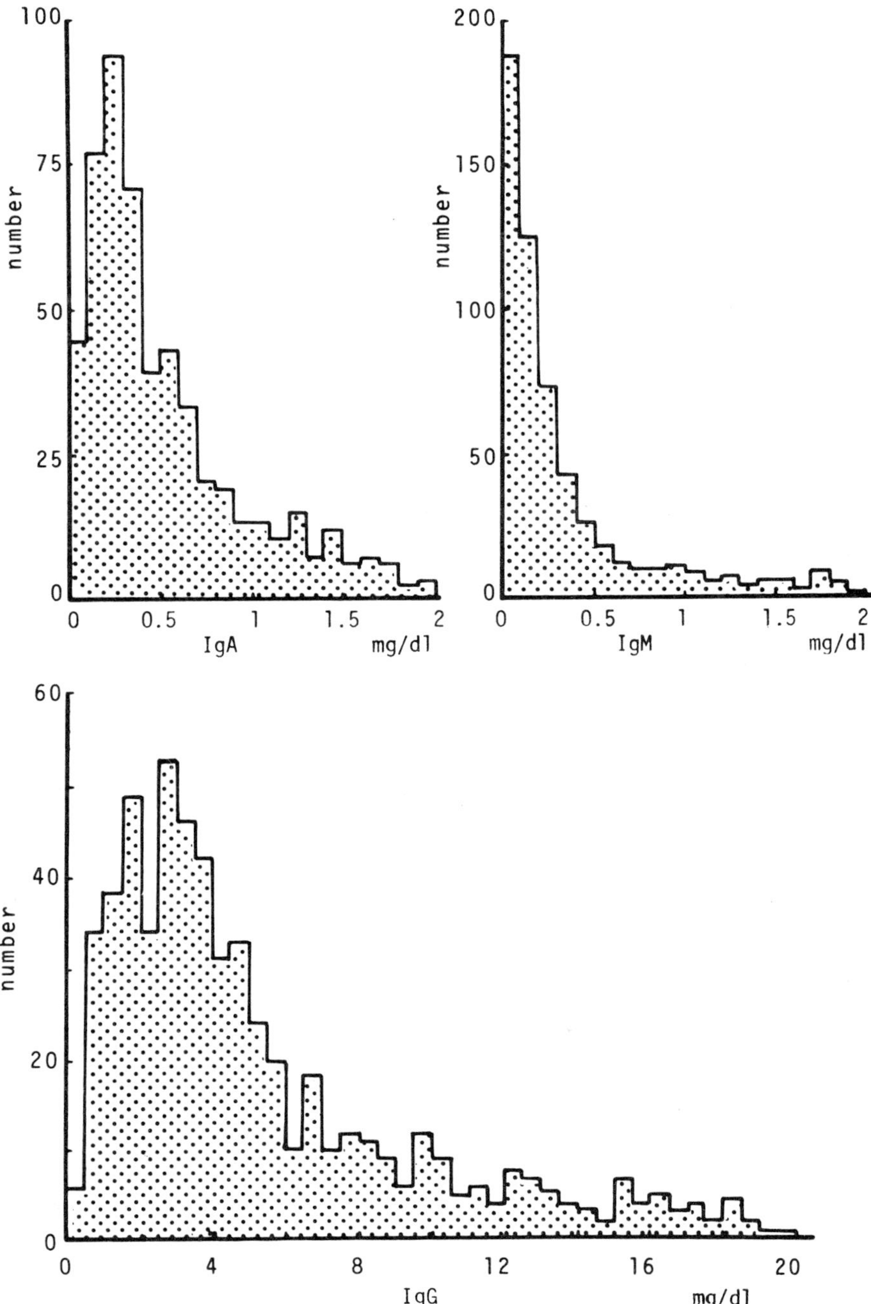

Fig. 2 Frequency distributions of IgG, IgA and IgM in CSF
determined in our Laboratory

Table 2   Ig levels in CSF specimens obtained in our hospital

	IgG (mg/dl)	IgA (mg/dl)	IgM (mg/dl)
Glioblastoma	0.40 - 364.90	2.17 - 49.79	1.00 - 15.85
Craniopharyngioma	0.44 - 104.30	0.03 - 6.80	0.04 - 4.80
Brain tumor	6.05 - 20.53	0.39 - 0.67	0.18 - 0.51
Spinal cord tumor	0.92 - 551.00	0.33 - 5.05	0.19 - 34.91
Arachnoid cyst	1.10 - 26.70	0.05 - 1.20	0.10 - 10.75
Spinabifida	0.31 - 16.17	0.03 - 2.03	0.01 - 2.36
Subarachnoid hemorrhage	178.20	23.11	2.03
Encephalitis	7.30 - 71.60	0.10 - 17.31	0.03 - 4.85
Meningitis	0.60 - 22.99	0.05 - 2.71	0.03 - 1.22
Polyneuropathy	5.44 - 22.81	0.78 - 2.27	0.14 - 0.51
Neuro-Beçhet	3.84 - 15.47	0.50 - 1.54	0.13 - 0.62
Guillain-Barré Syndrome	2.09 - 23.42	0.23 - 1.55	0.17 - 1.60

(a) bacterial and viral meningitis because of increased
passage of plasma proteins into CSF. (b) chronic infections
and brain tumors because of increased local production of
immunoglobulins, synthesized by the plasma cells infiltrating
the central nervous system.   (c) degenerative diseases such
as the Guillan-Barré syndrome.   (d) mechanical obstruction
because of progressive equilibration of static CSF with plasma.

Summary

This immunolatex method described here is found to be highly
sensitive and accurate for estimating immunoglobulins in CSF.
The system is quick and simple and thus suitable for routine
laboratory use.

Reference

1. Fritz R.B. and Rivers S.L.: J. Immunol. 108, 108-111(1972).

UNIQUE PROPERTIES OF BOVINE UBIQUITIN

Esterase Activity and Acetylation as judged by Electrophoresis

Hajime Matsumoto
Department of Physiology II, Azabu University, Sagamihara,
Kanagawa, 229, Japan

Naoyuki Taniguchi
Biochemistry Laboratory, Cancer Institute, Hokkaido University
School of Medicine, Sapporo, 060, Japan

Introduction

Ubiquitin is a polypeptide with a molecular weight
approximately 8,500. It was first isolated from bovine thymus
(1), but subsequently found to be widely distributed in
animals, yeast, bacteria and higher plants (1-3). During the
course of studies on carbonic anhdrases of ruminant and
non-ruminant tissues (4-8), we found a peptide which was
identified as ubiquitin. We have found that ubiquitin
exhibits an esterase activity toward p-nitrophenyl esters and
is concomitantly acetylated as judged by electrophoresis.
Ubiquitin also possesses a weak CO_2 hydration activity and this
is inhibited by acetazolamide.

Results

1)   Purification of ubiquitin from the bovine erythrocytes.
     The packed erythrocytes (800 ml) were hemolyzed by
     repeated freezing and thawing. Hemoglobin was removed by

a mixture of chloroform and ethanol. After dialysis against water in Spectrapore 3 membrane, the dialyzate was lyophilized and 2.5 g of solids were obtained. This material was further purified by chromatography on DEAE cellulose and repeated gel filtrations on Sephadex G-50. The purified material (40 mg) gave a single but relatively broad band in the PAGE experiment, in the presence or absence of SDS. Staining of the gel with bromothymol blue in the presence of CO_2 revealed a very weak but clearly discernable hydration activity. Acetazolamide (1 x 10^{-4}M) inhibited the reaction.

2) Enzymatic activities. The CO_2 hydration activity was 262 Wilbur and Anderson (9) units/mg and this was completely inhibited by acetazolamide at a molar ratio of acetazolamide to ubiquitin of 100:1. The release of the p-nitrophenol from PNPA at pH 7.7 and 30° was linear with respect to time. Using p-nitrophenyl acetate PNPA in potassium phosphate buffer as a substrate, an activity of 0.013 μmol/min/mg was found. This gives a turnover number of 116.

3) Effect of p-nitropheny esters on the ubiquitin hydrolysis. The rate of hydrolysis of three different p-nitrophenyl esters by ubiquitin was determined. The reactivity appeared to decrease in the order of increasing length of acyl groups (Fig.1).

4) Acetylation of ubiquitin by PNPA. In addition to the enzymatic reaction, we have observed that the ubiquitin was modified upon reacting with PNPA as judged by the appearance of more anodically migrating electrophoretic components. With increased time of reaction, there was an increase in the amounts of these components (Fig.2).

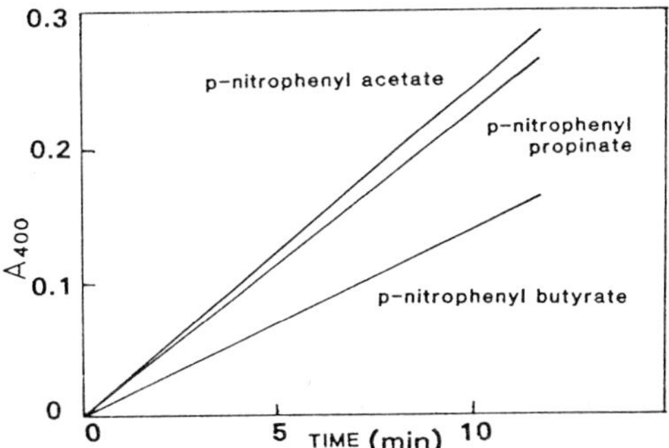

Fig.1 Rates of hydrolysis of various p-nitrophenyl esters by
ubiquitin. The reaction mixture contained 25 μmoles of
K-phosphate buffer, pH 6.5-8.01 or Tris-sulfate buffer, pH
8.4-8.6, 1.2 μmoles of PNPA and 0.06 μmole of ubiquitin in a
volume of 250 μl. The reaction was followed
spectrophotometrically at 400 nm, and corrections were made for
the absoption due to p-nitrophenol and the nonenzymatic
hydrolysis of PNPA.

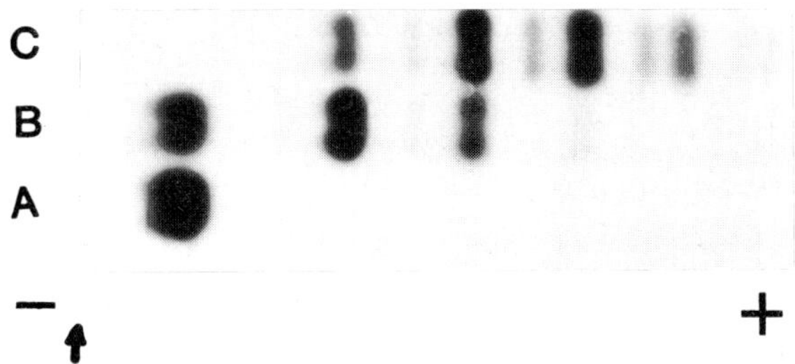

Fig.2 Results for electrophoresis of ubiquitin (0.25 nmol) in
12% polyacrylamide gel following reaction with PNPA
(ubiquitin:PNPA=1:20 molar ratio). The times of reaction at
37° are 0 (lane A), 60 min (lane B) and 24 h (lane C).
Electrophoresis was performed in 0.025 M Tris-glycine buffer,
pH 8.3 for 3.5 h at room temperature. The point of sample
application was indicated by the arrow.

Discussion

These facts indicate that the ubiquitin in addition to its

unique function of ATP-dependent proteolysis (3) also has

esterase activity toward p-nitrophenyl esters. Although the physiological importance of the esterase activity of ubiquitin, if any, remains to be determined, it may be of physiological significance in providing a pathway for the transfer of acetyl groups in vivo. The presence of a true substrate in vivo needs to be explored along with other aspects of the CO_2 hydration and esterase activity.

Acknowledgements

This work was supported by grant-in aid for scientific research from the Ministry of Education, Science and Culture, Japan and the Naito Foundation Research Grant for 1981.

References

1. Goldstein, G.: Nature (Lond.) <u>247</u>, 11-14 (1974).
2. Goldstein, G., Scheid, M., Hammerling, U., Boyse, E.A., Schlesinger, D.H. Niall, H.D.:
   Proc.Natl.Acad.Sci.U.S.A. 72, 11-15 (1975).

3. Ciechanover, A., Hod, Y., Hershko, A.:
   Biochem.Biophys.Res.Commun. 81, 1100-1105 (1978).

4. Deutsch, H.F., Funakoshi, S., Fujita, T., Taniguchi, N., Hirai, H.: J.Biol.Chem. 247, 4497-4502 (1972).

5. Taniguchi, N., Kondo, T., Ishikawa, N., Ohno, H., Takakuwa, E., Matsuda, I.: Anal.Biochem. 72, 144-152 (1976).

6. Taniguchi, N., Kondo, T., Ishikawa, N.:
   Biochem.Biophys.Res.Commun. 85, 952-958 (1978).

7. Kondo, T., Taniguchi, N., Taniguchi, K., Matsuda, I., Murao, M.: J.Clin.Invest. 62, 610-617 (1978).

8. Matsumoto, H., Fujioka, H., Obara, Y., Taniguchi, N.: Comp.Biochem.Physiol. 71, 317-320 (1982).

9. Wilbur, K.M., Anderson, N.G.: J.Biol.Chem. 176, 147-154 (1948).

Speech of Thanks to the Organizers

If you have a good sense of humour, you may enjoy the following saga. You may have to acquire some of the rich imagination
of the people of the Far East. Medical studies and work give
intimate contact with the marvellous life process of the human
creature. I have for several years lived with "Isoelectric
Focusing", a young fragile talent twenty years ago. It occurs
to me that there are some phenomenal similarities with the
development of a human and "my son Ampholine". He was born
after nine months in the spring nineteen years ago. Like all
newborn creatures he was sensitive during his first months and
I had to work hard for his survival. Little was then known
about him and his capabilities. However, within a few years
an increasing number of scientists were impressed by Ampholine
aiding in a rapidly growing number of applications, some of
them of wide interest. Ampholine consisting of several hundred
molecular species, each comprising myriads of molecules, having to follow certain rules of mutual interactions and order
when taking part in a separation experiment - like body cells
and their molecules. A lively fascinating phenomenon, almost
unbelievable, but functioning as all well organised organisms.
Even when young Ampholine was able to help several scientists
to many good and some splendid achievements. If this had not
been so, many of us would not have been here in Tokyo for this
conference nor would several scientists have just the projects
and occupations they have today.

"Mother Focusing" is not living in monogamy. During the latter
years the observers know of the birth of one "child" called
"Reoelectrolysis" and another one called "Immobiline". In
Comparison to Ampholine, Immobiline seems somehow rigid.
Ampholine being multi-disciplined for years serving both iso-
tachophoresis and increasingly 2-dimensional electrophoresis,

is now grown up, vigorous, powerful and capable to offer
plentiful, continued and exciting achievements for those who
utilise his services. Hopefully this will continue for many
years to come. I look forward with enthusiasm.

Written by Olof Vesterberg in Tokyo before sunrise May 12th,
1983. Presented as thanks to the organizers at the end of the
conference.

TWO-DIMENSIONAL ELECTROPHORESIS FOR "PROTEIN MAP" AND FOR "POLYPEPTIDE MAP"

T. Okuyama and T. Manabe

Dept. of Chemistry, Fac. of Science, Tokyo Metropolitan Univ., Setagaya-ku, Tokyo, Japan

There are two types of 2-D electrophoresis method, depending on "with" or "without" chaotropic reagent. The former method relying on the random coil structure of protein is called as "Polypeptide Map" method and the latter method relying on the natural structure or the higher dimensional structure of protein should be called as "Protein Map" method. The differences between the two methods were tested with the addition of 8M urea, mercaptoethanol, sodium dodecyl sulfate or/and nonionic detergent.

The protein map method was useful for the analysis of protein complexes, such as lipoproteins or haptoglobin-hemoglobin complexes and also protein polymers such as myeloma immunoglobulin polymers or haptoglobin polymers and further useful for the analysis of various isoenzymes. Thus the method is more useful for the physiological or clinical purposes and the conventional "Polypeptide Map" is useful for genetic studies.

STUDY ON THE COLORIMETRIC ANALYSIS OF TWO DIMENSIONAL
P.A.G. ELECTROPHORESIS BY SILVER STAINING METHODS

S. Saneyoshi, M. Akahori, T. Hattori and T. Ohshima

Tokyo Laboratory of JOOKOO CO., LTD. Tokyo, Japan

There are various approaches to the analysis of Two
Dimensional Electrophoresis images. Recently, the study
of the color analysis on protein map by silver staining
method has become a center of attention.

The silver stain process is described by Adams and Sammons
with net charge, molecular weight and color. In addition
to a characteristic position (orthogonal coordinates) on
a gel and an intensity, the spot staining is a characteristic
hue as well, usually of various shades of yellow, red, green
and blue. This multi dimensional analysis will permit
greater resolution, accuracy and reliability of either
automatic or manual analysis, which is attainable via
existing monochrome approaches.

However, usual method of color analysis is often unreliable
and uncertain. In order to remove this defect, a new
colorimetric instrument for electrophoresis use is designed,
which displays the digital color information. Based on
advanced color science, direct digital reading of tristi-
mulus value XYZ (red, green and blue), CIE chromaticity and
color difference delta E* by CIELAB system (Recommendation
of Commission Internationale de l'Eclairage) can be measured
immediately. Mutual relationship between color difference
and molecular weight was sufficiently estimated from the
result of measurement.

TWO-DIMENSIONAL ELECTROPHORESIS OF RAT AND RABBIT LIVER PLASMA MEMBRANES

A. Rahimi-Pour, D. Ratanasavanh, M.M. Galteau and G. Siest

Lab. Biochem. Pharmacol. ERA CNRS 698, Nancy I Univ.,
Nancy, France

We have used the technique of Two-Dimensional Electrophoresis for analysis of rat and rabbit liver plasma
membrane proteins and peptides. We have adapted the sample
preparation conditions as well as those for electrophoretic
separations. The membranes are isolated as described by
Neville (1).

The conditions of isoelectric focusing and electrophoresis
were adapted to plasma membranes. Isoelectric focusing is
carried out in the presence of an ampholine mixture
(pH 3,5-10 and 5-7) with 700 V for 15 hours. The second
dimension was performed on polyacrylamide gradient gel
(11 and 14%).

The technique of solubilisation of membrane proteins
mainly determine the quality of electrophoretic separation.

To determine optimal conditions of membrane solubilisation
we studied the effect of some chemical agents and detergents:
urea, nonidet Np-40, sodium dodecylsulfate (SDS), sodium
deoxycholate (DOC)... The data obtained confirmed the need
for use of urea for membrane solubilisation. However we
have assayed SDS and DOC. The use of DOC at a final concentration of 0.5% permitted to considerably increase the
number of spots in the electrophoretic pattern.

Thus using silver staining technique we could detect more
than 300 proteins and peptides in these membranes. The
electrophoretic patterns of plasma membranes of rat and
rabbit liver have a large number of peptides in common.

Now, we are trying to identify some of these proteins, after
their transfer on nitrocellulose, by specific "antigen-antibody" reaction.

(1) J.D.M. NEVILLE
    Isolation of an organ specific protein antigen from
    cell surface membrane of rat liver
    Biochem. Biophys. Acta, 1968, <u>154</u>, 540-552

CELLULAR ELECTROPHORETIC TECHNIQUES SHOW PLASMA ABNORMALITIES IN MULTIPLE SCLEROSIS

C.H. Tamblyn, R.L. Swank, G.V.F. Seaman

Dept. of Neurology, Oregon Health Sciences University, Portland, Oregon, U.S.A.

Analytical particle electrophoresis has been used in tests for the early diagnosis of multiple sclerosis (MS). The addition of linoleic acid (LA) to a suspension of fresh red blood cells (RBC) from MS patients in Medium 199 reduced their mean electrophoretic mobility (EPM). We found no significant change in the mobility of normal fresh RBC after the addition of LA unless they had undergone prior incubation in blood group compatible plasma from MS patients. When saline washed RBC from MS patients were incubated in normal plasma they then acquired normal electrophoretic properties in the fresh RBC test for MS.

While working toward the elucidation of the mechanism of this test we found that glutaraldehyde fixed RBC suspended in phosphate buffered saline (PBS), had an increased EPM after the addition of LA, with the mobility of RBC from MS patients still remaining lower than that of normal RBC. Polystyrene latex particles (PSL) were also used as carrier particles by incubating them in plasma from either MS patients or healthy subjects. The EPM of the plasma coated particles increased after the addition of LA, with the mobility of MS plasma cated PSL being less than those coated in normal plasma. This effect is similar to that found when RBC are used as carrier indicator particles. However, the results with PSL demonstrates that these electrophoretic phenomena originate from a plasma component abnormality (possibly a deficiency) in the blood of MS patients, rather than an intrinsic cell membrane defect. A group of MS patients in the early phases of the disease were given plasma infusions during acute exacerbations. Not only did they undergo neurological improvement, but their fresh RBC EPM test for MS went from positive into the normal range and remained there for 30 to 60 days after the infusion.

USE OF THREE ISOENZYMES FOR CHARACTERIZATION OF CELL CULTURE SPECIES BY AGAROSE ELECTROPHORESIS

J. Ziegenmeyer and C. Lepp

Corning Medical Development Laboratory, East Walpole, MA 02032, U.S.A.

Reports of cross contamination of tissue cultures have appeared frequently in recent years. The problem is quite common, with an estimated 25% of cultures having incorrectly designated identities, and occurs in laboratories and cell banks in several countries. We have studied the use of isoenzyme electrophoresis for routine monitoring of cell culture identity. In this method a combination of several isoenzymes from distinct alleles is used to describe the uniqueness of a given cell line. We describe the use of three isoenzymes to help characterize cultures.

This method separates the cytoplasmic enzymes purine nucleoside phosphorylase (NP), malate dehydrogenase (MD) and asparate aminotransferase (AST) by electrophoresis in a thin-layer agarose gel for 25 minutes at 23 V/cm. The locations of the enzymes are marked by an insoluble formazan produced after reaction _in situ_ with an enzyme-specific reagent. The migration distances of enzymes vary in different species. These differences can be used to identify the cell species.

The reproducibility of migrations is most affected by the time, voltage and temperature of electrophoresis and must be controlled. A standard is used to correct for minor differences in experimental conditions. Interlaboratory results are repeatable to 2mm(1SD). The standard chosen is an extract of a mouse lymphoma line. A control is used to help verify that separations have occurred properly and

to serve as a marker for the most commonly encountered
contaminant, e.g. HeLa.

This system has allowed us to distinguish among thirteen
cell lines derived from different species with at least
95% confidence. The result is unaffected by the tissue
of origin or by strains within the species. With proper
calibration the system can be used to assign species
identities to unknown cell lines.

DEVELOPMENT OF 3 ISOENZYME PROCEDURES FOR CHARACTERIZATION OF CELL CULTURE SPECIES BY AGAROSE ELECTROPHORESIS

C.A. Lepp, J.M. Ziegenmeyer

Corning Medical Development Laboratory, East Walpole,
MA 02032, U.S.A.

Contamination of continuously grown cell culture lines
by other species of cells has been shown to be a widespread problem. Various methods to detect contamination
exist but can be difficult to use and interpret. Electrophoretic methods exist but are difficult to put into general
practice. We have optimized and developed three colorimetric
reagents to easily distinguish between specific cell lines
using thin-layer agarose gel electrophoresis. Reagents for
purine nucleoside phosphorylase (EC 2.4. 2.1, NP) malate
dehydrogenase (EC 1.1.1.37, MD) and asparate aminotransferase (EC 2.6.1.1, AST) were formulated. Reagent for NP
consists of 50 mM inosine, 250 mM potassium phosphate,
1×10^3 units/L xanthine oxidase (EC 1.2.3.2), 0.2 mM
phenazine methosulfate (PMS), and 17 mM malate, 30 mM NAD,
0.012 mM EDTA, 0.18 mM PMS, and 12 mM MTT. The novel AST
reagent consists of 10 mM α-Ketoglutarate, 150 mM L-asparate,
125 mM NAD, 5 mM ADP, 0.04 mM pyridoxal-5-phosphate, 1.5×10^5
units/L glutamate dehydrogenase (EC 1.4.1.3), 0.18 mM PMS and
12 mM MTT. Each of these reagents makes use of the purple
formazan formed in the presence of the respective isoenzymes.
The reagents are each prepared in 130 mM Veronal buffer
pH 8.6. Separation is performed at 23 V/cm for 25 minutes
on thin-layer agarose plates, maintained at 4-10°C, containing 65 mM Veronal, 0.035% EDTA and 5% sucrose at pH 8.6.
One ml of reagent is then spread upon the gel and incubated
at 37°C, washed in dH_2O and dried at 60°C. The duration of
each step is 20 minutes.

Separations are clear, well resolved and easily readable. As low as 25 IU/L for each isoenzyme is detectable with 1 ul of sample. The upper limit of linearity is at least 1000 IU/L per zone. Reproducibility of migration is better than $\pm$ 2 mm under these conditions.

EVIDENCE FOR TWO NEW ALLELES IN THE COAGULATION FACTOR
XIII SUBUNIT B SYSTEM, FXIIIB* 4 AND FXIIIB* 6, REVEALED
BY IMMUNOFIXATION AGAROSE GEL ISOFOCUSING

P. Kühnl and W. Spielmann

Institute of Immunohematology, University Frankfurt,
Sandhofstr. 1, D-6000 Frankfurt/Main, FRG

Immunofixation agarose gel isoelectric focusing (IAGIF)
with monospecific coagulation factor XIII subunit B (S)
antiserum gave evidence for the existence of two new
alleles in this system, designated FXIIIB* 4 and FXIIIB* 6.
After 3h IAGIF of neuraminidase(CPN+)-treated serum samples
in a 1mm isogel agarose pH 5-7 on gelbond film at a cooling
temperature of 10°C at maximum settings of 1200V, 20mA,
10W, a 1:1 dilution of antibody was spread over the cathodal half of the gel, incubated 90 min at 37°C, pressed
under filter paper, washed out in saline overnight, dried
and stained. Whereas the B*4 and B*6 gene products (phenotypes B4-1, B4-2, and B6-1, confirmed by family studies)
were located <u>anodally</u> to B2 after 4h immunofixation agarose
gel electrophoresis (IAGE), IAGIF of CPN+ sera shifted the
B4 and B6 band in a more cathodal position. The pI of B2
became very similar to that of B1, suggesting that a) there
are three common alleles present in the major races b)
FXIIIB*B' (0.013 in Japan) is identical with B*2 in Caucasoids (0.095 in Germany) c) the p1 of B2 asialo-molecules
is decreased by the carrier ampholytes during IAGIF.

CROSSED IMMUNOELECTROPHORESIS FOR THE DETECTION OF SPLIT
PRODUCTS OF THE THIRD COMPLEMENT IN DENGUE HEMORRHAGIC
FEVER. I. OBSERVATION IN PATIENTS' PLASMA

V. Churdboonchart, N. Bhamarapravati and P. Futrakul

Dept. of Pathobiology, Faculty of Science, Mahidol
University, Bangkok, Thailand

Crossed Immunoelectrophoresis (CIEP) was applied to detect
the products of the third component of complement (C_3)
activation in plasma of patients who suffered from Dengue
Hemorrhagic Fever (DHF) which is caused by dengue viruses
types 1,2,3 and 4.

First dimension electrophoresis was performed using 5x10 cm
glass slides with 6 ml of 1% agarose in 0.01 M EDTA-Tris
buffer pH 8.6. Samples, 1 µl for each test, were placed in
to 1 mm punched wells. The electrophoresis was done by
using 10 v/cm in 4°C for 60 min. Second dimension electrophoresis was run perpendicularly in rabbit anti-C_3 serum
containing agarose and electrophoresis was performed by
using 2v/cm 4°C overnight. Normal human serum treated with
inulin, a known C_3 splitting agent, was used as positive
control in all tests.

C_3 split products (C3sp) were demonstrated in severely ill
patients clinically classified as the grade III and grade
IV, and rapidly disappeared during convalescent phase. The
appearance of C3sp correlated well with the signs of shock,
and this electropherogram of C3sp could be used as a parameter reflecting immunological activity in this disease.

Electrophoresis '82
Advanced Methods
Biochemical and Clinical Applications

**Proceedings of the International Conference
on Electrophoresis · Athens, Greece, April 21–24, 1982**
Editior   D. Stathakos
1983. 17 cm x 24 cm. XVI, 867 pages. Numerous illustrations.
Hardcover. DM 260,–; approx. US-$ 119.00 ISBN 3 11 008791 X

Electrophoresis '81
Advanced Methods
Biochemical and Clinical Applications

**Proceedings of the Third International Conference
on Electrophoresis
Charleston, South Carolina, USA, April 1981**
Editor   R. C. Allen
1981. 17 cm x 24 cm. XVIII, 1041 pages. Numerous illustrations.
Hardcover. DM 245,–; approx. US-$ 112.00 ISBN 3 11 008155 5

Electrophoresis '79
Advanced Methods
Biochemical and Clinical

**Proceedings of the Second International Conference
on Electrophoresis
Munich, Germany, October 15–17, 1979**
Editor   J. Radola
1980. 17 cm x 24 cm. XV, 858 pages. 361 figures.
Hardcover. DM 185,–; approx. US-$ 85.00 ISBN 3 11 008154 7

 **Walter de Gruyter · Berlin · New York**